Biology, Management, and Protection of Catadromous Eels

Biology, Management, and Protection of Catadromous Eels

Edited by

Douglas A. Dixon
3412 Hillview Avenue
Palo Alto, California 94304, USA

American Fisheries Society Symposium 33

Proceedings of the First International Symposium
Biology, Management, and Protection of Catadromous Eels
held at St. Louis, Missouri, USA
21–22 August 2000

American Fisheries Society
Bethesda, Maryland
2003

The American Fisheries Society Symposium series is a registered serial. Suggested citation formats follow:

Entire book

Dixon, D. A., editor. 2003. Biology, management, and protection of catadromous eels. American Fisheries Society, Symposium 33, Bethesda, Maryland.

Chapter within the book

Jessop, B. M. 2003. Annual variability in the effects of water temperature, discharge, and tidal stage on the migration of American eel elvers from estuary to river. Pages 3–16 *in* D. A. Dixon, editor. Biology, management, and protection of catadromous eels. American Fisheries Society, Symposium 33, Bethesda, Maryland.

Cover photography by Wendy E. Morrison.

© 2003 by the American Fisheries Society

All rights reserved. Photocopying for internal or personal use, or for the internal or personal use of specific clients, is permitted by AFS provided that the appropriate fee is paid directly to Copyright Clearing Center (CCC), 222 Rosewood Drive, Danvers, Massachusetts, 01923, USA; phone 978-750-8400. Request authorization to make multiple copies for classroom use from CCC. These permissions do not extend to electronic distribution or long-term storage of articles or to copying for resale, promotion, advertising, general distribution, or creation of new collective works. For such uses, permission or license must be obtained from AFS.

Printed in the United States of America on acid-free paper

Library of Congress Control Number: 2002110128
ISBN 1-888569-42-5
ISSN 0892-2284

American Fisheries Society
5410 Grosvenor Lane, Suite 110
Bethesda, Maryland 20814-2199
USA

Contents

Preface ix
Reviewer Acknowledgments xi
Symbols and Abbreviations xiii

Part I   Early Life History, Recruitment, and Habitat Utilization

Annual Variability in the Effects of Water Temperature, Discharge, and Tidal Stage on the Migration of American Eel Elvers from Estuary to River
Brian M. Jessop 3

Annual and Seasonal Variability in the Size and Biological Characteristics of the Runs of American Eel Elvers to Two Nova Scotia Rivers
Brian M. Jessop 17

Migration and Recruitment of Tropical Glass Eels to the Mouth of the Poigar River, Sulawesi Island, Indonesia
Takaomi Arai, Daniel Limbong, and Katsumi Tsukamoto 37

Upstream Migration by Glass Eels of Two *Anguilla* Species in the Hacking River, New South Wales, Australia
Bruce Pease, Veronica Silberschneider, and Trudy Walford 47

Contrasting Use of Daytime Habitat by Two Species of Freshwater Eel *Anguilla* spp. in New Zealand Rivers
Don J. Jellyman, Marty L. Bonnett, Julian R. E. Sykes, and Peter Johnstone 63

Response of Otolith Sr:Ca to a Manipulated Environment in Young American Eels
Richard T. Kraus and David H. Secor 79

Estuarine Habitat Use by Hudson River American Eels as Determined by Otolith Strontium:Calcium Ratios
Wendy E. Morrison, David H. Secor, and Philip M. Piccoli 87

Distribution, Relative Abundance, and Habitat Use of American Eel *Anguilla rostrata* in the Virginia Portion of the Chesapeake Bay
Patrick J. Geer 101

Size and Age of American Eels Collected from Tributaries of the Virginia Portion of Chesapeake Bay
Stephen J. Owens and Patrick J. Geer 117

Recruitment of American Eels in the Richelieu River and Lake Champlain:
Provision of Upstream Passage as a Regional-Scale Solution to a Large-Scale Problem
  Richard Verdon, Denis Desrochers, and Pierre Dumont 125

Upstream Migratory Movements of American Eel *Anguilla rostrata* between the
Beauharnois and Moses-Saunders Power Dams on the St. Lawrence River
  Richard Verdon and Denis Desrochers 139

Studies of Upstream Migrant American Eels at the Moses-Saunders Power Dam
on the St. Lawrence River near Massena, New York
  Kevin J. McGrath, Denis Desrochers, Carole Fleury, and Joseph W. Dembeck IV 153

PART II EEL FISHERIES

Effect of Changes in Growth and Eel Pot Mesh Size on American Eel Yield
per Recruit Estimates in Upper Chesapeake Bay
  Julie A. Weeder and James H. Uphoff, Jr. 169

Effect of Harvest on Size, Abundance, and Production of Freshwater Eels
Anguilla australis and *A. dieffenbachii* in a New Zealand Stream
  Benjamin L. Chisnall, M. L. Martin, and Brendan J. Hicks 177

Enhancement and Management of Eel Fisheries Affected by Hydroelectric
Dams in New Zealand
  Jacques Boubée, Ben Chisnall, Erina Watene, Erica Williams, David Roper, and Alex Haro 191

Age, Growth, and Catch-Related Data of Yellow Eel *Anguilla anguilla* (L.) from
Lakes of the Erne Catchment, Ireland
  Milton A. Matthews, Derek W. Evans, Charles A. McClintock, and Christopher Moriarty 207

A Review of Eel Fisheries in Ireland and Strategies for Future Development
  Christopher Moriarty 217

The Exploitation of the Migrating Silver American Eel in the St. Lawrence River
Estuary, Quebec, Canada
  Guy Verreault, Pierre Pettigrew, Rémi Tardif, and Gontrand Pouliot 225

Estimation of the Population Size, Exploitation Rate, and Escapement of Silver-Phase
American Eels in the St. Lawrence Watershed
  François Caron, Guy Verreault, and Eric Rochard 235

An Estimation of American Eel Escapement from the Upper St. Lawrence River
and Lake Ontario in 1996 and 1997
  Guy Verreault and Pierre Dumont 243

Eel Fishing in the Great Lakes/St. Lawrence River System During the 20th Century:
Signs of Overfishing
  Jean A. Robitaille, Pierre Bérubé, Serge Tremblay, and Guy Verreault 253

PART III SPAWNING MIGRATION AND PROTECTION

Lunar Cycles of American Eels in Tidal Waters of the Southern Gulf
of St. Lawrence, Canada
  David K. Cairns and Peter J. D. Hooley 265

Do Stocked Freshwater Eels Migrate? Evidence from the Baltic Suggests "Yes"
  *Karin E. Limburg, Håkan Wickström, Henrik Svedäng, Mikael Elfman,
  and Per Kristiansson* 275

Life History Patterns of Japanese Eel *Anguilla japonica* in Mikawa Bay, Japan
  Wann-Nian Tzeng, Jen-Chieh Shiao, Yoshiaki Yamada, and Hideo P. Oka 285

Downstream Movement of Mature Eels in a Hydroelectric Reservoir
in New Zealand
  Erina M. Watene, Jacques A. Boubée, and Alexander J. Haro 295

Surface and Midwater Trawling for American Eels in the St. Lawrence River
  *Kevin J. McGrath, Joseph W. Dembeck IV, James B. McLaren, Alan A. Fairbanks,
  Kevin Reid, and Stephen J. Cluett* 307

Differentiating Downstream Migrating American Eels *Anguilla rostrata*
from Resident Eels in the St. Lawrence River
  Kevin J. McGrath, Julie Bernier, Scott Ault, Jean-Denis Dutil, and Kevin Reid 315

Development of Hydrosonic Telemetry Technologies Suitable for Tracking
American Eel Movements in the Vicinity of a Large Hydroelectric Project
  Kevin J. McGrath, Scott Ault, Kevin Reid, David Stanley, and Fred Voegeli 329

Behavioral Study of Downstream Migrating Eels by Radio-Telemetry
at a Small Hydroelectric Power Plant
  Caroline Durif, Claude Gosset, Jacques Rives, François Travade, and Pierre Elie 343

Simulated Effects of Hydroelectric Project Regulation on Mortality
of American Eels
  *Alex Haro, Theodore Castro-Santos, Kevin Whalen, Gail Wippelhauser,
  and Lia McLaughlin* 357

Evaluation of Angled Bar Racks and Louvers for Guiding Silver Phase
American Eels
  *Stephen V. Amaral, Frederick C. Winchell, Brian J. McMahon,
  and Douglas A. Dixon* 367

Review of Research and Technologies on Passage and Protection
of Downstream Migrating Catadromous Eels at Hydroelectric Facilities
  William A. Richkus and Douglas A. Dixon 377

Preface

Worldwide, catadromous eel populations are in apparent decline. The reasons are unknown, but may include oceanic influences, pollution, over–fishing, natural predation and disease, and direct and indirect migration impacts caused by water resource projects. Apparent declines in eel abundance have triggered management concerns on the part of domestic and international fisheries managers. These concerns have elevated interest in eel biology, the sources of human impacts on eel populations, and methods to mitigate them. As such, information on eel life history, habitat requirements, protection needs and methods, and approaches toward stock management are rapidly evolving. Opportunities for information exchange and dialog, therefore, are required such that management and protection needs and measures can be effectively identified. This book addresses that need and represents a collection of peer–reviewed papers based on the symposium on Biology, Management, and Protection of Catadromous Eels held at the American Fisheries Society (AFS) annual meeting in St. Louis, Missouri, 21–22 August 2000.

The symposium successfully brought together those with a common interest in eels. Symposium participants came from Canada, France, Sweden, England, Ireland, Germany, Taiwan, Japan, New Zealand, Australia, and the United States. Key topics presented and discussed at the symposium and addressed in this book include:

- New tools and techniques for studying life history and migration patterns
- Life history characteristics relevant to population dynamics
- Recruitment and upstream migratory behavior of early life stages
- Maturation cues and downstream movement of silver eels
- Eel fisheries, stock management, supplementation and restoration
- Upstream and downstream passage and protection technologies

The book is organized into three parts: (1) early life history, recruitment and habitat utilization; (2) eel fisheries; and (3) spawning migration and protection. I hope that this book will provide the most recent, relevant technical information to benefit those involved in eel management, conservation, and protection. Furthermore, I hope that this book provides important insight into the additional research needs for this highly fascinating family of fishes. A particular research need, a topic notably absent at the symposium and in the papers herein, is reproductive behavior and spawning success on their oceanic spawning grounds. In the North Atlantic, not since the work of Kracht and Tesch (1981) has research been conducted in the Sargasso Sea spawning area of European and American eel. Similar research is limited for the Pacific, although Ishikawa et al. (2001) and Kimura et al. (2001) have made recent notable reports on spawning occurrence and larval transport. In fact, as Jellyman and Tsukamoto (2002) note, of the 15 species of catadromous eel around the world, the spawning grounds of only the European, American, and Japanese eel are known with reasonable certainty. Essentially, catadromous eel spawning is a "black box" into which mature eels disappear and from which glass eels emerge. Research such as the genetic analysis by Wirth and Bernatchez (2001), which challenges the panmixia concept, and Kimura et al. (2001), which examines how physical processes affect larval transport of the Japanese eel, allow us brief glimpses into the box. Until we have a greater understanding of spawning behavior and physical processes that influence spawning success and larval transport, we will not know if the observed declines are real and result from manageable impacts, or whether they represent natural variability that result from global scale processes, such as those that affect the North Atlantic Ocean (Attrill and Power 2002; Dickson et al. 2002; Pershing 2001; Ottersen et al. 2001)

Finally, the symposium and papers reflect an enormous effort by many individuals and organizations. For co–sponsorship of the original symposium, I express my appreciation to AFS and its North Central Division. Development of the symposium and selection of papers for presentation was supported by William (Bill) Richkus of Versar Inc., Kevin McGrath of the New York Power Authority, and Alex Haro of the USGS–Conte Anadromous Fish Research Center. Bill Richkus also served as symposium co–moderator. Laura Cameron of AFS has been a most excellent and patient copy editor and extremely diligent toward completion of this book—in fact, she remained as the copy editor despite a job change during the book's development. Completion of this book involved sustained and extensive effort by all of the authors, who were aided by the thoughtful and constructive reviews and comments of many others, as subsequently acknowledged. I am grateful to all these individuals for the diligence and patience they have shown in bringing this project to fruition.

Douglas A. Dixon, Ph.D
EPRI (Electric Power Research Institute)

References

Attrill, M. J., and M. Power. 2002. Climatic influence on marine fish assemblage. Nature 417: 275–278

Dickson, B., I. Yashayaev, J. Meinke, B. Turrell, S. Dye, and J. Holfort. 2002. Rapid freshening of the deep North Atlantic Ocean over the past four decades. Nature 416: 832–836

Ishikawa, S., and 14 co–authors. 2001. Spawning time and place of the Japanese eel *Anguilla japonica* in the North Equatorial Current of the western North Pacific Ocean. Fisheries Science 67: 1097–1103

Jellyman, D., and K. Tsukamoto. 2002. First use of archival transmitters to track migrating freshwater eels *Anguilla dieffenbachii* at sea. Marine Ecology Progress Series 233: 207–215

Kimura, S., T. Inoue, and T. Sugimoto. 2001. Fluctuation in the distribution of low–salinity water in the North Equatorial Current and its effect on the larval transport of the Japanese eel. Fisheries Oceanography 10(1): 51–60

Kracht, R., and F–W. Tesch. 1981. Progress report on the eel expedition of R.V. 'Anton Dohrn' and R.V. 'Friedrich Heinke' to the Sargasso Sea 1979. Environmental Biology of Fishes 6(3/4): 371–375

Ottersen, G., B. Planque, A. Belgrano, E. Post, P. C. Reid, and N. C. Stenseth. 2001. Ecological effects of the North Atlantic Oscillation. Oecologia 128: 1–14

Pershing, A. J. 2001. Oceanographic responses to climate in the Northwest Atlantic. Oceanography 14(3): 76–82

Wirth, T., and L. Bernatchez. 2001. Genetic evidence against panmixia in the European eel. Nature 409: 1037–1040

Reviewer Acknowledgments

The following individuals assisted with reviewing manuscripts:

Steve Amaral
Takaomi Arai
Scott Ault
Aurore Baisez
David Booth
Jacques Boubée
Cédric Briand
Glen Cada
David Cairns
François Caron
John Casselman
Martin Castonguay
Ted Castro-Santos
Ben Chisnall
Willem Dekker
Denis Desrochers
Pierre Dumont
Caroline Durif
Jean-Denis Dutil
Terry Euston
Rejéan Fortin
Pat Geer
Bob Graham
Alex Haro
Brendan Hicks
Paul Jacobson
Don Jellyman
Brian Jessop
Desmond Kahn
Yutaka Kawakami
Lisa Kline
Brian Knights
Richard Kraus
Brandon Kulik

Patrick Lambert
Steve LaPan
Karin Limburg
Javier Lobon-Cervia
Karl Lundstrom
Alastair Mathers
Milton Matthews
Henry Maxwell
Kevin McGrath
Lachlan McKinnon
Christopher Moriarty
Wendy Morrison
Ken Oliviera
Tsugo Otake
Steve Owens
Bruce Pease
Russel Poole
Bill Richkus
Jean Robitaille
Dave Secor
Ned Taft
Peter Todd
François Travade
Gilles Tremblay
Wann-Nian Tzeng
Richard Verdon
Guy Verreault
Asbjørn Vøllestad
Erina Watene
Julie Weeder
Stuart Welsh
Håkan Westerberg
Håkan Wickström
Gayle Zydlewski Barbin

Symbols and Abbreviations

The following symbols and abbreviations may be found in this book without definition. Also undefined are standard mathematical and statistical symbols given in most dictionaries.

A	ampere	hp	horsepower (746 W)
AC	alternating current	Hz	hertz
Bq	becquerel	in	inch (2.54 cm)
C	coulomb	Inc.	Incorporated
°C	degrees Celsius	i.e.	(id est) that is
cal	calorie	IU	international unit
cd	candela	J	joule
cm	centimeter	K	Kelvin (degrees above absolute zero)
Co.	Company		
Corp.	Corporation	k	kilo (10^3, as a prefix)
cov	covariance	kg	kilogram
DC	direct current; District of Columbia	km	kilometer
D	dextro (as a prefix)	l	levorotatory
d	day	L	levo (as a prefix)
d	dextrorotatory	L	liter (0.264 gal, 1.06 qt)
df	degrees of freedom	lb	pound (0.454 kg, 454g)
dL	deciliter	lm	lumen
E	east	log	logarithm
E	expected value	Ltd.	Limited
e	base of natural logarithm (2.71828...)	M	mega (10^6, as a prefix); molar (as a suffix or by itself)
e.g.	(exempli gratia) for example		
eq	equivalent	m	meter (as a suffix or by itself); milli (10^{-3}, as a prefix)
et al.	(et alii) and others		
etc.	et cetera	mi	mile (1.61 km)
eV	electron volt	min	minute
F	filial generation; Farad	mol	mole
°F	degrees Fahrenheit	N	normal (for chemistry); north (for geography); newton
fc	footcandle (0.0929 lx)		
ft	foot (30.5 cm)	N	sample size
ft³/s	cubic feet per second (0.0283 m³/s)	NS	not significant
g	gram	n	ploidy; nanno (10^{-9}, as a prefix)
G	giga (10^9, as a prefix)	o	ortho (as a chemical prefix)
gal	gallon (3.79 L)	oz	ounce (28.4 g)
Gy	gray	P	probability
h	hour	p	para (as a chemical prefix)
ha	hectare (2.47 acres)	p	pico (10^{-12}, as a prefix)

Pa	pascal	USA	United States of America (noun)
pH	negative log of hydrogen ion activity	V	volt
ppm	parts per million	V, Var	variance (population)
qt	quart (0.946 L)	var	variance (sample)
R	multiple correlation or regression coefficient	W	watt (for power); west (for geography)
		Wb	weber
r	simple correlation or regression coefficient	yd	yard (0.914 m, 91.4 cm)
		α	probability of type I error (false rejection of null hypothesis)
rad	radian		
S	siemens (for electrical conductance); south (for geography)	β	probability of type II error (false acceptance of null hypothesis)
SD	standard deviation	Ω	ohm
SE	standard error	μ	micro (10^{-6}, as a prefix)
s	second	$'$	minute (angular)
T	tesla	$''$	second (angular)
tris	tris(hydroxymethyl)-aminomethane (a buffer)	$°$	degree (temperature as a prefix, angular as a suffix)
UK	United Kingdom	%	per cent (per hundred)
U.S.	United States (adjective)	‰	per mille (per thousand)

PART I

Eary Life History, Recruitment, and Habitat Utilization

Annual Variability in the Effects of Water Temperature, Discharge, and Tidal Stage on the Migration of American Eel Elvers from Estuary to River

BRIAN M. JESSOP

*Department of Fisheries and Oceans, Bedford Institute of Oceanography,
Post Office Box 1006, Dartmouth, Nova Scotia B2Y 4A2, Canada*

Abstract.—Elvers of the American eel *Anguilla rostrata* were trapped at the mouth of the East River, Chester, Nova Scotia, in order to examine migration habits. They were counted daily through out the migration run (May to mid-July) during the years 1996–1999. Nightly tidal height, river water temperature and discharge, and the difference in temperature between bay and river all influenced daily elver abundance during upstream migration at the river mouth in some years and for some portions of the elver run. The particular temporal pattern and interaction of these environmental variables in any year may explain the annual differences in variables found to be of significant effect on elver upstream movement and the seasonal pattern in that effect. After accounting for the collinearity among environmental variables, it appears that increasing tidal height acts to deliver increasing quantities of elvers to the river mouth, relative to their availability in the estuary. River temperature (10–12°C) may then act initially as a gating factor to the start of upstream migration, while river discharge controls the subsequent rate of upstream movement.

Introduction

Elvers of the American eel *Anguilla rostrata* annually migrate into eastern North American estuaries and streams during late winter and spring, earlier in the southern part of their range than in the northern part (Helfman et al. 1987). Commercial fisheries for elvers, where permitted in North America, typically occur at the mouths of rivers near the head of tide and have varied in intensity with market demand during the 1990s (Jessop 1997, 1998; ASMFC 2000). The influence of tidal phase and elver semidiurnal vertical migrations on the selective tidal stream transport of elvers through the estuary towards the river have been examined by McCleave and Kleckner (1982), McCleave and Wippelhauser (1987), and Wippelhauser and McCleave (1987). The effects of environmental conditions, such as tidal phase, estuarine salinity and temperature, and river discharge and temperature, on the movement of elvers from the upper estuary and into the river have been of scientific interest. Where fisheries occur, the movement of elvers is also of interest to fishers whose catch may vary in response to these conditions. Martin (1995) noted that "large differences in latitude, season, experimental protocol, and species studied have precluded generalizations concerning the importance of each of these factors on the onset of elver migration." Some studies have examined catch per unit effort (CPUE; Dekker 1986; Ciccotti et al. 1995), others have used elver counts (Jellyman and Ryan 1983; Hvidsten 1985; Sorensen and Bianchini 1986), while Martin (1995) used otolith-based daily age estimates from samples of elvers. Some, such as Jellyman and Ryan (1983), Sorensen and Bianchini (1986), and Ciccotti et al. (1995), have considered the time series nature of their data, but others have not. Most studies examined only one or two of the environmental factors that might influence elver migration into freshwater, but the cumulative picture that emerges places importance on tidal stage, river water velocity (discharge), and river temperature. No studies have examined a suite of environmental factors for interannual variability in effect. This study examines the interannual variability over four years in the effect of water temperature (estuary and river), tidal height, and river discharge on the migration of American eel elvers into freshwater.

Study Area

The East River, Chester, enters the Atlantic Ocean on the northeast side of Mahone Bay at 44°35'16"N, 64°10'02"W, just south of the midway point on the Atlantic coast of Nova Scotia (Figure 1). It has a watershed area of 134.0 km², of which 10.5% is lake surface area. The water is of low pH (range 4.7–5.0) and high color, due to organic acids from bog areas in the drainage. The dominant fish species, by a factor of at least four, is the American eel (Watt et al. 1997).

The elevation of the East River drops about 1.1 m over a distance of 10.6 m (slope 0.11) between the small falls at the outlet of the pond-like widening of the river and the head of tide just upriver of the Highway 3 bridge (Figure 1). Most (about 0.6 m) of the vertical drop occurs at the waterfall or within 2–3 m of it. The presence of rapids and a small falls at the mouth of the river was a major factor in the selection of this river as a project site. A presumed impediment to elver movement upstream is created by the steep slope and associated high water velocities, particularly during high discharge. Additional details are given in Jessop (2000a).

Methods

Four Irish-style elver traps (O'Leary 1971; Jessop 2000a) were operated at the river mouth annually throughout the elver run in the years 1996 through 1999 (Figure 1). Two traps were sited on each side of the river just downstream of the small falls at the river mouth and at (lower two traps), or just upstream of, the uppermost limit of tidal influence (upper two traps) and over 15 m upriver of the salt water limit. The objective was to collect all elvers migrating upstream. The vertical drops across the fall line and associated high water velocities were expected to prevent upstream movement of elvers, except when water velocities declined with seasonally reducing discharge and where elvers could find convenient, nearshore paths around the water velocity barriers. All nearshore pathways were physically blocked wherever possible so as to force elvers back into the main stream and prevent such movement as much as possible. The possibility of elvers bypassing the falls via these routes was evaluated in 1998 and 1999 by setting tube traps (Jessop 2000a) at the edge of each stream bank just upstream of the falls.

Elver catches were estimated daily, with individual elvers counted when numbers were about 150 elvers in each trap. Otherwise, counts were estimated volumetrically, in aliquots of 50, 75, or 100 mL by graduated cylinder calibrated by six to nine counts at one or more of the aliquot

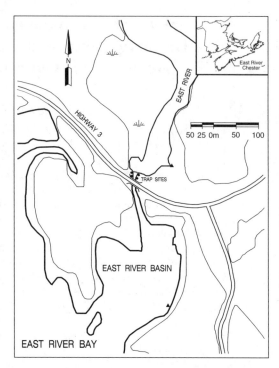

Figure 1. Map of the East River, Chester, Nova Scotia showing trap sites at the river mouth and thermograph and water level gauge sites (dark triangles).

volumes. Sometimes, other intermediate volumes were used and adjusted to the 50 mL calibration value (the slope of the calibration regression was such that the number of elvers at 100 mL was essentially double that at 50 mL). The graduated cylinder was calibrated twice during the run to account for the known decline in elver size during the run (Haro and Krueger 1988; Jessop 1998b). The total daily elver trap catch for each calibration period was estimated as $Y = \sum_i N_i \bar{y} + count$ where Y is the total daily trap catch, N_i is the number of aliquots at the ith volume (50, 75, 100 mL), $\bar{y}$ is the mean calibration count at that volume, and $count$ is the count of individual elvers (Cochran 1977). The daily trap counts for each calibration period were summed to estimate the total seasonal trap catch.

Water temperatures (to 0.1°C) were recorded every two hours by thermographs set in the East River, 80 m upriver of the falls, and at a wharf in the estuary about 0.3 km from the river mouth (Figure 1). As a proxy for river discharge (and velocity, which changes with site geomorphology), a water level logger measured the relative water level of the pond at the river mouth, hourly (to ± 2 mm). Mean daily water temperatures and levels were used in subsequent analyses. A second water level logger installed at the estuary wharf recorded hourly tidal water level fluctuation (to ± 8 mm). The maximum high tide at night was selected for subsequent analysis, rather than the diel high tide, because the traps were at, or upriver of, the high tide limit, depending upon stream discharge. Most elver movement is also at night (Tesch 1977; Dutil et al. 1988, 1989), although some daytime movement occurs later in the run (Jessop, B. M. unpublished data). Nightly high tide was moderately correlated with the diel high tide ($r = 0.54 - 0.73$, $P < 0.0001$) for the years 1998–1999, with the largest differences between the two time series occurring when the diel high tide occurred during the day.

The effects of environmental variables on the start and pattern of the elver migration were analyzed by multiple linear regression, according to a modification of the temperature and temperature-difference models proposed by Martin (1995). The temperature model was

(1) $\quad E = B_0 + B_1 T + B_2 H + B_3 M,$

where E = daily elver trap total count, B_0 = intercept, B_i = coefficient for each parameter, T = daily mean river water temperature, H = daily mean river gauge height (river level), and M = maximum tide height for the night preceding the elver count. The hypothesis proposed by Gandolfi et al. (1984), that elver migration peaked when sea and river temperatures become nearly equal, was also examined with the temperature-difference model proposed by Martin (1995). Thus, $\Delta T = (T_r - T_b)$ was substituted for T in model 1, where ΔT = the difference between mean daily river (T_r) and bay (T_b) water temperatures and was termed the temperature-difference model:

(2) $\quad E = B_0 + B_1 \Delta T + B_2 H + B_3 M.$

The environmental analysis was limited in each year to that time period in which more than 95% of the run occurred. The data analyzed in each annual time series began shortly before the start of the first wave of elvers, to various dates in each year, so as to permit an evaluation of potential changes in the effects of each environmental factor over the run. The chosen dates (Table 1; Figures 2–5) were those on which daily elver counts troughed between wave peaks, so that each date would contain one or more complete waves of elvers. The start date of each annual data series is the first date listed in Table 1 (e.g., 21 May in 1996) minus n days, where n = df + 1 (e.g., $n = 3 + 6 + 1 = 10$ d) and the end date is the final day listed (e.g., 23 June). Thus, in 1996, the data series analyzed extended from 11 May to 23 June.

Methods appropriate to regression on time series data were used (Rose et al. 1986; Wilkinson et al. 1996). The daily elver counts were logarithmically (base 10) transformed to increase normality of the count distribution. No zero counts occurred during the time periods examined, and no correction constant was required before transformation. In order to reduce, to the extent possible, the autocorrelation within each time series, to achieve independence of residuals and to avoid inflated correlations between the daily elver count and each environmental variable, and between environmental variables, each time series once to achieve stationarity (no time trend). Correlations between differenced values of the daily elver count and each environmental variable, and between environmental variables, were examined for lag effects (e.g., a delay of one or more days between occurrence of an environmental change and any effect on daily elver count or on another environmental variable).

Uncertainty exists as to whether linear or multiple linear regressions of differenced data should include a constant (y-intercept model) or not (regression through the origin or no-intercept model). Wilkinson et al. (1996) prefer regression with a constant (intercept model), unless prior knowledge requires a zero intercept, while Neter et al. (1996) indicate that differenced time series data should be analyzed with a no-intercept regression model. Although parsimonious models eliminate insignificant variables, and elimination of an insignificant constant produces a no-intercept model, use of an intercept model where the intercept is insignificant and differs from zero by only a small sampling error will have minor consequences (Neter et al. 1996). In all cases of the intercept regression model examined, the constant was not statistically significant ($P > 0.90$), and the pattern of variable significance was similar to that

Table 1. t-values and their significance, degrees of freedom (df), adjusted R^2 values and regression significance (P) for parameters of the multiple regression model $E = B_0 + B_1T + B_2H + B_3M$ where E = daily elver trap total count, B_0 = intercept, B_i = coefficient for each parameter, T = daily mean river water temperature (°C), H = daily river gauge height (river level, m), and M = maximum tide height (m) for the night preceding the elver count. The daily elver count has been logarithmically (base 10) transformed, all variable time series have been differenced and have been lagged as appropriate. In 1996, lags were applied to river temperature (–1 d), river height (–4 d), and tidal height (–3 d); in 1998, river temperature and height were each lagged –1 d. The adjusted R^2 is adjusted for the degrees of freedom.

Year	Date	t-values (Probability)			df	Adj. R^2	P
		River temperature	River height	Tide height			
1996	21 May	4.16 (0.006)	−2.11 (0.08)	−3.87 (0.008)	3,6	0.77	0.007
	1 June	1.94 (0.07)	1.78 (0.09)	−1.55 (0.14)	3,17	0.47	0.003
	6 June	2.69 (0.01)	0.74 (0.47)	−3.37 (0.003)	3,22	0.46	< 0.001
	11 June	2.76 (0.01)	1.13 (0.27)	−3.07 (0.005)	3,27	0.42	< 0.001
	23 June	2.59 (0.01)	2.12 (0.04)	−2.56 (0.01)	3,39	0.34	< 0.001
1997	30 May	−1.30 (0.26)	−0.79 (0.48)	0.05 (0.96)	3,4	0.25	0.29
	6 June	−0.93 (0.37)	−1.78 (0.09)	−0.22 (0.83)	3,11	0.08	0.29
	11 June	−0.84 (0.41)	−1.96 (0.07)	−0.13 (0.90)	3,16	0.10	0.21
	18 June	−0.85 (0.41)	−2.42 (0.02)	0.42 (0.68)	3,23	0.14	0.089
	27 June	−0.78 (0.44)	−2.79 (0.01)	0.65 (0.52)	3,32	0.14	0.051
1998	24 May	1.51 (0.15)	−1.39 (0.18)	−0.19 (0.85)	3,17	0.15	0.13
	6 June	1.71 (0.10)	−1.89 (0.07)	0.10 (0.92)	3,30	0.14	0.06
	19 June	2.49 (0.02)	−2.34 (0.02)	−0.11 (0.91)	3,43	0.24	0.002
	4 July	2.06 (0.04)	−2.16 (0.04)	−0.51 (0.61)	3,58	0.19	0.002
1999	13 May	4.62 (0.006)	1.13 (0.31)	1.28 (0.26)	3,5	0.72	0.025
	27 May	3.38 (0.003)	−0.84 (0.41)	1.46 (0.16)	3,27	0.43	0.003
	6 June	4.18 (< 0.001)	−0.86 (0.40)	1.48 (0.15)	3,29	0.45	< 0.001
	11 June	4.46 (< 0.001)	−0.98 (0.34)	1.76 (0.09)	3,34	0.46	< 0.001
	29 June	4.40 (< 0.001)	−1.48 (0.14)	0.86 (0.40)	3,52	0.37	< 0.001

produced by the no-intercept model. Thus, I have chosen to present results from the intercept model. Output from temperature models 1 and 2, with all variables included and with insignificant variables deleted (the "best" model), has been presented for completeness.

Residual plots of various types (e.g., residual on predicted value, autocorrelation, and cross-correlation function plots of residuals, and tests on residuals) revealed no serious violations of the assumptions underlying the use of the regression models. Studentized residual values, leverage measures, and Cook's D statistic indicated no unduly influential data points, and the Durbin-Watson statistic, which evaluates autocorrelation of residuals from the fitted regression, was within acceptable limits (around 2). Statistical significance was accepted at $\alpha = 0.05$.

Results

Temporal patterns in nightly tidal height, and river water temperature and level (discharge and velocity), varied greatly among years (Figures 2–5). The lunar periodicity of tidal water levels was relatively constant in pattern among years. In all years, as the elver run progressed seasonally, the river and bay water temperatures increased, while river water levels typically declined (except late spring in 1998). Elvers began actively moving upstream at river temperatures of 10–12°C, and the peak of the first wave of elvers occurred at temperatures of 11–16°C.

For the river temperature model, the influence on the daily elver count of tidal height, river temperature, and discharge varied among years and over the duration of the elver run (Table 1).

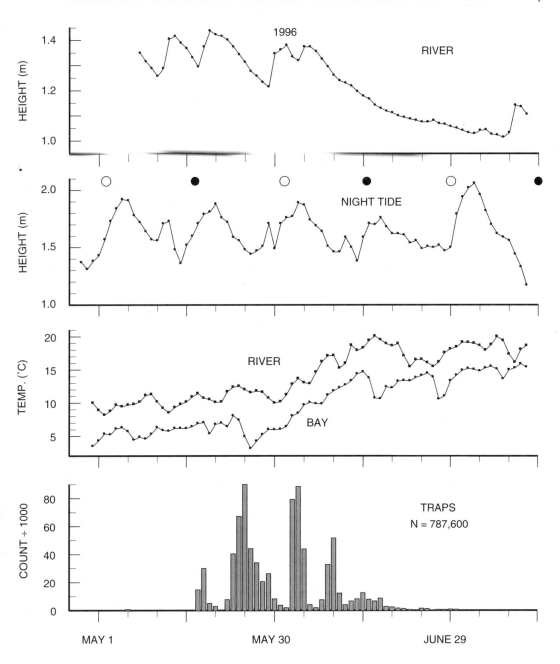

Figure 2. Daily river water levels, nightly tide heights, bay and river water temperatures and the trap catch of American eel elvers in the East River, Chester, 1996.

The sign of the *t*-value indicates whether the relation between the daily elver count and the environmental variable was positive or negative, and its magnitude indicates the relative importance of the variable. River temperature and discharge significantly affected the daily elver count at some period of the run in three of four years, while tidal height showed an effect only in 1996. In 1996, river water temperature and discharge, and tidal height showed significant effects either throughout most of the run, or at some point during the run. In 1999, only

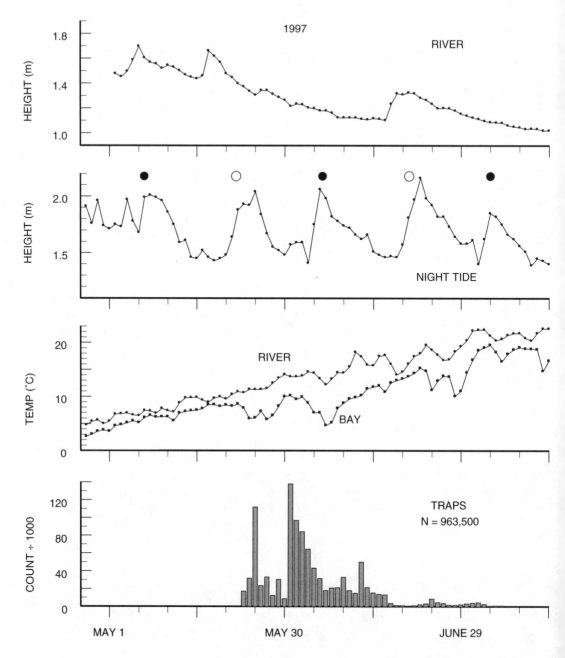

Figure 3. Daily river water levels, nightly tide heights, bay and river water temperatures and the trap catch of American eel elvers in the East River, Chester, 1997.

river water temperature showed a significant effect. When river temperature was significant in all seasonal time periods, such as 1996 and 1999, river discharge tended to be insignificant. The magnitude of the effect, as measured by the adjusted R^2, varied among years from 0.14 to 0.46, during the latter part of the elver run. However, the inclusion in the river temperature model of both river temperature and river height may be inappropriate because of collinearity. Despite differencing, highly significant ($P \leq 0.006$) correlations occurred in all years between

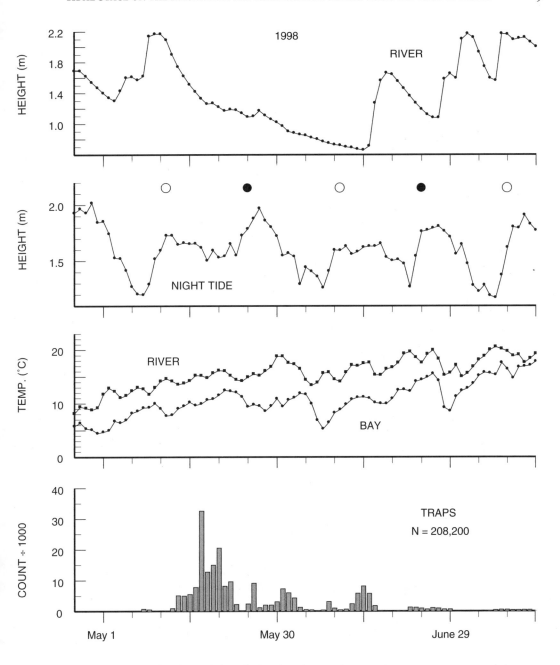

Figure 4. Daily river water levels, nightly tide heights, bay and river water temperatures and the trap catch of American eel elvers in the East River, Chester, 1998.

river water temperature and river level, with effect size (R^2) magnitudes ranging from 0.10 to 0.21. However, in all cases the condition index diagnostic for collinearity was low (typically less than two, rarely greater than six and tended to decline with increasing sample size or progression of the run), which suggests that collinearity was not a serious problem. Given that water temperature varied with water level, water temperature was dropped from the temperature model (equation 1) to produce the temperature-deleted model (Table 2). In three of four years, the daily

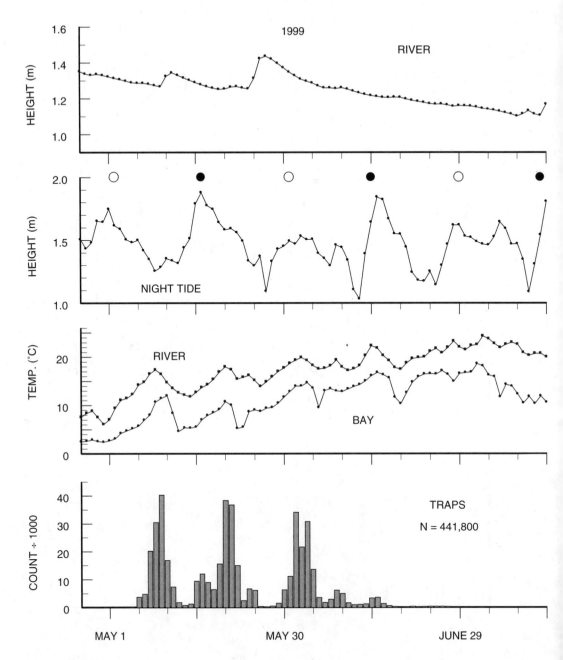

Figure 5. Daily river water levels, nightly tide heights, bay and river water temperatures and the trap catch of American eel elvers in the East River, Chester, 1999.

elver count decreased with increasing river discharge, while tide height showed no significant effect. In 1996, the daily elver count increased with increasing river height, and decreased with increasing nighttime tide level. Cross correlations between daily elver count and river discharge showed significant negative correlations at a lag of 0 d in three of four years, and positive correlations of marginal significance at negative lags of four to five days in two years. The significance of an environmental variable usually increased as the run progressed.

Table 2. t-values and their significance, degrees of freedom (df), adjusted R^2 values and regression significance (P) for parameters of the multiple regression model $E = B_0 + B_2H + B_3M$ where E = daily elver trap total count, B_0 = intercept, B_i = coefficient for each parameter, H = daily river gauge height (river level, m), and M = maximum tide height (m) for the night preceding the elver count. The daily elver count has been logarithmically (base 10) transformed, all variable time series have been differenced and have been lagged as appropriate. In 1996, lags were applied to river height (–4 d), and tidal height (–3 d); in 1998, river height was lagged –1 d. The adjusted R^2 is adjusted for the degrees of freedom.

| Year | Date | t-values (Probability) | | df | Adj. R^2 | P |
		River height	Tide height			
1996	21 May	0.37 (0.72)	–1.07 (0.32)	2,7	0.24	0.16
	1 June	3.07 (0.007)	–0.98 (0.34)	2,18	0.39	0.004
	6 June	2.32 (0.03)	–2.69 (0.013)	2,23	0.31	0.005
	11 June	2.61 (0.014)	–2.50 (0.018)	2,28	0.29	0.003
	23 June	2.96 (0.005)	–2.31 (0.026)	2,40	0.24	0.002
1997	30 May	0.39 (0.71)	1.50 (0.19)	2,5	0.15	0.29
	6 June	–1.65 (0.13)	0.41 (0.69)	2,12	0.09	0.22
	11 June	–1.85 (0.08)	0.59 (0.56)	2,17	0.11	0.14
	18 June	–2.36 (0.027)	0.86 (0.40)	2,24	0.15	0.052
	27 June	–2.72 (0.010)	0.77 (0.45)	2,33	0.15	0.026
1998	24 May	–1.78 (0.09)	–0.22 (0.82)	2,18	0.09	0.17
	6 June	–2.16 (0.04)	–0.03 (0.97)	2,31	0.08	0.10
	19 June	–3.14 (0.003)	–0.19 (0.85)	2,44	0.15	0.010
	4 July	–3.53 (0.001)	–0.70 (0.48)	2,59	0.15	0.003
1999	13 May	–0.41 (0.70)	–0.36 (0.73)	2,6	0.00	0.82
	27 May	–2.34 (0.03)	0.58 (0.57)	2,20	0.14	0.08
	6 June	–2.62 (0.014)	0.96 (0.35)	2,30	0.15	0.03
	11 June	–2.71 (0.010)	1.46 (0.15)	2,35	0.17	0.015
	29 June	–3.13 (0.003)	1.48 (0.15)	2,53	0.16	0.004

The temperature-difference model (equation 2) produced results with an annual pattern of significant effects on the daily elver count similar to those of the temperature model (equation 1), in three of four years, but with differences in the temporal patterns of significance for each variable (Tables 1 and 3). In 1999, river temperature was significant in the temperature model, while river discharge was significant in the temperature-difference model. The bay-river temperature difference was significant in the temperature-difference model in two of four years (1996, 1998). The bay-river water temperature difference and river temperature variables were significantly collinear in all years ($r = 0.27$–$0.68, P \leq 0.016$ in all years). In three of four years, the bay-river water temperature difference and river height were not significantly correlated ($P = 0.12 – 0.32$), although in 1997 the correlation was significant ($P = 0.018$).

Discussion

Elvers of the American eel began their upstream migration into the East River, Chester, between early and late May during the years 1996–1999. Elvers were present in the upper estuary for at least several weeks prior to upstream migration into freshwater (Jessop 1998a, 2000b). In all years, elvers began actively moving upstream when stream water temperatures reached 10–12°C. These temperature values are similar to the 11°C threshold for American eels suggested by Helfman et al. (1984) for Georgia, 10–15°C (peak migration at 14°C) for Rhode Island (Sorensen and Bianchini 1986), and 10–12°C for New Brunswick (Smith 1955; Groom 1975). The first wave of upstream migration peaked at water temperatures ranging from 11°C to 16°C. Stream water temperatures of less than 10–12°C evidently have a gating effect on elver upstream movement.

Table 3. t-values and their significance, degrees of freedom (df), adjusted R^2 values and regression significance (P) for parameters of the multiple regression model $E = B_0 + B_1 \Delta T + B_2 H + B_3 M$ where E = daily elver trap total count, B_0 = intercept, B_i = coefficient for each parameter, ΔT = difference between daily mean river and estuary water temperatures (°C), H = daily river gauge height (river level, m), and M = maximum tide height (m) for the night preceding the elver count. The daily elver count has been logarithmically (base 10) transformed, all variable time series have been differenced and have been lagged as appropriate. In 1996, lags were applied to temperature difference (–1 d), river height (–4 d), and tide height (–3 d); in 1998, temperature difference was lagged –1 day and height was lagged –1 day. The adjusted R^2 is adjusted for the degrees of freedom.

| Year | Date | t-values (Probability) | | | df | Adj. R^2 | P |
		Temperature difference	River height	Tide height			
1996	21 May	1.45 (0.20)	–0.16 (0.88)	–1.47 (0.19)	3,6	0.34	0.15
	1 June	0.53 (0.61)	2.64 (0.017)	–1.00 (0.33)	3,17	0.37	0.013
	6 June	1.57 (0.13)	1.99 (0.060)	–2.63 (0.015)	3,22	0.35	0.005
	11 June	2.14 (0.042)	2.10 (0.045)	–2.59 (0.015)	3,27	0.37	0.002
	23 June	2.33 (0.025)	2.70 (0.010)	–2.55 (0.015)	3,39	0.32	< 0.001
1997	30 May	–1.02 (0.37)	0.79 (0.47)	1.73 (0.16)	3,4	0.16	0.36
	6 June	0.24 (0.82)	–1.55 (0.15)	0.42 (0.68)	3,11	0.01	0.40
	11 June	0.36 (0.72)	–1.81 (0.090)	0.61 (0.55)	3,16	0.07	0.26
	18 June	0.08 (0.94)	–1.97 (0.060)	0.84 (0.41)	3,23	0.12	0.12
	27 June	0.35 (0.73)	–2.33 (0.026)	0.79 (0.44)	3,32	0.13	0.06
1998	24 May	2.16 (0.046)	–1.66 (0.11)	–0.20 (0.84)	3,17	0.24	0.054
	6 June	2.32 (0.028)	–1.99 (0.055)	–0.09 (0.93)	3,30	0.20	0.023
	19 June	3.16 (0.003)	–2.96 (0.005)	–0.15 (0.88)	3,43	0.29	0.0004
	4 July	3.63 (0.001)	–3.38 (0.001)	–0.40 (0.69)	3,58	0.29	0.0001
1999	13 May	0.03 (0.97)	–0.38 (0.72)	–0.31 (0.76)	3,5	0.00	0.95
	27 May	–0.48 (0.63)	–2.24 (0.036)	0.56 (0.57)	3,27	0.09	0.19
	6 June	–0.74 (0.46)	–2.55 (0.016)	1.01 (0.32)	3,29	0.12	0.08
	11 June	–0.54 (0.59)	–2.58 (0.014)	1.53 (0.14)	3,34	0.13	0.049
	29 June	–0.83 (0.41)	–3.02 (0.004)	1.60 (0.12)	3,52	0.13	0.015

The significant effect of stream water temperature throughout the elver migration in 1996 and 1999, later in the run in 1998, and absence of a temperature effect in 1997, differs from observations in other studies. Jellyman and Ryan (1983) concluded that temperatures early and late in the season, below about 12°C, are of most importance as a minimum threshold for migration. Sorensen and Bianchini (1986) found no relation between elver movement and water temperature above 12°C, while Martin (1995) noted a river temperature effect only during the start of the run and near the end. However, Sorensen and Bianchini (1986) did not examine water level (tidal or river). Martin (1995) did not consider the time series nature of the data, which requires adjustment for autocorrelation within variables, stationarity of trend (seasonal water temperatures tend to increase while water levels decrease) and collinearity among variables, all of which could introduce potentially serious bias into his analysis and the conclusions drawn from it. The collinearity of river water temperature and level, also noted by Jellyman and Ryan (1983), confounds interpretation of the specific effects of each variable. Collinearity reduces the precision of the estimated regression coefficients and can vary the effect of each predictor variable depending upon which are included in the model; regression coefficients can markedly change depending upon the variable composition (Neter et al. 1996). Where collinearity exists, the more dependant variable, or the one with less biological rationale, is often dropped from the regression model. Water level was retained as the primary variable because rainfall was the common factor that increased water level and decreased water temperature. Water level also acts directly upon elver movement by increasing the difficulty of upstream migration as a result of increased water velocity (McCleave 1980; Barbin and Krueger 1994) while variability in water temperatures above 10–12°C has little effect on elver movement (Jellyman and Ryan 1983; Sorensen and Bianchini 1986). Dropping water temperature from the water temperature model increased the significance of stream discharge in most cases, but did not change the influence of discharge in all years or of tide height during most of the run in 1996.

Tide height is clearly an important environmental factor on the upstream movement of elvers because of its role in their selective tidal stream transport to the head of tide (McCleave and Kleckner 1982; McCleave and Wippelhauser 1987; Wippelhauser and McCleave 1987). The wave pattern in daily elver abundance is largely due to tidal effects (Ciccotti et al. 1995). However, in the zone of interaction between tidal effect and stream discharge, annual variability in the particular pattern of interaction of these variables may account for the annual variability in their observed statistical significance. Thus, tidal height was only of significant effect during 1996, while stream discharge was significant in all years. The decrease in elver count with increasing tide height (negative effect) in 1996 results from a lagged effect of tide at the trap sites and the interaction with discharge. Tides lagged three days from the count day were declining from a peak, four to five days earlier. Of 14 major waves of elvers during the four years observed, only four waves (29%) occurred within two days of a full or new moon. The effect of a major wave of elvers between high tide periods would tend to reduce the overall effect of tide in the multiple regression analysis. Such a pattern could result from the resumption, after an initial delay due to high discharge, of upstream movement of a tidally-deposited wave of elvers following decreased river discharge. Martin (1995) concluded that tide height was a significant factor influencing the start of upstream migration in both the temperature and temperature-difference models. Sorensen and Bianchini (1986) observed a tidal effect on elver abundance only in a section of stream upriver of the mouth, rather than at the river mouth, but their river mouth study site was itself some distance up an estuary and only weakly influenced by the tide.

River discharge significantly controlled elver daily abundance in all years in the temperature-deleted and temperature-difference models. In three of four years, increasing discharge resulted in decreased daily elver counts. The increased daily elver count with increasing discharge in 1996, when tide height was a significant factor, is explainable by the significant correlation between night tide and river level at lags of one day and, less significantly, five days. These correlations are believed to be coincidental and to illustrate the effects of interaction among unrelated variables. In 1996, river level was significantly in negative correlation with daily elver counts at a lag of zero days, and in positive correlation at a lag of negative four days. Thus, daily elver counts may increase four to five days after a decrease in river discharge in some years, depending upon the tidal cycle.

Elvers undergo a period of physiological adaptation in the estuary prior to active migration upstream (Deelder 1958; Tesch 1977; Sorensen and Bianchini 1986; Haro and Krueger 1988). Any delay to upstream migration caused by such an adaptation period might extend the migration period and reduce the absolute daily abundance of elvers, but should have no effect on the pattern of movement by actively migrating elvers. The nightly catch of the commercial dip-net fishery in the upper estuary, just downstream of the elver traps, was highly correlated in timing (lag one to three, rarely five to seven days) and abundance with the trap catch (Jessop 1998b, 2000b). The degree of lag reflected the river discharge. Elvers not in an active-migration mode may hold position further down the estuary.

The effect on the daily elver count, of the difference in water temperature between bay and river, was significant at some period during the elver run in two of four years. This contrasts with the absence of any effect noted by Martin (1995), who concluded that river temperatures that were higher than bay temperatures were preferred by elvers. In all years, the mean daily river temperature during elver migration was always higher, if only slightly more so, than the bay temperature. Although the preference for elvers for warmer, rather than cooler, temperatures in long-term experiments (Tongiorgi et al. 1986; Chen and Chen 1991) may have some effect on upstream migration, the observations by other studies (Jellyman and Ryan 1983; Sorensen and Bianchini 1986) that water temperatures above 10–12°C have little effect on elver movement seem more relevant. The former observations accord with the observation by this study that river water temperature significantly influences daily elver abundance; the latter observations accord with the view that water temperature may be highly correlated with river discharge, and that discharge is the primary factor of influence when water temperatures are above the threshold value. Also, the collinearity between bay-river temperature difference and river temperature and between river temperature and discharge, a clearly influential factor, would tend to bias the relationship, possibly overstating the significance of the bay-river temperature difference. Neither Martin (1995), nor this study, found support for the hypothesis of Gandolfi et al. (1984) that elver migration peaked when bay and river temperatures were nearly equal.

Conclusions drawn early in the run about the significance of the effect of an environmental variable on the daily elver count may be increasingly less reliable at shorter time spans, and should be made cautiously. Thus, the frequent insignificance of a t-value at low degrees of freedom is equivocal, because it may simply be impossible to detect an effect of low magnitude, but of known biological importance (e.g., there is low power to correctly reject the null hypothesis).

In summary, nightly tidal height, river water temperature and discharge, and the difference in temperature between bay and river, all influenced daily elver abundance during upstream migration at the river mouth in some years and for some portions of the elver run. The particular temporal pattern and interaction of these environmental variables in any year may explain the annual differences in variables found to be of significant effect on elver upstream movement and the seasonal pattern in that effect. Given such variability, it is difficult to generalize from past studies that examined only one year or different numbers of variables or differed in other important ways (Martin 1995). After accounting for the collinearity among environmental variables, it appears that increasing tidal height acts to deliver increasing quantities of elvers to the river mouth, relative to their availability in the estuary. River temperature may then act initially as a gating factor to the initiation of upstream migration, while river discharge controls the subsequent rate of upstream movement.

Acknowledgments

I thank all those who have contributed over the years to the fieldwork for this project: J. Baltzer, N. Caron, D. Hasselman, J. Orser, and B. Zisserson. Thanks also to C. Harvie for statistical advice and for reviewing the manuscript and to D. Cairns, K. Oliveira, and K. Robichaud-LeBlanc for reviewing the manuscript.

References

ASMFC (Atlantic States Marine Fisheries Commission) 2000. Fishery management plan for American eel. Atlantic States Marine Fisheries Commission, Washington, D.C.

Barbin, G. P., and W. H. Krueger. 1994. Behaviour and swimming performance of elvers of the American eel, *Anguilla rostrata*, in an experimental flume. Journal of Fish Biology 45:111–121.

Chen, Y. L., and H. Y. Chen. 1991. Temperature selections of *Anguilla japonica* (L.) elvers, and their implications for migration. Australian Journal of Marine and Freshwater Research 42:743–750.

Ciccotti, E., T. Ricci, M. Scardi, E. Fresi, and S. Cataudella. 1995. Intraseasonal characterization of glass eel migration in the River Tiber: space and time dynamics. Journal of Fish Biology 47:248–255.

Cochran, W. G. 1977. Sampling techniques. John Wiley and Sons, New York.

Deelder, C. L. 1958. On the behavior of elvers (*Anguilla vulgaris* Turt.) migrating from the sea into freshwater. Journal du Conseil International pour l'Exploration de la Mer 24:135–146.

Dekker, W. 1986. Regional variation in glass eel catches; an evaluation of multiple sample sites. Vie Milieu 36:251–254.

Dutil, J. D., A. Giroux, A. Kemp, G. Lavoie, and J. P. Dallaire. 1988. Tidal influence on movements and on daily cycle of activity of American eels. Transactions of the American Fisheries Society 117:488–494.

Dutil, J. D., M. Michaud, and A. Giroux. 1989. Seasonal and diel patterns of stream invasion by American eels (*Anguilla rostrata*) in the northern Gulf of St. Lawrence. Canadian Journal of Zoology 67:182–188.

Gandolfi, G., M. Pesaro, and P. Tongiorgi. 1984. Environmental factors affecting the ascent of elvers, *Anguilla anguilla* (L.), into the Arno River. Oebalia 10:17–35.

Groom, W. 1975. Elver observations in New Brunswick's Bay of Fundy Region. Resource Development Branch, N. B. Department of Fisheries, Fredericton, New Brunswick, Canada.

Haro, A. J., and W. H. Krueger. 1988. Pigmentation, size and migration of elvers (*Anguilla rostrata* (LeSueur)) in a coastal Rhode Island stream. Canadian Journal of Zoology 66:2528–2533.

Helfman, G. S., E. L. Bozeman, and E. B. Brothers. 1984. Size, age, and sex of American eels in a Georgia river. Transactions of the American Fisheries Society 113:132–141.

Helfman, G. S., D. E. Facey, L. S. Hales, Jr., and E. L. Bozeman, Jr. 1987. Reproductive ecology of the American eel. Pages 42-56 in M. J. Dadswell, R. J. Klauda, C. M. Moffitt, R. L. Saunders, R. A. Rulifson, and J. E. Cooper, editors. Common strategies of andromous and catadromous fishes. American Fisheries Society, Symposium 1, Bethesda, Maryland.

Hvidsten, N. A. 1985. Ascent of elvers (*Anguilla anguilla* L.) in the stream Imsa, Norway. Report of the Institute of Freshwater Research, Drottningholm 62:71–74.

Jellyman, D. J., and C. M. Ryan. 1983. Seasonal migration of elvers (*Anguilla* spp.) into Lake Pounui, New Zealand, 1974–1978. New Zealand Journal of Marine and Freshwater Research 17:1–15.

Jessop, B. M. 1997. An overview of European and American eel stocks, fisheries, and management issues. Pages 6-20 in R. H. Peterson, editor. The American eel in Eastern Canada: stock status and management strategies. Proceedings of Eel Management Workshop, January 13–14, 1997. Quebec City, Q.C. Canadian Technical Report of Fisheries and Aquatic Sciences No. 1296.

Jessop, B. M. 1998a. The management of, and fishery for, American eel elvers in the Maritime Provinces, Canada. Bulletin Français de la Pêche et de la Pisciculture 349:103–116.

Jessop, B. M. 1998b. Geographical and seasonal variation in biological characteristics of American eel elvers in the Bay of Fundy area and on the Atlantic coast of Nova Scotia. Canadian Journal of Zoology 76:2722–2185.

Jessop, B. M. 2000a. Estimates of population size, and instream mortality rate of American eel elvers in a Nova Scotia river. Transactions of the American Fisheries Society 129:514–526.

Jessop, B. M. 2000b. Size, and exploitation rate by dip net fishery, of the run of American eel (*Anguilla rostrata*) elvers in the East River, Nova Scotia. Dana 12:51–65.

Martin, M. H. 1995. The effects of temperature, river flow, and tidal cycles on the onset of glass eel and elver migration into freshwater in the American eel. Journal of Fish Biology 46: 891–902.

McCleave, J. D. 1980. Swimming performance of European eel [*Anguilla anguilla* (L.)] elvers. Journal of Fish Biology 16:445–452.

McCleave, J. D., and R. C. Kleckner. 1982. Selective tidal stream transport in the estuarine migration of glass eels of the American eel (*Anguilla rostrata*). Journal du Conseil Permanent International pour l'Exploration de la Mer 40:262–271.

McCleave, J. D., and G. S. Wippelhauser. 1987. Behavioral aspects of selective tidal transport in juvenile American eels. Pages 138-150 in M. J. Dadswell, R. J. Klauda, C. M. Moffitt, R. L. Saunders, R. A. Rulifson, and J. E. Cooper, editors. Common strategies of andromous and catadromous fishes, American Fisheries Society, Symposium 1, Bethesda, Maryland.

Neter, J., M. H. Kutner, C. J. Nachtsheim, and W. Wasserman. 1996. Applied linear statistical models. 4th edition Irwin. Chicago, Illinois.

O'Leary, D. 1971. A low head elver trap developed for use in Irish rivers. EIFAC (European Inland Fisheries Advisory Committee) Technical Paper 14.

Rose, K. A., J. K. Summers, R. A. Cummins, and D. G. Heimbuch. 1986. Analysis of long-term ecological data using categorical time series regression. Canadian Journal of Fisheries and Aquatic Sciences 43:2418–2426.

Smith, M. W. 1955. Control of eels in a lake by preventing the entrance of young. Canadian Fish Culturist 17:13–17.

Sorensen, P. W., and M. L. Bianchini. 1986. Environmental correlates of the freshwater migration of elvers of the American eel in a Rhode Island brook. Transactions of the American Fisheries Society 115:258–268.

Tesch, F. W. 1977. The eel: biology and management of Anguillid eels. Chapman and Hall, London, Enlgand.

Tongiorgi, P., L. Tosi, and M. Balsamo. 1986. Thermal preferences in upstream migrating glass-eels of *Anguilla anguilla* (L.). Journal of Fish Biology 33:721–733.

Watt, W. D., P. J. Zamora, and W. J. White. 1997. Electrofishing data from a monitoring program designed to detect the changes in acid toxicity that are expected to result from a reduction in the long-range transport of acid pollutants into Nova Scotian salmon rivers. Canadian Data Report of Fisheries and Aquatic Sciences No. 1002.

Wilkinson, L., G. Blank, and C. Gruber. 1996. Desktop data analysis with SYSTAT. Prentice Hall, Upper Saddle River, New Jersey.

Wippelhauser, G. S., and J. D. McCleave. 1987. Precision of behaviour of migrating juvenile American eels (*Anguilla rostrata*) utilizing selective tidal stream transport. Journal du Conseil Permanent International pour l'Exploration de la Mer 44: 80–89.

Annual and Seasonal Variability in the Size and Biological Characteristics of the Runs of American Eel Elvers to Two Nova Scotia Rivers

BRIAN M. JESSOP

*Department of Fisheries and Oceans, Bedford Institute of Oceanography,
Post Office Box 1006, Dartmouth, Nova Scotia B2Y 4A2, Canada*

Abstract.—The recruitment of American eel *Anguilla rostrata* elvers to the East River, Sheet Harbor, Nova Scotia, Canada annually varied from 101,500 to 467,400 during the years 1990–1999, with no temporal trend. In the more southerly and smaller East River, Chester, recruitment annually varied from 432,400 to 1,419,000 elvers between 1996 and 2000. Annual elver counts in each river were highly correlated ($r = 0.998$, $P = 0.002$, $N = 4$), perhaps due to the effects on coastal distribution of elvers of the southwestward flowing Nova Scotia Current. Upstream migration typically occurred between early May and mid-July, beginning about 13 days earlier in the East River, Chester, than in the East River, Sheet Harbor. Two to six, usually three or four, waves or modes of daily abundance occurred over the elver run with an interval between waves of five to eight days, reflecting the tidal cycle and effects of river discharge. Sample sizes of about 1,500 elvers collected systematically throughout the run provided estimates of the seasonal population mean elver length and weight that generally differed little from estimates that were adjusted by the weekly count frequency. Annual mean length, weight, and condition index varied significantly among years and rivers. Annual variability in mean length was small relative to that for weight and condition and possibly of less biological importance. Elver mean length declined during the first six to nine weeks of the run in 13 of 16 annual observations, weight declined in all years, and the residual index of elver condition declined in 15 of 16 cases. The biological meaning should be carefully considered of statistically significant effects with small effect sizes (degree to which an effect exists) and of insignificant effects with moderate or large effect sizes.

Introduction

Large numbers of elvers of the catadromous American eel *Anguilla rostrata* enter the estuaries of eastern North American rivers during late winter and spring (Tesch 1977; Helfman et al. 1987; ASMFC 2000). Most elvers are believed to migrate into and up streams and rivers where they reside until the onset of sexual maturation and their return to the sea to spawn.

Variability in abundance and biological characteristics, whether geographic, annual or seasonal, is typical of diadromous fishes but has been minimally described for the catadromous American eel, particularly the elver phase. Vladykov (1966, 1970) and Haro and Krueger (1988) examined the latitudinal cline of increasing elver mean length along the Atlantic coast of North America and seasonal changes in size and pigmentation. Groom (1975), Hutchison (1981), Dutil et al. (1989), Jessop (1998a, b) and EPRI (1999) document various aspects of the geographic, seasonal, and occasionally annual, trends and/or changes in one or more aspects of elver length, weight, condition, and pigmentation stage. Dutil et al. (1989) also examined annual variability in elver relative abundance while Able and Fahay (1998) reported the seasonal timing and relative abundance of an elver run. Jessop (1997) provided annual estimates of elver run size based on trap catches, Jessop (1998b) examined seasonal timing and relative abundance on a regional basis and Jessop (2000a) estimated run size and seasonal mortality rate. Recent Canadian (Peterson 1997) and American (EPRI 1999) surveys of eel stock status and biology and the U.S. Interstate Fishery Management Plan for American Eel (ASMFC 2000) have noted the limited information available on the elver phase and have recommended additional research to address knowledge gaps so as to assist development of appropriate management decisions.

This study examines the annual variability in run size and timing and seasonal (May–July) mean lengths, weights, and condition indices of the elver runs to the East River, Sheet Harbor, Nova Scotia for the years 1989–1999, and to the East River, Chester, for the years 1996–2000. European eels *A. anguilla* have declined in elver recruitment and mean length since the 1970s (Desaunay and Guerault 1997). A similar relation might be hypothesized for North American elvers given the concern about the status of the eel population (EPRI 1999; ASMFC 2000).

The geographic pattern in elver catch and catch-per-unit-fishing effort (CPUE) by the commercial fishery suggests that the pattern and relative abundance of elvers distributed to different geographic areas may be annually consistent (Jessop 1998b). Thus, it can be hypothesized that annual elver counts in the two East Rivers will vary synchronously in relative abundance. Any synchrony in annual elver abundance between rivers in adjacent geographic areas might be influenced by environmental factors such as the Nova Scotia Current that runs from north-east to south-west along the Atlantic coast of Nova Scotia and is largely driven by the discharge from the St. Lawrence River (Smith 1989; Jessop 1998b). Further, annual river discharge, as modified by monthly mean precipitation during the January–June period when elvers are moving shoreward and into rivers, might have an influence on elver run size (Jellyman and Ryan 1983). Hvidsten (1985) observed a correlation between the annual size of the run of European eel elvers in the Imsa River and the cumulative number of degree-days exceeding 11°C as did Vøllestad and Jonsson (1988) with mean June–July water temperatures. Similar hypotheses were examined by correlation of the annual counts in the East River, Sheet Harbor with annual discharge from the St. Lawrence River, monthly mean precipitation, and seasonal river water temperatures.

Study Area

The East River, Sheet Harbor, (ERSH) is located on the Atlantic coast of Nova Scotia; the East River, Chester, (ERC) is located about 130 km to the southwest. Both rivers have low pH (range 4.7–5.0) and are moderately colored by organic acids (40–130 relative units) (Watt 1986; Farmer et al. 1988; Watt et al. 2000). Water temperatures increase from about 8–10°C in late April to 20–24°C in mid-July in both rivers. The ERSH drains an area of 526 km^2 and mean discharge ranges from about 23.5 m^3/s in May to 8.6 m^3/s in July (Environment Canada 1991). A 3-m high, vertical face, concrete barrier dam is located at the head of tide. The ERC drains 134 km^2 with discharge ranging from 0.5 to 4.8 m^3/s between 28 May and 20 September 1999 (Jessop 2000b). The stream mouth drops about 1.1 m over a distance of 10.6 m (slope 0.11) between the small falls (0.6 m) at the outlet and the high tide mark. Additional details about the ERSH and ERC sites may be found in Jessop (1995, 2000a, c).

Methods

Elvers were collected by Irish-type elver traps (O'Leary 1971) set up just at or upstream of the head of tide (Jessop 1998b, 2000a). In the ERSH, one elver trap was set at each side of the low head dam at the head of tide. In the ERC, two elver traps were set at each side of the river just downstream of the low falls at the river mouth. The elver traps were annually operated between about the beginning of May and mid-July. The objective was to enumerate and obtain representative life history data on a subsample of all elvers migrating upstream in each river as part of an annual elver-index monitoring program. The run was declared over when catches had declined to about 100 elvers per week. Elver trap catches were estimated daily. All elvers were counted in 1989 and 1990. Catches exceeding about 1,000 elvers were estimated with a calibrated 500-mL measuring cup between 1991 and 1993. After 1993, total catch was estimated volumetrically with a calibrated graduate cylinder with a 1 mm^2 mesh bottom for water drainage. Beginning in 1996, the calibration was repeated midway through the run so as to account for the decline in elver length and weight during the run (Jessop 1998a). Catch estimation procedures followed Jessop (2000b, c) with the exception that the total daily elver trap catch for each calibration period ($k = 1,2$) was estimated as $Y = S_i N_i \bar{y} + count$ where Y is the total daily trap catch, N_i is the number of aliquots at the ith volume (50, 75, 100 mL), $\bar{y}$ is the mean calibration count at that volume, and *count* is the count of individual elvers (Cochran 1977). The daily trap counts for each calibration period were summed to estimate the total trap catch. The variance of the estimated trap catch for each calibration period (C_k) was estimated as

$$S_C^2 = \sum_i \frac{N_i^2 S_i^2}{n_i}$$

where N_i is the number of aliquots at the ith calibration volume, S_i^2 is the variance of the calibration for that volume, and n_i is the number of counts for that volume (Cochran 1977). The standard error and 95% confidence interval for the estimated total trap catch was estimated as in Jessop (2000c).

In four of five years, an analysis of covariance (ANCOVA) of the regressions of calibration count on volume for the ERC data indicated a significant difference ($P < 0.0001$) between the adjusted mean counts for each calibration period. In 1996, there was no significant difference ($P = 0.37$) between calibration counts at 50 mL (the only measurement volume used that year) because the time between calibrations (seven days) was short relative to the three to four weeks in other years.

Dip net fisheries for commercial or aquacultural purposes occurred in each river during the years 1996–1998 (the market collapsed in 1999 and 2000). Daily commercial fisheries catches (kg), reported by logbook for each river, were converted on a weekly basis to numbers of elvers, weekly catches were summed to give the season total commercial catch, and a 95% confidence interval for the total catch was estimated following the procedures in Jessop (2000c).

When commercial fisheries occurred, the total count = fishery count + trap count. Except for 1989, the annual total or run count is believed to be a close underestimate of the true run size. A single trap that was not fully effective was used during 1989, the first year of operation at ERSH, and the run was probably largely underestimated. The low head dam in the ERSH is believed, based on observation, to be an effective barrier to upstream elver movement. In the ERC, the effectiveness as a barrier to elver movement of the low falls at the river mouth was enhanced by active measures at the stream edge to prevent their upstream movement and the accuracy of the run estimates is believed to be high (Jessop 2000b). Confidence intervals for the total run (fishery plus trap catch) were estimated in the standard manner after estimating the standard error of the total run as:

$$S_{Tot} = \sqrt{S_F^2 + S_T^2 + 2rS_FS_T},$$

where S_T^2 and S_F^2 are the variances of the fishery and trap catches and the final term adjusts for the covariance (r often about 0.6) between weekly fishery and trap catches.

Samples of elvers were collected systematically throughout the runs in each river but sampling protocols varied over time and between rivers. In the ERSH from 1989 to 1991, nonselective samples of up to 50 elvers per day, as available, were collected five to seven days per week. During 1992, up to 60 elvers per day were collected three days per week every second day, Monday to Friday. From 1993–1996, up to 30 elvers per day were collected three days per week and from 1997 to 1999, up to 30 elvers per day were collected five days per week. For the years 1989–1994, the elver samples were killed in 5% formalin and remained in the formalin for up to two hours before processing for biological data; from 1995 onwards, elvers were killed in 5% formalin and immediately processed. Elvers shrink in length and gain in weight when preserved in formalin (Jessop 1998a, 2001); consequently, the years in which elvers were preserved (1989–1994) were analyzed separately from those in which they were measured fresh (1995–1999). Elver condition was reported only for those years in which elvers were unpreserved. In the ERC, samples of up to 50 elvers per day were collected three days per week and were measured fresh immediately after killing in 4% formalin. Elvers were measured for total length (TL, to 0.1 mm with calipers), weight (to 0.01 g) after blotting dry, and classified as to pigment stage following Jessop (1998a).

Annual discharge (1990–1998) from the St. Lawrence River was obtained from the Ocean Circulation Division, Fisheries and Oceans Canada, Dartmouth, Nova Scotia and monthly mean precipitation (no discharge data are available) for the ERSH at Malay Falls was obtained for 1990–1999 from the Environment Canada Weather Service, Bedford, Nova Scotia. Water temperatures for the ERSH were recorded by thermograph (to 0.1°C) on a one or two hour frequency in 1990, 1992, 1996, and 1997 and in other years were taken daily in mid-morning with a mercury or digital thermometer.

Although the importance of a treatment effect is often evaluated by the degree of statistical significance at a chosen probability level, typically $\alpha = 0.05$, statistical significance does not necessarily imply biological importance (Kirk 1996; Johnson 1999). The potential biological importance of an observed effect can be evaluated by estimates of effect magnitude (degree to which a phenomenon is present or to which the null hypothesis is believed false), including measures of strength of association (e.g., correlation coefficient and coefficient of determination) and effect size (standardized mean difference) (Cohen 1988; Kirk 1996). Effect size can also be conceptually defined as the significance-test statistic divided by the sample size (Tatsuoka 1993). Differences in annual run lengths and weights of an elver run to a river, as estimated from weekly systematic samples, and from those samples adjusted by the weekly count (theoretically more accurate) and among annual means in elver length, weight, and condition were evaluated by measures of effect magnitude. Other statistical relations were also evaluated for effect magnitude. Effect size analyses are uncommon in fisheries studies but Myers (1997) used effect size in the meta-analysis of recruitment variation in fish populations.

Annual population mean lengths and weights for the ERSH were estimated first from weekly sample values then were estimated by adjusting the weekly sample frequency distributions by the weekly counts. The magnitude of the difference between sample-based and count-adjusted means was evaluated by Glass's g' measure of effect size, the standardized mean difference between treatment groups, where $g' = (\bar{Y}_c - \bar{Y}_s)/S_c$ and $\bar{Y}_c$ is the count-adjusted mean value, $\bar{Y}_s$ is the sample-based mean, and S_c is the standard deviation of the count-adjusted mean (Kirk 1996). The effect size of the difference between annual estimates of elver condition was estimated by Hedge's g, where $g = (\bar{Y}_{max} - \bar{Y}_{min})/S_{pooled}$ and the $\bar{Y}$ values are the maximum and minimum condition values to be compared and S_{pooled} is the error mean square from the ANCOVA of the annual weight-length regressions. Cohen (1988) provides guidelines for the interpretation of the magnitude of the experimental effect size g' and g values, where 0.2 is a small effect, 0.5 is a medium effect, and 0.8 is a large effect.

Annual and seasonal variability in mean elver length and weight, for each river and for formalin-preserved and fresh measurement groups, was evaluated by analysis of variance (ANOVA) and the Tukey-Kramer multiple comparison test. The effect magnitude of the differences among annual means in elver length and weight was evaluated by the correlation coefficient (r), which is a measure of the strength of association (Cohen 1988; Kirk 1996). Effect magnitude values for r of 0.1 are defined as small, 0.3 as medium and 0.5 as large (Cohen 1988).

Seasonal variability in weekly mean lengths and weights was examined visually by box and whisker plots and statistically by ANOVA. The magnitude of change in mean weekly length and weight over the period of the run characterized by decline (typically weeks 1 to 7–9, after which growth became evident) was evaluated by the correlation coefficient (r).

Annual and weekly sample length distributions typically showed slight nonnormality (positive skewness, occasional extended tails) as assessed by normal probability plots while weight distributions showed moderate positive skewness. Annual length and weight samples showed significant ($P < 0.01$; statistical significance was set at $\alpha = 0.05$) heterogeneity of variances (F_{max} test; Sokal and Rohlf 1981), mainly as a consequence of large ($N = 450$–$2,821$) sample sizes, and a slight tendency for variances to increase with the mean. Weekly lengths and weights, within a given year, typically had homogeneous variances. Elver lengths and weights were logarithmically (base 10) transformed to meet the assumptions of normality of distribution and homogeneity of variances that underlie linear regression, ANOVA, and ANCOVA. Independence within and among annual sample lengths and weights was assumed. An index of mean elver condition (a measure of weight relative to length, indicative of well-being) was estimated by the mean weights adjusted to a common (the overall mean) length by ANCOVA (Sokal and Rohlf 1981) of annual weight-length regressions (Cone 1989; Springer et al. 1990). Comparisons of mean elver condition among years were evaluated by Tukey-Kramer multiple comparison tests. An index of individual elver condition was estimated by the residuals from the annual weight-length regression and plotted by week to examine the seasonal pattern in the condition of individual elvers (Jakob et al. 1996; Sutton et al. 2000).

The large number and ovoid distribution of elver weight-length data points in each annual data set presents three problems during analysis. At large sample sizes (median $N = 1,020$, range 450–2,821 for ERSH; median = 1,374, range 1,181–1,549 for ERC) statistical significance may not imply biological significance because very small effect sizes become statistically significant.

Unequal numbers of observations among the range of observed sizes tends to bias regression parameters (Ricker 1975) and may lead to heterogeneity of variances at each length (Sokal and Rohlf 1981). Consequently, representative annual weight-length regressions for elvers, for comparing seasonal elver condition among years, were based upon random subsamples of up to 60 elvers per 5 mm length interval, e.g., 50.0–54.9 mm, selected from the total sample. Annual weight-length regressions were thun based on sample sizes ranging from 158 to 247 elvers for the ERSH and 246–268 elvers for the ERC. Where the requirement for homogeneity of slopes among annual data was not met, the data were subdivided into groups of years having homogeneous slopes before further analysis. The homogeneity among years of weight-length regression slopes was evaluated by ANCOVA F-test of the interaction between treatment (year) and covariate (total length) (Wilkinson et al. 1996). Estimates of weight at a given length from the weight-length regressions were back-transformed from logarithmic values following Ricker (1975, page 275).

Results

Elver Abundance and Run Timing

Between 1989 and 1999, the annual estimate of American eel elvers migrating to the ERSH ranged from 10,700 elvers in 1989, 467,400 elvers in 1997 (Table 1; Figure 1). If the 1989 elver abundance estimate is omitted as a large underestimate, the annual abundance over 10 years varied 4.6-fold, ranging from 101,500 elvers in 1995, to 467,400 elvers in 1997. No temporal trend in abundance is evident ($F = 0.19$, df = 1,9, $P = 0.67$). In the ERC, annual abundance varied threefold over the five years from 1996 to 2000, ranging from 432,400 elvers in 1998, to 1,419,000 elvers in 1997. The elver run in the ERSH annually began (daily count exceeding 50 elvers) between 13 May and 1 June while in the ERC it began between 4 May and 22 May, about 13 days earlier (range 4–28 days). Once the elver run began, it developed rapidly (Figure 2). In the ERSH, the dates by which 5% of the run had entered the river ranged among years from May 6–28 while 95% of the run had entered by 22 June to 15 July. The modal duration between the fifth and 95th percentiles of the run was 42 days. In the ERC, the dates at which 5% of the run had entered the river ranged among years from 7 to 23 May while 95% of the run had entered by 8–23 June. Between the 5th and 95th percentiles of the run, the modal duration was 29.5 days. In both rivers, the elver run essentially ended (daily count less than 50 elvers) between 6 July and 30 July, most often between 10 July and 19 July.

In the ERSH and the ERC, the annual elver run consisted of two to six, usually three or four, waves or modes of elver abundance. The first run mode occurred from two to 15 days (median five days) after the start of the run. Indications of the wave pattern can be seen in Figure 2 where a reduced slope indicates a trough

Table 1. Annual estimates, with 95% confidence intervals (CI), of the run of American eel elvers to the East River, Sheet Harbour, and the East River, Chester, Nova Scotia.

Year	East River, Sheet Harbour		East River, Chester	
	Estimate	95% CI	Estimate	95% CI
1989	10,700[a]			
1990	218,300			
1991	376,000			
1992	219,200			
1993	134,100			
1994	309,900	± 10,900		
1995	101,500	± 1,600		
1996	336,500	± 11,800	1,138,100	± 28,100
1997	467,400	± 8,500	1,419,000	± 58,900
1998	109,200	± 2,500	432,400	± 9,900
1999	134,600	± 600	441,800	± 9,800
2000			791,200	± 17,300

a. The run size was greatly underestimated due to operational problems.

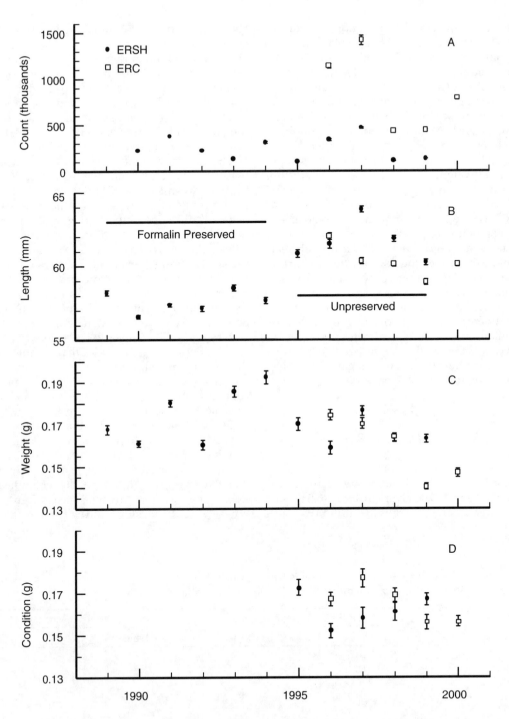

Figure 1. Annual run size (A), mean length (B), mean weight (C), and index of condition (D), with 95% confidence intervals, for American eel elvers from the East Rivers, Chester and Sheet Harbor, 1990–2000 (the 1989 run size was excluded as an underestimate). Elvers from the East River, Sheet Harbor, were preserved in formalin prior to measurement during 1989–1994. Condition was estimated as the mean weight, adjusted to the overall mean length of 60.97 mm, from weight-length regressions of the logarithmically (base 10) transformed data then back-transformed for presentation.

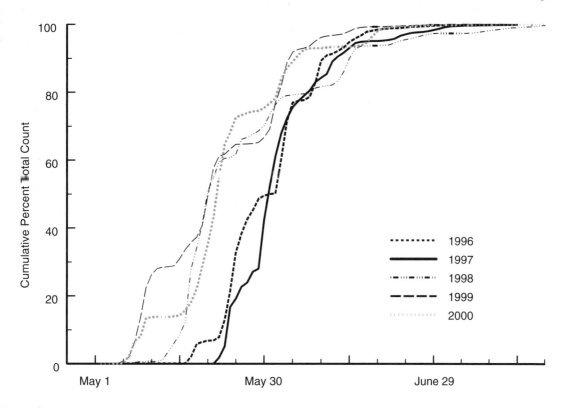

Figure 2. Seasonal progression of the annual run of American eel elvers to the East River, Chester, as indicated by the cumulative percentage of the daily count for the years 1996–2000.

between waves. The largest daily mode of elvers typically entered with the first wave of elvers (eight of eleven years in ERSH, two of five years in ERC), less frequently with the second wave. The interval between the first and second waves of elvers entering the river ranged from 5 to 25 days in the ERSH. Seven of 11 years had an interval between waves of five to eight days, and 3 of 11 years had an interval of 13–15 days. In the ERC, the period between the first two waves of elvers ranged from four to 11 days.

Annual elver counts in the two rivers were highly correlated ($r = 0.998$, $P = 0.002$, $N = 4$). Annual elver counts in the ERSH were not significantly correlated with annual ($r = 0.34$, $P = 0.33$, $N = 10$) or with January–June ($r = 0.21$, $P = 0.55$, $N = 10$) discharge from the St. Lawrence River. Nor were they correlated with January–June precipitation in the local area ($r = 0.04$, $P = 0.91$, $N = 10$) or with the number of degree-days above 11°C ($r = 0.17$, $P = 0.63$, $N = 10$,) or with mean June water temperatures ($r = 0.13$, $P = 0.73$, $N = 10$). No significant autocorrelations or cross-correlations were found at any lag in the time series of annual elver counts and environmental variables. Annual elver counts were not significantly correlated with mean elver length for either the ERSH ($r = 0.80$, $P = 0.11$, $N = 5$) or the ERC ($r = 0.61$, $P = 0.28$, $N = 5$).

Elver Length, Weight, and Condition

Estimates of the mean annual elver length and weight generally differed little whether based on systematic samples or on samples adjusted by count frequency. The effect size g' of the comparison of the two estimation methods was typically smaller for length than for weight, ranging from less than 0.01–0.41 ($N = 15$, median = 0.08) for elver length and from 0.02 to 0.57 (median = 0.13) for weight. For elver length, only one of 15 cases had a g' value exceeding 0.20 while for weight, 5 of 15 cases had a g' value exceeding 0.20, with one case exceeding 0.5. Seasonal sample sizes larger than about 1,500 elvers produced effect sizes less than 0.3 (small) for both mean length and weight. Sample sizes less than 1,500 elvers may produce effect sizes less than 0.3 but may also reach about 0.6 (large).

Annual weight distributions of elvers were often more skewed than were length distributions. For the ERSH, the median was typically about 60–65 mm for elver length (unpreserved) and 0.13–0.20 g for elver weight (Figure 3) while for the ERC, the respective values were 59–62 mm and 0.14–0.17 g (Figure 4). Outliers of the length and weight frequency distributions were more frequent at the higher side. Annual mean lengths of elvers from the ERSH were all significantly shorter for the years (1989–1994) during which they were measured after preservation than when they were measured fresh (1995–1999) (Table 2; Figures 1B, 3A). The annual mean weights of preserved elvers were higher (0.175 g) than for unpreserved elvers (0.166 g) but there was much overlap in mean weights between groups (Table 2; Figures 1C, 3B). In the ERSH and ERC, annual sample mean elver lengths and weights varied significantly ($P < 0.0001$) among years whether they were preserved or measured fresh (Table 3). The effect magnitude of the difference in mean lengths and weights among years, as measured by r, ranged from 0.17 to 0.35 (Table 3).

Annual mean elver lengths, measured fresh, were significantly smaller in the ERC than in the ERSH in three of four years (1997–1999) but were larger in 1996 (Figure 1B). Mean annual elver weights were smaller in the ERC than in the ERSH in two of four years, larger in one year and similar in one year (Figure 1C). The annual pattern of change in elver lengths and weights varied between rivers but both rivers experienced a decline in length and weight during the years 1997–1999 (Figures 1B, 1C).

Elver condition (weight adjusted to the overall mean elver length of 60.97 mm) varied significantly among years for each river (Figure 1D), as can be roughly judged by the degree of no overlap of the 95% confidence intervals. The comparison of condition among years and between rivers could not be made by ANCOVA and multiple-comparison test where the assumption of homogeneous regression slopes was not met. For the ERC the annual weight-length regression slopes (Table 4) were homogeneous ($F = 1.66$, df = 3,1007, $P = 0.17$) for the years 1996–1999 but were heterogeneous when the year 2000 was included ($F = 3.06$, df = 4, 1249, $P = 0.016$). Elver condition in the ERC varied among the years 1996–1999 ($F = 24.4$, df = 3,1010, $P < 0.0001$), such that years without a letter in common were significantly different (1997, 0.177z; 1998, 0.169zy; 1996, 0.167y; 1999, 0.156x). For the ERSH the annual (1995–1999) weight-length regression slopes were heterogeneous ($F = 15.6$, df = 4,1037, $P < 0.0001$) except among the pairs of years 1995 and 1996 ($F = 0.87$, df = 1,375, $P = 0.35$) and 1997 and 1999 ($F = 0.72$, df = 1,462, $P = 0.40$), which had homogeneous slopes. The mean condition of elvers from ERSH was significantly higher in 1995 (0.172 g) than in 1996 (0.152 g) and higher in 1999 (0.167 g) than in 1997 (0.158 g). The interaction of elver condition at various lengths when annual weight-length regression slopes are nonhomogeneous is illustrated in Figure 5 by the crossing of the lines connecting annual condition values at the lengths examined. Elver mean condition was higher in the ERC than in the ERSH for the years 1996–1998 and lower in 1999 (Figure 1D).

In the ERSH elver mean lengths varied significantly ($P < 0.01$) among weeks in 9 of 11 years (probability of annual variability in weekly mean lengths = 0.82) and weights varied significantly ($P < 0.004$) in all years (Table 5). Specifically, elver mean lengths declined significantly ($P < 0.01$) over the first six to nine weeks of the run in seven of 11 years (probability of decline in weekly mean lengths = 0.64), declined after an initial increase in length in 1 year, and varied with no trend in three years. In the ERSH, elver weights declined in all years although minimally so in two years. Elver weight increased late in the run in some years. The weekly variation in mean length was of generally small, occasionally medium, effect with r values ranging from 0.12 to 0.35, and of small to large effect for weight, with r values from 0.13 to 0.55 (Table 5). In the ERC, elver mean lengths and weights declined seasonally in all years with r values ranging from 0.21 to 0.42 for length (small to medium effect) and from 0.45 to 0.56 for weight (medium to large effect).

Individual elver condition, as represented by the residual from the annual weight-length regression, declined significantly through the first five to nine weeks of the run in 15 of 16 years of data (probability of annual decline = 0.94) from the ERSH and ERC (Table 5). For those years in which the elver run continued beyond the decline-in-condition phase, elver mean condition leveled out or increased while the variability

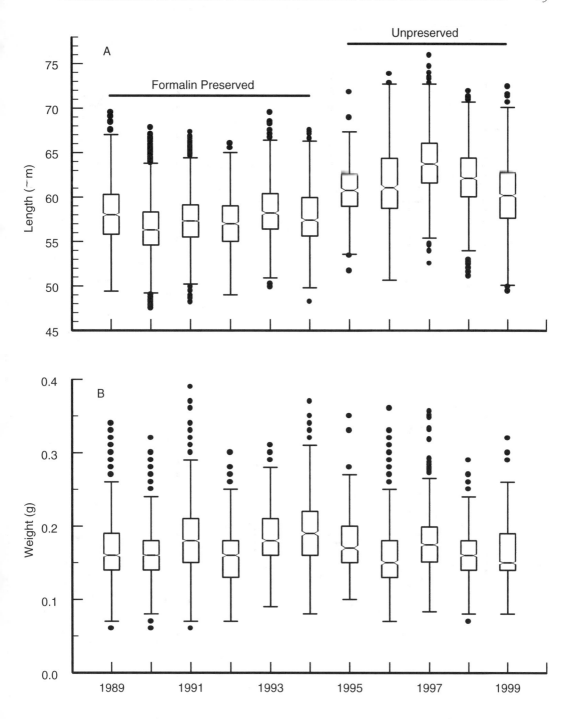

Figure 3. Box plots of the annual length and weight frequency distributions of American eel elvers from the East River, Sheet Harbor, 1989–1999. Annual sample sizes ranged from 450 elvers in 1995, to 2,821 elvers in 1991 (median 1,020 elvers). Notches indicate the 95% confidence interval about the median; the box encloses the middle 50% of the data; whiskers indicate the data range except where outliers (solid dot) occur. Length and weight data were from preserved samples during 1989–1994.

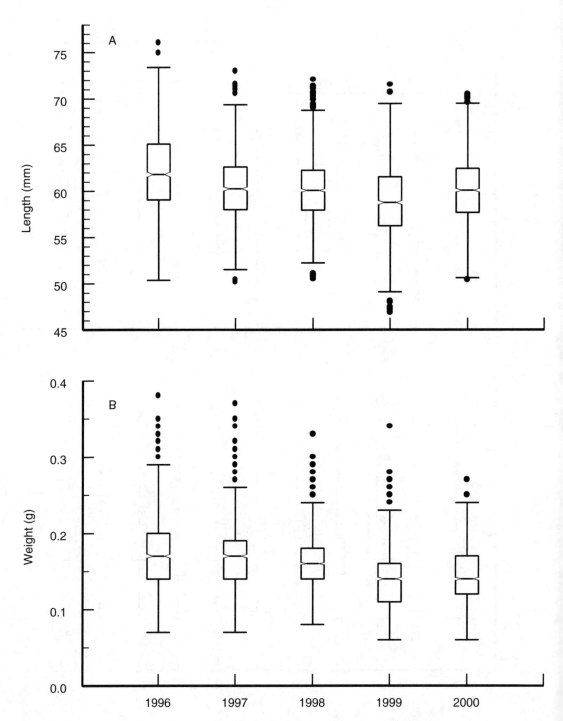

Figure 4. Box plots of the annual length and weight frequency distributions of American eel elvers from the East River, Chester, 1996–2000. Annual sample sizes ranged from 1,181 elvers in 1997, to 1,641 elvers in 2000 (median 1,446 elvers). Notches indicate the 95% confidence interval about the median; the box encloses the middle 50% of the data; whiskers indicate the data range except where outliers (solid dot) occur.

Table 2. Annual mean lengths (mm) and weights (g) of American eel elvers from the East Rivers, Sheet Harbour and Chester. Means from each river have been arranged in ascending order; those without a letter in common are significantly different at α = 0.05. The annual sample sizes (*N*) apply to both length and weight for the year indicated.

	East River, Sheet Harbour										
Length	56.57z	57.12y	57.36y	57.66y	58.20x	58.52x	60.25w	60.84v	61.49vu	61.84u	63.84t
Year	1990	1992	1991	1994	1989	1993	1999	1995	1996	1998	1997
Weight	0.159z	0.160z	0.161z	0.163zy	0.164zy	0.168yx	0.170yx	0.176xw	0.180w	0.186v	0.192v
Year	1996	1992	1990	1999	1998	1989	1995	1997	1991	1993	1994
N	2,156	1,004	2,821	868	1,316	819	1,320	450	694	1,106	1,020
	East River, Chester										
Length	58.90z	60.13y	60.14y	60.34y	62.02x						
Year	1999	1998	2000	1997	1996						
Weight	0.140z	0.147y	0.164x	0.170v	0.174v						
Year	1999	2000	1998	1997	1996						
N	1,446	1,549	1,614	1,181	1,301						

Table 3. Analysis of variance *F*, degrees of freedom (df), *P*, and correlation coefficient (*r*) values with associated scale of effect magnitude from the comparison of annual mean American eel elver lengths and weights (base 10 logarithm transformed) measured either fresh or after preservation in 5% formalin. Data from the East River, Sheet Harbour (ESHS) are for the years 1989–1994 (preserved) and 1995–1999 (fresh) and from the East River, Chester (ERC) are for 1996–2000.

River	Year-group	ANOVA	*F*	df	*P*	*r*	Effect size
ERSH	preserved	length x year	80.0	5, 8,978	< 0.0001	0.20	small
	fresh	length x year	158.8	4, 4,585	< 0.0001	0.35	medium
	preserved	weight x year	133.6	5, 8,978	< 0.0001	0.26	small
	fresh	weight x year	36.2	4, 4,585	< 0.0001	0.17	small
ERC	fresh	length x year	124.3	4, 7,086	< 0.0001	0.26	small
	fresh	weight x year	209.4	4, 7,086	< 0.0001	0.33	medium

Table 4. Annual parameter values for weight-length regressions for American eel elvers (measured fresh) from the East Rivers, Sheet Harbour and Chester. The data were random subsamples of up to 60 elvers per 5-mm length interval, e.g., 50.0–54.9 mm, selected from the total annual sample.

Year	N	r^2	Slope	95% CI	Intercept	95% CI
			East River, Sheet Harbour			
1995	158	0.63	3.0533	2.6846–3.4221	–6.2200	–6.8784 to –5.5616
1996	221	0.73	3.2794	3.0164–3.5425	–6.6793	–7.1502 to –6.2084
1997	219	0.43	2.3462	1.9855–2.7068	–4.9978	–5.6845 to –4.3470
1998	202	0.33	1.7512	1.4089–2.0936	–3.9283	–4.5409 to –3.3157
1999	247	0.74	2.5126	2.3233–2.7018	–5.2665	–5.6030 to –4.9300
			East River, Chester			
1996	268	0.74	2.7833	2.5828–2.9838	–5.7520	–6.1110 to –5.3930
1997	246	0.63	2.8372	2.5604–3.1140	–5.8257	–6.3182 to –5.3332
1998	246	0.67	2.5162	2.2952–2.7372	–5.2695	–5.6627 to –4.8763
1999	255	0.71	2.8665	2.6402–3.0928	–5.9313	–6.3325 to –5.5301
2000	244	0.82	3.1084	2.9241–3.2927	–6.3582	–6.6861 to –6.0304

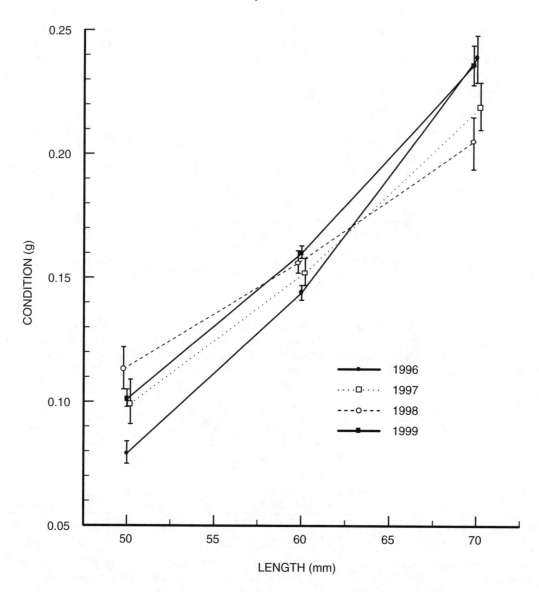

Figure 5. Index of condition, with 95% confidence intervals, at different lengths (50, 60, 70 mm) for American eel elvers from the East River, Sheet Harbor, 1996–1999. Condition was estimated as the mean weight at a given length from annual weight-length regressions of the logarithmically (base 10) transformed data and back-transformed for presentation. For each year, condition indices at different lengths were connected to illustrate the interaction effect of nonhomogeneous weight-length regression slopes.

about the mean increased. Thus, for example, during 1999 in the ERC the decline in the residual index of individual condition during the first 6 weeks of the run ($F = 28.8$, df = 1, 137, $P < 0.0001$, $r = 0.41$) was followed by a net leveling of mean condition and by an increased variance in condition (Figure 6).

Discussion

Elver Abundance and Run Timing

Annual estimates of the run of American eel elvers to the ERSH showed no temporal trend between 1990 and 1999. The cause of the decline in

Table 5. Analysis of variance results, by year, from the examination of weekly variability and trend in mean length and weight (base 10 logarithm transformed) of American eel elvers from the East Rivers, Sheet Harbour and Chester. The correlation coefficient r indicates the strength of association between mean length or weight and week. Samples at the end of the run showing a change in trend have been omitted.

					Length (mm)						Weight (g)			
Year	Weeks	Trend	N	df	F	P	r	Effect size	Trend	F	P	r	Effect size
					East River, Sheet Harbour									
1989	1–9	None[a]	1,192	7, 1,184	5.0	<0.0001	0.17	small	decline	21.9	<0.0001	0.34	medium
1990	1–7	decline	1,954	6, 1,947	12.6	<0.0001	0.19	small	decline	34.8	<0.0001	0.31	medium
1991	1–7	decline	2,180	6, 2,173	13.1	<0.0001	0.19	small	decline	76.1	<0.0001	0.42	medium
1992	1–6	decline	1,004	5, 998	3.1	0.009	0.12	small	decline	63.5	<0.0001	0.49	medium
1993	1–7	decline	529	6, 522	8.3	<0.0001	0.29	small	decline	9.6	<0.0001	0.31	medium
1994	1–8	decline	718	7, 710	13.6	<0.0001	0.34	medium	decline	43.0	<0.0001	0.55	large
1995	1–7	decline	450	6, 443	4.3	<0.0001	0.23	small	decline	11.9	<0.0001	0.37	medium
1996	1–7	none	619	6, 612	1.2	0.31	0.11	small	decline	9.2	<0.0001	0.29	small
1997	3–8	decline	763	5, 757	11.1	<0.0001	0.26	small	decline	10.1	<0.0001	0.25	small
1998	3–8	none	660	5, 654	2.0	0.07	0.12	small	decline	12.1	<0.0001	0.29	small
1999	1–8	decline	1,170	7, 1,162	3.4	0.001	0.14	small	decline	3.0	0.004	0.13	small
					East River, Chester									
1996	1–9	decline	1,201	8, 1,192	7.8	<0.0001	0.22	small	decline	39.5	<0.0001	0.46	medium
1997	1–9	decline	1,181	8, 1,172	31.6	<0.0001	0.42	medium	decline	66.5	<0.0001	0.56	large
1998	1–8	decline	1,112	7, 1,114	7.2	<0.0001	0.21	small	decline	41.6	<0.0001	0.45	medium
1999	1–7	decline	1,000	6, 993	7.3	<0.0001	0.20	small	decline	48.0	<0.0001	0.47	medium
2000	1–9	decline	1,200	8, 1,191	19.8	<0.0001	0.34	medium	decline	51.0	<0.0001	0.50	large

a. No trend although length varied significantly among weeks.

commercial eel fishery catches in many parts of eastern North America, and in some juvenile abundance indices, is uncertain (Peterson 1997; ASMFC 2000; EPRI 1999). It may reflect a decline in elver recruitment induced by marine environmental effects or human activity such as growth or recruitment overfishing or obstructed upstream access or a combination of these factors.

Whether the presently observed level of recruitment is low or high relative to historic levels is unknown. Few, and short (less than 12 years), elver time series exist for North America. There are several juvenile eel indices (EPRI 1999) but their relation to elver recruitment is uncertain. Elver recruitment to the Atlantic coast of Nova Scotia and New Jersey may have been relatively stable during the 1990s (Jessop 1997; EPRI 1999). The constant, predominantly male sex ratio for silver eels in Rhode Island since the 1980s implies that elver recruitment may also have been stable, or at least high, relative to the available habitat, because high elver densities produce male-dominated sex ratios (Krueger and Oliveira 1999). The relatively stable recruitment of European eel elvers during the 1990s in some index rivers was low relative to abundances during the 1960s–1980s but comparable to abundances in the 1930s and 1940s (Moriarty 1990; ICES 2000). Any decline in elver recruitment in North America since the 1980s may have been restricted to the Gulf of St. Lawrence (Castonguay et al. 1994a, b). More specifically, decline may have been largely restricted to the lower Gulf of St. Lawrence (Nova Scotia, New Brunswick, Prince Edward Island; Chaput et al. 1997) and the upper Gulf of St. Lawrence (Quebec; Castonguay et al. 1994a, b). The northern Gulf of St. Lawrence (north shore of Quebec east of the Saguenay River) may have been less affected (Caron and Verreault 1997). The strong correlation between the juvenile eel index at the Moses-Saunders Dam and the total annual flow through the locks at the downstream Beauharnois Dam may compromise the

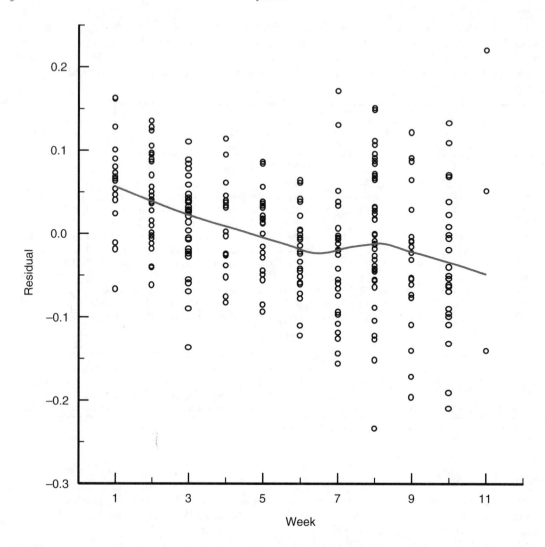

Figure 6. Seasonal pattern in the index of condition of American eel elvers from the East River, Chester, 1999. The index of individual elver condition was estimated as the residual from the weight-length regression of the logarithmically (base 10) transformed data. The line is loess smoothed.

reliability of the index as an indicator of natural juvenile eel recruitment to the upper St. Lawrence River, but not to Lake Ontario (EPRI 1999; Desrochers and Fleury 1999).

Despite the reduced elver recruitment to Europe since the 1980s, there is no evidence that the level of recruitment is insufficient to minimally stock the available habitat although the availability of recruitment and habitat may be geographically mismatched (Moriarty and Dekker 1997). Elver abundances vary geographically along the Atlantic coast of Nova Scotia and in the Bay of Fundy (Jessop 1998b). The geographic variability in oceanographic conditions provides reason to expect geographic variability in elver recruitment to other regions of the Atlantic coast of North America (Smith 1989; Castonguay et al. 1994b) as occurs along the European coast (Moriarty and Dekker 1997). The high acidity, low productivity and relatively small size of most Atlantic coastal Nova Scotia streams (Watt 1986; Watt et al. 2000) and high

(> 0.99; Jessop 2000b), possibly density-dependent, elver mortality rates (Vøllestad and Jonsson 1988) may be such that the present elver recruitment exceeds habitat carrying capacity. The high proportion (65%) of male silver eels in the ERC also implies a high elver density relative to the available habitat (Oliveira et al. 2001).

Annual elver recruitment to the ERC exceeded that to the ERSH and began earlier in the year. The high correlation between recruitment to the two rivers implies a consistent, environmentally mediated, pattern in geographic distribution. The distance between rivers with correlated run sizes has practical implications in the number and location of the index rivers required to monitor North American elver recruitment (e.g., ASMFC 2000).

Synchrony in annual elver abundance and run timing may result from the combined effects of several factors. The southwest flow of the Nova Scotia Current may affect the coastal distribution of elvers, the mid-shelf location of channels and basins in the Scotian Shelf may allow earlier shoreward access to the ERC by elvers utilizing the deep shoreward current flow in these channels, and consistent geographic and temporal differences in coastal and stream water temperature patterns may influence run timing (Smith 1989; Jessop 1998b). Although the correlation between annual discharge from the St. Lawrence River, which contributes greatly to the Nova Scotia Current, and elver run size to the ERSH was not significant, the correlation coefficient r of 0.34 indicated a medium effect size and possible biological importance. The absence over 10 years of any significant correlation between elver run size and seasonal total precipitation and a negligible effect size ($r = 0.04$) implies that stream discharge had little impact on run size. This contrasts with the conclusion by Jellyman and Ryan (1983), based on four years of data, that higher rainfall during the migratory period produced a larger elver run to a New Zealand stream. A significant correlation between the elver run and the number of degree-days greater than 11°C was noted by Hvidsten (1985) and with mean June–July water temperatures by Vøllestad and Jonsson (1988) for the Imsa River, Norway. The absence of such correlations for the ERSH may indicate a difference among regions in such a temperature effect given the similarity in temperature response by European and American eel elvers (Tesch 1977; Helfman et al. 1984; Sorensen and Bianchini 1986). The Imsa is one of the warmest streams in Norway, which may partially account for its relative success in elver recruitment (Hvidsten 1985) while the ERSH is typical of streams along the Eastern Shore of Nova Scotia. Although no significant relation was found between elver run size and mean length for either river, perhaps because of the short available time series, a simultaneous decrease was found for European elvers between the 1970s and 1990s (Desaunay and Guerault 1997).

Environmental conditions offshore and across the continental shelf may substantially control the abundance and general timing of the elver run to a river (Jessop 1998a, b, this study). However, the daily abundance and rate of movement of elvers into freshwater is controlled by environmental conditions inshore and at the river mouth (Sorensen and Bianchini 1986; Haro and Krueger 1988; Martin 1995). Waves of elver abundance with periods of 5–11 and 13–15 days during stream ascent are consistent with the interactions of selective tidal stream transport and tidal phase (McCleave and Kleckner 1982; Jessop this volume) and stream discharge and temperature (Haro and Krueger 1988; Martin 1995; Jessop 1998a). Waves of elver abundance are typical of European (Cantrelle 1984; Böetius and Böetius 1989; Ciccotti et al. 1995), Pacific (Jellyman 1979; Tzeng 1984) and American eels (Martin 1995; Jessop 1997; 1998a).

The estimation of mean population lengths and weights for the run of elvers to a stream by systematic samples was biased little by daily abundance that varied greatly over the run rather than being relatively uniform. The generally small effect size g' less than about 0.2) or small-to-medium (less than 0.3) effect size of the difference in run mean elver lengths and weights, as estimated by sample means and sample means adjusted by the elver count, implies that sample means provide an acceptable estimate of population mean values. Seasonal sample sizes of at least 1,500 elvers will generally ensure an effect size for differences between means sufficiently small to be of little biological concern. Small differences in mean lengths and weights between estimation methods may be statistically significant at $\alpha = 0.05$ because of the large sample (over one thousand elvers) sizes involved. The relative abundance of outliers in the upper range of the frequency distributions of length and weight may result from the natural occurrence of unusually large elvers and from the difficulty of readily distinguishing large elvers, when highly pigmented, from small juvenile eels in this size range (Jessop 1998a).

Some consistencies occurred between rivers (larger lengths and weights in the East River, Sheet Harbor) and among years (declining in three of four years) in the pattern of annual differences in population mean elver lengths and weights in the East Rivers, Chester and Sheet Harbor. Mean elver condition was higher in the ERC than in the ERSH in three of four years. Such consistencies may reflect the geographic and temporal consistency of local oceanographic conditions. The pattern of increasing elver size with increasing latitude over the 140 km separating the two rivers matches the geographic cline of increasing elver length, and presumably weight, from south to north along the Atlantic coast of North America (Vladykov 1966; Haro and Krueger 1988). However, it more likely results from regional effects of the Nova Scotia Current on the coastal distribution of elvers rather than continental-scale effects by the Gulf Stream (Jessop 1998a, b). Annual variability in elver mean lengths and weights for each river, while statistically significant, was generally of small effect magnitude (r values ranged from 0.17 to 0.35), suggesting that such variability is of little biological importance. A low effect magnitude (r) may be influenced by the interaction of high annual variability in individual elver length and weight and in the magnitude of their seasonal decline, thereby masking the degree to which both factors may be of biological importance. European elvers show similar geographic and temporal differences in biometrics (Desaunay and Guerault 1997).

Elver Length, Weight, and Condition

The declines in elver mean length, weight and condition during the elver run documented for single years (Haro and Krueger 1988; Jessop 1998a) can be expected to occur in almost all years and to varying degree. Seasonal effects accounted for more of the decline in elver weight (21–31% of variance accounted for) than of length (4–18%). This difference may be due to the wider variability of elver weights ($CV\%$ range = 19.5–26.5) than of lengths ($CV\%$ range = 4.6–6.6) throughout the annual run and the greater physiological effects on weight (e.g., starvation, energetic cost of swimming) during the estuarine adjustment period (Tesch 1977; Cantrelle 1984; Jessop 1998a). Statistically significant declines in seasonal mean length had low to medium effect size f values and may be of less biological importance than the declines in weight. The effect size f values for weight were often (nine of sixteen years) large (exceeding 0.4) and are probably of biological significance since biological significance for large effect sizes has been reported in other studies (Kirk 1996; Arft et al. 1999). Note that weighing a sample of 50 elvers to the nearest 0.01 g provides a potential 5% measurement difference (effect size g' of about 0.3 or small-to-medium effect) on an elver of 0.20 g with very little change in true weight. Measuring elvers to the nearest 0.1 mm provides a potential 0.2% measurement difference (effect size g' of less than 0.01) and is clearly trivial in effect.

The seasonal decline in elver length, weight, and condition during stream entrance has been linked to changes in physiology and behavior interacting with oceanic, estuarine and stream environmental conditions (Callamand 1943; Cantrelle 1984; Haro and Krueger 1988; Böetius and Böetius 1989; Jessop 1998a, b). The consequences of seasonal declines in length, weight, and condition on subsequent elver growth and mortality are poorly understood (Jessop 1998b, 2000b). Reliable estimates of seasonal parameters require sampling throughout the run. The increased elver mean weight sometimes observed at the end of the run may arise from the resumption of growth or the recent arrival of elvers with high condition although their advanced pigmentation suggests the former.

The use of an index of weight adjusted to a common length by ANCOVA to describe elver condition among different rivers, years, or even weeks within a year is problematic because of the requirement for equal regression slopes (Bolger and Connolly 1989; Cone 1989; Springer et al. 1990; Jakob et al. 1996). American eel elvers from different rivers and years have a common genetic basis for their weight-length relationship (Helfman et al. 1987) but the effects of sampling variability, seasonal decline in length and weight, and local environmental conditions may create differences among empirical weight-length relations. At 60.7 mm, the mean annual length over 10 years of data (unpreserved) from both East

Rivers (range 58.9–63.8 mm), the effect size g of the maximum difference in 1996–1999 annual mean elver condition for the East River, Sheet Harbor, was 0.65. This medium-large effect size indicates probable biological importance for the variability in annual mean condition. A comparison of elver condition at the overall mean length is biased, but minimally so, to the degree that the weight-length slopes differ, with the bias increasing at the extremes of the length distribution. The resulting interaction (change in relative magnitude or order) in weight at different lengths requires caution when interpreting differences in elver condition among stocks (Bolger and Connolly 1989). The comparison of elver condition from rivers in different geographic regions will be further biased to the extent that the latitudinal increase in mean elver length (Haro and Krueger 1988) increases the difference in mean length of elvers from the regions of interest. Also, the assumption that changes in wet body weight relative to length reflect changes in physiological condition, particularly fat content, may be weak for elvers as it is for Atlantic salmon *Salmo salar* parr (Sutton et al. 2000).

The use of a residual index of condition proved useful for examining the seasonal change in individual elver condition. Elvers that may have resumed growth following the transitional period of stream entrance and elvers that may be showing signs of starvation and potential mortality could be identified by the increased variance in the residual index of condition and the positive or negative residual value. Residual variation may be interpreted as environmentally induced variation in condition given that the genetic variation in body shape is assumed to be small between individuals, as is plausible for the American eel (Helfman et al. 1987; Sutton et al. 2000). A residual index is not appropriate for comparing condition among populations unless a common weight-length regression can be estimated (Jakob et al. 1996; Sutton et al. 2000).

Further evaluation is required of the meaning of effect sizes in relation to various aspects of elver biology. The meaning of statistically significant results should be evaluated because not all will be of probable biological importance. The biological relevance of an effect size may ultimately be determined by the nature of the study and the researcher's judgment (Cohen 1988).

Acknowledgments

I thank all those who have contributed over the years to the fieldwork for these projects. Thanks also to D. Cairns and C. Harvie for reviewing the draft manuscript and to the anonymous reviewers for their helpful comments.

References

Able, K. W., and M. P. Fahay. 1998. *Anguilla rostrata* (Lesueur) American eel. Pages 38–41 *in* The first year in the life of estuarine fishes in the middle Atlantic bight. Rutgers University Press, New Brunswick, New Jersey.

Arft, A. M., et al. 1999. Responses of tundra plants to experimental warming: meta-analysis of the international tundra experiment. Ecological Monographs 69:491–511.

ASMFC (Atlantic States Marine Fisheries Commission) 2000. Interstate fishery management plan for American eel. Fishery Management Report No. 36, Atlantic States Marine Fisheries Commission, Washington, D.C.

Böetius, I., and J. Böetius. 1989. Ascending elvers, *Anguilla anguilla*, from five European localities. Analyses of pigmentation stages, condition, chemical composition and energy reserves. Dana 7:1–12.

Bolger, T., and P. L. Connolly. 1989. The selection of suitable indices for the measurement and analysis of fish condition. Journal of Fish Biology. 34:171–182.

Callamand, O. 1943. L'Anguille Européenne: les bases physiologique de sa migration. Annales de l'Institut Océanographique 21:361–440.

Cantrelle, I. 1984. Les populations de civelles d'*Anguilla anguilla* L. en migration dans l'estuaire de la Gironde. Vie et Milieu 34:109–116.

Caron, F., and G. Verreault. 1997. Estimation du stock d'anguille d'Amerique (*Anguilla rostrata*) argentée en dévalaison dans le bas Saint-Laurent et son taux d'exploitation en 1996. Pages 94–105 *in* R. H. Peterson, editor. The American eel in Eastern Canada: stock status and management strategies. Proceedings of Eel Management Workshop, January 13–14, 1997, Quebec City, Q.C. Canadian Technical Report of Fisheries and Aquatic Sciences No. 1296.

Castonguay, M., P. V. Hodson, C. M. Couillard, M. J. Ekersley, J.-D. Dutil, and G. Verreault. 1994a. Why is recruitment of the American eel, *Anguilla rostrata*, declining in the St. Lawrence river and gulf? Canadian Journal of Fisheries and Aquatic Sciences 51:479–488.

Castonguay, M., P. V. Hodson, C. Moriarty, K. F. Drinkwater, and B. M. Jessop. 1994b. Is there a role of ocean environment in American and European eel decline? Fisheries Oceanography 3:197–203.

Chaput, G., A. Locke, and D. Cairns. 1997. Status of American eel (*Anguilla rostrata*) from the southern Gulf of St. Lawrence. p. 69-93. *In* Peterson, R. H. (editor.) The American eel in Eastern Canada: stock status and management strategies. Proceedings of Eel Management Workshop, January 13–14, 1997, Quebec City, Q.C. Canadian Technical Report of Fisheries and Aquatic Sciences No. 1296.

Ciccotti, E., T. Ricci, M. Scardi, E. Fresi, and S. Cataudella. 1995. Intraseasonal characterization of glass eel migration in the River Tiber: space and time dynamics. Journal of Fish Biology 47:248–255.

Cochran, W. G. 1977. Sampling techniques. John Wiley and Sons, New York.

Cohen, J. 1988. Statistical power analysis for the behavioural sciences. 2nd edition. Erlbaum, Hillsdale, New Jersey.

Cone, R. S. 1989. The need to reconsider the use of condition indices in fishery science. Transactions of the American Fisheries Society 118:510–514.

Desaunay, Y., and D. Guerault. 1997. Seasonal and long-term changes in biometrics of eel larvae: a possible relationship between recruitment variation and North Atlantic ecosystem productivity. Journal of Fish Biology 51(Supplement A):317–339.

Desrochers, D., and C. Fleury. 1999. Passe migratoire à anguille (*Anguilla rostrata*) au barrage de Chambly et étude de la migration des anguilles juvéniles du Saint-Laurent. [par] MILIEU Inc., [pour] Hydraulique et Environnement, Groupe Production, Hydro-Quebec, Canada.

Dutil, J.-D., M. Michaud, and A. Giroux. 1989. Seasonal and diel patterns of stream invasion by American eels (*Anguilla rostrata*) in the northern Gulf of St. Lawrence. Canadian Journal of Zoology 67:182–188.

EPRI (Electric Power Research Institute, Inc.) 1999. American eel (*Anguilla rostrata*) scoping study: a literature and data review of life history, stock status, population dynamics, and hydroelectric impacts. TR-111873. Palo Alto, California.

Environment Canada. 1991. Historical streamflow summary, Atlantic Provinces. Inland Waters Directorate, Water Survey of Canada, Ottawa, Canada.

Farmer, G. J., D. K. MacPhail, and D. Ashfield. 1988. Chemical characteristics of selected rivers in Nova Scotia during 1982. Canadian Manuscript Report of Fisheries and Aquatic Sciences 1961.

Groom, W. 1975. Elver observations in New Brunswick's Bay of Fundy region. Department of Fisheries and Environment, Fredericton, New Brunswick, Canada.

Haro, A. J., and W. H. Krueger. 1988. Pigmentation, size and migration of elvers (*Anguilla rostrata* [LeSueur]) in a coastal Rhode Island stream. Canadian Journal of Zoology 66:2528–2533.

Helfman, G. S., E. L. Bozeman, and E. B. Brothers. 1984. Size, age, and sex of American eels in a Georgia river. Transactions of the American Fisheries Society 113:132–141.

Helfman, G. S., D. E. Facey, L. S. Hales, Jr., and E. L. Bozeman, Jr. 1987. Reproductive ecology of the American eel. Pages 42–56 *in* M. J. Dadswell, R. J. Klauda, C. M. Moffitt, R. L. Saunders, R. A. Rulifson, and J. E. Cooper, editors. Common strategies of anadromous and catadromous fishes. American Fisheries Society, Symposium 1, Bethesda, Maryland.

Hutchison, S. J. 1981. Upstream migration of the glass-eel (*Anguilla rostrata*) in Nova Scotia—1981. Nova Scotia Department of Fisheries, Manuscript and Technical Report No. 81-02.

Hvidsten, N. A. 1985. Ascent of elvers (*Anguilla anguilla* L.) in the Stream Imsa, Norway. Report—Institute of Fresh-water Research, Drottningholm, Norway.

ICES (International Council for the Exploration of the Sea). 2000. Report of the EIFAC/ICES Working Group on Eels, Silkeborg, Denmark, 20-24 September. 1999. ICES CM 2000/ACFM:6.

Jakob, E. M., S. D. Marshall, and G. W. Uetz. 1996. Estimating fitness: a comparison of body condition indices. Oikos 77:61–67.

Jellyman, D. J. 1979. Upstream migration of glass-eels (*Anguilla* spp.) in the Waikato river. New Zealand Journal of Marine and Freshwater Research 13:13–22.

Jellyman, D. J., and C. M. Ryan. 1983. Seasonal migration of elvers (*Anguilla* spp.) into Lake Pounui, New Zealand, 1974-1978. New Zealand Journal of Marine and Freshwater Research 17:1–15.

Jessop, B. M. 1995. *Ichthyophthirius multifilis* in elvers and small American eels from the East River, Nova Scotia. Journal of Aquatic Animal Health 7:54–57.

Jessop, B. M. 1997. American eel elvers and their fishery in the Scotia-Fundy area of Atlantic Canada: An overview. Pages 134–143 in R. H. Peterson, editor. The American eel in Eastern Canada: stock status and management strategies. Proceedings of Eel Management Workshop, January 13–14, 1997, Quebec City, Q.C. Canadian Technical Report of Fisheries and Aquatic Sciences 1296.

Jessop, B. M. 1998a. Geographical and seasonal variation in biological characteristics of American eel elvers in the Bay of Fundy area and on the Atlantic coast of Nova Scotia. Canadian Journal of Zoology 76:2172–2185.

Jessop, B. M. 1998b. The management of, and fishery for, American eel elvers in the Maritime Provinces, Canada. Bulletin Français de la Pêche et de la Pisciculture 349:103–116.

Jessop, B. M. 2000a. Estimates of population size, and instream mortality rate of American eel elvers in a Nova Scotia river. Transactions of the American Fisheries Society 129:514–526.

Jessop, B. M. 2000b. The biological characteristics of, and efficiency of dip-net fishing for, American eel elvers in the East River, Chester, Nova Scotia. 1999. Department of Fisheries and Oceans Document No. 2000-01, Halifax, Nova Scotia, Canada.

Jessop, B. M. 2000c. Size, and exploitation rate by dip net fishery, of the run of American eel, *Anguilla rostrata* (LeSueur), elvers in the East River, Nova Scotia. Dana 12:51–65.

Jessop, B. M. 2001. Change in length, weight, and condition of American eel *Anguilla rostrata* elvers preserved in 4% and 10 % formalin. Canadian Technical Report of Fisheries, and Aquatic Sciences No. 2339.

Johnson, D. H. 1999. The insignificance of statistical significance testing. Journal of Wildlife Management 63:763–772.

Kirk, R. E. 1996. Practical significance: a concept whose time has come. Educational and Psychological Measurement 56:746–759.

Krueger, W. H., and K. Oliveira. 1999. Evidence for environmental sex determination in the American eel, *Anguilla rostrata*. Environmental Biology of Fishes 55:381–389.

Martin, M. H. 1995. The effects of temperature, river flow, and tidal cycles on the onset of glass eel and elver migration into freshwater in the American eel. Journal of Fish Biology 46:891–902.

McCleave, J. D., and R. C. Kleckner. 1982. Selective tidal stream transport in the estuarine migration of glass eels of the American eel (*Anguilla rostrata*). Journal du Conseil Internationale pour l'Exploration de la Mer 40: 262–271.

Moriarty, C. 1990. European catches of elver of 1928-1988. Internationale Revue Gestalt Hydrobiologie 75:701–706.

Moriarty, C., and W. Dekker editors. 1997. Management of the European eel. Fisheries Bulletin (Dublin) 15.

Myers, R. A. 1997. Recruitment variation in fish populations assessed using meta-analysis Pages 451–467 in R. C. Chambers and E. A. Trippel, editors. Early life history and recruitment in fish populations. Chapman and Hall, London, England.

Oliveira, K., J. D. McCleave, and G. S. Wippelhauser. 2001. Regional variation, and the effect of lake:river area on sex distribution of American eels. Journal of Fish Biology 58:942–952.

O'Leary, D. 1971. A low head elver trap developed for use in Irish rivers. EIFAC (European Inland Fisheries Advisory Committee) Technical Paper 14.

Peterson, R. H. (Editor). 1997. The American eel in Eastern Canada: stock status and management strategies. Proceedings of Eel Management Workshop, January 13–14, 1997, Quebec City, QC. Canadian Technical Report of Fisheries and Aquatic Sciences 1296.

Ricker, W. E. 1975. Computation and interpretation of biological statistics of fish populations. Fisheries Research Board of Canada Bulletin 191.

Smith, P. C. 1989. Seasonal and interannual variability of current, temperature and salinity off southwest Nova Scotia. Canadian Journal of Fisheries and Aquatic Sciences 46 (Suppl. 1):4–20.

Sokal, R. R., and F. J. Rohlf. 1981. Biometry. 2nd edition. W. H. Freeman, New York.

Sorensen, P. W., and M. L. Bianchini. 1986. Environmental correlates of the freshwater migration of elvers of the American eel in a Rhode Island brook. Transactions of the American Fisheries Society 115:258–268.

Springer, T. A., B. R. Murphy, S. Gutreuter, R. O. Anderson, L. E. Miranda, D.C. Jackson, and R. S. Cone. 1990. Properties of relative weight and other condition indices. Transactions of the American Fisheries Society 119:1048–1058.

Sutton, S. G., T. P. Bult, and R. L. Haedrich. 2000. Relationships among fat weight, body weight, water weight, and condition factors in wild Atlantic salmon parr. Transactions of the American Fisheries Society 129:527–538.

Tatsuoka, M. 1993. Effect size. Pages 461–479 *in* G. Keren and C. Lewis, editors. A handbook for data analysis in the behavioural sciences: methodological issues. Erlbaum, Hillsdale, New Jersey.

Tesch, F. W. 1977. The eel: biology and management of anguillid eels. Chapman and Hall, London, England.

Tzeng, W. N. 1984. An estimation of the exploitation rate of *Anguilla japonica* immigrating into the coastal waters off Shuang Chi River, Taiwan. Bulletin of the International Zoological Academy Sinica 23:173–180.

Vladykov, V. D. 1966. Remarks on the American eel (*Anguilla rostrata* LeSueur). Sizes of elvers entering streams; the relative abundance of adult males and females; and the present economic importance of eels in North America. Internationale Vereinigung für Theoretische und Angewandte Limnologie Verhandlungen 16:1007–1017.

Vladykov, V. D. 1970. Elvers of the American eel (*Anguilla rostrata*) in the Maritime Provinces. Pages 7–31 *in* V. D. Vladykov. Progress Reports Nos. 1-5 of the American eel (*Anguilla rostrata*) studies in Canada. Department of Fisheries and Forestry, Ottawa, Canada.

Vøllestad, L. A., and B. Jonsson. 1988. A 13-year study of the population dynamics and growth of the European eel *Anguilla anguilla* in a Norwegian river: evidence for density-dependent mortality, and development of a model for predicting yield. Journal of Animal Ecology 57:983–997.

Watt, W. D. 1986. The case for liming some Nova Scotia salmon waters. Water, Air, and Soil Pollution 31:775–789.

Watt, W. D., C. D. Scott, and P. Mandell. 2000. Acid toxicity levels in Nova Scotian rivers have not declined in synchrony with the decline in sulfate levels. Water, Air, and Soil Pollution 118:203–229.

Wilkinson, L., G. Blank, and C. Gruber. 1996. Desktop data analysis with SYSTAT. Prentice Hall, Englewood Cliffs, New Jersey.

Migration and Recruitment of Tropical Glass Eels to the Mouth of the Poigar River, Sulawesi Island, Indonesia

TAKAOMI ARAI

Otsuchi Marine Research Center, Ocean Research Institute, the University of Tokyo, 2-106-1, Akahama, Otsuchi, Iwate 028-1102, Japan

DANIEL LIMBONG

Faculty of Fisheries and Marine Science, Sam Ratulangi University, Manado 95115, Indonesia

KATSUMI TSUKAMOTO

Ocean Research Institute, the University of Tokyo, Minamidai, Nakano, Tokyo 164-8639, Japan

Abstract.—In order to determine the inshore migration mechanisms and early life history of the tropical anguillid eels, *Anguilla celebesensis*, *A. marmorata*, and *A. bicolor pacifica*, species composition and early life history based on otolith microstructure and microchemistry in Sr:Ca of glass eels of each species migrating to an Indonesian river were examined. Each of the three species occurred throughout the season with fluctuating abundance. *A. celebesensis* and *A. marmorata* occurred throughout almost the entire year in 1997 and 1998, with a peak in June. *A. bicolor pacifica* showed the same immigration pattern in 1997; however, in 1998 they were captured in only seven months (January to March, June, September, October, and December). The age of glass eels at metamorphosis and recruitment to the Poigar River were found to be almost constant throughout the year; age at metamorphosis and age at recruitment in *A. celebesensis*, *A. marmorata*, and *A. bicolor pacifica* were 88, 128, and 141 days, and 109, 155, and 173 days, respectively. The estimated spawning months in the 1997 sample, ranged throughout the year in *A. celebesensis* and *A. marmorata*; however, although the estimated spawning months of *A. bicolor pacifica* were intermittent throughout the year, this was likely due to the limited number of otoliths of this species that were examined (year-round spawning is assumed). All year-round spawning and inshore migration with a constant age at recruitment suggest year-round spawning migration of silver eel of each species in tropical rivers, a life history characteristic markedly different than temperate anguillid species.

Introduction

Considerable knowledge has been accumulated on the inshore migration and early life history of glass eels of numerous temperate anguillid species, such as *Anguilla japonica* (Matsui 1952; Tzeng 1985; Tsukamoto 1992; Cheng and Tzeng 1996; Arai et al. 1997), *A. anguilla* (Deelder 1958; Gandolfi et al. 1984; Lecomte-Finiger 1992; Arai et al. 2000a), *A. australis* and *A. dieffenbachii* (Jellyman 1977; 1979; Sloane 1984; Arai et al. 1999a; Marui et al. 2001), and *A. rostrata* (Sorensen 1986; Tongiorgi et al. 1986; Martin 1995; Wang and Tzeng 1998; Arai et al. 2000a). These reports have revealed the species composition and migration timing of glass eels, as well as environmental factors affecting the onset of glass eel migration, such as temperature, salinity, tidal cycles, and moon phase. In addition, spawning area, birth date, larval age and growth, timing of metamorphosis, and age at recruitment to the coastal waters have also been revealed.

Compared to temperate anguillid species, however, knowledge of the inshore migration of tropical species is only at a rudimentary level (Tabeta et al. 1976a; Arai et al. 1999b, c, d, 2001). Furthermore, in tropical areas some anguillid species are sympatric, and thus their exact species identification is difficult because intra-species variation of the morphological key characters considerably overlap between species (Tabeta et al. 1976a). For this reason, species identification and composition have not been well established in the tropical species.

We collected glass eel samples over the course of two years from the mouth of the Poigar River, north Sulawesi Island, Indonesia, where Arai et al. (1999b) precisely determined the species composition using morphological characteristics and mitochondrial DNA (mtDNA) analysis. In order to establish a basic foundation for ecological study of tropical eels, we examined the species composition and early life history parameters such as age at metamorphosis and recruitment, and hatching date of tropical anguillid glass eels.

Methods

Field Sampling and Species Composition Analyses

Glass eels were collected at the mouth of the Poigar River, north Sulawesi Island, Indonesia, in 1997 and 1998 (Figure 1). Sampling was conducted with triangular scoop nets quantitatively every two hours in a day at each new moon. They were caught in a 10 m transect along the beach within 1.5 m from shore using two triangular scoop nets (mouth 0.3 m², 1 mm mesh). The nets were fished simultaneously at depths of 25–50 cm in 10 replicate passes. The glass eels sampled for species composition analysis were fixed in 10% formalin immediately after collection, and the number of specimens was counted.

Samples (2–116 specimens in each month) for otolith microstructure and microchemistry analyses were also collected at night (20:00–02:00 h) just after each quantitative sampling. These samples were preserved in 99% ethanol immediately after collection. In months when only a few specimens were collected (February, March, September and December), all otoliths were examined (2–10 specimens), whereas in each of the other months (January, April to August, October and November), 30 specimens were randomly selected from each sample bottle. Otolith analyses were performed on a total of 272 specimens.

Total length (TL), predorsal, ano-dorsal and preanal length of the specimens were measured to the nearest 0.1 mm. Pigmentation stage was determined according to the method of Bertin (1956). All specimens were identified as short-finned eel, *A. bicolor pacifica* or as one of the two long-finned eels, *A. celebesensis* and *A. marmorata* according to the methods of Ege (1939), Castle and Williamson (1974) and Arai et al. (1999b).

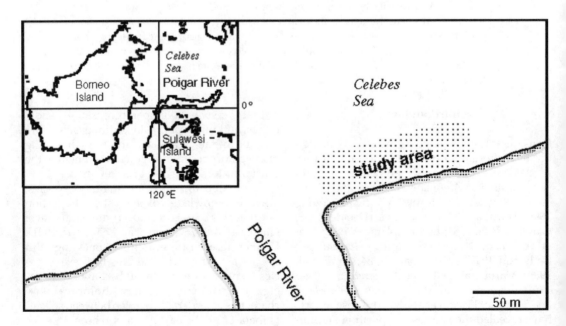

Figure 1. Location of Poigar River, Sulawesi Island, Indonesia and the study area from which glass eels were collected.

Early Life History Analysis

Sagittal otoliths were extracted from each fish and, after measuring the radius, embedded in epoxy resin (Struers, Epofix) and mounted on glass slides. Otoliths were ground to expose the core using a grinding machine equipped with a diamond cup-wheel (Struers, Discoplan-TS) and further polished with 6 μm and 1 μm diamond paste on an automated polishing wheel (Struers, Planopol-V). They were then cleaned in an ultrasonic bath and rinsed with deionized water pending subsequent examinations.

For electron microprobe analysis, 112 otoliths (54 specimens of *A. celebesensis*, 43 specimens of *A. marmorata* and 15 specimens of *A. bicolor pacifica*) were given a carbon coating by high vacuum evaporator. Strontium (Sr) and calcium (Ca) concentrations were measured along the longest axis of the otolith using a wavelength dispersive X-ray electron microprobe (JEOL JXA-733) as described in Arai et al. (1997). Accelerating voltage and beam current were 15 kV and 7 nA, respectively. The electron beam was focused on a point about 1 μm in diameter, spacing measurements at 1 μm intervals. Each data point represents the average of three measurements (each counting time: 4.0 seconds). Microprobe measurement points, which were seen as burn depressions, were assigned to otolith growth increments that were examined as described below. The average of successive data of Sr and Ca concentrations pooled in every 10 successive growth increments were used for the life history transect analysis.

Following the electron microprobe analysis, otoliths were re-polished to remove the coating, etched with 0.05 M HCl and vacuum coated with Pt-Pd in an ion-sputterer for examination with a scanning electron microscope (SEM, Hitachi S-4500). Otoliths of 160 specimens (135 specimens of *A. celebesensis* and 25 of *A. marmorata*) that were not used for electron microprobe analysis were also etched at the ground surfaces and coated by the same procedure for SEM examination. SEM photographs at various magnifications (150×, 180×, 1000×, and 1500×) were used for counting the number of growth increments and measuring increment widths. The longest "radius" from the core to edge in the ground otolith surface was regarded as the otolith radius along which increment width were measured. The averages of every 10 succeeding ring widths between the hatch check (Umezawa et al. 1989) of about 10 μm in diameter and the otolith edge were used for otolith growth analysis. Based on the findings of Arai et al. (2000b) and Sugeha et al. (2001a) that the growth rings in *A. celebesensis* and *A. marmorata* were formed daily, the equivalent rings in *A. bicolor pacifica* were also considered to be formed daily.

Results

Species Composition

Glass eels occurred at the Poigar estuary throughout the course of the study (Figure 2). The peak season of inshore migration was June in both years, while total catches of glass eels were approximately 17 times higher in 1997 (21,741 specimens) than 1998 (1,243 specimens). All specimens showed undeveloped pigmentation with only a few melanophores in the caudal region, or at the skull, caudal and rostral regions of the body. These were classified as being in the glass eel stage (VA or VB in pigmentation stage as per Bertin [1956]).

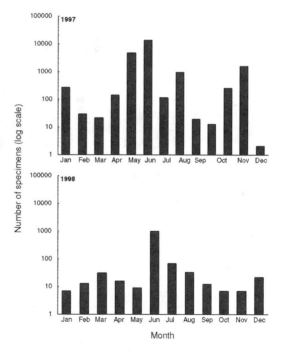

Figure 2. Monthly catches of glass eels in new moon days at the mouth of the Poigar River throughout 1997 (top) and 1998 (bottom).

The three species, *A. celebesensis*, *A. marmorata*, and *A. bicolor pacifica*, occurred throughout the season with fluctuating abundance (Figure 3). *A. celebesensis* dominated (87% in 1997, and 85% in 1998) the collections, occurring almost throughout the year with a peak in June for both years. *A. marmorata* was the second most dominant (12% in 1997, and 13% in 1998), and was seen throughout the year with peak in June in both years. *A. bicolor pacifica* comprised the lowest percentage of the collections of the three species (1% in 1997, and 2% in 1998). The immigration pattern of *A. bicolor pacifica* in 1997 was the same as the two other species, except the abundance peak was observed in August. In 1998, however, *A. bicolor pacifica* was only observed during the months of January to March, June, September, October, and December.

Mean total lengths throughout the year were significantly different among the three species (analysis of variance (ANOVA); $P < 0.005 - 0.0001$) in each year, except for between *A. marmorata* and *A. bicolor pacifica* in 1998 ($P > 0.05$). Mean total lengths ($\pm$ SD) for each species in 1997

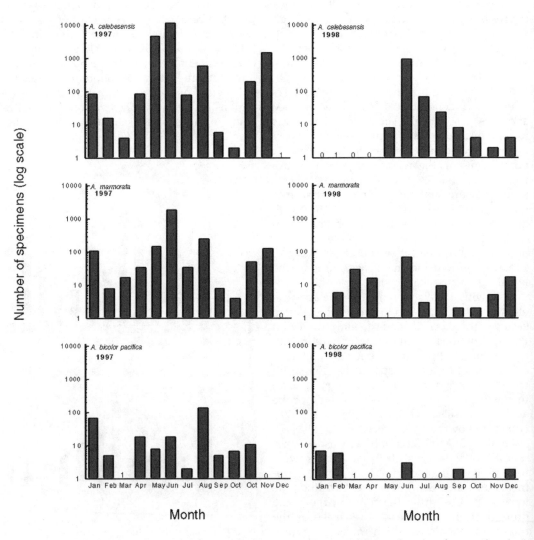

Figure 3. Monthly catches of *A. celebesensis* (top), *A. marmorata* (middle) and *A. bicolor pacifica* (bottom) glass eels at the mouth of the Poigar River in 1997 (left) and 1998 (right).

and 1998, respectively were: *A. celebesensis* (49.1 ± 2.2 mm and 49.4 ± 1.7 mm), *A. marmorata* (51.1 ± 2.0 mm and 51.7 ± 1.7 mm) and *A. bicolor pacifica* (49.6 ± 2.0 mm and 50.7 ± 2.0 mm). We found no significant difference in mean total length in each species between 1997 and 1998 ($P > 0.05$).

Otolith Microstructure and Microchemistry

All otoliths of the glass eels of each species displayed a similar growth pattern and Sr:Ca along the core to edge life history transect (Figure 4). The drastic increase in otolith increment width were exactly overlapped with the sharp drop of otolith Sr:Ca in all species examined. Otolith increment widths increased between the hatch check and age 20–40 days in each species (first phase), thereafter becoming constant or gradually decreasing where average widths were 0.41–0.47 μm (second phase). Beyond the second phase, increment width increased sharply to a maximum, averaging 2.81–3.02 μm (third phase), followed by a rapid drop (fourth phase). Sr:Ca, averaging 8.3–9.6 × 10^{-3} in the core, slightly dropped around the end of the first phase of otolith growth. Sr:Ca subsequently increased to a maximum level, averaging 15.7–17.2 × 10^{-3} during the second phase of otolith growth. A marked decrease was found thereafter toward the edge, which was coincident with the beginning of third phase.

Metamorphosis and Recruitment

Based on the otolith microstructure and microchemistry analyses, the onset of metamorphosis in *A. celebesensis*, *A. marmorata* and *A. bicolor pacifica* was estimated to occur at the mean (± SD) age 88 days (± 9.8 d), 128 days (± 15.2 d) and 141 days (± 20.9 d), respectively (Figure 5). Mean age of onset of metamorphosis was significantly different between the species (ANOVA, $P < 0.01$–0.0001). Mean age of onset of metamorphosis did not significantly differ between months of collection in each species (ANOVA, $P > 0.1$).

Mean age at recruitment of *A. celebesensis*, *A. marmorata* and *A. bicolor pacifica* varied significantly (ANOVA, $P < 0.01$) and were (mean ± SD) 109 ± 10.9 days, 155 ± 14.8 days, and 173 ± 20.9 days, respectively (Figure 6). Mean age at recruitment did not significantly differ between months of collection in each species (ANOVA, $P > 0.1$).

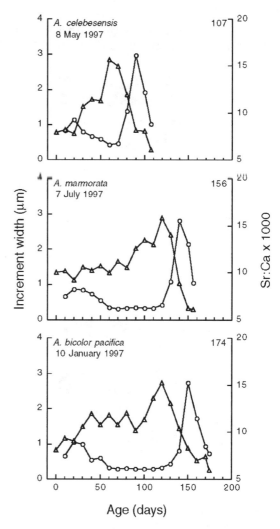

Figure 4. Tropical *Anguilla* spp. Profiles of otolith incremental widths from the core to the edge (circle) and otolith Sr:Ca measured with a wavelength dispersive electron microprobe from the core to the edge (triangle). Each point represents the average of data for every 10 days. Numbers at the upper right indicate age (days). Specimens were collected at the mouth of the Poigar River in 1997.

The estimated hatching month of *A. marmorata*, back-calculated from their sampling date and age, occurred throughout the year (July 1996 to June 1997; Figure 7). Hatching month of *A. celebesensis* was estimated to occur almost throughout the year (September 1996 to August 1997, except May and November). However, due

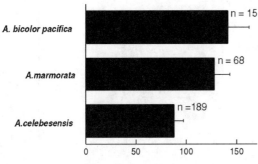

Figure 5. Mean age (± SD) at metamorphosis of *A. celebesensis*, *A. marmorata* and *A. bicolor pacifica*, as estimated from glass eel otolith microstructure analysis.

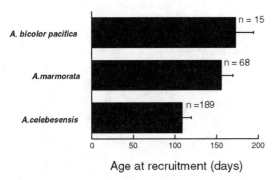

Figure 6. Mean age (± SD) at recruitment of *A. celebesensis*, *A. marmorata* and *A. bicolor pacifica*, as estimated from glass eel otolith microstructure analysis.

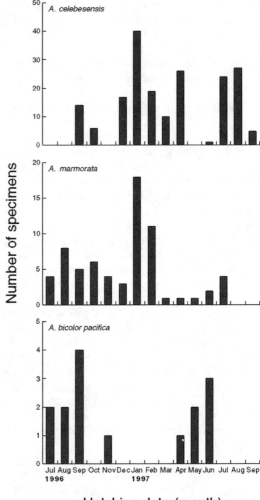

Figure 7. Hatching month of *A. celebesensis* (top), *A. marmorata* (middle) and *A. bicolor pacifica* (bottom), as revealed by aging of otoliths from 272 glass eels caught at the mouth of the Poigar River in 1997.

to limited presence in collections, hatching month of *A. bicolor pacifica* was estimated to occur during only seven months (July 1996 to June 1997, except for October, and December through March).

Inshore Migration

The relationship between the age at metamorphosis and the age at recruitment clearly showed that glass eels that underwent metamorphosis at an earlier age tended to migrate or recruit to inshore areas at a younger age (Fisher's Z-transformation, $P < 0.0001$; Figure 8).

Discussion

The pattern in otolith microstructure and microchemistry observed in the tropical species of this study, i.e., the sudden increase in increment width with a corresponding decrease in Sr:Ca is quite similar to the patterns in temperate eels *A. japonica* (Otake et al. 1994; Arai et al. 1997), *A. australis* (Arai et al. 1999a), *A. rostrata* and *A. anguilla* (Arai et al. 2000a), and *A. dieffenbachii* (Marui et al. 2001). Otake et al. (1997) ascertained

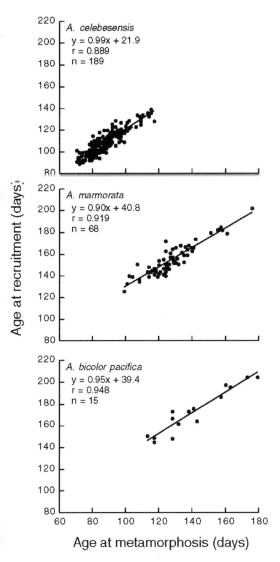

Figure 8. Relationship between age at metamorphosis (days) and age at recruitment (days) in *A. celebesensis* (top), *A. marmorata* (middle) and *A. bicolor pacifica* (bottom) glass eels caught at the mouth of the Poigar River in 1997.

crease in otolith increment widths and sharp decrease in Sr:Ca heralded the onset of metamorphosis. Therefore, early life history transect in otoliths of the tropical eels examined in this study can be interpreted as follows: first phase for preleptocephalus, second phase for leptocephalus, third phase for metamorphosis stage and fourth phase for glass eel.

The mean age at metamorphosis in the three tropical species we studied ranged from 88 to 141 days, and the mean age at recruitment was estimated to occur from 109 to 173 days. Mean age at metamorphosis and recruitment differed significantly between species. It is noteworthy, however, that the duration of metamorphosis stage was a constant 16–19 days in each of the three species. Arai et al. (1997, 1999c, d) suggested that metamorphosis was completed before the maximum peak of otolith increment width, because some specimens had no maximum peak in increment width and the following decreasing phase (fourth phase). Therefore, the duration of metamorphosis is likely to be almost identical, at most 16–19 days for each species.

We also observed, based on the relationship between the age at metamorphosis and the age at recruitment, that glass eels that underwent metamorphosis at an earlier age tended to migrate to inshore or estuarine waters at a younger age. Tsukamoto and Umezawa (1994), Arai et al. (1997, 1999a, 2000a), and Marui et al. (2001) have found the same phenomenon in the temperate eels, *A. japonica*, *A. australis*, *A. rostrata*, *A. anguilla*, and *A. dieffenbachii*. Therefore, this early life history strategy also appears to be common in anguillid fishes.

Spawning of *A. celebesensis* and *A. marmorata* were found to occur throughout the year. Spawning of *A. bicolor pacifica*, however, was estimated to occur in only seven months of the year possibly due to the early life history analyses being carried out only five months of *A. bicolor pacifica* samples. Because *A. bicolor pacifica* glass eels were found to occur throughout the year at the mouth of the Poigar River in 1997 and because the duration of leptocephalus stage and age at recruitment were found to be constant throughout the year, the spawning season may also occur throughout the year in this species. In the temperate species, spawning occurs over a limited period; i.e., February to April in *A. rostrata* (McCleave et al. 1987), March to June in *A. anguilla* (McCleave et al. 1987), April to

the same phenomenon in *Conger myriaster*. Coincidence of timings between rapid increase in increment width and sharp decrease in otolith Sr:Ca appears to be a common pattern in anguillid fishes. Arai et al. (1997) concluded, by comparing microstructure and microchemistry analyses in fully grown *A. japonica* leptocephalus and glass eel that the timing of rapid in-

November in *A. japonica* (Tsukamoto 1990), August to December in *A. dieffenbachii* (Marui et al. 2001), and October to February in *A. australis* (Arai et al. 1999a; Marui et al. 2001). The difference in the spawning seasons between tropical species and temperate species might be due to the difference in timing of their seaward migration of maturing adult eels. Those in the temperate species occur over a limited period, August to November in *A. rostrata* (Hain 1975), August to December in *A. anguilla* (Haraldstad et al. 1985), August to December in *A. japonica* (Matsui 1952), August to October in *A. dieffenbachii* (Jellyman 1987) and February to April in *A. australis* (Jellyman 1987). In the tropical species examined in this study, based on estimated year round spawning, spawning migrations may be occurring throughout the year, although no direct information are available on migration timing of maturing adult eels in these tropical species. The year-round spawning in tropical species and constant larval growth throughout the year might extend the recruitment period to the estuarine habitats throughout the year, as found in previous studies (Tabeta et al. 1976b; Arai et al. 1999b; Arai 2000, Sugeha et al. 2001b).

The early life history parameters found in all tropical species overlap with the range of those in a temperate species, *A. japonica* (Cheng and Tzeng 1996; Arai et al. 1997), although the former was a little shorter than those of *A. anguilla*, *A. rostrata*, *A. australis* and *A. dieffenbachii* (Lecomte-Finiger 1992; Wang and Tzeng 1998; Arai et al. 1999a; Marui et al. 2001). *A. celebesensis*, *A. marmorata* and *A. bicolor pacifica* leptocephali were defined to take about three to six months to migrate from their spawning area to the estuarine waters. *A. bicolor bicolor* was also found to take the same period in our previous study (Arai et al. 1999c). In temperate eels, which migrate via oceanic current systems (Schmidt 1922, 1925; Tsukamoto 1992), the duration of oceanic migration seems to be related to the distance and complexity of the current systems between the spawning areas and their destinations to freshwater habitats.

According to Jespersen (1942), spawning areas of tropical eels including *A. celebesensis* and *A. marmorata* found in north Sulawesi Island are possibly situated in Celebes Sea, Sulu and Molucca Seas, close to the area sampled in this study (Figure 1). Therefore, the rather long migration period of those species relative to short distance between growth habitat and spawning area may due to the complicated local current systems around north Sulawesi Island. The occurrence of tropical leptocephali of various sizes including even preleptocephalus to metamorphosing stages and glass eel in the waters off Sumatra (Jespersen 1942) support this supposition. This situation is quite different from those of temperate eels, suggesting that the migration mechanisms of tropical eel larvae is not so simple as those of temperate eels which can be interpreted as simple transportation by stable current systems. These observations suggest that the ancestral eel would start its diadromous migration from a local short distance movement in complex currents in tropical coastal waters and evolve it to a long distant migration of temperate eels well established in a subtropical gyre in both hemispheres.

Acknowledgments

We are grateful to the students of Aquatic Resources Management Laboratory, Sam Ratulangi University, for their assistance during the field study. We also express our appreciation for comments by Donald J. Jellyman and Tsuguo Otake, and an anonymous reviewer on earlier drafts of this manuscript. This work was supported in part by Grants-in-Aid Nos. 07306022, 07556046, 08041139 and 08456094 from the Ministry of Education, Science, Sports and Culture, Japan; Research for the Future Program No. JSPS-RFTF 97 L00901 from the Japan Society for the Promotion of Science; Eel Research Foundation from Nobori-kai; Research Foundation from Touwa Shokuhin Shinkoukai.

References

Arai, T., T. Otake, and K. Tsukamoto. 1997. Drastic changes in otolith microstructure and microchemistry accompanying the onset of metamorphosis in the Japanese eel *Anguilla japonica*. Marine Ecology Progress Series 161: 17–22.

Arai, T., T. Otake, D. J. Jellyman, and K. Tsukamoto. 1999a. Differences in the early life history of the Australasian shortfinned eel, *Anguilla australis* from Australia and New Zealand, as revealed by otolith microstructure and microchemistry. Marine Biology 135:381–389.

Arai, T., J. Aoyama, D. Limbong, and K. Tsukamoto. 1999b. Species composition and inshore migration of the tropical eels, *Anguilla* spp., recruiting to the estuary of the Poigar River, Sulawesi Island. Marine Ecology Progress Series 188: 299–303.

Arai, T., D. Limbong, T. Otake, and K. Tsukamoto. 1999c. Metamorphosis and inshore migration of tropical eels, *Anguilla* spp., in the Indo-Pacific. Marine Ecology Progress Series 182: 283–293.

Arai, T., T. Otake, D. Limbong, and K. Tsukamoto. 1999d. Early life history and recruitment of the tropical eel, *Anguilla bicolor pacifica*, as revealed by otolith microstructure and microchemistry. Marine Biology 133:319–326.

Arai, T. 2000. Ecological study on the inshore migration of the eels, *Anguilla* spp. in Sulawesi Island. Doctoral dissertation. The University of Tokyo, Tokyo, Japan.

Arai, T., T. Otake, and K. Tsukamoto. 2000a. Timing of metamorphosis, and larval segregation of the Atlantic eels *Anguilla rostrata* and *A. anguilla*, as revealed by otolith microstructure, and microchemistry. Marine Biology, 137:39–45.

Arai, T., D. Limbong, and K. Tsukamoto. 2000b. Validation of otolith daily increments in the tropical eel, *Anguilla celebesensis*. Canadian Journal of Zoology 78:1078–1084.

Arai, T., D. Limbong, T. Otake, and K. Tsukamoto. 2001. Recruitment mechanisms of tropical eels, *Anguilla* spp., and implications for the evolution of oceanic migration in the genus *Anguilla*. Marine Ecology Progress Series 216:253–264.

Bertin, L. 1956. Eels —a biological study. Cleaver-Hume Press Ltd, London, England.

Castle, P. H. J., and G. R. Williamson. 1974. On the validity of the freshwater eel species *Anguilla ancestralis* Ege from Celebes. Copeia 2:569–570.

Cheng, P. W., and W. N. Tzeng. 1996. Timing of metamorphosis and estuarine arrival across the dispersal range of the Japanese eel *Anguilla japonica*. Marine Ecology Progress Series 131:87–96.

Deelder, C. L. 1958. On the behavior of elvers (*Anguilla vulgaris* Turt.) migrating from the sea into fresh water. Journal du Conseil, Conseil Permanent International pour l'Exploration de la Mer 24:135–146.

Ege, V. 1939. A revision of the Genus *Anguilla* Shaw. Dana Report 16 (13):8–256.

Gandolfi, G., M. Pesaro, and P. Tongiorgi. 1984. Environmental factors affecting the ascent of elvers, *Anguilla anguilla* (L.), into the Arno River. Oebalia 10:17–35.

Hain, J. H. W. 1975. The behavior of migratory eels, *Anguilla rostrata*, in response to current, salinity and lunar period. Helgoländer Meeresuntersuchungen 27:211–233.

Haraldstad, O., L. A. Vollestad, and B. Jonsson. 1985. Descent of European silver eels, *Anguilla anguilla* L., in a Norwegian watercourse. Journal of Fish Biology 26:37–41.

Jellyman, D. J. 1977. Invasion of a New Zealand freshwater stream by glass eels of two *Anguilla* spp. New Zealand Journal of Marine and Freshwater Research 11:193–209.

Jellyman, D. J. 1979. Upstream migration of glass eels (*Anguilla* spp.) in the Waikato River. New Zealand Journal of Marine and Freshwater Research 31:13–22.

Jellyman, D. J. 1987. Review of the marine life history of Australasian temperate species of *Anguilla*. Pages 276–285 *in* M. J. Dadswell, R. J. Klauda, C. M. Moffitt, R. L. Saunders, E. F. Rulifson, and J. E. Cooper, editors. Common strategies of anadromous and catadromous fishes. American Fisheries Society, Symposium 1, Bethesda, Maryland.

Jespersen, P. 1942. Indo-Pacific leptocephali of the Genus *Anguilla*. Dana Report 22:1–128.

Lecomte-Finiger, R. 1992. Growth history and age at recruitment of European glass eels (*Anguilla anguilla*) as revealed by otolith microstructure. Marine Biology 114:205–210.

Martin, M. H. 1995. The effects of temperature, river flow, and tidal cycles on the onset of glass eel and elver migration into freshwater in the American eel. Journal of Fish Biology 46: 891–902.

Marui, M., T. Arai, M. J. Miller, D. J. Jellyman, and K. Tsukamoto. 2001. Comparison of early life history between New Zealand temperate eels, and Pacific tropical eels revealed by otolith microstructure, and microchemistry. Marine Ecology Progress Series 213:273–284.

Matsui, I. 1952. Morphology, ecology and culture of the Japanese eel. Journal of Shimonoseki College of Fisheries 2:1–245.

McCleave, J. D., R. C. Kleckner, and M. Castonguay. 1987. Reproductive sympatry of American and European eels and implications for migration and taxonomy. Pages 286–297 *in* M. J. Dadswell, R. J. Klauda, C. M. Moffitt, R. L. Saunders, E. F. Rulifson, and J. E. Cooper, editors. Common strategies of anadromous and catadromous fishes. American Fisheries Society, Symposium 1, Bethesda, Maryland.

Otake, T., T. Ishii, M. Nakahara, and R. Nakamura. 1994. Drastic changes in otolith strontium/calcium ratios in leptocephali and glass eels of

Japanese eel *Anguilla japonica*. Marine Ecology Progress Series 112:189–193.

Otake, T., T. Ishii, T. Ishii, M. Nakahara, and R. Nakamura. 1997. Changes in otolith strontium: calcium ratios in metamorphosing *Conger myriaster* leptocephali. Marine Biology 128: 565–572.

Schmidt, J. 1922. The breeding places of the eel. Philosophical Transactions of the Royal Society of London, B: Biological Sciences 211: 178–208.

Schmidt, J. 1925. The breeding places of the eel. Smithsonian Institution Annual Report 1924:279–316.

Sloane, R. D. 1984. Invasion and upstream migration by glass eels of *Anguilla australis* Richardson and *A. reinhardti* Steindachner in Tasmanian freshwater streams. Australian Journal of Marine and Freshwater Research 35: 47–59.

Sorensen, P. W. 1986. Origins of the freshwater attractant (s) of migrating elvers of the American eel *Anguilla rostrata*. Environmental Biology of Fishes 17:185–200.

Sugeha, H. Y., A. Shinoda, M. Marui, T. Arai, and K. Tsukamoto. 2001a. Validation of otolith daily increments in the tropical eel *Anguilla marmorata*. Marine Ecology Progress Series 220:291–294.

Sugeha, H. Y., T. Arai, M. J. Miller, D. Limbong, and K. Tsukamoto. 2001b. Inshore migration of the tropical eels *Anguilla* spp. recruiting to the Poigar River estuary on north Sulawesi Island. Marine Ecology Progress Series 221:233–243.

Tabeta, O, T. Takai, and I. Matsui. 1976a. The sectional counts of vertebrae in the anguillid elvers. Japanese Journal of Ichthyology 22:195–200.

Tabeta, O., T, Tanimoto, T. Takai, I. Matsui, and T. Imamura. 1976b. Seasonal occurrence of anguillid elvers in Cagayan River, Luzon Island, the Philippines. Bulletin of the Japanese Society of Scientific Fisheries 42:421–426.

Tongiorgi, P., L. Tosi, and M. Balsamo. 1986. Thermal preferences in upstream migrating glasseels of *Anguilla anguilla*. Journal of Fish Biology 28:501–510.

Tsukamoto, K. 1990. Recruitment mechanism of the eel, *Anguilla japonica*, to the Japanese coast. Journal of Fish Biology 36:659–671.

Tsukamoto, K. 1992. Discovery of the spawning area for the Japanese eel. Nature (London) 356:789–791.

Tsukamoto, K., and A. Umezawa. 1994. Metamorphosis: a key factor of larval migration determining geographic distribution and speciation of eels. Proceedings of fourth Indo-Pacific fish conference, Bangkok, Thailand.

Tzeng, W. N. 1985. Immigration timing and activity rhythms of the eel, *Anguilla japonica*, elvers in the estuary of northern Taiwan, with emphasis on environmental influences. Bulletin of the Japanese Society of Fisheries Oceanography 47:11–28.

Umezawa, A., K. Tsukamoto, O. Tabeta, and H. Yamakawa. 1989. Daily growth increments in the larval otolith of the Japanese eel, *Anguilla japonica*. Japanese Journal of Ichthyology 35:440–444.

Wang, C. H., and W. N. Tzeng. 1998. Interpretation of geographic variation in size of American eel *Anguilla rostrata* elvers on the Atlantic coast of North America using their life history and otolith aging. Marine Ecology Progress Series 168:35–43.

Upstream Migration by Glass Eels of Two *Anguilla* Species in the Hacking River, New South Wales, Australia

Bruce Pease, Veronica Silberschneider, and Trudy Walford

*Fisheries Research Institute, New South Wales Fisheries,
Cronulla, New South Wales 2230, Australia*

Abstract.—In order to determine seasonal recruitment patterns of the two local eel species in southeastern Australia, glass eels were sampled from the Hacking River catchment, from May 1998 to June 2000. Samples were collected at an estuarine site 7 km upstream from the ocean, and 5 km further upstream at a tidal barrier, which forms the upper boundary of the estuary. Extensive overlap of the two species was observed among seasons during the initial passive recruitment of glass eels into the estuary using flood tides. The more temperate species *Anguilla australis* recruited to the estuary from March through August, while the more tropical species *Anguilla reinhardtii* recruited to the estuary all year round, with annual peaks between January and August. Inter-annual variability in the timing of recruitment and size at recruitment to the estuary was high for both species. Glass eels of both species appeared to delay their upstream migration at the freshwater interface and accumulate below the tidal barrier. There was little overlap of species composition among seasons during the secondary active migration into fresh water. Thus, the two-stage migration process appears to provide a mechanism for temporally isolating the recruitment of the two *Anguilla* species into freshwater habitats in this region. This process may have important implications for ecological interactions and competition between the two species. Along with significant ecological implications for the populations of these two species, the process provides a potential mechanism for reducing mixed species catches in the local glass eel fishery.

Introduction

Two species of catadromous anguillid eels occur in abundance and are harvested commercially on the east coast of Australia. *Anguilla reinhardtii*, known locally as the speckled longfin eel (longfin), is a predominantly tropical species found in the coastal catchments of eastern Australia, from Cape York to Tasmania (Beumer 1996). It is also known to occur in New Guinea, Solomon Islands, New Caledonia, Lord Howe Island, and New Zealand (Schmidt 1928; Ege 1939; Allen 1991; Jellyman et al. 1996). *Anguilla australis*, known locally as the shortfin eel (shortfin), is a predominantly temperate species found in the coastal catchments of eastern Australia from southern Queensland (Caboolture River) to Tasmania (Beumer 1996). This species also occurs in New Caledonia, Norfolk Island, Lord Howe Island, and New Zealand (Schmidt 1928; Ege 1939; Dijkstra and Jellyman 1999).

Both species are believed to spawn in the southwestern Pacific north of New Caledonia (Aoyama et al. 1999). The leptocephalus larvae are carried to the east coast of Australia by the East Australian Current (Jespersen 1942; Castle 1963; Jellyman 1987; Beumer and Sloane 1990), where they metamorphose into postlarval glass eels, which recruit to eastern Australian estuaries. In Australia, juvenile and adult longfins are most abundant in the coastal catchments of Queensland and New South Wales (NSW), while shortfins are most abundant in Victoria and Tasmania. Both species are relatively abundant in NSW, where their distributions overlap.

Extensive fisheries for all postlarval life history stages of anguillid eels have developed worldwide (FAO 1999a, 1999b). Three life history stages of the two Australian eel species are commercially harvested in Australia. Glass eels of both species are harvested in limited quantities (< 500 kg annually) from estuaries for use as seed stock in a developing intensive eel aquaculture industry (Beumer and Harrington 1980; McKinnon and Gooley 1998; Gooley et al. 1999). Shortfin elvers (juveniles 70–200 mm) are harvested in Tasmania for extensive eel aquaculture. Sub-adults (commonly referred to as yellow eels) of both species are captured in a wild harvest fishery, which exports frozen eel meat and live eels to Europe and Asia (Kailola et al. 1993).

An understanding of glass eel recruitment provides important biological information for the management of these fisheries. Studies of glass eel recruitment in Australia have been limited primarily to aggregate compilations of seasonal occurrence, derived from relatively few samples collected sporadically over a range of catchments. None of these studies have addressed the issue of competitive recruitment between the two species within the NSW overlap zone.

Schmidt (1928) and Ege (1939) provide the earliest descriptions of glass eel collections based on four samples, each collected from different catchments during different years. Sloane (1984) collected glass eel samples from freshwater areas within six catchments in Tasmania, with an irregular sampling protocol during four years to describe seasonal glass eel recruitment patterns of both species at the southern end of their Australian range. Less than 3% of the glass eel occurrence data that Beumer and Sloane (1990) compiled from all earlier studies was collected from NSW, which comprises almost half of the coastline area studied. In 1995, a survey of glass eels in southern Queensland (Queensland Department of Primary Industries 1995) showed that both species recruited to each of the six estuaries sampled in the northern part of their Australian range. Most recently, McKinnon and Gooley (1998) and Gooley et al. (1999) conducted surveys of shortfin glass eels recruiting to the estuarine waters of 26 catchments in NSW, Victoria and Tasmania from 1994 to 1996. However, catches of longfins were not reported and less than 6% of the samples were collected from NSW.

The objective of our study was to document and compare the temporal and spatial characteristics of migration from estuarine into fresh waters by glass eels of the two Australian eel species in the middle of their distributional overlap zone in eastern Australia. Seasonal occurrence, abundance, size, and pigmentation characteristics of each species were compared among sampling sites in the estuary and at the freshwater transition zone. Relationships between these biological characteristics of migrating glass eels and key environmental factors, including tidal flow, lunar periodicity, water temperature and salinity, were also examined.

Methods

Study Sites

Sampling was conducted at two sites in the Hacking River catchment (Figure 1). The estuarine portion of this catchment consists of a relatively small, drowned river valley type estuary (Roy 1984) called Port Hacking, which is located within the central estuarine bioregion of NSW (Pease 1999). The Hacking River and several small streams flow into this marine dominated estuary, which has full tidal exchange (mean spring tidal range of 1.32 m) through a permanently open entrance. The estuary extends 12 km upstream to a causeway, which forms a tidal barrier across the Hacking River.

This small catchment (180 km^2) has a mean annual freshwater discharge of 69×10^6m^3 (Bell and Edwards 1980). The northern shore of Port Hacking forms the southern boundary of the Sydney suburban area; however, most of the catchment lies within the Royal National Park. Therefore, the waters of the Hacking catchment are relatively pristine (Hacking River Catchment Management Committee 1997; Environment Protection Authority 1997). While the entire catchment is closed to commercial fishing, recreational fishing is permitted and is very popular in these waters, with sub-adult longfins occasionally caught by anglers. There is no recreational harvest of glass eels.

Site one (Figure 1) was located 7 km up the estuary from the ocean, at the mouth of the Hacking River channel. The channel at this site narrows to approximately 50 m wide and 5 m deep at mean low water (MLW), with a sand bottom and strong tidal currents.

Site two (Figure 1) was located at the estuary/fresh water interface formed by the Audley Causeway. The existing concrete causeway was constructed in 1951, but there has been a tidal barrier at this site since 1882. It forms a low-head dam 90 m long and 2.4 m high at MLW. The channel immediately below the causeway is approximately 80 m wide and 5 m deep (MLW), with a boulder substrate. There is no tidal current at this site.

Field Sampling Protocol

The study was conducted over two years, from May 1998 through April 2000. During the first year, samples were collected at both sites every two weeks, near the new and full moon peri-

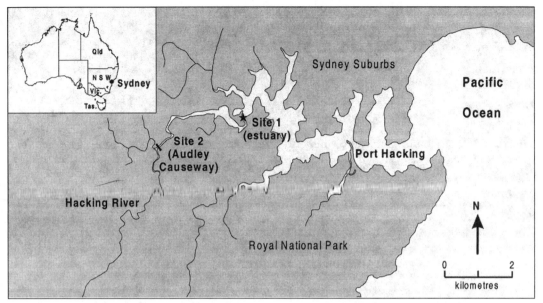

Figure 1. Map showing sampling sites in the Port Hacking estuary and Hacking River.

ods. Based on the work of McKinnon and Gooley (1998), samples were collected within five days after the new or full moon, when low tide occurred near sunset. During the second year, samples were again collected biweekly, except during the period from September through January, when sampling only occurred at new moon periods. Site 1 was sampled on two nights during each lunar period from February to August, and only on one night during all other lunar periods. Samples at Site 2 were collected on one night for each of the lunar sampling periods.

Samples were collected at Site 1, hereafter referred to as the "estuary," with a fyke net similar to the stow net described by Weber (1986). It has two wings, each of which is 11 m in length with an effective fishing depth of 2 m. The wings and bunt are constructed of 2 mm (stretched) mesh nylon netting. The detachable cod end is constructed from nylon material with a mesh size of less than 0.5 mm. A rigid aluminum bycatch reduction grid with a bar spacing of 0.5 cm was fitted in the mouth of the net, as described by Broadhurst and Kennelly (1996). Preliminary trials showed that a bycatch reduction device was needed to prevent the net from becoming clogged with large volumes of jellyfish.

At low tide the net was set on the south side of the channel to fish the incoming tide. The inshore wing was attached to a star-picket at the edge of the channel and the outer wing was anchored 11 m offshore. The catch was removed from the cod end every hour during the first four hours of the incoming tide. The entire incoming tide was not fished because preliminary trials showed that catches declined rapidly after the fourth hour, when the tidal current dropped and puffer fish *Tetractenos glaber* often chewed holes in the cod end in order to eat some of the catch.

Each hourly catch was sorted, using a series of screens, and processed in the field. If the hourly catch of glass eels was less than 100 individuals, they were all counted and weighed (total wet weight in grams). In compliance with NSW animal ethics legislation, the glass eels were then euthanased in a solution of 100 mg/L benzocaine in ambient water, before being preserved in 95% ethanol. If the catch was greater than 100 individuals, the total wet weight was recorded and a subsample of approximately 100 individuals was processed as above. The remainder of the sample was released and species composition of the subsample was used to calculate the total number of each species in the hourly sample.

Actual tide height at the start and end of each nightly sampling period was recorded from a tide gauge 50 m east of the sampling station at Site 1. Temperature and salinity were recorded every 10 minutes during the nightly sampling period with a Yeo-Kal 606SDL data logger placed 1 m below MLW at the nearby tide gauge. Tidal flow next to the net was measured every hour with a General Oceanics 2030 flowmeter.

Three stations were sampled at Site 2. These stations will be referred to as "below the causeway," "climbing the causeway" and "above the causeway." All three stations were sampled on one night during the same lunar periods described for Site 1.

Glass eels below and above the causeway were sampled with artificial habitat collectors using the sampling protocol described by Silberschneider et al. (2001). The collectors were based on a design initially developed by Phillips (1972) and later modified by Montgomery and Craig (1997) for collecting lobster puerulus larvae. Each collector consisted of one PVC panel (61 cm × 35 cm × 0.4 cm) with 25 evenly spaced tufts of fiber 25 cm long. Lead weights were attached to each corner of the panel and a single rope was attached for retrieving the collector. All collectors were initially "aged" by hanging them vertically in the water column for a minimum of two months. Between each lunar sampling period, the collectors were stored, tufts facing up, on the shallow (< 2 m) seabed near the NSW fisheries wharf in Port Hacking in order to maintain algal growth on the fibers.

At low tide on one afternoon of each lunar sampling period, collectors were positioned above and below the causeway. In order to sample below the causeway, two collectors were placed on the seabed, tufts facing up, within 1–2 m of the base of the causeway at a depth of approximately 1 m (MLW). Above the causeway, collectors were placed 1–2 m from the top of the causeway at a depth of approximately 1 m. The collectors were retrieved at both locations, shortly after low tide on the following morning and placed into a plastic tub. Each collector was shaken at least 20 times (tufts facing down) above the plastic tub to remove the glass eels.

Glass eels were also collected with small rectangular aquarium dip nets during the night-time incoming tide as they swam up the drain at the base of the causeway or climbed up the moist zone above the water level in the drain. Preliminary sampling during the entire period of the incoming tide showed that glass eels attempted to climb the causeway primarily during the three hour period starting approximately two hours after low tide, when the tidal level approached within 5 cm of the bottom of the causeway drain. All dip net samples for this study were collected during this three hour period. All glass eels collected at Site 2 were euthanased in 100 mg/L benzocaine and preserved in 95% ethanol. The water temperature above the causeway was measured once per month by the Port Hacking Catchment Management Committee.

Laboratory and Data Analyses

Glass eel species were identified based on the relative position of the dorsal fin and vent (Schmidt 1928). The total length of each individual was measured to the nearest millimeter and the stage of pigmentation was assessed according to the scale outlined by Struberg (1913). Glass eels in this study are defined as the eel-shaped postlarvae of *A. australis* and *A. reinhardtii* that have not yet become fully pigmented. To simplify the analysis of pigmentation data, only the eight abbreviated stages, first used by Ege (1939), are reported. These simplified stages (1–8) are derived from Struberg's stages VA-VB, VIAI, VIAII1-VIAII2, VIAII3-VIAII4, VIAIII1-VIAIII3, VIAIV1-VIAIV2, VIAIV3-VIAIV4 and VIB, respectively.

Only data from hours two through four were used for calculating mean or maximum daily catches at Site 1. Glass eel abundances in samples collected during new moon periods in the estuary and climbing the causeway were compared with abundances at these respective stations during full moon periods, using paired-sample t-tests (Zar 1974). Data from samples collected during new and full moon periods of the same month were paired, using only data from months when more than ten glass eels were collected. Relationships between glass eel abundance (log transformed) of each species and maximum tidal flow at the estuary site, on days when more than 10 glass eels were caught, were analyzed using product-moment correlation coefficients (Zar 1974). All glass eels that were collected and processed from each station were used in the analysis of lengths and pigmentation stages. Student's t statistic was used to test hypotheses about mean lengths and the Kolmogorov-Smirnov D statistic was used to test hypotheses about the frequency distribution of the ordinal pigmentation stages (Zar 1974). Correlations between length and pigmentation stage were analyzed with the nonparametric Gamma statistic (STATISTICA version 5 statistical software), rather than the product-moment correlation coefficient because of the ordinal nature of the pigmentation data and the high number of tied ranks.

Results

The maximum daily catches of each species collected at each sampling station during each month of the study period are summarized in Figure 2. The actual magnitude of catches cannot be compared among stations, where differing sampling methods were used. However, temporal patterns were examined within and between sites by observing time periods when

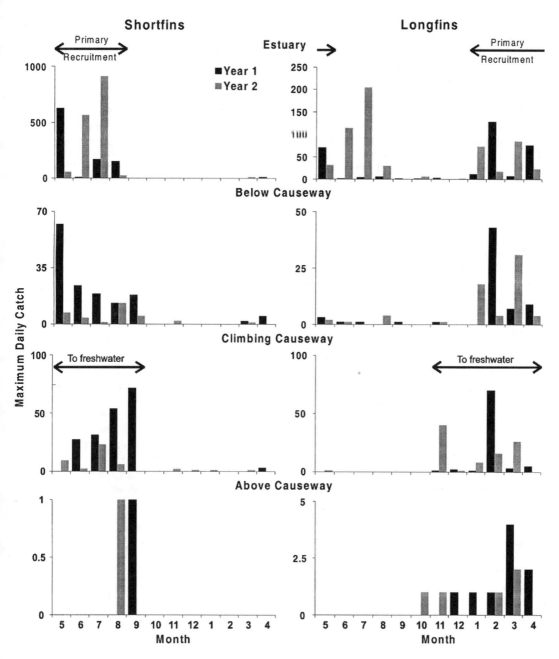

Figure 2. Maximum daily glass eel catches at the sampling stations during each month of the study period. Daily unit of effort at each sampling station: 1) estuary = hours 2, 3, and 4 of evening incoming tide using stow net, 2) below causeway = two habitat collectors per night, 3) climbing causeway = one person dip netting per incoming tide per night, and 4) above causeway = two habitat collectors per night.

relatively large catches occurred. Within these periods when large catches occurred, there was extensive inter-annual variability in the occurrence of maximum monthly catches of both species at all stations.

Relatively large catches (> 10 glass eels per month) of shortfins were consistently obtained from May through August in the estuary during both years, indicating that this is the primary period of shortfin recruitment to the estuary. Shortfins were found at the causeway from May until September during both years and it is assumed that the catches of glass eels climbing the causeway during this period are indicative of periods when shortfins are moving from the estuary to fresh water. Only two shortfins were caught above the causeway, however, this does indicate that some glass eels of this species successfully negotiate the causeway, particularly near the end of the recruitment period in August and September. Very few shortfins were collected at any of the stations between September and May.

Relatively large catches (> 10 glass eels per month) of longfins were obtained from January through May in the estuary during both years, indicating that this is the primary period of annual longfin recruitment. During year 2, large catches were also obtained from the estuary during the period from June through August. These catches during the shortfin recruitment period may be atypical and indicate that inter-annual variability in the timing of longfin recruitment is much greater than the variability of shortfin recruitment. Catches below the causeway were high in February and March during both years and catches of longfins climbing the causeway were obtained from November through April. Very few longfins were found at the causeway from May through October. As with the shortfins, very few longfins were collected above the causeway, but catches of glass eels climbing the causeway are probably indicative of periods when longfins are moving to fresh water.

Table 1 provides a summary of annual abundance indices and species composition at each sampling station. Both species were more abundant in the estuary during the second year, but shortfins were much less abundant at the causeway during that year. Shortfins also outnumbered longfins in the estuary during both years, but the proportion of longfins consistently increased at the upstream stations.

Average monthly water temperatures and salinities in the estuary during the study period are plotted in Figure 3, along with freshwater temperatures above the causeway. It was found that a horizontal line positioned at the y-intercept of the average estuary temperature in May (18.5°C), which occurs during the transition period from longfin recruitment to shortfin recruitment, seasonally bisected the temperature curves. The primary period of shortfin recruitment to the estuary and freshwater during both years occurred when water temperatures were at or below 18.5°C, whereas the primary period of longfin recruitment to the estuary and freshwater occurred when water temperatures rose above 18.5°C. The recruitment of longfins to the estuary during the coldwater period (June to August) in year 2 is not consistent with this pattern. Declines in estuarine salinities during both the coldwater recruitment period for shortfins and the warmwater recruitment period for longfins were caused by increased freshwater discharge from the Hacking River during these periods in both years.

Freshwater temperatures above the causeway were consistently lower than the estuarine

Table 1. Annual abundance index ("Abundance") of each species and proportion of longfins (%) at each sampling station during years one and two of the study. The abundance index is the sum of the average daily catches for each month of the year.

	Year 1			Year 2		
	Abundance		Longfins (%)	Abundance		Longfins (%)
	Shortfin	Longfin		Shortfin	Longfin	
Estuary	373	143	28	510	306	37
Below Causeway	76	37	32	21	44	68
Climbing Causeway	236	137	36	50	94	65
Above Causeway	1	5	83	1	4	80

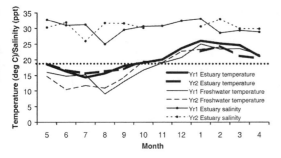

Figure 3. Average monthly water temperature and salinity at the estuary site and monthly temperature of fresh water above the causeway during the study period. Horizontal dotted line indicates average estuarine temperature in May of both years (18.5°C).

water temperatures during the coldwater period when shortfins were climbing the causeway and moving to freshwater, except for one occurrence of warmer water during July in year 1. Freshwater temperatures during the warmwater period when longfins climbed the causeway and recruited to freshwater (November to April) were consistently similar to the concurrent estuarine water temperatures.

Maximum daily glass eel catches during biweekly semilunar periods are summarized for the four month season that each species concurrently recruited to the estuary and climbed the causeway (Figure 4). Maximum tidal flows (m/s) recorded at the estuary site are also plotted concurrently with the catch data at that site. The timing of peak recruitment/movement events of both species varied greatly between years at both sites. There was no significant correlation between glass eel catches (log transformed) of each species and maximum tidal flow in the estuary on days when a minimum of ten glass eels were caught ($r = 0.20$, $p > 0.2$ for shortfins and $r = 0.37$, $p > 0.1$ for longfins). However, most of the largest recruitment events at both sites occurred during new moon periods. The largest catches in seven out of the eight seasonal graphs occurred during new moon periods and in seven out of the eight seasonal graphs, three out of the four highest catches occurred during new moon periods. Paired-sample t-tests showed that catches of both shortfins and longfins in the estuary were significantly greater during new moon periods ($t = 1.90$, $p < 0.04$ for shortfins and $t = 3.01$, $p < 0.005$ for longfins). Catches of shortfins climbing the causeway were not significantly greater during new moon periods ($t = 0.23$, $p > 0.4$) due to one recruitment event during a full moon period in September of year 1 (Figure 4). When the data for this month was excluded from the analysis, the remaining catches were significantly higher during new moon periods ($t = 3.45$, $p < 0.02$). Catches of longfins climbing the causeway were significantly greater during new moon periods ($t = 5.50$, $p < 0.02$).

There appears to have been a delay of at least two weeks between the time migrants were sampled in the estuary and the time they started climbing the causeway (Figure 4). The largest estuarine catch of the year 1 shortfin season occurred in May, but glass eels of this species did not start climbing the causeway until June. Also during year 1, shortfin recruitment to the estuary ended in August but glass eels were still climbing the causeway in September (Figure 2). During year 2, the largest peak in the number of shortfins climbing the causeway occurred at the end of July, after the largest catches in the estuary were collected in June and early July (Figure 4). Longfin catches do not show as much evidence of a delay. Other than the delay between the small peaks in January of year 1 at the estuary, and the first peak in the number of longfins climbing the causeway during the full moon period in February, most peak catches from both areas generally occurred at the same time. However, there was a three month delay between the atypical recruitment of longfins to the estuary during June through August in year 1, and the first indication that longfins were climbing the causeway in November (Figure 2). During the atypical recruitment from June through August, and the subsequent three month delay period, very few longfins were found below the causeway, indicating either dispersal to other parts of the estuary or a very high mortality rate. During the primary periods that shortfins and longfins moved to fresh water, they were found in relatively high numbers below the causeway, indicating that they were at the causeway during periods when delays may have occurred, and were not dispersed to other parts of the estuary.

In Figure 5, the pigmentation stages of each species are summarized monthly during the period they were found concurrently in the estuary and climbing the causeway. Pigmentation

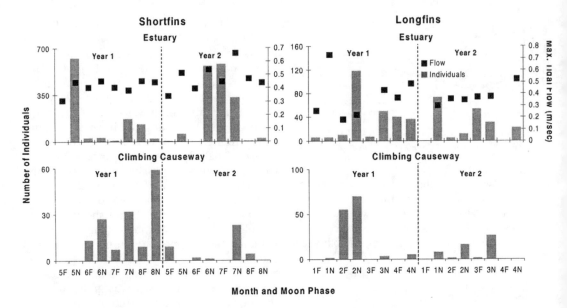

Figure 4. Top graphs show maximum daily catches (during hours 2, 3, and 4) of glass eels and maximum tidal flow recorded during the evening sampling period during each biweekly sampling period in the estuary during the period that each species was caught at the estuary and climbing the causeway. Bottom graphs show daily catch (one dip net sample per evening incoming tide) of glass eels climbing the causeway. On the x axis of all graphs: Numeric calendar month and moon phase where: F = Full moon and N = New moon.

data for the two years were pooled after the Kolmogorov-Smirnov test revealed that the frequency distribution of pigmentation stages in year 1 was not significantly different from the distribution in year 2 ($D = 0.02$, $p > 0.50$ for shortfins and $D = 0.11$, $p = 0.10$ for longfins). Spatial and temporal patterns in the frequency distribution of pigmentation stages during the recruitment seasons were generally similar for both species (Figure 5). The majority of glass eels sampled in the estuary had stage 1 pigmentation, while the majority of glass eels that were climbing the causeway had stage 3–6 pigmentation. The frequency distributions of pigmentation stages of all glass eels from both study years were significantly different between locations (Sites 1 and 2) for each species ($D = 0.75$, $p < 0.001$ for shortfins and $D = 0.72$, $p < 0.001$ for longfins). For both species at both locations, there was a progressive increase in the proportion of higher stages as the recruitment season progressed (Figure 5).

Length frequencies of each glass eel species captured from the estuary, were summarized by month during the period they were found concurrently in the estuary and climbing the causeway (Figure 6). The data for each year is shown separately, because t-tests of all length frequency data from both sites revealed that the annual mean length during year 2 was significantly greater than the annual mean length during year 1 for each species ($t = 17.96$, $p < 0.001$ for shortfins and $t = 20.01$, $p < 0.001$ for longfins). The mean length of shortfins was also significantly greater than the mean length of longfins in the estuary during both years ($t = 12.46$, $p < 0.001$ in year 1 and $t = 19.75$, $p < 0.001$ in year 2). Figure 6 shows that the lengths of both species were generally greater in year 2 during each month that they recruited to the estuary. The monthly shortfin length-frequency distributions also followed a seasonal pattern of variability, with the largest glass eels recruiting in the middle of the season, during June and July, of both years. There was very little seasonal variability of monthly longfin length-frequency distributions within the primary recruitment period. However, the mean lengths of longfins that recruited to the

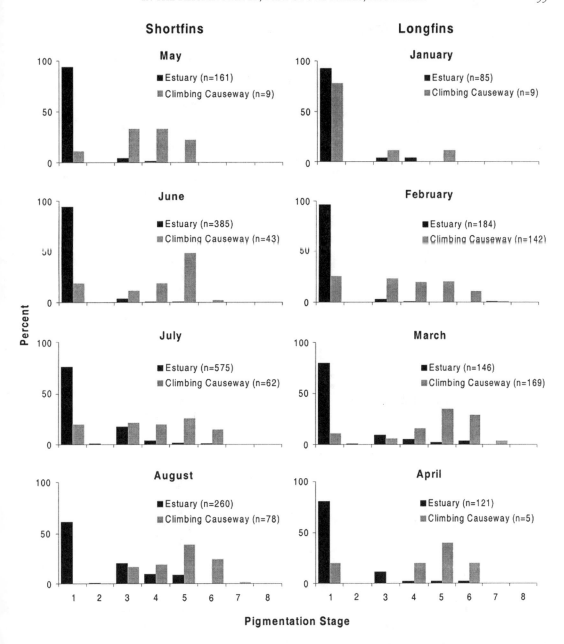

Figure 5. Frequency of glass eel pigmentation stages in catches of both species from the estuary and climbing the causeway, during each month that high catches were obtained concurrently in both areas (data for years one and two pooled).

estuary out of season in June, July and August of year 2 were more similar to the mean lengths of larger longfins that recruited five months later, during the following season ($t = 1.99$, $p = 0.05$), than to the mean lengths of longfins that recruited one to two months earlier, during the previous season ($t = 9.51$, $p < 0.001$). The longfins climbing the causeway out of season in November were also more similar in length to those climbing the causeway during the next season ($t = 1.98$, $p = 0.05$), than to those climbing during the first year ($t = 6.47$, $p < 0.001$).

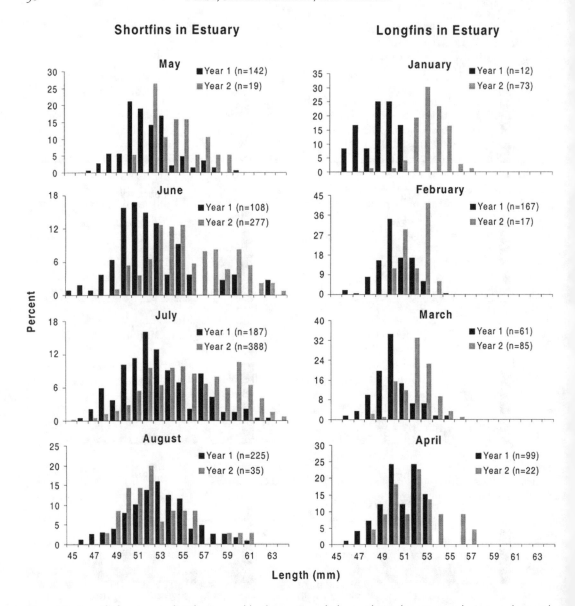

Figure 6. Length frequency distributions of both species of glass eels in the estuary during each month that high catches were obtained concurrently in the estuary and climbing the causeway.

Further t-tests showed that mean lengths of longfins at Site 1 were significantly lower than those at Site 2 during both years ($t = 8.51$, $p < 0.001$ in year 1 and $t = 2.42$, $p = 0.02$ in year 2). Mean lengths of shortfins were also significantly lower at Site 2 during year 2 ($t = 8.51$, $p < 0.001$), but not during year 1 ($t = 0.32$, $p = 0.75$). Longfin lengths during both years were also significantly correlated (inversely) with pigmentation stage (Gamma = -0.21, $p < 0.03$ in year 1 and Gamma = -0.19, $p < 0.01$ in year 2). Shortfin lengths were not correlated with pigmentation stages during either year (Gamma = -0.01, $p > 0.8$ in year 1, Gamma = -0.08, $p > 0.06$ in year 2).

Discussion

The recruitment migration of both eel species into this catchment on the central coast of NSW appears to follow the two stage process which has been previously described in detail for *Anguilla anguilla* in Europe (Deelder 1970; Tesch 1977; Gascuel 1986). According to this conceptual model, glass eels first move into the estuaries with the selective use of night-time flood tides. Upon reaching the saltwater/freshwater transition zone there is a delay, while they undergo the necessary physiological changes to adapt to fresh water. Then there is a secondary migration into the fresh waters of the upper catchment by actively swimming against the flow. It appears that the temporal details of this migration process can differ substantially between the two Australian eel species, and these seasonal differences temporally isolate recruitment of the two species into freshwater habitats.

Shortfins consistently migrated into the estuary from May through August, whereas longfins migrated from January through August (Figure 2). Longfin recruitment overlapped with peak shortfin recruitment in May of year 1, and July of year 2. Beumer and Sloane (1990) indicated that longfins recruit to tropical and sub-tropical (latitudes 20–29°S) estuaries all year round and shortfins recruit to temperate (35–44°S) estuaries all year round. Between these regions (Port Hacking is at 34°S), inter-annual variability of longfin recruitment is high, as indicated by the major recruitment events during the second shortfin season. This high variability also results in extensive recruitment overlap for the two species in the estuary. However, the recruitment period for shortfins in this study was clearly not all year round, and longfins did not recruit substantially between September and December.

Freshwater discharge and water temperature are thought to be important factors influencing the recruitment of European (Tongiorgi et al. 1986; Tosi et al.1990), Japanese (Chen et al. 1994), and Australian (McKinnon and Gooley 1998) glass eels to estuaries. Periods of increased freshwater discharge, as indicated by lower salinities (Figure 3), occurred during both longfin and shortfin primary recruitment periods. The peak monthly recruitment of both species in year 2 occurred on the same month that the lowest salinity was recorded during that year. Based on findings for European and American eels, it is likely that Australian glass eels are attracted into estuaries by organic chemoattractants (Creutzberg 1961; Tesch 1977; Sorensen 1986; Tosi et al. 1990; Sola 1995; Sola and Tongiorgi 1996) in the fresh water discharge from the estuaries. However, given the fact that anguillid eels have extremely sensitive olfactory organs (Tesch 1977), the relative volume of freshwater discharge may be unimportant.

Shortfins consistently recruited to the estuary when mean monthly water temperatures were below 18.5°C (Figure 3) and longfins consistently recruited when mean monthly water temperatures were above 18.5°C during both years. In Victoria, McKinnon and Gooley (1998) collected their highest catches of shortfins when water temperatures were 10–14°C. However, the recruitment of longfins when water temperatures were lower than 18.5°C in year 2 of this study indicates that recruitment to the estuary may not be directly related to water temperature. Jellyman (1977), Sloane (1984) and Beumer and Sloane (1990) also found no relationship between temperature and timing of glass eel recruitment.

The high inter-annual variability of longfin recruitment periods and species ratios, along with the significant inter-annual variability of body length of both species, indicates that glass eel recruitment to the estuary is probably dictated primarily by oceanographic conditions associated with the currents that deliver leptocephali to the continental shelf adjacent to the estuary. Oceanographic conditions off central NSW are complex (Hamon 1965; Cresswell 1987). The East Australian Current, which most researchers believe is carrying leptocephali from their spawning grounds to the east coast of Australia (Jespersen 1942; Castle 1963; Jellyman 1987; Beumer and Sloane 1990; Aoyama et al. 1999), often breaks away from the coast north of Sydney and forms eddies that may last from several days to a year (Cresswell 1987). At different times during the year, the ocean waters off Port Hacking may be under the influence of: the warm East Australian Current, one of the eddies encountering the coast, or a cold inshore counter current. Possibly the glass eels that recruited during year 2, including the longfins that recruited during the year 2 shortfin season, were all significantly longer than the glass eels that recruited during the previous year because they had been caught in such an eddy for a period of time. Alternative explanations are that growth conditions (primarily temperature) in the current or the track of the current (distance traveled) is highly variable between years. Further temporal studies of glass eel recruitment in this region are needed to identify factors related to inter-annual recruitment variability.

European (Creutzberg 1961) and American (McCleave and Kleckner 1982) glass eels have been shown to move upstream in estuaries by selectively using tidal currents. All glass eels collected in the estuary during this study were captured as they moved upstream with the flood tide shortly after dusk. Studies in Europe (Deelder 1958; Tesch 1977) and New Zealand (Jellyman 1979) have found that high catches of glass eels are often obtained at semilunar intervals near both new and full moon and relate this primarily to the local cycle of spring and neap tides with associated tidal currents. However, upstream movements in Port Hacking were not significantly correlated with the magnitude of the tidal flow (Figure 4). Instead, significantly greater numbers of both species recruited during new moon periods. Antunes (1994) and de Casamajor et al. (1996) also found that estuarine catches of European glass eels were highest during new moon periods. The estuarine waters of Port Hacking are generally very clear. It may be that the low light conditions at night during new moon periods are an important cue for glass eel movement in this region. Semilunar movement patterns may predominate in areas where cloud cover and/or turbid waters often occur.

This study also provides some evidence that both species delay their migration below the freshwater transition zone at the causeway. One indication of a delay was the significantly higher frequency of later pigmentation stages at the causeway than at the estuary site. Laboratory experiments with European (Struberg 1913) and New Zealand (Jellyman 1977) species have shown that the rate of pigmentation development increases with increasing water temperatures, and may vary with other environmental parameters. Struberg (1913) found that most of the *A. anguilla* in his experiments had developed from stage 1 to stages 3 and 4 in 12 days at 7.5–10°C. Jellyman found that it took shortfins *A. australis* three weeks to develop from stage 1 to stage 4 at 13.5°C. It is assumed that glass eels in Port Hacking could easily travel the 5 kilometers from Site 1 to Site 2 within two to three days, indicating a possible delay time of around two weeks. A delay of at least two weeks for shortfins was also indicated by tracking peak biweekly catches during the primary recruitment season from the estuary, to climbing the causeway (Figure 4).

There is no available data for rate of pigmentation development in Australian longfins, and there are few instances of an obvious delay between peak catches in the estuary and climbing the causeway (Figures 2 and 4). It may be that the delay is shorter for longfins and they develop pigmentation more rapidly than the shortfins during the warmwater period when they primarily recruit. However, some evidence of a delay for longfins is the fact that their mean lengths were significantly lower at the causeway than at the estuary site during both years. This suggests that they are generally not feeding and are shrinking until they make the transition to freshwater. The significant negative correlation between length and pigmentation stage of longfins supports the hypothesis that the shrinkage and pigmentation processes are taking place concurrently over a period of time. Shortfins were only significantly shorter at the causeway on one of the two study years, and length was not correlated with pigmentation stage, indicating that they may start feeding earlier than the longfins.

The data from the causeway (Figure 2) provides some interesting evidence that longfin migration to fresh water can be temporally isolated from the migration of shortfins to fresh water. Longfins moved to fresh water almost exclusively during the warmwater period from November to April, while shortfins moved almost exclusively during the coldwater period from May to September. Researchers have shown that water temperature is a key factor regulating movement of Japanese (Hiyama 1952), European (Creutzberg 1961; Tesch 1977; Gandolfi et al. 1984; Tongiorgi et al. 1986; Tosi et al. 1988; Tosi et al. 1990), and American (Martin 1995) glass eels from the estuary to fresh water. In some instances, glass eels have been shown to prefer fresh water at a different temperature from their ambient estuarine waters, while in other instances they apparently prefer fresh water that is a similar temperature to the estuary.

The fresh water of the Hacking River is typically several degrees colder than the average estuarine water temperatures during the period when shortfins move into fresh water (Figure 3). It is highly probable that the colder river water provides an important cue for this temperate species during the freshwater transition phase of their migration. This movement also occurred most often during new moon periods.

During the warmwater period when longfins move to fresh water, the estuarine and river waters are at a similar temperature. It appears that an important cue for this more tropical species to move to fresh water may be water temperatures greater than 18.5°C, with very little difference between estuarine and river water temperatures. Good evidence for this is the fact that the longfins that recruited to the estuary during June to August of year 2 did not appear to climb the causeway during this coldwater period. Longfins did not start climbing the causeway until November, when the conditions described above first occurred during year 2. During the primary period that longfins moved into fresh water each year, both estuarine and freshwater temperatures were very similar, and at or near the seasonal peak. As with shortfins, this movement also occurred most often during new moon periods.

This work verifies the two-stage nature of glass eel recruitment in southeastern Australia, and indicates that movement into fresh water in the middle of the distributional overlap zone is temporally isolated for the two species. Consistent changes in the species composition of glass eels migrating upstream from the estuary through the freshwater transition zone (Table 1) provide evidence that this recruitment process may have important implications for ecological interactions and competition between the two species in this region. These processes also have important management implications regarding the optimum time and location for commercially harvesting glass eels of each species. Mixed species catches may be minimized by fishing at the estuary/freshwater boundary.

Acknowledgments

This project was supported by the Fisheries Research and Development Corporation (FRDC) of Australia as part of FRDC Project 97/312: "Assessment of eastern Australian glass eel stocks and associated aquaculture." Veronica Silberschneider was also supported by an Australian Postgraduate Award for work related to her Ph.D. thesis. We thank Dave Pollard, Rick Fletcher, David Booth, and Ben Chisnall for providing valuable comments on the manuscript.

References

Allen, G. R. 1991. Field guide to the freshwater fishes of New Guinea, G. R. Allen. Christensen Research Institute, Madang, Papua New Guinea.

Antunes, C. 1994. The seasonal occurrence of glass eels (*Anguilla anguilla* L.) in the Rio Minho between 1991 and 1993 (North Portugal). Internationale Revue der Gesumpten. Hydrobiologie 79(2):287–294.

Aoyama, J. et al. 1999. Distribution and dispersal of anguillid leptocephali in the western Pacific Ocean revealed by molecular analysis. Marine Ecology Progress Series 188:193–200.

Bell, F. C. and A. R. Edwards. 1980. An environmental inventory of estuaries and coastal lagoons in New South Wales. Total Environment Centre, Sydney, Australia.

Beumer, J. P. 1996. Freshwater eels. Pages 39–43 *in* R. McDowell, editor. Freshwater fishes of southeastern Australia. Reed, Sydney, Australia.

Beumer, J. P. and D. J. Harrington. 1980. Techniques for collecting glass-eels and brown elvers. Australian Fisheries 39(8):16–22.

Beumer, J. P. and R. D. Sloane. 1990. Distribution and abundance of glass-eels *Anguilla* spp. in East Australian waters. Internationale. Revue der Gesumpten Hydrobiologie. 75(6):721–736.

Broadhurst, M. K and S. J. Kennelly. 1996. Rigid and flexible separator panels in trawls that reduce the by-catch of small fish in the Clarence River prawn-trawl fishery, Australia. Marine and Freshwater Research 47:991–998.

Castle, P. H. J. 1963. Anguillid leptocephali in the southwest Pacific. Victoria University of Wellington. Zoology Publication 33, Wellington, New Zealand.

Chen, Y. L., H. Y. Chen, and W. N. Tzeng. 1994. Reappraisal of the importance of rainfall in affecting catches of *Anguilla japonica* elvers in Taiwan. Australian Journal of Marine and Freshwater Research 45:185–190.

Cresswell, G. 1987. The east Australian current. CSIRO Marine Lab., Information Sheet No. 3, Hobart, Australia.

Creutzberg, F. 1961. On the orientation of migrating elvers (*Anguilla vulgaris* Turt.) in a tidal area. Netherlands Journal of Sea Research 1(3):257–338.

de Casamajor, M. N., N. Bru, and P. Prouzet. 1996. Influence of environmental conditions on glass-eel (*Anguilla anguilla* L.) catchability on the Ardour estuary. EIFAC/ICES Working Party on Eel. Ijmuiden, The Netherlands.

Deelder, C. L. 1958. On the behaviour of elvers (*Anguilla vulgaris* Turt.) migrating from the sea into fresh water. Journal du Counseil International pour l'Exploration de la Mer 24(1):135–146.

Deelder, C. L. 1970. Synopsis of biological data on the eel *Anguilla anguilla* (Linnaeus, 1758). FAO Fisheries Synopsis 80:1–73.

Dijkstra, L. H. and D. J. Jellyman. 1999. Is the subspecies classification of the freshwater eels *Anguilla australis australis* Richardson and *A-a-schmidtii* Phillipps still valid? Marine and Freshwater Research 50(3):261–263.

Environment Protection Authority. 1997. Proposed interim environmental objectives for NSW waters. Sydney, central coast and Illawarra catchments. Environment Protection Authority, Chatswood, Australia.

Ege, J. 1939. A revision of the genus *Anguilla* Shaw. A systematic, phylogenetic and geographical study. Carlsberg Foundation, Dana-Report 16, Copenhagen, Denmark.

FAO (Food, and Agriculture Organisation). 1999a. Aquaculture production statistics 1988–1997. Data and Statistics Unit, Fishery Information, Circular 815(11), Rome, Italy.

FAO. 1999.b. Yearbook of Fishery Statistics - Commodities. Data and Statistics Unit, Fishery Information No. 85, Rome, Italy.

Gandolfi, G., M., Pesaro, and P. Tongiorgi. 1984. Environmental factors affecting the ascent of elvers, *Anguilla anguilla* (L.), into the Arno River. Oebelia 10:17–35.

Gascuel, D. 1986. Flow-carried and active swimming migration of the glass eel (*Anguilla anguilla*) in the tidal area of a small estuary on the French Atlantic coast. Helgolander Meeresuntersuchungen 40:321–326.

Gooley, G. J. et al. 1999. Assessment of juvenile eel resources in southeastern Australia and associated development of intensive eel farming for local production. Department Natural Resources and Environment. Marine and Freshwater Resources Institute, Final Report to Fisheries Research and Development Corporation No. 94/067, Melbourne, Australia.

Hacking River Catchment Management Committee. 1997. Hacking River Catchment. A pollution source inventory. Hacking River Catchment Management Committee, Sydney, Australia.

Hamon, B. V. 1965. The east Australian current, 1960-1964. Deep-Sea Research 12:899–921.

Hiyama, Y. 1952. Thermotaxis of eel fry in stage of ascending river mouth. Japanese Journal of Ichthyology 2:22–30.

Jellyman, D. J. 1977. Invasion of a New Zealand freshwater stream by glass-eels of the two *Anguilla* spp. New Zealand Journal of Marine and Freshwater Research 2:193–209.

Jellyman, D. J. 1979. Upstream migration of glass eels (*Anguilla* spp.) in the Waikato River. New Zealand Journal of Marine and Freshwater Research 13(1):13–22.

Jellyman, D. J. 1987. Review of the marine life history of Australasian temperate species of *Anguilla*. Pages 276–285 *in* M. J. Dadswell, R. J. Klauda, C. M. Moffitt, R. L. Saunders, R. A. Rulifson, and J.E. Cooper, editors. Common strategies of anadromous and catadromous fishes. American Fisheries Society, Symposium 1, Bethesda, Maryland.

Jellyman, D. J., B. L. Chisnall, L. H. Dijstra, and J. A. T. Boubee. 1996. First record of the Australian longfinned eel, *Anguilla reinhardtii*, in New Zealand. Marine and Freshwater Research 47:1037–1040.

Jespersen, P. 1942. Indo-Pacific leptocephalids of the genus *Anguilla*. Systematic and biological studies. The Carlsberg Foundation, Dana-Report 22, Copenhagen, Denmark.

Kailola, P. J. et al. 1993. Australian fisheries resources. Bureau of Resource Sciences, Department of Primary Industries and Energy, and the Fisheries Research and Development Corporation, Canberra, Australia.

Martin, M. H. 1995. The effects of temperature, river flow, and tidal cycles on the onset of glass eel migration into fresh water in the American eel. Journal of Fish Biology 46:891–902.

McCleave, J. D. and R. C. Kleckner. 1982. Selective tidal stream transport in the estuarine migration of glass eels of the American eel (*Anguilla rostrata*). Journal du Counseil International pour l'Exploration de la Mer 40: 262–271.

McKinnon L. J. and G. J. Gooley. 1998. Key environmental criteria associated with the invasion of *Anguilla australis* glass eels into estuaries of south-eastern Australia. Bulletin Francaise de la Peche et de la Pisciculture 349:117–128.

Montgomery, S. S. and J. R. Craig. 1997. A strategy for measuring the relative abundance of pueruli of the spiny lobster *Jasus verreauxi*. Pages 574–578 *in* D. A. Hancock, D.C. Smith, A. Grant, and J. P. Beumer, editors. Developing and Sustaining World Fisheries Resources, The State of Science and Management, Second World Fisheries Congress, CSIRO Publishing, Brisbane, Australia.

Pease, B. C. 1999. A spatially oriented analysis of estuaries and their associated commercial fisheries in New South Wales, Australia. Fisheries Research 42:67–86.

Phillips, B. F. 1972. A semi-quantitative collector of the puerulus larvae of the western rock lobster *Panulirus longipes cygnus* George (Decapoda, Palinuridea). Crustaceana 22:147–154.

Queensland Department of Primary Industries. 1995. A survey on the distribution and composition of glass eel populations *Anguilla* spp. in south-east Queensland 15 April to 27 August 1995. Aquaculture Information Series, Brisbane, Australia.

Roy, P. S. 1984. New South Wales estuaries: their origin and evolution. 99–121 *in* B. G. Thom, Editor. Coastal Geomorphology in Australia. Academic Press, Sydney, Australia.

Silberschneider, V., B. C., Pease, and D. J. Booth. 2001. A novel artificial habitat collection device for studying resettlement patterns in anguillid glass eels. Journal of Fish Biology 58:1359–1370.

Schmidt, J. 1928. The freshwater eels of Australia. Records of the Australian Museum 5(16):179–210.

Sloane, R. D. 1984. Invasion and upstream migration by glass-eels of *Anguilla australis australis* Richardson and *A. reinhardtii* Steindacher in Tasmanian freshwater streams. Australian Journal of Marine and Freshwater Research 35:47–59.

Sola, C. 1995. Chemoattraction of upstream migrating glass eels *Anguilla anguilla* to earthy and green odorants. Environmental Biology of Fishes 43:179–185.

Sola, C. and P. Tongiorgi. 1996. The effect of salinity on the chemotaxis of glass eels, *Anguilla anguilla*, to organic earthy and green odorants. Environmental Biology of Fishes 47:213–218.

Sorensen, P. W. 1986. Origins of the freshwater attractant(s) of migrating elvers of the American eel, *Anguilla rostrata*. Environmental Biology of Fishes 17(3):185–200.

Struberg, A. C. 1913. The metamorphosis of elvers as influenced by outward conditions. Meddelelser Kommission for Havundersogelser, Serie Fiskeri 4:1–11.

Tesch, F. W. 1977. The eel. Chapman and Hall, London, England.

Tongiorgi, P., L. Tosi, and M. Balsamo. 1986. Thermal preferences in upstream migrating glass eels of *Anguilla anguilla* (L.). Journal of Fish Biology 28:501–510.

Tosi, L., L., Sala, C. Sola, A. Spampanato, and P. Tongiorgi. 1988. Experimental analysis of the thermal and salinity preferences of glass-eels, *Anguilla anguilla* (L.), before and during the upstream migration. Journal of Fish Biology 33:721–733.

Tosi, L., A., Spampanato, C. Sola, and P. Tongiorgi. 1990. Relation of water odour, salinity and temperature to ascent of glass eels, *Anguilla anguilla* (L.): a laboratory study. Journal of Fish Biology 36:327–340.

Weber, M. 1986. Fishing method and seasonal occurrence of glass eels (*Anguilla anguilla* L.) in the Rio Minho, west coast of the Iberian peninsula. Vie et Milieu 36(4):243–250.

Zar, J. H. 1974. Biostatical analysis. Prentice-Hall, Englewood Cliffs, New Jersey.

Contrasting Use of Daytime Habitat by Two Species of Freshwater Eel *Anguilla* spp. in New Zealand Rivers

Don J. Jellyman, Marty L. Bonnett, and Julian R. E. Sykes

National Institute of Water and Atmospheric Research Ltd., Riccarton, Christchurch, New Zealand

Peter Johnstone

AgResearch, Private Bag 50034, Mosgiel, New Zealand

Abstract.—The daytime microhabitat of shortfinned eels *Anguilla australis* and longfinned eels *A. dieffenbachii* was studied in four physically contrasting rivers in New Zealand. Within each waterway, individual sites were selected to provide extensive combinations of water depths and velocities; sites averaged 3.3 m² although actual size depended upon the visual uniformity of depth, velocity, substrate and cover. All sites (549) were sampled by electrofishing and 81% contained eels. Of the 2816 eels captured, 55% were shortfinned eels (56–840 mm TL) and 45% longfinned eels (66–1035 mm TL). Log-linear models were used to derive habitat associations per 100 mm length groups of each species, with eels greater than 500 mm treated as a single class. For all sites combined, models explained between 25 and 63% of total sample variance; for smallest size groups of both species, substrate features and the presence of algae were highly significant, but for larger size groups, shade and cover were more important. There were marked differences in habitat preference curves of depth and velocity both intraspecifically (between different size groups) and interspecifically, with juveniles (< 100 mm) of both species preferring slow (< 0.5 m/s) and shallow (< 0.2 m deep) reaches, while larger (> 500 mm) eels were strongly associated with cover (undercut banks, weed, instream debris) and deeper (> 0.3 m), but slow-flowing water. There were marked changes in habitat preferences of shortfins below and above 200 mm, and for longfins below and above 300 mm, probably associated with shifts from living within substrates to open water. While factors like closeness to the sea and invertebrate abundance were of importance to similar-sized eels in different rivers, there was extensive substitutability of similar categories of habitat variables indicating a high degree of ecological adaptability in both species.

Introduction

New Zealand has two species of freshwater eel that are widespread and abundant, the shortfinned eel *Anguilla australis* and the longfinned eel *Anguilla dieffenbachii*. Both species have broad habitat differences: shortfinned eels (shortfins) prefer slow flowing, lowland, and coastal waters, while longfinned eels (longfins) penetrate further inland and prefer faster flowing water and stony substrates (McDowall 1990), but the two frequently co-exist (Glova et al. 1998). Because of this coexistence, and the fact that both species are nocturnally active (Glova and Jellyman 2000), there is potential for interspecific competition for food and space. To decrease the potential for such negative interactions, ecological theory would suggest that some segregating mechanisms must exist (Ross 1986).

Some species-specific differences in diet have been demonstrated for eels of comparable sizes in lakes (Jellyman 1989). Although there have been a number of studies of the diet of both species of eel in flowing water habitats (e.g., Cairns 1942; Burnet 1969; Hopkins 1970; Sagar and Glova 1998), none has compared the diet of varying size groups of both species. Glova et al. (1998) recorded some interspecific differences in the distribution and habitat associations of eels of different sizes; their analysis indicated no differences in habitat associations for eels greater than or equal to 300 mm, so eels larger than this were treated as a single group. However, as Glova et al. (1998) averaged habitat variables over 10 m lanes of varying width, it was not possible to generate 'point source' habitat preferences. Therefore the present study investigated habitat associations on much smaller spatial scales to determine whether there was any evidence of segregation by species and/or size. Three hypotheses were investigated: firstly, that habitat preferences of both species would vary with size (intraspecific variation); secondly, for eels of the same size, habitat preferences would differ between species (interspecific variation); and thirdly, that habitat preferences of similar sized groups of each species would be similar for different rivers (intraspecific spatial variation). For all three hypotheses, the habitat observed was that used during daytime for concealment. It was recognized that while eels might make extensive foraging movements at night, the type and amount of daytime cover might determine the density of larger eels (Burnet 1952).

Methods

To determine whether habitat use was consistent between physically different waterways, three contrasting river types where both species of eel were known to be present were selected for sampling. All three rivers (Figure 1) had good access along their entire lengths. The waterway types were: a small coastal stream (single thread, hard substrate): Pigeon Bay Stream (east coast South Island); a medium-sized braided lowland river (braided, hard substrate): Ashburton River (east coast South Island); and a small lowland tributary (single thread, entrenched, soft substrate): Firewood Creek, itself a tributary of the Waipa River, that is in turn a lower tributary of the Waikato River (west coast, North Island).

In practice, a second medium sized lowland river (Ashley River, east coast South Island—braided, hard substrate) was substituted for the Ashburton River as the latter river was experiencing unseasonably low flows. A total of 549 sites were sampled; the main physical features of the waterways are summarized in Table 1. For simplicity, hereafter all waterways are generically referred to as "rivers."

Sampling

Within each river, five separate reaches were selected for electrofishing, with up to 30 sites sampled per reach. Reaches were distributed

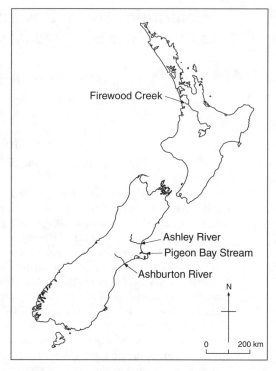

Figure 1. Sampling locations

equidistantly throughout the catchments except that two reaches were chosen in the lower sections of rivers where eel density was known (Pigeon Bay Stream) or suspected of being higher than farther inland, and where habitat variability was greatest. For each reach, individual sites were selected according to two criteria—visual uniformity of habitat (depth, velocity, substrate, and cover), and representation on a velocity and depth matrix to ensure that as wide a range of these combinations was sampled as possible.

Water temperature and conductivity were measured once at each reach, but at each site, a number of physical factors were measured. Most of these factors were chosen on the basis of experience gained from a previous study of habitat features affecting eel distribution (Glova et al. 1998). Factors recorded were:

- Distance upstream (m, measured from the site to entry to the sea from 1:50 000 scale topographic maps) and expressed as both actual distance and relative distance. The latter measure was used to accommodate Firewood Creek (a tributary 110 km upstream of the mouth of the Waikato

River)—relative distance ignored distance from the confluence with the Waikato River to the sea and was expressed relative to the creek length alone
- Water type (run, pool, riffle, backwater)
- Area sampled (m^2)
- Average depth (m): mean of a minimum of three measurements
- Average water column velocity (m/s): mean of a minimum of three measurements at 0.6 of the depth below the surface with an electronic flowmeter
- Average streambed velocity (m/s): mean of a minimum of three measurements
- Dominant substrate type (using the standard measures of Wolman 1954)
- Compaction (using a qualitative scale of 1–5 where 1 = firmly cemented together, and 5 = very loose)
- Embeddedness (using a qualitative scale of 1–5 where 1 = < 5% and 5 = > 75% embedded in fines)
- Substrate depth (m, measured with uniform pressure of a graduated metal rod)
- Stream width (m), and distance to nearest bank (m) and nearest riffle (m)
- Overhead shade (% of site shaded by sun when directly overhead)
- Bank type (% of site bounded by banks—both true right and true left banks; categories being no bank, vertical, sloped, collapsed, undercut). For undercut bank, the average horizontal extent of the undercut was also measured
- Surface cover (% of site covered by surface material, e.g., surface sticks/branches, floating aquatic plants or leaves, willow roots if not reaching to the bottom); the dominant "type" of cover was also recorded (e.g., branches, leaves, aquatic plants)
- Debris cover (% of streambed covered by debris clusters, e.g., sunken sticks/logs, willow root mats if reaching to the bottom); the dominant "type" of debris was also recorded (e.g., roots, sticks, logs)
- Algal cover (% of streambed covered by either encrusting algae, or filamentous algae)
- Aquatic plant cover (% of streambed covered by submerged or emergent plants such as grasses, water cress *Rorippa nasturtium-aquaticum*, etc); the dominant plant "type" was also recorded (e.g., grass, cress, rush, reed)
- Aquatic invertebrate abundance (using a qualitative scale of 0–5 where 0 = completely absent, 5 = abundant); the dominant taxon was also recorded (e.g., snails, mayflies, caddis).

Sampling was completed during stable flow periods in summer 1999 (25 January–10 February) using a battery-powered electrofishing machine (pulsed DC). The minimum sampling area of each site was 1 m^2. If the area was consistent for depth, velocity, substrate and cover, then this area was often extended to 5 m^2 and occasionally more, to a maximum of 150 m^2. On the few occasions that an eel was not completely within a given sampling area, it was included in the catch for that area provided more than half its length was within the area. At each site, electrofishing was generally carried out in a downstream direction with eels being captured in a fine-mesh (0.2 cm mesh) hand-held stopnet. However, if the substrate was comprised mainly of silt, fishing was carried out in an upstream direction so that any disturbed silt did not obscure the area being sampled. Captured eels were anesthetized with 2-phenoxyethanol, identified by species, measured (total length, to 1 mm), and released downstream of the site. For consistency, the same observer always recorded habitat measurements. Further sites within the same reach were always sampled in an upstream direction to avoid "contamination" of sites by any released eels.

Data were assembled in an Access database, imported into Excel, and then into Genstat (Genstat 5 Committee 1993), or Systat (Wilkinson 1990) for statistical analysis. Associations of eel abundance (no. m^{-2}) with habitat variables were principally analyzed by generalized linear models (McCullagh and Nelder 1989). Analysis of the count data were with a logarithmic link and Poisson errors. Qualitative factors were compactness, embeddedness, and invertebrate abundance. Eels were separated into 100 mm size classes although all eels greater than or equal to 500 mm were regarded as a single class. To determine habitat associations of any particular size-group, abundance data from all sites, including those where eels were absent, were used.

Scatter plots were used to determine possible relationships between the eel abundance and other variables and factors. From these results, together with experience and results from previous studies (Glova et al. 1998), likely variables were tested against abundance of particular size classes of each species—for variables with $P \approx 0.05$, their influence was tested by seeing the response of the model when they were dropped. The influence of qualitative variables like invertebrates (six classes) were tested similarly as the model produced P values for each class rather than for the variable as a single entity. A parsimonious principle (Aitkin 1978) was adopted to reduce models to as few significant variables as possible ($P < 0.05$)—because of the number of variables included, interactions were not included.

Models used data from all sites combined (per species and size-class). Comparisons of selected size classes were made between rivers where numbers ($n > 50$) allowed. Finally, based on results from the habitat preference curves, shortfins were separated into "small" (< 200 mm) or "big" ($\geq$ 200 mm), while longfins were separated as "small" (< 300 mm) and "big" ($\geq$ 300 mm). These size groups were then used to examine any intra- and interspecific associations.

Habitat Preference Curves

Preference curves for water depth and velocity, and substrate size, are used by the in-stream flow incremental methodology (IFIM: Bovee 1982) to assess the influence of varying flows on the availability of fish habitat. To determine these preferences for each 100 mm size-group of both species, the percentage of fish use of each habitat variable (depth, column velocity or substrate) was defined as the average fish density in the various classes of that variable (e.g., 0.1 m classes for depth) expressed as a percentage of the sum of the average fish densities over all habitat classes.

$$p_i = \frac{\bar{d}_i}{\Sigma \bar{d}_i} \times 100$$

$i = 1, n$

where p_i = percentage fish-use in habitat class i, $\bar{d}_i$ = average fish density in habitat class i over all sites and n = the number of habitat classes.

The frequency curves of habitat used and habitat available were generated from kernel smoothed density distributions (Silverman 1986). Preference curves were derived by dividing the frequency of use by the frequency of available habitat (Bovee 1986). Curves were scaled to a maximum preference of one by dividing by the maximum ordinate.

Results

Sites Sampled and Sizes and Numbers of Eels Captured

A total of 549 sites was sampled (Table 1), with the proportions of the different water types sampled varying among rivers. The area of each individual site sampled ranged from 1.0 to 150 m², with a mean of mean 3.30 m² (SE 0.30). Depths for all sites ranged from 0.04 to 1.02 m, and column velocities from 0.0 to 1.39 m/s. Table 2 shows the numbers of eels captured per 100 mm size classes. The length range for shortfins was 56–840 mm (mean length = 176 mm), and 66–1035 mm (mean length = 335 mm) for longfins. Size-class 1 (< 100 mm) dominated shortfin catches in both Pigeon Bay Stream and the Ashley River, but was virtually absent from longfin catches. Few shortfins greater than or

Table 1. Summary of some physical characteristics of the waterways, including the numbers of each water type sampled. Water temperatures and conductivities are means of those recorded during the survey.

	Pigeon Bay Stream	Ashburton River	Ashley River	Firewood Creek
Catchment area (km²)	12	1593	1287	52
River length (km)	6	90	97	13
Mean flow (m³/s)	0.04	11	15	0.35
Water temperature (°C)	16.9	16.4	17.4	19.0
Conductivity (μs/cm)	211	83	97	106
Water type sampled (no)				
Runs	39	38	76	67
Riffles	49	18	37	31
Pools	66	0	12	65
Backwaters	6	10	23	12

Table 2. Numbers of shortfin and longfin eels per 100 mm size groups sampled from Pigeon Bay Stream, the Ashburton and Ashley Rivers, and Firewood Creek.

Eel size group (mm)	Shortfin					Longfin				
	Pigeon Bay Stream	Ashburton River	Ashley River	Firewood Creek	Total	Pigeon Bay Stream	Ashburton River	Ashley River	Firewood Creek	Total
< 100	227	2	150	34	413	7	–	–	–	7
100–199	179	3	97	134	413	81	2	59	29	171
200–299	25	10	8	81	124	63	1	58	23	145
300–399	7	18	6	61	92	44	1	70	21	139
400–499	–	12	6	28	46	17	3	85	46	151
500–599	1	13	1	13	28	6	1	18	42	67
600–699	–	5	–	3	8	1	1	2	12	16
700–799	–	4	1	1	6	3	–	1	3	7
800–899	–	3	–	–	3	2	–	–	3	5
> 900	–	–	–	–	–	3	–	1	1	5
Total	439	70	269	355	1133	227	9	294	183	713

equal to 500 mm were captured except in the Ashburton River and Firewood Creek

Water Type

To study the influence of water type on the size distribution of both species, the sites with the best overall representation of shortfins and longfins were examined; i.e., Firewood Creek for shortfins and the Ashley River for longfins. Shortfins less than 100 mm were found mainly in riffles, but most eels larger than this were found in pools. The data for all shortfins showed similar trends except that highest densities of eels greater than or equal to 500 mm were from backwaters (mainly from the Ashburton River). Similarly, longfins less than 300 mm from the Ashley River were found in riffles with larger eels found in pools; data from all sites were similar.

Associations with Habitat Variables

Although alternative measures of distance (absolute or relative distance) and water velocity (mean column or mean bottom velocity) were recorded, tests of the model indicated that relative distance and bottom velocity were the more appropriate measures, and these were adopted. The exception was the use of column velocity to generate habitat suitability curves, as this is the standard measure of velocity for IFIM calculations (Bovee 1982).

Results from the log linear models for each size-group of both species and all sites combined (Table 4) were all highly significantly ($P < 0.001$). Although the models did not always identify the same criteria for adjacent size groups, some patterns were apparent—for example, depth of substrate was highly important for shortfins less than 100 mm but compaction was the substrate feature of importance for shortfins 100–199, 200–299, and 300–399 mm. For the smallest categories of shortfins there was an inverse relationship with distance to the nearest bank meaning that these eels did not show a bankside distribution; likewise, they had a negative association with relative distance indicating that they were found in the lower reaches of rivers. Similarly, mean depth had a negative association, indicating a preference by small shortfins for shallow water; other significant variables were encrusting algae, invertebrate abundance, depth of substrate (shortfins < 100 mm) and substrate compaction. For most size categories of shortfins, specific cover items (debris plant or surface cover) were not important.

For both species, there was a tendency for a higher percentage of the variance to be explained with increasing size of eels, indicating that the habitat associations of eels were more predictable with increasing size. For the larger size categories of shortfins (shortfin 400–499, ≥ 500 mm) shade and plant cover were of importance. Substrate size was also significant although the relationship was negative, indicating

Table 3. Density of eels (no/100m^2) per 100 mm size groups per water type sampled from Firewood Creek (shortfin), Ashley River (longfin), and all sites. N = number of eels.

Eel size group (mm)	Run	Pool	Riffle	Backwater	N	Run	Pool	Riffle	Backwater	N
					Shortfin					
		Firewood Creek						All sites		
< 100	5.8	3.9	14.5	0	34	13.9	21.3	39.7	21.9	413
100–199	22.3	18.9	43.6	7.8	134	17.8	23.3	32.3	18.0	415
200–299	11.2	18.5	6.7	13.7	82	5.6	10.3	4.0	5.1	124
300–399	5.3	15.4	6.7	7.8	60	4.5	7.7	2.0	3.5	91
400–499	2.7	7.7	2.2	3.9	29	2.0	3.7	0.6	3.5	46
> 500	2.1	4.2	0	2.0	16	2.9	2.1	0	4.7	44
N	93	178	66	18	355	320	391	277	145	1133
					Longfin					
		Ashley River						All sites		
< 100	0	0	0	0	0	0.1	0	1.7	0	7
100–199	12.1	1.1	20.3	2.0	61	8.2	1.2	30.3	1.2	173
200–299	9.2	3.2	11.1	8.1	56	8.5	3.0	16.2	4.7	144
300–399	9.2	20.2	3.7	2.7	72	7.3	10.3	7.4	2.3	141
400–499	9.2	28.6	1.8	1.3	84	7.3	15.0	3.1	1.6	151
> 500	1.8	6.9	0	2.0	21	3.7	10.7	1.1	2.7	97
N	117	111	42	24	294	240	230	211	32	713

a preference for finer sediments. Slow velocity was important for shortfins 300–499 mm.

For the smallest two size categories of both species (< 200 mm for shortfins, < 300 mm for longfins), upstream distance was significant (a negative relationship indicating a lowland distribution); encrusting algae and invertebrate abundance were also important, as were aspects of substrate like compaction, embeddedness and substrate size. In contrast with shortfins greater than or equal to 400 mm, the relationship between substrate size and abundance of the smallest shortfins (< 100 mm) and longfins (< 200 mm) was positive indicating a preference for larger substrates; velocity was also important for these longfins. With increasing size, the most significant factors associated with the distribution of longfins greater than 300 mm were components of cover (i.e., debris, plant and surface cover). Again aspects of substrate (compaction and embeddedness) were important for longfin up to 500 mm. Undercut banks were highly significant for longfins greater than or equal to 300 mm and there was a small negative association with bottom velocity.

Of the eight types of surface cover recorded, the most frequently encountered were willow (Salix spp., 62% of all assessments) followed by floating grass (20%). Debris clusters were dominated by accumulations of wood (64%) followed by willow root mats (22%). Surface plant cover was dominated by floating grasses, (50%) followed by watercress (31%). Aquatic macrophytes were present at only 12% of sites sampled, all in Firewood Creek. Caddisfly larvae dominated the invertebrate communities (56% of sites), followed by snails, (34%) and mayfly larvae (7%); caddisfly larvae were dominant in all the waterways sampled with the exception of Firewood Creek where snails were the main species encountered.

To compare habitat variables associated with similar-sized eels from different rivers, samples where n greater than 50 were selected (Table 5). Shortfins less than 100 mm from Pigeon Bay Stream and the Ashley River generally showed similar associations, with eel abundance being negatively associated with relative distance, while algae and invertebrate abundance were strongly positively associated with eel abundance. For shortfins 100–199 mm, comparisons were possible for three rivers (no single variable was common to all three rivers, but 6 of the 16 variables associated with this size-class were

Table 4. Summary of habitat variables that were significantly positively or negatively related ($P < 0.05$) to the abundance of eels of different size classes. (*/- = $P < 0.05$; **/- - = $P < 0.01$; ***/- - - = $P < 0.001$). Df = degrees of freedom; % Dev = percent variance explained by model; P = overall significance of model

Habitat variable	Shortfin						Longfin				
	<100	100–199	200–299	300–399	400–499	>500	100–199	200–299	300–399	400–499	>500
Distance to nearest riffle			+								
Distance to nearest bank	- -		-				-				-
Elevation											
Gradient											
Relative distance	- - -	-	+				- -	-			
Mean depth	- - -	- - -								+	+
Width											-
Shade		-		-	+++		+				
Algae—encrusting	+++	+++		+			+++	+			
Algae—filamentous		++	+	+				+			
Invertebrate abundance	+++	+++			++		++	+++			
Debris cover										+++	+++
Plant cover					+++			+++			
Surface cover				+				+++		+++	+++
Depth of substrate	+++			++							
Compaction	+++	+++	++	+++			+++	+++	+++	+++	
Embeddedness							+++	+++	++	+++	
Dominant substrate size	+		+		-	- -	+++				
Bank—vertical			+								
—sloped									+		
—collapsed								+++			
—undercut									+++	+++	+++
—no bank											
Bottom velocity				-	- -		+++	+			-
Df.	16	22	13	15	7	12	19	19	19	13	12
%Dev	50	28	25	41	36	63	30	24	41	54	46
P	<0.001	<0.001	<0.001	<0.001	<0.001	<0.001	<0.001	<0.001	<0.001	<0.001	<0.001

common to two of the three rivers). Important features were shallow water, the presence of encrusting algae and aquatic invertebrates, and aspects of the substrate. Bank features were important in the Ashley River, but not at either of the other two sites.

For the smallest longfin size-class, 100–199 mm, features common to both Pigeon Bay Stream and the Ashley River were relative distance, extent of algae, and aspects of substrate; again, bank features were important in the Ashley River. For longfins 200–299 mm, comparisons were possible for the same two rivers, but the only features in common were invertebrate abundance and substrate size. The percent deviance explained was much lower for Pigeon Bay Stream (31%) than the Ashley River (57%), with the only highly significant variable ($P < 0.001$) for Pigeon Bay Stream being water velocity. In contrast, distance, shade, substrate compaction and size, and bank features were all highly significant variables for 200–299 mm longfins in the Ashley River.

Associations of "small" and "large" eels of both species were explored by linear regression of the abundance of the particular combination

Table 5. Summary of habitat variables that were significantly positively or negatively related ($P < 0.05$) to the abundance of shortfin and longfin eels of similar size classes from different sites. PBS = Pigeon Bay Stream; AR = Ashley River; FWC = Firewood Creek. (*/- = $P < 0.05$; **/- - = $P < 0.01$; ***/- - - = $P < 0.001$). Df = degrees of freedom; % Dev = percent variance explained by model; P = overall significance of model.

	Shortfin < 100		Shortfin 100–199			Longfin 100–199		Longfin 200–299	
	PBS	AR	PBS	AR	FWC	PB	AR	PB	AR
Distance to nearest riffle		- - -							
Distance to nearest bank			+		+				
Elevation									
Gradient									
Relative distance	- - -	- - -				- -	-		- - -
Mean depth			- - -	- - -		- - -			
Width		- - -	-	- -					
Shade		+++	- - -			++			+++
Algae—encrusting	+++	+++		++	+	+++	+		++
Algae—filamentous	+++		++	+++		++			
Invertebrate abundance	+++	+++	+++		+++		++	++	+
Debris cover									
Plant cover		++		++					
Surface cover					- -				
Depth of substrate		++			++				
Compaction	+++		+++			++	+++		+++
Embeddedness				+++			++		+
Dominant substrate size			- - -			+		+	+++
Bank–vertical							+++		+++
—sloped		+		- - -			+++		+++
—collapsed							+++		+++
—no bank					- - -				
Bottom velocity		- - -				+	+++	+++	
N									
Df	13	14	12	12	10	11	16	6	17
% Deviance	35	97	30	60	23	62	61	31	57
P	<0.001	<0.001	<0.001	<0.001	<0.001	<0.001	<0.001	<0.001	<0.001

of size groups. The only significant relationship was for small shortfins and small longfins (Table 6); neither of the intraspecies associations (small versus large shortfins and small versus large longfins) was significant.

Habitat Preferences

The smallest size-group of shortfins (< 100 mm) showed a marked preference for shallow water (Figure 2). With increasing size, there was a shift towards preferring water 0.5–1.0 m deep. Similarly, the smallest size-group of longfins, 100–199 mm, showed a preference for water less than 0.25 m deep, with larger eels again preferring deeper water. Smallest shortfins had a preference for water with a velocity of less than 1.0 m/s; the size-group 100–199 mm showed little preference except for an avoidance of water swifter than 1.25 m/s. Larger size groups of shortfins showed a marked preference for water up to 0.5 m/s or less. In contrast, longfins up to 300 mm showed a preference for faster water,

Table 6. Associations of small and large shortfin and longfin eels. P values from linear regressions; sample number in brackets

	Small longfin	Big shortfin	Big longfin
Small shortfin	<0.001 (84)	0.69 (68)	0.85 (70)
Small longfin		0.44 (26)	0.74 (52)
Big shortfin			0.70 (54)

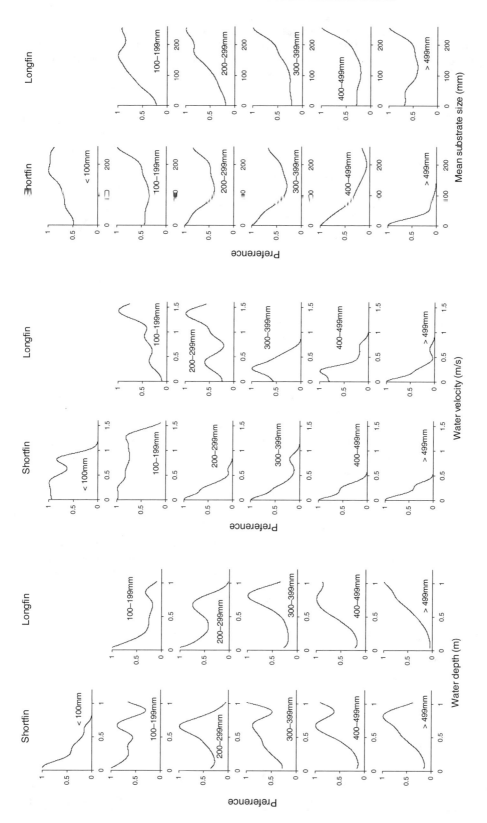

Figure 2. Habitat preference curves (water depth, water velocity, and substrate size) for shortfin and longfin eels (by 100 mm size groups).

with the preferred velocity being between 1.2 and 1.4 m/s; above this size, there was a marked preference for velocities less than 0.5 m/s. Shortfins up to 200 mm showed a slight preference for coarse substrate, but with increasing size there was a significant shift towards fine substrate. Small longfins showed a preference for coarse substrate that was reflected in all but the largest size-class.

In the present study, eel abundance was not necessarily significantly associated with depth, velocity and substrate size, the variables used in the IFIM assessments (Bovee 1982; Jowett and Richardson 1995). Therefore, to test the influence of these variables on eel abundance, log-linear models were developed using these factors alone (Table 7). Depth was almost always significant to both species, either positively (increasing depth) for largest size groups, or negatively (decreasing depth) for smaller size groups; likewise, velocity was generally significant to both species. Substrate was of lesser importance to shortfins, and was unimportant to longfins.

Discussion

Co-occurrence of Species

From a review of the distribution and relative abundance of freshwater fish in New Zealand, Minns (1990) noted that the longfin eel was the most common native species in lotic environments, with shortfin eels the fourth most common; both eel species showed one of the strongest associations of any lotic fish species cluster. Recent information on the national distribution of freshwater eels in New Zealand (from NIWA's Freshwater Fish Database, FFDB), shows that eels are present at a high proportion (51%) of all sites ($n = 14{,}718$); of the 7,522 sites where eels have been recorded, both species co-occur at 22% of these sites.

Longfins are ubiquitous, occurring in a wide range of habitats throughout New Zealand. When discussing their distribution in South Westland, Taylor (1988) noted that apart from an avoidance of swamps, longfins were recorded from all other habitats. Longfins dominate inland populations (McDowall and Taylor 2000), but the lack of shortfins in these areas may be more a feature of lack of suitable habitat than lack of ability to penetrate inland by this species (McDowall 1993). Results from the FFDB tend to confirm this as longfins predominate in all inland habitats except swamps where 73% of records of eels ($n = 44$) were for shortfins. In the present study, small shortfins were strongly associated with small longfins, no doubt partly because of their restricted spatial distribution within rivers. Both these groups were highly unlikely to coexist with large longfins and, to a lesser extent, large shortfins. While this can be described in terms of changing habitat requirements with size, it is also likely to indicate an avoidance of larger predatory eels (Cairns 1942; Jellyman 1989) as removal of large longfins has been associated with increased density of smaller eels (Chisnall and Hicks 1993).

Size Distribution

The size distributions of shortfins in Pigeon Bay Stream and the Ashley River were typical of those expected for a catchment entering the sea and with regular recruitment; i.e., a predominance of smallest size groups and a progressive

Table 7. Relationships between the abundance of eels of different size classes and depth, column velocity, and mean substrate size. (*/- = $P < 0.05$; **/- - = $P < 0.01$; ***/- - - = $P < 0.001$). Df = degrees of freedom; % Dev = percent variance explained by model; P = overall significance of model

Habitat variable	Shortfin						Longfin				
	< 100	100–199	200–299	300–399	400–499	> 500	100–199	200–299	300–399	400–499	> 500
Depth	- - -	- - -	+	+		+++	- - -	- -	+++	+++	+++
Velocity	- - -	- - -	- - -	- - -	-		+++		-	-	- - -
Substrate			+		- - -	- - -					
Df.	3	3	3	3	3	3	3	3	3	3	3
%Dev	11	5	8	7	18	27	17	4	6	14	22
P	<0.001	<0.001	<0.001	<0.001	<0.001	<0.001	<0.001	<0.001	<0.001	<0.001	<0.001

decline in numbers of larger eels. A similar size distribution to those in the present study was recorded by Glova et al. (1998) for a small North Island stream. The lower proportion of shortfins less than 100 mm from Firewood Creek is attributable to the distance upstream (110 km) of this Waikato River tributary. Likewise, relatively few age-0 eels are present at Karapiro Dam (Jellyman 1977; Beentjes et al. 1997), 130 km upstream on the Waikato River.

Direct comparisons can be made between the size distributions of both species of eel from Pigeon Bay Stream recorded by Glova et al. (1998) and the present study. For shortfins, although the relative proportions of eels in the smallest two size classes (< 100, 100–199 mm) differ, both studies recorded that eels less than 200 mm contributed the bulk of shortfin samples (92 and 94% respectively). Similarly the proportion of longfins less than 100 mm was low in both studies, being 3% and 4% respectively. While the absence of this size-class of longfins from Firewood Creek could be largely explained by distance inland (110 km), the explanation does not apply to the Ashley River that enters directly to sea. From the proportions of larger eels recorded in the present study and that of Glova (1988), the Ashley River would be classified as principally a "longfin river"; however, this is not apparent from glass eel recruitment where Jellyman et al. (1999) recorded only 4% of glass eels were longfins.

Temporal variability in the strength of glass eel recruitment is a feature of *Anguilla* spp. (e.g., Moriarty 1986), and is to be expected in a species where adults undertake an extensive marine spawning migration and returning larvae are mainly passively transported on surface ocean currents. Unfortunately, northern hemisphere species of *Anguilla* have shown substantially reduced recruitment over recent years (e.g., Moriarty and Dekker 1997; Richkus and Whalen 1999; Haro et al. 2000); reasons are uncertain, although loss of habitat and overfishing have both been suggested. Similarly the virtual absence of small longfins in the present study gives rise to some concern about the well being of this species. Such concern is reinforced by results of conceptual modeling (Hoyle and Jellyman in press) that indicate there is insufficient escapement of migratory female longfin to maintain stocks; this is not surprising given the extreme longevity of the species where generation times for slow-growing stocks can exceed 90 years (Jellyman 1995) and prespawning fish are subject to extensive commercial fishing.

Although shortfins numerically dominated three of the four waterways sampled, there were few greater than or equal to 300 mm in either Pigeon Bay Stream or the Ashley River. Reasons for this are unknown, but as males do not mature until a minimum of 380 mm (Jellyman and Todd 1982), seaward migration of males is not a contributing factor. Bonnett and Jellyman (1996) suggested it was possible that there was a subsequent migration back to sea of juvenile shortfins that had not found suitable habitat in the stream they initially entered. Proof of this would require an extensive tagging program, or analysis of otolith microchemistry.

All four rivers in this study were commercially fished by fyke nets, although there are no catch records to determine the number and size of eels removed. Fyke nets are size selective (Berg 1990; Naismith and Knights 1990), meaning that the number of eels greater than 400–500 mm in these waterways will have been affected by fishing. Being ecologically dominant, large longfins appear particularly susceptible to fyke netting (Hoyle and Jellyman in press), and the fact that only 14% of all longfins exceeded 500 mm may largely reflect this selective removal. Certainly, large longfins were prominent in precommercial catches of Burnet (1952) where dominant size groups were 760–840 mm. Recent catch sampling by Beentjes (1999) has also demonstrated the presence of substantial numbers of female longfins greater than 750 mm in areas that are seldom commercially fished.

Habitat Associations

Few studies of *Anguilla* spp. have investigated habitat associations. However, distance from the sea is known to be important in explaining the distribution of size and abundance on various species of *Anguilla* (e.g., *A. anguilla*, Naismith and Knights 1993; Lobon-Cervia et al. 1995; *A. rostrata*, Smogor et al. 1995; *A. australis*, Sloane 1984; *A. dieffenbachii*, Hayes et al. 1989). In a study of the habitat associations of both New Zealand species, Glova et al. (1998) found that distance from the sea was important in describing the distribution of shortfins less than 300 mm, but not of similar-sized longfins. In the present study, upstream distance was associated with the distribution of eels less than 300 mm of both species.

Most studies involving habitat utilization of New Zealand eels have concluded that eel habitat use changes with increasing size. This has often been described in relation to water types. For example, Jowett and Richardson (1995) recorded higher proportions of eels of both species (size unspecified) in riffles than runs. Both the present study and that of Glova et al. (1998) found that the abundance of longfins less than 300 mm was greatest in riffles. While our study found that shortfins less than 200 mm inhabited riffles, Glova et al. (1998) found them evenly distributed across all habitat types. Glova et al. (1998) found that eels greater than 300 mm showed no consistent habitat preferences, but in the present study these eels were clearly associated with pools, a similar result to that from a previous study (Glova 1988).

Various sizes have been suggested when changes in habitat use occur. Burnet (1952) found that for longfins, changes took place between 300 and 350 mm. In adopting a similar size, 300 mm, Glova et al. (1998) considered that with growth, interstitial spaces within the substrates provided inadequate cover for eels, necessitating a shift to larger cover items like undercut banks and debris clusters. Their analysis suggested that this transition occurred at lengths greater than 300 mm. Subsequently, Glova (1999) experimentally demonstrated differences in cover and habitat preferences for differing size groups of small eels (< 300 mm) of both species tested separately and together, but he did not test eels greater than 300 mm.

Results from the present habitat association model and the distribution of eels by water type, indicated that shortfins less than 200 mm tended to have different habitat associations than shortfins larger than this. Likewise, the habitat associations of longfins varied with size, but principally occurred for fish greater than 300 mm. Small shortfins (< 200 mm) were found in shallow downstream reaches of open water, with high densities of algae and invertebrates, where the substrate was deep and loosely compacted. Habitat associations for large shortfins ($\geq$ 300 mm) were less well defined, but included shade, the presence of algae and invertebrates, fine and deep substrate, plant cover and slow water velocity. Habitat preferences for small longfins were rather similar to those for small shortfins—downstream reaches with high densities of algae and invertebrates, and coarse but loosely compacted sediments. For large longfins, algae and invertebrate abundance were not important, but substrate features (compaction and embeddedness) were very important—a consistent feature for large longfins was the importance of cover, variously in the form of debris, plant, surface, or undercut bank. These overall size-related differences in preferred habitat were also evident in the habitat suitability curves; i.e., for shortfins, there was a shift in preferences between 200 and 300 mm, while for longfins the main change occurred between 300 and 400 mm.

Habitat preference curves in the present study show significant differences to those derived from other studies (Glova and Duncan 1985; Jowett and Richardson 1995), mainly because neither of these previous studies differentiated eels by length. Preference curves for longfins from a stony, braided South Island river (Glova and Duncan 1985) indicated that longfins preferred shallow water (0.1–0.5 m), moderate velocities (0.2–1.5 m/s), and coarse substrate. Results for longfins from a study of 34 rivers nationally (Jowett and Richardson 1995) were generally similar to those of Glova and Duncan (1985) except that the species had a preference for fast water—in contrast, shortfins preferred slower velocity as (0–1.0 m/s) and fine substrate. The present study showed marked changes in depth preferences between small and large eels of both species (small eels preferring shallow water and larger eels preferring deeper water); further changes were from high to low velocities for small and big longfins respectively, and a change from coarse to fine substrate for large shortfins.

When the components used in IFIM studies, water depth, velocity, and substrate size, were the only factors used in models of eel abundance, not all showed as significant variables (Table 7). For example, substrate size did not appear in any of models for longfins. However, results from models using all habitat variables (Table 4) indicated that substrate size was only of significance to longfins 100–199 mm, and that the substrate features of importance were compaction and embeddedness. The change in the relationship between depth and abundance of both species (negative for small eels then changing to positive for larger eels), confirmed the size-related shifts in habitat discussed above.

The models suggested that invertebrate abundance was of considerable importance for small eels of both species, of some importance for large shortfins, but of no importance for larger longfins. These outcomes are consistent with dietary studies of both species. Thus, small eels are invertebrate carnivores (Cairns 1942; Hopkins 1970; Jellyman 1989; Sagar and Glova 1998), and probably consume a significant proportion of their food from within these substrate (Sagar and Glova 1998) rather than foraging in the open where they would be more vulnerable to predation (Glova and Jellyman 2000). Larger shortfins also eat a high proportion of invertebrates (Jellyman 1989), but are largely piscivorous (Cairns 1942; Jellyman 1989).

The importance of cover to large longfins was demonstrated by Burnet (1952) who found that high densities of larger eels were associated with high levels of cover. As eels are nocturnally active, daytime cover provides areas of subdued lighting, but can also serve as an area for stalking and ambushing prey fish species (Cairns 1942; Burnet 1952). In their study of habitat use by both New Zealand species, Glova et al. (1998) noted that eels greater than 300 mm were strongly associated with cover, with longfins using a greater variety (macrophytes, undercut bank, instream debris, shade) than shortfins (principally riparian cover). The ability of shortfins to use riparian cover would partly explain the observation of Hicks and McCaughan (1997) that densities of shortfins in pastoral areas substantially exceed those of longfins. Their observation that densities of longfins were similar in all land-use types (native forest, exotic forest, pasture) is consistent with the observation of Glova et al. (1998) that the longfin "is a more generalist species" than the shortfin. Further evidence of this comes from Hayes et al. (1989) where longfins were a component of all four fish assemblages established in a North Island, New Zealand, river.

Comparisons of habitat variables used by eels of similar size in the different rivers were only possible for small eels. As the habitat associations of larger eels were more predictable than those for smaller eels, differences between rivers would be expected to be more pronounced for the smaller fish. Despite differences in the specific variables used by eels in different rivers, there was substantial commonality for "categories" of variables; i.e., the presence of algae was almost universally important, although this was variously encrusting algae or filamentous algae. Likewise, substrate features were important, usually the degree of compaction or embeddedness, but sometimes depth of substrate or dominant substrate size. It was concluded that while variables like distance from the sea and abundance of invertebrates were of primary importance, there was a degree of substitutability for other variables—thus it was the presence of algae that was more important than whether the algae was encrusting or filamentous. Such observations support the conclusion of Glova (1999) that juvenile eels show "behavioral plasticity" and an ability to adapt to utilize available cover even when this cover is not the preferred type.

Conclusions and Management Implications

The first hypothesis that habitat preferences of each species would vary with size proved correct. Both species showed shifts in habitat associations consistent with a shift from living within substrate interstices to utilizing instream features like debris and undercut banks, although the size when this change took place differed between the species. The second hypothesis of interspecific differences in habitat for eels of similar size was also accepted, although differences were more apparent for larger eels than smaller eels. The third hypothesis was that habitat preferences of similar sized groups of each species would be similar for different rivers—results for this hypothesis were inconclusive. In practice, as rivers were physically different, the habitats they provided were correspondingly different; while eels showed preferences for particular "categories" of habitat, the specific type of habitat used varied between rivers.

The study demonstrated that, in order to maintain a full size range of both species of eel, a wide range of habitat types is required. Although their ecological plasticity is one of the reasons that *Anguilla* species are so widespread (Helfman et al. 1987), both shortfins and longfins have definable habitat preferences that vary both interspecifically and intraspecifically (with size). For juveniles, features of the substrate are very important, but larger eels require more complex cover like debris clusters and undercut banks; the availability of such cover will largely determine the density of large eels.

Previous habitat suitability curves (Glova and Duncan 1985; Jowett and Richardson 1995) have not differentiated between different size classes of eel. Given the habitat shifts that both species undergo, the preference curves generated by the present study will allow for more accurate assessments of eel habitat in future IFIM habitat surveys.

Acknowledgments

We thank our colleagues Ben Chisnall and Greg Kelly for their assistance in the field. Helpful discussions on statistical techniques were held with Eric Graynoth (NIWA) and David Baird (AgResearch). Thanks to Paul Sagar, National Institute of Water and Atmospheric Research Ltd., Christchurch, and two external referees, Ken Oliveira and Javier Lobon-Cervia, for helpful comments on the manuscript. The research was funded by the New Zealand Foundation for Research, Science and Technology Contract CO 1605.

References

Aitkin, M. 1978. The analysis of unbalanced cross-classifications. Journal of the Royal Statistical Society 141, Part 2:195–223.

Beentjes, M. P. 1999. Size, age, and species composition of commercial eel catches from South Island market sampling, 1997-98. National Institute of Water and Atmospheric Research, New Zealand. Technical Report 51.

Beentjes, M. P., B. L. Chisnall, J. A. T. Boubée, and D. J. Jellyman. 1997. Enhancement of the New Zealand eel fishery by elver transfers. New Zealand Fisheries Technical Report 45.

Berg, R. 1990. The assessment of size-class proportions and fisheries mortality of eel using various catching equipment. International Revue gesamten Hydrobiologie 75:775–780.

Bonnett, M. L., and D. J. Jellyman. 1996. Dealing with slippery customers—what happens to large shortfinned eels in streams? National Institute of Water and Atmospheric Research Ltd., Water and Atmosphere 4(3):26–27.

Bovee, K. D. 1982. A guide to stream habitat analysis using the instream flow incremental methodology. United States Fish and Wildlife Service, Cooperative Instream Flow Group, Instream Flow Information Paper 12.

Bovee, K. D. 1986. Development and evaluation of habitat suitability criteria for use in the instream flow incremental methodology. United States Fish and Wildlife Service, Cooperative Instream Flow Group. Instream Flow Information Paper 21.

Burnet, A. M. R. 1952. Studies on the ecology of the New Zealand longfinned eel, *Anguilla dieffenbachii* Gray. Australian Journal of Marine and Freshwater Research 3:32–63.

Burnet, A. M. R. 1969. A study of the inter-relation between eels and trout, the invertebrate fauna and the feeding habits of the fish. New Zealand Marine Department. Fisheries Technical Report 36.

Cairns, D. 1942. Life-history of the two species of freshwater eel in New Zealand. II. Food and inter-relationships with trout. New Zealand Journal of Science 23:132–148.

Chisnall, B. L., and B. J. Hicks. 1993. Age and growth of longfinned eels (*Anguilla dieffenbachii*) in pastoral and forested streams in the Waikato River basin, and in two hydro-electric lakes in the North Island, New Zealand. New Zealand Journal of Marine and Freshwater Research 27: 317–332.

Genstat 5 Committee. 1993. Genstat 5 Release 3 reference manual. Oxford, Clarendon Press.

Glova, G. J. 1988. Fish density variations in the braided Ashley River, Canterbury, New Zealand. New Zealand Journal of Marine and Freshwater Research 22:9 15.

Glova, G. J. 1999. Cover preference tests of juvenile shortfinned eels (*Anguilla australis*) and longfinned eels (*A. dieffenbachii*) in replicate channels. New Zealand Journal of Marine and Freshwater Research 33:193–204.

Glova, G. J., and M. J. Duncan. 1985. Potential effects of reduced flows on fish habitats in a large braided river, New Zealand. Transactions of the American Fisheries Society 114:165–181.

Glova, G. J., and D. J. Jellyman 2000. Size-related differences in diel activity of two species of juvenile eel (*Anguilla*) in a laboratory stream. Ecology of Freshwater Fish 9:210–218.

Glova, G. J., D. J. Jellyman, and M. L. Bonnett. 1998. Factors associated with the distribution and habitat of eels (*Anguilla* spp.) in three New Zealand lowland streams. New Zealand Journal of Marine and Freshwater Research 32: 255–269.

Haro, A., W. Richkus, K. Whalen, A. Hoar, W.-D. Busch, S. Lary, T. Brush, and D. Dixon. 2000. Population decline of the American eel: implications for research, and management. Fisheries 25(9):7–16.

Hayes, J. W., J. R. Leathwick, and S. M. Hanchett. 1989. Fish distribution patterns and their association with environmental factors in the Mokau River catchment, New Zealand. New Zealand

Journal of Marine and Freshwater Research 23:171–180.

Helfman, G. S., D. J. Facey, L. S. Hales, and E. L. Bozeman, Jr. 1987. Reproductive ecology of the American eel. Pages 42–56 *in* M. J. Dadswell, R. J. Klauda, C. M. Moffitt, R. L. Saunders, R. A. Rulifson, and J. E. Cooper, editors. Common strategies of anadromous and catadromous fishes. American Fisheries Society, Symposium 1, Bethesda, Maryland.

Hicks, B. J., and H. M. C. McCaughan. 1997. Land use, associated eel production, and abundance of fish and crayfish in streams in Waikato, New Zealand. New Zealand Journal of Marine and Freshwater Research 31:635–650.

Hopkins, C. L. 1970. Some aspects of the bionomics of fish in a brown trout nursery stream. New Zealand Marine Department, Fisheries Research Bulletin 4.

Hoyle, S. D., and D. J. Jellyman In press. Longfin eels need reserves: modelling the impacts of commercial harvest on stocks of New Zealand eels. Marine and Freshwater Research.

Jellyman, D. J. 1977. Summer upstream migration of juvenile freshwater eels in New Zealand. New Zealand Journal of Marine and Freshwater Research 11:61–71.

Jellyman, D. J. 1989. Diet of two species of freshwater eel (*Anguilla* spp.) in Lake Pounui, New Zealand. New Zealand Journal of Marine and Freshwater Research 23:1–10.

Jellyman, D. J. 1995. Longevity of longfinned eels *Anguilla dieffenbachii* in a New Zealand high country lake. Ecology of Freshwater Fish 4:106–112.

Jellyman, D. J., B. L. Chisnall, M. L. Bonnett, and J. R. E. Sykes. 1999. Seasonal arrival patterns of juvenile freshwater eels (*Anguilla* spp.) in New Zealand. New Zealand Journal of Marine and Freshwater Research 33:249–262.

Jellyman, D. J., and P. R. Todd. 1982. New Zealand freshwater eels: their biology and fishery. New Zealand Ministry of Agriculture and Fisheries, Fisheries Research Division Information Leaflet 11.

Jowett, I. G., and J. Richardson. 1995. Habitat preferences of common, riverine New Zealand native fishes and implications for flow management. New Zealand Journal of Marine and Freshwater Research 29:13–23.

Lobon-Cervia, J., C. G. Utrilla, and C. G. Rincon. 1995. Variations in the population dynamics of the European eel *Anguilla anguilla* (L.) along the course of a Cantabrian river. Ecology of Freshwater Fish 4:17–27.

McCullagh, P., and J. A. Nelder. 1989. Generalized linear models. Chapman and Hall, London, England.

McDowall, R. M. 1990. New Zealand freshwater fishes: a natural history and guide. Auckland, Heinemann-Reed.

McDowall, R. M. 1993. Implications of diadromy for the structuring and modelling of riverine fish communities in New Zealand. New Zealand Journal of Marine and Freshwater Research 27:453–462.

McDowall, R. M., and M. J. Taylor. 2000. Environmental indicators of habitat quality in a migratory freshwater fish fauna. Environmental Management 25:357–374.

Minns, C. K. 1990. Patterns of distribution and association of freshwater fish in New Zealand. New Zealand Journal of Marine and Freshwater Research 24:31–44.

Moriarty, C. 1986. Variations in elver abundance at European catching stations from 1938 to 1985. Vie Milieu 36:233–235.

Moriarty, C., and W. Dekker. 1997. Management of the European eel. Fisheries Bulletin (Dublin) 15.

Naismith, I. A., and B. Knights. 1990. Studies of sampling methods and of techniques for estimating populations of eels, *Anguilla anguilla* L. Aquaculture and Fisheries Management 21:357–367.

Naismith, I. A., and B. Knights. 1993. The distribution, density and growth of the European eel, *Anguilla anguilla*, in the freshwater catchment of the River Thames. Journal of Fish Biology 42:217–226.

Richkus, W., and K. Whalen. 1999. American eel (*Anguilla rostrata*) scoping study. A literature and data review of life history, stock status, population dynamics and hydroelectric impacts. Palo Alto, California, EPRI. TR-111873.

Ross, S. T. 1986. Resource partitioning in fish assemblages: a review of field studies. Copeia 1986:352–388.

Sagar, P. M., and G. J. Glova. 1998. Diel feeding and prey selection of three size classes of shortfinned eel (*Anguilla australis*) in New Zealand. Marine and Freshwater Research 49:421–428.

Silverman, B. W. 1986. Density estimation for statistics and data analysis. Chapman and Hall, London, England.

Sloane, R. D. 1984. Distribution and abundance of freshwater eels (*Anguilla* spp.) in Tasmania. Australian Journal of Marine and Freshwater Research 35:463–470.

Smogor, R. A., P. L. Angermeier, and C. K. Gaylord. 1995. Distribution and abundance of American eels in Virginia streams: tests of null models across spatial scales. Transactions of the American Fisheries Society 124:789–803.

Taylor, M. J. 1988. Features of freshwater fish habitat in South Westland and the effect of forestry practices. New Zealand Ministry of Agriculture and Fisheries, Freshwater Fisheries Report 97.

Wilkinson, L. 1990. SYSTAT: the system for statistics. SYSTAT Inc., Evanston, Illinois.

Wolman, M. G. 1954. A method of sampling coarse river-bed material. American Geophysical Union 36:951–956.

Response of Otolith Sr:Ca to a Manipulated Environment in Young American Eels

RICHARD T. KRAUS AND DAVID H. SECOR

*University of Maryland Center for Environmental Science,
Chesapeake Biological Laboratory, One Williams Street,
Solomons, Maryland 20688, USA*

Abstract.—There has been increased use of otolith composition data to track eel *Anguilla* spp. migrations in coastal and estuarine environments. Numerous studies have used strontium (measured as Sr:Ca) to infer salinity related habitat use, yet the method remains largely unverified. It is not known whether the otolith Sr:Ca is primarily related to ambient salinity, or whether this relationship is confounded by temperature or growth. We manipulated experimental rearing environments of young American Eels *Anguilla rostrata* to determine the amount of variability in otolith Sr:Ca related to salinity, temperature, and growth; and estimate the lag time in response of otolith chemistry to changes in salinity. Our results suggest that otolith Sr:Ca in eels can be used to discriminate broad scale life history periods of fresh, brackish, and ocean habitat occupation, but finer scale interpretations cannot be supported. In addition to salinity, temperature, and growth effects, further study is needed concerning the influence of dietary sources of Sr on eel otolith composition.

Introduction

Important anguillid eel fisheries exist worldwide but primarily occur for three species: *Anguilla japonica*, *A. anguilla*, and *A. rostrata*. Each species is comprised of a single panmictic population that has a broad geographical and estuarine range during the juvenile life history stage (i.e., glass eel and elver stages). Eels occupy a variety of habitats in which there is considerable potential for variable reproductive contribution to the population (Tsukamoto et al. 1998). To better understand the population dynamics of different eel species, it is important to identify divergent life cycles and quantify significant differences in vital population rates between major habitat types.

One method that provides retrospective inference of habitat use is microprobe analysis of otolith strontium:calcium ratios (Sr:Ca). Otolith Sr:Ca is expected to be proportional to Sr:Ca in the water (Bath et al. 2000), and in estuaries salinity represents a proxy for Sr:Ca in the water, which typically increases from freshwater to marine habitats. Observed variability in eel otolith Sr:Ca suggests that time spent in fresh, brackish, and ocean waters can be differentiated using microprobe-based chemical analyses in *A. rostrata*, *A. anguilla*, and *A. japonica* (Tsukamoto and Arai 2001; Morrison et al., Limburg, and Tzeng et al., all this volume); however, predictive relationships between salinity and otolith Sr:Ca have not been determined. Although otolith Sr is positively related to ambient salinity for several tested fishes (Secor and Rooker 2000), temperature and growth may also be important in regulating the deposition of Sr and Ca in fish otoliths (Mugiya and Tanaka 1995; Kalish 1989; Radtke 1989; Campana 1999).

The assumption that otolith Sr:Ca is positively correlated with salinity has been examined for *A. japonica* (Tzeng 1996; Kawakami et al. 1998), but not for other Anguillidae. Further, Casselman (1982) observed differences in otolith Sr:Ca between *A. rostrata* and *A. anguilla*, and suggested that uptake of Sr into otoliths may vary between these species. The objective of our study was to experimentally quantify the relationship between otolith Sr:Ca and ambient salinity, temperature and growth in young elver-stage (< 7 g) *A. rostrata* through controlled rearing experiments.

Methods

Young eels were obtained from Anguilla Culture Technology in Hopewell, Virginia. Eels were captured at the glass eel stage, primarily by fishers in Canada and Maine, and cultured in fresh water. Two experiments were conducted to evaluate effects of environmental salinity and temperature on otolith Sr:Ca in young *A. rostrata*. In the first (short-term) experiment, randomly selected eels were acclimated to five salinities (0.5, 5, 10, 15, 25) and three temperatures (16°C, 20°C, 28°C) and held for 53 days. These treatments were chosen to represent the ranges of temperature and salinity that are potentially encountered by estuarine juveniles during the growing season. Two replicates were run simultaneously. In the second (cycle) experiment, randomly selected tanks of eels were switched from high (15) to low salinity (0.5) or vice versa to compare with control treatments with no salinity change. The cycle experiment was replicated twice at two temperatures (20°C and 28°C) and lasted for 74 days. Salinity was changed on experimental day 53. The durations in both the short-term and the cycle experiments allowed sufficient otolith growth at all treatment levels to conduct the microprobe analyses. Demarcation of experimental otolith growth was accomplished with *in vivo* marking with alizarin complexone (50 ppm, by immersion for 24 hours). *Ad libitum* feedings were conducted daily with BioDiet Starter, 1 mm food pellets (from Bio-Oregon, Inc. of Warrenton, Oregon).

Microchemical Analysis

Sagittal otoliths were embedded in epoxy resin (Spurr 1969). Longitudinal thin sections were mounted on petrographic slides, ground to the widest plane, and polished to a smooth appearance at 50× magnification. Chemical analysis was accomplished with Wavelength Dispersive X-ray Spectrometry (WDS) using a JEOL JXA-8900 electron probe microanalyzer for the elements, Sr and Ca. Calcite and strontianite standards were used. A nominal beam diameter was used to scan an area 25 μm^2 with an accelerating voltage of 25 kV and current of 20 na. Peak Sr count time was 30 seconds, and peak count time for Ca was 10 seconds. For the short-term experiment, points were oriented along the periphery of otolith sections. For otolith sections from the cycle experiment, transects with evenly spaced points at 10 μm intervals were analyzed along a major growth axis from the core region to either the rostral or postrostral margin, whichever direction appeared to have the most uniform surface at 600× magnification. Subsequent examination of slides with epifluorescent microscopy was used to view the alizarin complexone mark(s) and assign probe points to the preexperimental or experimental regions of otolith.

Additional chemical analyses were performed on the experimental eel diet and on water samples from the various salinities using a quadrapole inductively coupled plasma mass spectrometer (ICP-MS). Approximately 0.5 g of food was placed into a 20 mL Teflon vial along with 3 mL of concentrated HNO_3 (J.T. Baker, Optima Grade) and 1 mL of concentrated HCl (J.T. Baker, Optima Grade). The vials were tightly capped and placed in a drying oven at 60°C overnight. The samples were diluted with Q-water prior to analysis by ICP-MS. The final dilution was dependent on the sample concentration of Ca and Sr.

Statistical Analyses

In the short-term experiment, the mean of Sr:Ca values for multiple points (n = 2–10) between the alizarin mark and the edge were used to represent the Sr:Ca response by each eel to experimental treatments. A regression approach was used to model the effects of salinity, temperature and growth (as measured by an index; see below) on otolith Sr:Ca. Individual eels were considered as experimental units, and within-tank covariances between individuals were modeled using compound symmetry structure in the residual error matrix (Littell et al. 1996). Individual eel growth was indexed as the ratio of the increment of otolith growth during the experimental period to the entire otolith diameter (measured along ventral-dorsal axis). The index required the assumption that otolith and somatic growth were proportional, which we tested by comparing otolith weight and wet body weight for a subsample of eel otoliths. The relationship was significant: otolith weight in mg = 0.091 + 0.069 × wet body weight in g (r^2 = 0.77, p < 0.001). The Sr:Ca response was log transformed to satisfy homogeneous variance assumptions, while temperature, salinity and growth variables were centered to reduce multicollinearity. Treatment effects on arcsine transformed survival data were analyzed using mixed model analysis. In the cycle experiment, nested analysis of variance (ANOVA) was used to compare the mean Sr:Ca response from the area between the first and second marks (corresponding to the initial salinity) with the area between the second mark and the edge (corresponding to the final salinity). The four cycle treatments (0–0, 15–15, 0–15, and 15–0) were nested within two temperatures (20°C and 28°C).

Results

Calcium and strontium in the tank water were positively correlated with experimental salinity levels (Figure 1; Ca, $p < 0.0001$, Sr; $p < 0.0001$). Calcium varied from 1.4 to 206.5 ppm and strontium varied from 0.53 to 6.11 ppm. Slopes of the regressions for calcium and strontium were linear, indicative of conservative mixing across the salinity gradient. There was little variation in the Sr:Ca of the water among the tanks with 5 ppt salinity or greater (mean = 0.031, CV = 8%), and higher and more variable Sr:Ca was observed at the 0.5 ppt treatment (mean = 0.091, CV = 44%). The concentrations of strontium and calcium in the diet used in these experiments were orders of magnitude higher than in the tank water concentrations: 140 and 33,590 ppm, respectively. Dietary concentrations of Sr and Ca are consistent with levels measured by Limburg (1995) with ICP-MS in an artificial diet for American Shad (Sr = 158 and Ca = 19,180 ppm). However, another study that measured Sr and Ca in several artificial diets with atomic absorption spectrophotometry found much lower values: Sr ranged from 1.8 to 7.8 ppm, and Ca ranged from 658 to 5040 ppm (Hoff and Fuiman 1995).

In the short-term experiment (Table 1; Figure 2), Sr:Ca increased significantly with temperature (DDF = 16, F = 4.73, p = 0.04) and salinity (DDF = 14.7, F = 18.43, p = 0.0007). No significant temperature-salinity interaction was detected. Otolith Ca (measured as calcium oxide, CaO) remained relatively constant across transect points within otoliths and across treatments (mean = 53.9%, s.e. = 0.09); observed otolith Sr:Ca ranged from 0.0019 to 0.0082. No significant effect on otolith Sr:Ca due to growth was detected, and no treatment effects on survival were detected.

In the cycle experiment, nested comparisons were made between the two marked regions, but no significant differences were found at any of the four treatment levels within either temperature treatment (20°C or 28°C). Mean otolith Sr:Ca was significantly different between the temperature treatments (DDF = 44, F = 5.0, p = 0.03), and within each temperature level, the mean otolith Sr:Ca was significantly different between the four possible cycle treatments (DDF = 44, F = 6.4, $p < 0.001$). At 20°C, mean otolith Sr:Ca ratio in the 0–0 salinity cycle was significantly lower than both the 15–15 salinity cycle (df = 44, t = –3.62, p = 0.02) and the 0–15 salinity cycle (df = 44, t = –3.45, p = 0.04), using Bonferroni adjusted probability values. At 28°C, mean otolith Sr:Ca in the 15–15 salinity cycle was significantly higher than the 0–15 salinity cycle (df = 44, t = 4.19, p = 0.004). Where significant differences were observed, the sign was consistent with our expectations; however, some of the comparisons in which significant differences were not observed led to contradictory results. For example, at the 20°C/15–0 cycle, observed Sr:Ca was variable and not significantly different from any of the other cycle treatments.

Discussion

The observed magnitude and variability of otolith Sr:Ca at the different salinity treatments in this study were consistent with otolith Sr:Ca observed in *A. japonica* reared in a similar range of salinities (Tzeng and Tsai 1994; Kawakami et al. 1998). Contrary to our expectations, a direct relationship between Sr:Ca in the tank water and that in the otilith was not indicated by these results, an outcome which warrants further investigation. Based on the observed variability, inferences from otolith Sr:Ca in wild eels appears to be limited to discriminating periods of freshwater from saltwater habitat occupation. However, a linear extrapolation (to salinities from 25 to 30 ppt) of the trend that we observed suggests that additional information might support discrimination between periods of brackish and ocean habitat occupation. There are three possible confounding influences on this relationship due to temperature and growth effects, diet composition, and the temporal resolution of the electron microprobe.

In our experiments, lower temperatures were associated with decreased otolith Sr:Ca. This result contrasts to the expected inverse relation between ambient temperature and otolith Sr:Ca attributed to molecular kinetics (Radtke 1989). Temperature effects observed in this study may be confounded due to a small, yet significant correlation between temperature and the growth index (r = 0.46, p = 0.002). However, the correlation is driven by a low growth index at the 16°C treatment; the growth index did not increase between 20°C and 28°C. Slow growth due to

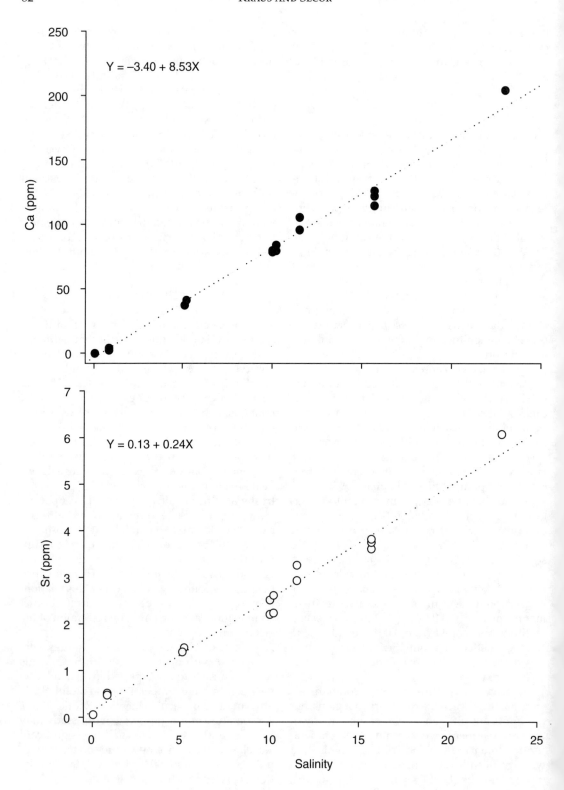

Figure 1. Strontium (lower panel) and calcium (upper panel) concentrations in water versus salinity level in rearing tanks for juvenile American Eel holding experiments.

Table 1. Effects of temperature and salinity on American Eel otolith Sr:Ca (by weight percent). Model coefficients and tests for significance were determined using a mixed model analysis to account for within tank correlation among individual fish responses. Degrees of freedom were determined using a Satterthwaite approximation. s.e. = standard error.

Effect	Estimate	s.e.	df	t-statistic	p-value
Intercept	−5.4	0.039	14.6	−143.5	<0.001
Temperature	0.026	0.009	17	2.9	0.0101
Salinity	0.019	0.004	14.7	4.3	0.0007

reduced metabolism at 16°C may have resulted in a decreased uptake of strontium that would account for the lower otolith Sr:Ca.

Pre-experimental otolith Sr:Ca averaged 0.002, which is consistent with otolith Sr:Ca in eels from freshwater regions of Hudson River (Morrison et al. these proceedings), and slightly less than, but within the range observed in freshwater eels from Maine (Arai et al. 2000). The higher mean Sr:Ca levels we found in the freshwater treatments (0.0037) may be an artifact of the experimental conditions related to stress or diet, but may also result from high Sr:Ca in water observed at this treatment level. By

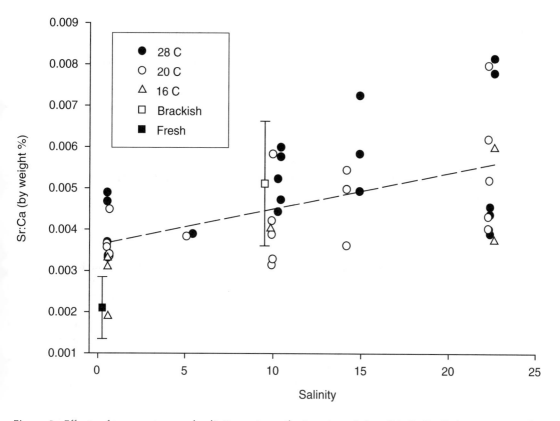

Figure 2. Effects of temperature and salinity on juvenile American Eel otolith Sr:Ca. Points represent the mean otolith Sr:Ca for a given eel plotted against the mean salinity recorded in the tank. Different symbols represent the three temperature treatments indicated in the key. Approximately 30% of the variability in Sr:Ca was accounted for by salinity (see Table 1 for regression results). For comparison, symbols with 95% confidence bars represent measurements on the marginal increment of wild Hudson River eels captured at fresh and brackish water sites (Morrison et al., this volume).

comparison, the mean otolith Sr:Ca observed in the freshwater experimental treatment is also greater than the means observed in wild individuals from three other species in freshwater: 0.0034 in *A. celebesensis* and 0.0028 in *A. marmorata* (Arai et al. 1999a), and 0.0034 in *A. australis* (Arai et al. 1999b).

Diet could have affected otolith Sr:Ca through increased uptake of strontium relative to calcium. We expect that dietary and water borne sources of calcium did not affect otolith Sr:Ca because calcium is a highly regulated ion. Calcium concentrations are similar throughout the otolith's microstructure, and past experimental studies have documented that otolith calcium is unresponsive to variation in direct or intestinal exposure of calcium (Berg 1968; Farrell and Campana 1996). In contrast, high levels of strontium in the diet relative to the ambient tank water could have elevated the intercept of the regression relationship (Table 1; Figure 2), explaining the higher otolith Sr:Ca observed for the 0.5 salinity treatment. Because discrimination of strontium by direct absorption is higher than by intestinal absorption (Berg 1968; Farrell and Campana 1996), it is not clear how dietary Sr would affect the slopes in the regression. The linear increase in otolith Sr:Ca with experimental salinity and the conservative mixing behavior of ambient strontium lends support to the assumption that changes in otolith Sr:Ca are in part associated with salinity changes. Our results suggest further study of diet related factors is needed.

Electron microprobe point size (25 μm^2) for the Sr:Ca analysis in these experiments roughly integrated growth over 5–25 days on the otolith, depending on the plane in the otolith section and the size and growth history of the individual. Thus, otolith precipitation rates ranged from 0.2 to 1 $\mu m/d$ along the axes that we analyzed. High otolith Sr:Ca variability coupled with coarse temporal resolution in the microprobe analysis precluded evaluation of a fine scale lag response to ambient salinity changes. However, we believe that the lack of significant change in otolith Sr:Ca between the cycle phases was more likely a function of the high variability of otolith Sr:Ca among individual eels rather than improper temporal scaling. There were some significant differences in otolith Sr:Ca observed among the four different cycle treatments; however, some of the multiple comparisons gave contradictory results, and further study into the sources of variability in otolith Sr:Ca among individuals is warranted.

In summary, the results of the short-term experiment indicated a linear increase in Sr:Ca with salinity, although the absolute relationship should not be applied directly to field observations until further study has determined the contribution of dietary strontium to otolith strontium in eels. In addition, the variability in otolith Sr:Ca suggests that while broad scale inference of fresh, brackish, and ocean habitat use appears valid, finer scale interpretation of estuarine life history movements may not be supported by this method.

Acknowledgments

We acknowledge Phil M. Piccoli for providing expertise and assistance with the microprobe analyses, and we thank Wendy Morrison for the use of her data in this paper. We also thank Debbie Connell for conducting the ICP-MS analyses. The comments and recommendations of Lisa Kline and Shingo Kimura on an earlier draft are also appreciated. This work was funded by the National Science Foundation (OCE-9812069). This is contribution 3592 of the University of Maryland Center for Environmental Science, One Williams Street, Solomons, Maryland 20688.

References

Arai, T., D. Limbong, T. Otake, and K. Tsukamoto. 1999a. Metamorphosis and inshore migration of tropical eels *Anguilla* spp. in the Indo-Pacific. Marine Ecology Progress Series. 182:283–293.

Arai, T., T. Otake, D. J. Jellyman, and K. Tsukamoto. 1999.b. Differences in the early life history of the Australasian shortfinned eel *Anguilla australis* from Australia and New Zealand, as revealed by otolith microstructure and microchemistry. Marine Biology. 135:381–389.

Arai, T., T. Otake, and K. Tsukamoto. 2000. Timing, and metamorphosis, and larval segregation of the Atlantic eels *Anguilla rostrata* and *A. anguilla*, as revealed by otolith microstructure, and microchemistry. Marine Biology. 137:39–45.

Bath, G. E., S. R. Thorrold, C. M. Jones, S. E. Campana, J. W. McLaren, and J. W. H. Lam. 2000. Strontium and barium uptake in aragonitic otiliths of marine fish. Geochimica et Cosmochimica Acta 64:1705–1714.

Berg, A. 1968. Studies on the metabolism of calcium and strontium in the freshwater fish. I. –Relative contribution of direct and intestinal absorption. Memorie dell'Istituto Italiano di Idrobiologia 23:161–196.

Campana, S. E. 1999. Chemistry and composition of fish otoliths: pathways, mechanisms and applications. Marine Ecology Progress Series 188:263–297.

Casselman, J. M. 1982. Chemical analyses of the optically different zones in eel otoliths. Pages 74–82 in K. H. Loftus, editor. Proceedings of the 1980 North American Eel Conference. Ontario Fisheries, Technical Report Series, No. 4. Maple, Ontario, Canada.

Farrell, J. and S. E. Campana. 1996. Regulation of calcium and strontium deposition on the otoliths of juvenile tilapia, Oreochromis niloticus. Comparative Biochemistry and Physiology A-Physiology 115:103–109.

Hoff, G. R. and L. A. Fuiman. 1995. Environmentally induced variation in elemental composition of Red Drum (Sciaenops ocellatus) otoliths. Bulletin of Marine Science 56:578–591.

Kalish, J. M. 1989. Otolith microchemistry: validation of the effects of physiology, age and environment on otolith composition. Journal of Experimental Marine Biology and Ecology 132:151–178.

Kawakami, Y., N. Mochioka, K. Morishita, T. Tajima, H. Nakagawa, H. Toh, and A. Nakazono. 1998. Factors influencing otolith strontium/calcium ratios in Anguilla japonica elvers. Environmental Biology of Fishes 52:299–303.

Limburg, K. 1995. Otolith strontium traces environmental history of subyearling American Shad, Alosa sapidissima. Marine Ecology Progress Series 119:25–35.

Littell, R. C., G. A. Milliken, W. W. Stroup, and R. D. Wolfinger. 1996. SAS System for Mixed Models. SAS Institute Inc., Cary, North Carolina.

Mugiya, Y., and S. Tanaka. 1995. Incorporation of water-borne strontium into otoliths and its turnover in the goldfish Carassius auratus: effects of strontium concentrations, temperature, and 17b-estradiol. Fisheries Science (Japan) 61:29–35.

Radtke, R. L. 1989. Strontium-calcium concentration ratios in fish otoliths as environmental indicators. Comparative Biochemistry and Physiology A-Physiology 92:189–193.

Secor, D. H., and J. R. Rooker. 2000. Is otolith strontium a useful scalar of life cycles in estuarine fishes? Fisheries Research 46:359–371.

Spurr, A. R. 1969. A low-viscosity resin embedding medium for electron microscopy. Journal of Ultrastructure Research 26:31–43.

Tsukamoto, K., Nakai, I., and Tesch, W. V. 1998. Do all freshwater eels migrate? Nature 396: 635–636.

Tsukamoto, K. and T. Arai. 2001. Facultative catadromy of the eel Anguilla Japonica between freshwater and seawater habitats. Marine Ecology Progress Series 220:265–276.

Tzeng, W. N. 1996. Effects of salinity and ontogenetic movements on strontium:calcium ratios in the otoliths of the Japanese eel, Anguilla japonica Temminck and Schlegel. Journal of Experimental Marine Biology and Ecology 199:111–122.

Tzeng, W. N. and Y. C. Tsai. 1994. Changes in otolith microchemistry of the Japanese eel, Anguilla japonica, during its migration from the ocean to the rivers of Taiwan. Journal of Fish Biology 45:671–683.

Estuarine Habitat Use by Hudson River American Eels as Determined by Otolith Strontium:Calcium Ratios

WENDY E. MORRISON*

National Oceanic and Atmospheric Administration, Biogeography Program, 1305 East West Highway, SSMC4, 9th Floor, Silver Spring, Maryland 20910, USA
Corresponding author: wendy.morrison@noaa.gov

DAVID H. SECOR

University of Maryland Center for Environmental Science, Chesapeake Biological Laboratory, Post Office Box 38, Solomons, Maryland 20688, USA.

PHILIP M. PICCOLI

Department of Geology and Center for Microanalysis, University of Maryland, College Park, Maryland 20742, USA

Abstract.—The migration histories of individual American eels (*Anguilla rostrata*) in the Hudson River, New York were investigated using otolith microchemistry. Strontium:calcium ratios were used as a proxy for salinity; mean levels ± SD estimated for fresh water and mesohaline (5–18 ppt) habitats were 0.002 ± 0.0005 and 0.005 ± 0.0007, respectively. Yellow-phase eels captured in fresh water showed no evidence of having previously resided in brackish water after metamorphosis into the elver stage. Yellow-phase eels captured at mesohaline sites showed two distinct migration modes, either spending their entire life in brackish water, or having a period of fresh water residency for the first 2–19 years before migrating into brackish water. Size-at-age analysis suggested that eels that spent their yellow-phase in brackish water grew faster and matured earlier than those that spent this portion of life in fresh water.

Introduction

Interest in the biology of the American eel *Anguilla rostrata* has been influenced by recent increase in demand and economic value, and a decline in U.S. landings (Figure 1; National Marine Fisheries Service 2000; Haro et al. 2000). Fisheries managers, therefore, are concerned about the sustainability of the resource and have noted a need for life history research to support a fishery management plan for the species (Atlantic States Marine Fisheries Commission 2000). Growth is highly variable in this species and one life history hypothesis is that productivity and growth of the yellow eel stage are greater in downstream estuarine or brackish water habits than in upstream fresh water locations (Helfman et al. 1987). An untested assumption of this hypothesis is that once a juvenile eel has settled into a particular habitat, it does not undertake migrations between brackish and fresh water habitats. Further, if eels do migrate between these habitats, this may affect their growth.

Yellow eels have high fidelity to home ranges over short time periods (Bozeman et al. 1985; Ford and Mercer 1986; LaBar and Facey 1983), and remain in the same location during a given season over several years (Oliveira 1997; Morrison 2002). Eels use selective tidal stream transport to return to an area after being displaced (Parker and McCleave 1997), but the mechanism for locating a site is not yet completely understood (Tesch 1977). Helfman et al. (1987) suggested that eels living in fresh water have a large home range, with possible seasonal migrations, while brackish water eels have smaller home ranges without seasonal migrations.

To the best of our knowledge, no study has looked at long term (years), long distance (> 10 km) dispersal by yellow American eels in estuaries. It has been suggested that glass eels entering the estuary metamorphose into elvers and yellow eels, and then undertake slow upstream

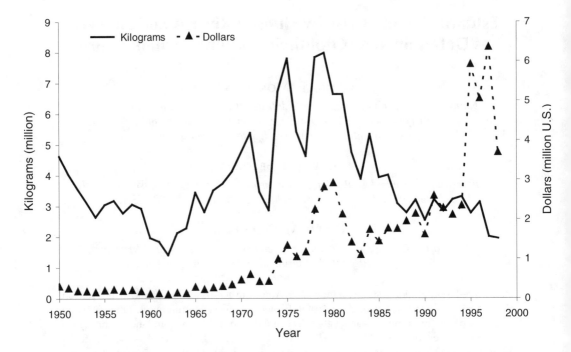

Figure 1. Atlantic and Gulf state American eel landings 1950–1998. Source: National Marine Fisheries Service, Statistics Division (http://www.st.nmfs.gov/st1/commercial/landings/annual_landings.html).

migrations that may last many years (Haro and Krueger 1991). Helfman et al. (1987) hypothesized that eels could minimize their vulnerability to predation by spending an initial period of their life in downriver productive areas (favoring rapid growth rate) before migrating upriver into areas where growth is slower.

This study investigates the large-scale migrations of yellow eels through the use of otolith microchemistry (Tzeng et al. 1997; Secor and Rooker 2000). Sr:Ca ratios were used to analyze the salinity history of yellow eels caught throughout the tidal portion of the Hudson River, New York. We tested the hypothesis that eels occupied only a single salinity zone (fresh or brackish water) during their entire yellow eel stage. Consequences of modes of yellow eel habitat use on growth were also evaluated.

Methods

The Hudson River Estuary (Figure 2) is characterized as a long, straight, and relatively deep basin formed by glaciers during past glaciation. The River is tidal from its mouth at Manhattan Island to Troy Dam, 255 km upriver. The salt front is usually found near Yonkers (river km 25) in high flow months (late winter to early spring), and moves upriver to near Newburgh (river km 100) when the water flow decreases during summer months (Dovel et al. 1992). Contamination of the sediment and fauna of the Hudson River by polychlorinated biphenyls (PCBs) led to a ban on harvesting of American eels in 1976 that remains in place today. The closure provides a unique opportunity to study stock dynamics in the absence of significant exploitation

During summers in 1998–1999, yellow eels were captured in standard 1-m long double funnel eel pots (13 × 13 mm mesh) that were baited with menhaden *Brevoortia tyrannus* and soaked overnight. A sample of 29 brackish water eels (27 females and 2 males) was collected at river km 20 near George Washington Bridge (GWB) in 1998 and 1999. The river at this site is relatively narrow (1.5 km) and salinity is variable, but typically ranges from 13 to 24 ppt (Figure 3). Pots were deployed in shallow (2–4 m) fringe banks adjacent to the river's main channel.

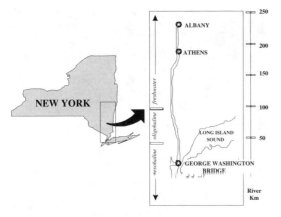

Figure 2. Collection sites on the Hudson River, New York. Mesohaline is defined as between 5 and 18 ppt, oligohaline between 0.5 and 5 ppt, and fresh water less than 0.5 ppt.

Total length (TL) of eels in the sample ranged from 31 to 69 cm, and their ages ranged from 5 to 22 years (aging method is described below). Fourteen female eels were collected from fresh water locations near Athens (AT) and Albany (ALB) (river km 190 and 240, respectively) in 1998. Pots were deployed in shallow banks near shore in 3–10 m depth. Eels ranged between 41 and 67 cm TL and 12–25 years in age.

Captured eels were euthanatized in MS-222, and frozen until otoliths could be removed. Sagittal otoliths were extracted, soaked briefly (3 minutes) in 10% hypochlorite solution, rinsed in deionized water and stored until dry. Right and left otoliths were randomly selected, and embedded in spur epoxy resin. Transverse sections were cut using a wafering saw, and then polished using wetted carborundum paper and alumina powder until the primordium was exposed (see Secor et al. 1991). Sections were sonicated in a deionized water bath, air dried, and coated with approximately 300 angstrom of carbon using standard thermal evaporation techniques. Microprobe analysis was performed on a JEOL JXA-8900 electron probe microanalyzer using wavelength dispersive spectrophotometry. Strontium (Sr) and calcium (Ca) concentrations were measured for a series of point analyses, i.e., a "life history transect," from the

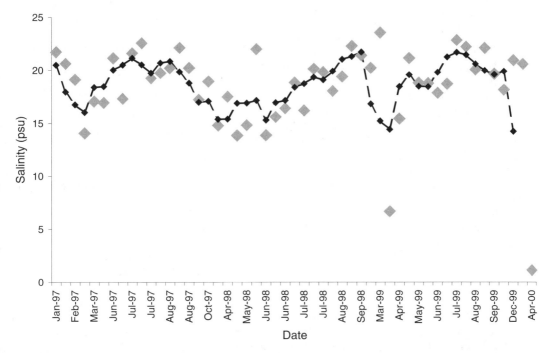

Figure 3. Hudson River bottom salinity and three point moving average at George Washington Bridge (January 1997–April 2000). Data from Naji Yao, Marine Sciences Section of New York City Department of Environmental Protection, 59–17 Junction Boulevard, 10th Floor, Flushing, New York, 11373.

core of the otolith to the otolith's edge, usually on an axis along one of the sulcal ridges. Standards used for the analysis were calcite and strontianite. The electron beam operating conditions were adapted from Tzeng et al. (1997), with a beam current of 20 nA, and an accelerating voltage of 25 kV. The beam was rastered across an area 5 × 3 µm for 30 seconds (peak) and 5 seconds (background) for Sr, and 10 seconds (peak) and 5 seconds (background) for Ca. Sr:Ca ratios were measured as mass fraction ratio. For aging purposes, otoliths were then cleaned and etched with a solution of 5% toluidine blue and 2% EDTA adapted from Graynoth (1999). Under this treatment, transmitted light exposed blue opaque zones that were enumerated and assigned to the microprobe point in closest proximity (Figure 4). For this study, the "transition" check (Michaud et al. 1988; Cieri and McCleave 2000) was assigned to age-1, and the following opaque zone identified to age-2 individuals. Therefore, ages are presented as actual age-classes, and not as age after metamorphosis (see Morrison 2002). Growth patterns were interpreted from length-at-age relationships. Length data were normally distributed, permitting the use of a standard t-test when comparing groups; however, age data required the use of the nonparametric Kruskal–Wallis test.

Otolith Sr:Ca as a scalar of estuarine habitat use (i.e., salinity) has been applied to life history investigations of Japanese eel *A. japonicus* (Tzeng 1996; Kawakami et al. 1998a; 1998b), and European eel *A. anguilla* (Tzeng et al. 1997), but has not been previously used for American eels. To provide field-based evidence to support the expected positive relationship between otolith strontium and ambient salinity, Sr:Ca in the most recently formed part of the otolith (otolith edge) was compared between eels captured in fresh and brackish water. We assumed that the outermost edge of the otolith represented the recent history of the eel, and should reflect the salinity of the water in which the eel was captured.

Results

Sr:Ca ratios at the otolith edge were significantly lower for eels collected in fresh water than those collected in brackish water ($P < 0.001$). Mean Sr:Ca ratios ± SD estimated for fresh water and brackish water habitats were 0.002 ± 0.0005 and 0.005 ± 0.0007, respectively.

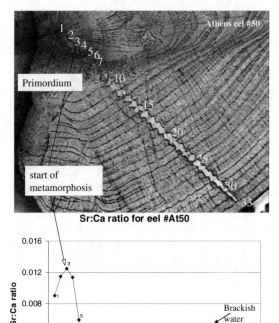

Figure 4. Sr:Ca ratios for eel otoliths collected at Athens (river km 190), from the core to the otolith margin. Microprobe points (white squares 5 × 3 µm, 15–20 µm between points) on the otolith are shown (top), and corresponding Sr:Ca ratios shown (bottom). In this plot, microprobe points have been assigned ages (years) on the x-axis. Primordium (embryonic stage) and start of metamorphosis are labeled. Sr:Ca ratios that correspond to fresh water and brackish water are 0.002 and 0.005 respectively.

Life history transects (from the otolith core to the edge) indicated all eels collected at the fresh water sites, AT and ALB, entered fresh water within one year after metamorphosis, and remained in fresh water until capture. There was no evidence of extended brackish water habitat use before the eels migrated upriver (Figure 5). Three eels showed a slight decrease in Sr:Ca ratio over a five year period (Figure 6, a76) that could be construed as a slow entry into fresh water, but the magnitude of the decline was within the range of oscillations (Figure 6, at22, at59, a61) present in other eels that showed an overall pattern of fresh water habitat use. Eels collected near GWB showed two distinct

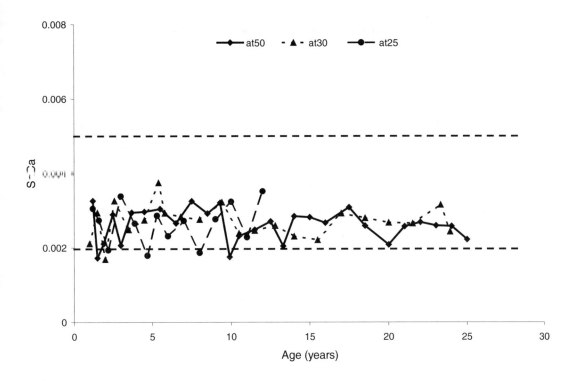

Figure 5. Sr:Ca life history transects (Sr:Ca ratio versus age) for eels collected at Athens (river km 190) and Albany (river km 240) that have spent their entire yellow eel stage in fresh water. Plots start at the elver metamorphic check. Microprobe points are 3 × 5 μm, with 20 μm between points. Plots were randomly selected from all the eels showing this mode of habitat use (n = 16). All eels were female. Horizontal dashed lines represent mean Sr:Ca ratios at the otolith edge for eels collected in fresh (0.002) and brackish water (0.005) habitats.

migration patterns. Over half (65%, $n = 29$) of the eels entered fresh water immediately after metamorphosis, remained there for 2–19 years, then migrated back downriver into brackish water for an additional 2–9 years before they were captured (Figure 7). The majority of the eels classified into this mode remained in fresh water for two to four years (78%), with two eels (17%) remaining 5 or 6 years, and one remaining 19 years in fresh water before returning to brackish water habitats. Approximately one-third of the eels captured in brackish water (35%) showed no record of entering fresh water, and appeared to have spent their entire life in brackish water (Figure 8). Both of the male eels analyzed showed this mode of habitat use. Results showed intra-annual fluctuations in Sr:Ca ratios that could be indicative of seasonal migration, or if the eels remain in place, of seasonal variations in river discharge (Figures 5–8). The amplitude of the fluctuations varied between eels.

Ages and lengths were compared between the eels exhibiting the mixed and brackish modes of habitat utilization. The sample of eels from the fresh water collections was biased towards large individuals (Figure 9), and therefore was not compared with the other modes. No significant differences in length ($t = -0.86$, df = 26, $P = 0.40$) and age (Kruskal-Wallis rank test: $x^2 = 2.6$, df = 1, $P = 0.10$) were detected between eels exhibiting the mixed and brackish dispersal modes. However, a power analysis indicates that with a larger sample size (combined $n = \geq 90$) differences in age would be evident. While the results are insignificant, observed differences in age suggest that yellow eels showing evidence of fresh water residency have slower growth rates than eels that spent

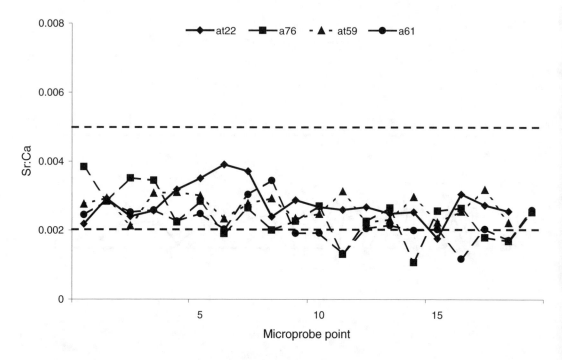

Figure 6. Sr:Ca life history transects for eels collected at Athens (river km 190) and Albany (river km 240) that have spent their entire yellow eel stage in fresh water. Sr:Ca ratio was plotted against microprobe point to emphasize the decreasing trend in eel a76. Plots start at the elver metamorphic check. Microprobe points are 3 × 5 μm, with 20 μm between points. Horizontal dashed lines represent mean Sr:Ca ratios at the otolith edge for eels collected in fresh (0.002) and brackish water (0.005) habitats.

their entire life in brackish water (Figure 9). Comparison of habitat use by age showed that by age six, most mixed mode individuals had moved into brackish water habitats (Figure 10).

Analysis of the Sr:Ca ratios prior to the elver metamorphic check showed similar amplitude and pattern to other *Anguilla* species (Otake et al. 1994; 1997; Tzeng et al. 1997; Wang and Tzeng 2000; Arai et al. 2000), with high Sr:Ca ratios (> 0.012) near the primordium, increasing ratio during the leptocephalus stage, and decreasing Sr:Ca during metamorphosis into the glass eel stage (Figure 4).

Discussion

The difference between Sr:Ca levels in the otolith edge for yellow eels captured in fresh water and brackish water yellow eels supports the use of Sr:Ca to investigate the estuarine habitat use by American eels. Sr:Ca ratios measured for American eels were similar to those measured for other Anguillid species. Japanese eels *A. japonica* captured in fresh water exhibited Sr:Ca ratios between 0.003 and 0.004 (Tzeng and Tsai 1994). Tzeng et al. (1997) reported that European eels *A. anguilla* had ratios near 0.006 for brackish water and 0.0003 for fresh water. These minor discrepancies could reflect differences between the species or analytical techniques of the investigators. Studies on laboratory-reared eels yielded similar but less precise results. For laboratory-reared Japanese eels, Kawakami et al. (1998a) and Tzeng (1996) observed mean Sr:Ca ratios of 0.004 (0.002–0.007) for fresh water and 0.005 (0.004–0.006) for 1/3 seawater. In addition, Kraus and Secor (this volume) found higher otolith Sr:Ca levels for American eel juveniles reared in fresh water (0.002–0.005) than those observed in this study. They attributed the difference to dietary sources of Sr in the experiment. It is possible that dietary Sr caused elevated otolith Sr:Ca ratios in laboratory studies of other *Anguilla* species. Consistent with other *Anguilla*

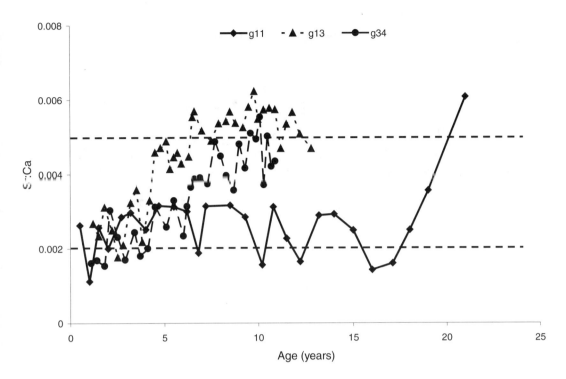

Figure 7. Sr:Ca life history transects (Sr:Ca ratio versus age) for eels collected at George Washington Bridge (river km 20) that spent 2–19 years of their yellow eel stage in fresh water before returning to brackish water. Plots start at the elver metamorphic check. Microprobe points are 3 × 5 μm, with 20 μm between points. Plots were randomly selected from all the eels showing this mode of habitat use ($n = 19$). All eels were female. Horizontal dashed lines represent mean Sr:Ca ratios at the otolith edge for eels collected in fresh (0.002) and brackish water (0.005) habitats.

laboratory studies, Kraus and Secor (this volume) documented a positive relationship between experimental salinity and otolith Sr:Ca.

Similar to past microchemistry studies describing otolith Sr:Ca during the leptocephalus and glass eel stage (Otake et al. 1994, 1997; Tzeng et al. 1997), we observed an initial increase in Sr:Ca during the leptocephalus stage, followed by a rapid decline that was probably associated with leptocephalus metamorphosis. We note as others have that these trends in otolith Sr:Ca are independent to salinity. Thus, otolith Sr:Ca as a scalar of salinity should only be used for elver and yellow phase eel, where evidence supports such a relationship.

Three modes of habitat use were distinguished in this study. The first mode included those eels that utilized only fresh water during elver and yellow eel stages (fresh water mode). The eels appeared to have entered fresh water habitats immediately during glass eel or elver stages and remained there until capture (Figure 5). A limitation of the analysis is that continued upstream migration or seasonal migrations within fresh water cannot be detected and may be present. In contrast to these results, Tzeng et al. (2000), using microprobe analysis on European eels found evidence of an extended (4–7 years) upstream migration into fresh water habitats by yellow eels in fjord systems. Eels captured in brackish water habitats in the Hudson River showed two distinct modes of habitat use. Over half of the eels showed early fresh water habitat use for at least two years, followed by a downstream migration into environments with brackish salinities (mixed mode) (Figures 7 and 10). The third mode of habitat use showed no utilization of fresh water environments (brackish mode). Both male eels were classified into this mode, however, the paucity of male eels in

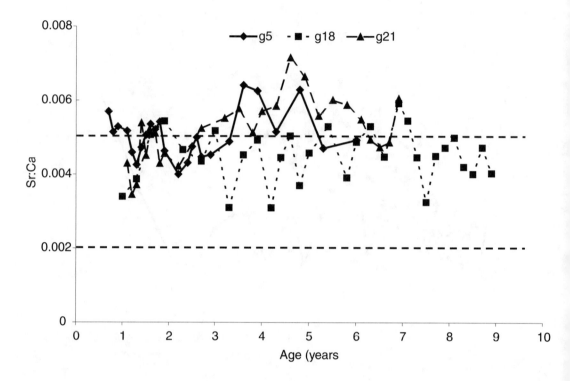

Figure 8. Sr:Ca life history transects (Sr:Ca ratio versus age) for eels collected at George Washington Bridge (river km 20) that have spent their entire yellow eel stage in brackish water. Plots start at the elver metamorphic check. Microprobe points are 3 × 5 μm, with 20 μm between points. Plots were randomly selected from all the eels showing this mode of habitat use ($n = 11$). All eels were female. Horizontal dashed lines represent mean Sr:Ca ratios at the otolith edge for eels collected in fresh (0.002) and brackish water (0.005) habitats.

this study and presumably, in the Hudson River estuary (Morrison 2002) does not allow for generalizations on gender specific habitat use.

Oscillations of Sr:Ca ratios were observed in the life history transects of this study (Figures 5–8). The causes are speculative, but may include: seasonal and yearly differences in fresh water inputs (e.g., freshets versus base flow); seasonal migrations within brackish water into and out of waters with varying salinity levels; seasonal differences in growth rates and their subsequent influence on strontium deposition rates (Sadovy and Severin 1994); and small analytical errors possibly due to scratches on the surface of the otolith section, or anomalies in the aragonite structure of the otoliths. The fresh water eels showed smaller fluctuations than the brackish-water eels, suggesting that measurement errors were not the cause of the fluctuations. The salinity at GWB is highly variable between times of high flow and low flow (Figure 3), suggesting that seasonal changes in salinity at this site could cause oscillations in Sr:Ca ratios for eels that resided within a limited home range.

Overall growth rates were slower for eels residing entirely in fresh water than for those using brackish water habitats. This was also shown on an annual and seasonal basis through mark recapture investigations on the Hudson River; annual growth rates were approximately twofold higher in brackish water habitats (Morrison 2002). Potential causes of higher growth in brackish water habitats include higher food abundance, better quality food, lower predation pressure, and less thermal and osmotic stress. The higher growth rates in brackish water support Helfman et al.'s (1987) generalization that productivity is higher in downriver brackish sites than upriver sites. This result suggests that

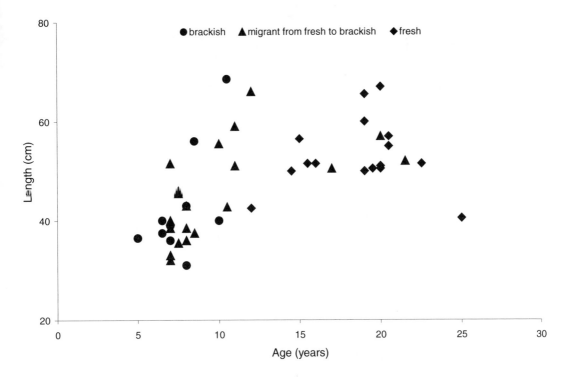

Figure 9. Total length-at-age for eels among all sites and habitat use modes. Circles represent eels that spent their entire yellow eel stage in brackish water, diamonds represent yellow eels that spent 2–19 years in fresh water before migrating to brackish water, and triangles represents eels that spent their entire yellow eel stage in fresh water.

the eels that utilized brackish water during their yellow eel stage grew faster, matured earlier, and emigrated as silver eels sooner than those that resided in fresh water throughout their yellow eel stage.

For this study, all eels captured in brackish water that were older than 11 years of age exhibited the mixed mode, confirming the hypothesis that freshwater residency may lead to later maturation. However, young eels captured at GWB (< 10 years old) showed similar size-at-age, regardless of whether time was previously spent in fresh water or not (Figure 9). This could suggest two hypotheses, either that early growth is comparable for eels located upriver and downriver, or that eels that resided in freshwater have a compensatory ability to later increase growth after entering brackish water areas. Alternatively, our sample size may have been too low to detect growth differences between these modes. The consequences of production differences (related to growth, mortality, and relative abundance) between the three habitat use modes defined here are important to issues of management and deserve additional research.

Several researchers have directly observed movements of older yellow eels between brackish and fresh water areas. Smith and Saunders (1955) documented a seasonal movement of eels into fresh water lakes from brackish water areas each fall. Casselman et al. (1997) documented large eels moving upriver through an eel ladder on the St Lawrence River from 1974 to 1987. Haro and Krueger (1991) observed that eels in the Annaquatucket River in Rhode Island were longer and older as he sampled upriver, prompting them to hypothesize that the eels slowly moved upriver with age. Our results reinforce the findings of Oliveira (1999), who investigated eels in the Annaquatucket River and found no

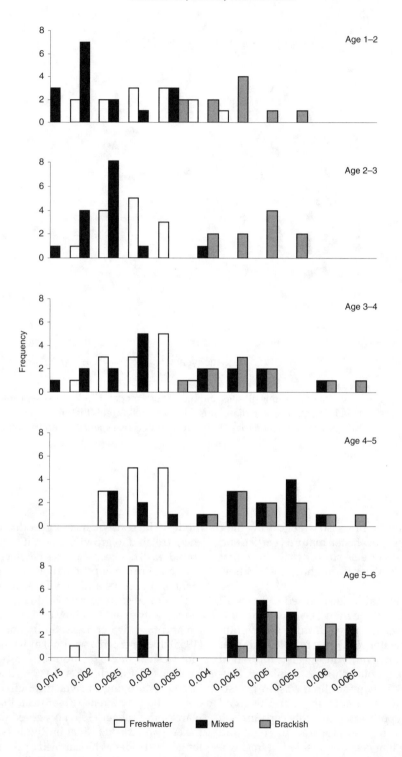

Figure 10. Mean Sr:Ca ratios in the first year (a), the second year (b), the third year (c), the fourth year (d), and the fifth year (e) after metamorphosis. Different fill patterns represent different dispersal modes: [□] = fresh water residency, [▨] = brackish water residency, and [■] = fresh water to brackish water migrant.

upstream movement of eels greater than 200 mm, and attributed the older ages found upriver to slower growth rate. The otolith microchemistry results from this study did not show compelling evidence for seasonal migrations between fresh and brackish-water in the Hudson River. Some of the eels captured in brackish water showed fluctuations in the Sr:Ca ratios that could be attributed to small seasonal migrations into waters of slightly lower salinity, but the eels did not appear to move all the way into fresh water. As mentioned above, these fluctuations could also be attributed to seasonal differences in salinity at one location.

In summary, our results support Helfman et al.'s (1987) hypothesis that growth and productivity are higher in downriver sites, but did not support his hypothesis that eels spend a portion of their lives in brackish water before moving upriver. Fish aggregations based upon divergent migration behaviors or habitat use within a genetic population have been termed contingents (Hjort 1914; Clark 1968; Secor 1999). Based on this definition, the eels within the Hudson River seem to contain three different contingents. There is remarkable convergence among the three temperate *Anguilla* species in their modes of habitat use (freshwater, mixed or brackish water) during their elver and yellow eel stages. In this symposium Tzeng (this volume) and Limburg et al. (this volume), reported similar contingent behaviors for *A. japonicus* and *A. anguilla* respectively. Secor (1999) hypothesized that the mechanism behind the different life-time modes of habitat use is dependent on early life divergence in growth rates. Growth rate differences observed in Hudson River eels were only observed after a given pattern of habitat use was undertaken. Therefore, our results suggest that growth rate was a consequence rather than a cause of variable migration patterns by Hudson River yellow eels.

Acknowledgments

We would like to acknowledge Steve Nack, Ron Ingold, and Scott Ingold for their perspectives and assistance in field work on the Hudson River. Troy Gunderson and Steve Larson assisted in the field and in the laboratory. We appreciate the help from the Riverkeeper in providing accommodations and laboratory space during field seasons. We appreciate comments from committee members Ed Houde and Chris Rowe. We also acknowledge the referee comments by Patrick Geer and Karin Limburg on an earlier draft of this paper. This research was supported by the Hudson River Foundation. This is contribution number 3623 of the University of Maryland Center for Environmental Science.

References

Arai, T., T. Otake, and K. Tsukamoto. 2000. Timing of metamorphosis, and larval segregation of the Atlantic eels *Anguilla rostrata* and *A. anguilla*, as revealed by otolith microstructure, and microchemistry. Marine Biology 137: 39–45.

Atlantic States Marine Fisheries Commission. 2000. Interstate Fishery Management Plan for American Eel. Fishery Management Report No. 36. Washington, D.C.

Bozeman, E. L., G. S. Helfman, and T. Richardson. 1985. Population size and home range of American eels in a Georgia tidal creek. Transactions of American Fisheries Society 114:821–825.

Casselman, J. M., L. A. Marcogliese, and P. V. Hodson. 1997. Recruitment index for the upper St. Lawrence River and Lake Ontario eel stock: a re-examination of eel passage at the R. H. Saunders hydroelectric generating station at Cornwall, Ontario, 1974–1995. Pages 161–169 *in* R.H. Peterson, editor. The American Eel in Eastern Canada: Stock Status and Management Strategies. Proceedings of Eel Workshop, January 13–14, 1997. Quebec City, Quebec, Canada.

Cieri, M. D., and J. D. McCleave. 2000. Discrepancies between otoliths of larvae, and juveniles of the American eel: is something fishy happening at metamorphosis? Journal of Fish Biology 57: 1189–1. 198.

Clark, J. 1968. Seasonal movements of striped bass contingents of Long Island Sound and the New York Bight. Transactions of American Fisheries Society 97:320–343.

Dovel, W. L., A. W. Pekovitch, and T. J. Berggren. 1992. Biology of shortnose sturgeon (*Acipenser brevirostrum*) in the Hudson River Estuary, New York. Pages 187–216 *in* Lavett C. Smith, editor. Estuarine Research in the 1980s. The Hudson River Environmental Society Seventh Symposium on Hudson River Ecology. State University of New York Press, Albany, New York.

Ford, T. E., and E. Mercer. 1986. Density, size distribution and home range of American eels, *Anguilla rostrata*, in a Massachusetts salt marsh. Environmental Biology of Fish 17:309–314.

Graynoth, E. 1999. Improved otolith preparation, ageing and back-calculation techniques for New Zealand freshwater eels. Fisheries Research 42:137–146.

Haro, A. J., and W. H. Krueger. 1991. Pigmentation, otolith rings, and upstream migration of juvenile American eels (*Anguilla rostrata*) in a coastal Rhode Island Stream. Canadian Journal of Zoology 69:812–814.

Haro, A., W. Richkus, K. Whalen, A. Hoar, W.-D. Busch, S. Lary, T. Brush, and D. Dixon. 2000. Population decline of the American eel: Implications for research, and management. Fisheries 25:7–16.

Helfman, G. S., D. E. Facey, J. L. Stanton Hales, and J. Earl L. Bozeman. 1987. Reproductive ecology of the American eel. Pages 42–56 *in* M. J. Dadswell, R. J. Klauda, C. M. Moffitt, R. L. Saunders, R. A. Rulifson, and J. E. Cooper, editors. Common strategies of anadromous and catadromous fishes. American Fisheries Society, Symposium 1, Bethesda, Maryland.

Hjort, J. 1914. Fluctuations in the great fisheries of Northern Europe. Conseil Permanent International pour L'Exploration de la Mer 20:1–228.

Kawakami, Y., N. Mochioka, K. Morishita, T. Tajima, H. Nakagawa, H. Toh, and A. Nakazono. 1998a. Factors influencing otolith strontium/calcium ratios in *Anguilla japonica* elvers. Environmental Biology of Fish 52:299–303.

Kawakami, Y., N. Mochioka, K. Morishita, H. Toh, and A. Nakazono. 1998.b. Determination of the freshwater mark in otoliths of Japanese eel elvers using microstructure and Sr/Ca ratios. Environmental Biology of Fish 53:421–427.

LaBar, G. W., and D. E. Facey. 1983. Local movements and inshore population sizes of American eels in Lake Champlain, Vermont. Transactions of American Fisheries Society 112:111–116.

Michaud, M., J.-D. Dutil, and J. J. Dodson. 1988. Determination of the age of young American eels, *Anguilla rostrata*, in fresh water, based on otolith surface area and microstructure. Journal of Fish Biology 32:179–189.

Morrison, W. E. 2002. Demographics, dispersal, and relative abundance of Hudson River yellow-phase American eels. Thesis. University of Maryland, College Park, Maryland.

National Marine Fisheries Service. 2000. Statistics, and Economics Division. Web site http://www.st.nmfs.gov/st1/commercial/landings/annual_landings.htm.

Oliveira, K. 1997. Movements and growth rates of yellow-phase American eels in the Annaquatucket River, Rhode Island. Transactions of American Fisheries Society 126:638–646.

Oliveira, K. 1999. Life history characteristics and strategies of the American eel, *Anguilla rostrata*. Canadian Journal of Fish and Aquatic Sciences 56:795–802.

Otake, T., T. Ishii, T. Ishii, M. Nakahara, and R. Nakamura. 1997. Changes in otolith strontium:calcium ratios in metamorphosing *Conger myriaster* leptocephali. Marine Biology 128:586–572.

Otake, T., T. Ishii, M. Nakahara, and R. Nakamura. 1994. Drastic changes in otolith strontium/calcium ratios in leptocephali and glass eels of Japanese eel *Anguilla japonica*. Marine Ecological Progress Series 112:189–193.

Parker, S. J., and J. D. McCleave. 1997. Selective tidal stream transport by American eels during homing movements and estuarine migration. Journal of Marine Biology 77:871–889.

Sadovy, Y., and K. P. Severin. 1994. Elemental patterns in red hind otoliths from Bermuda and Puerto-Rico reflect growth-rate, not temperature. Canadian Journal of Fish and Aquatic Sciences 51(1):133–141.

Secor, D. H., J. M. Dean, and E. H. Laban. 1991. Manual for otolith removal and preparation for microstructural examination. Univ. South Carolina Press, Columbia, South Carolina.

Secor, D. H. 1999. Specifying divergent migrations in the concept of stock: the contingent hypothesis. Fisheries Research 43:13–34.

Secor, D. H., and J. Rooker. 2000. Is otolith strontium a useful scalar of life cycles in estuarine fishes? Fisheries Research 46:359–371.

Smith, M. W., and J. W. Saunders. 1955. The American eel in certain fresh waters of the Maritime Provinces of Canada. Journal of Fish Research Board of Canada 12:238–269.

Tesch, E. W. 1977. The eel: Biology and management of anguillid eels. Chapman and Hall Ltd. Halsted Press, New York, New York.

Tzeng, W.-N. 1996. Effects of salinity and ontogenetic movements on strontium:calcium ratios in the otoliths of the Japanese eel, *Anguilla japonica* Temminck and Schlegel. Journal of Experimental Marine Biology and Ecology 199:111–122.

Tzeng, W. N., K. P. Severin, and H. Wickstrom. 1997. Use of otolith microchemistry to investigate the environmental history of European eel *Anguilla anguilla*. Marine Ecology Progress Series 149:73–81.

Tzeng, W. N., and Y. C. Tsai. 1994. Changes in otolith microchemistry of the Japanese eel, *Anguilla japonica*, during its migration from the ocean to the rivers of Taiwan. Journal of Fish Biology 45:671–683.

Tzeng, W. N., C. H. Wang, H. Wickstrom, and M. Reizenstein. 2000. Occurrence of the semicatadromous European eel *Anguilla anguilla* in the Baltic Sea. Marine Biology 137:93–98.

Wang, C. H., and W. N. Tzeng. 2000. The timing of metamorphosis, and growth rates of American, and European eel leptocephali: A mechanism of larval segregative migration. Fisheries Research 46:191–205.

Distribution, Relative Abundance, and Habitat Use of American Eel *Anguilla rostrata* in the Virginia Portion of the Chesapeake Bay

PATRICK J. GEER*

*Virginia Institute of Marine Science, Department of Fisheries Science,
Post Office Box 1346, Gloucester Point, Virginia 23062, USA*

Abstract.—The mid-Atlantic region of the U.S. typically has the highest landings of American eel *Anguilla rostrata* with the Chesapeake Bay area (Maryland and Virginia) comprising the bulk of the U.S. catch (63%). Eel landings during the period 1997–1999, however, have declined nearly 70% from the peak harvest of the mid-1980s. The Virginia Institute of Marine Science has conducted a fisheries independent trawl survey to monitored the Chesapeake Bay and its tributaries since 1955, with yellow eels a regular component of upriver catches. Data from this survey were used to create estimates of relative abundance for each major tributary in Virginia (James, York, and Rappahannock Rivers), as well as a comprehensive estimate, to determine if these downward trends in commercial landings are also evident in fisheries independent data. These estimates indicate below average catch rates (measured as a weighted geometric mean catch per trawl) from each river when compared against the long-term average (1979–1999). Overall, catch rates are down 50% from the long-term average, with a threefold decline from the high values seen in the late 1980s. These results support similar findings from other areas indicating a coast-wide decline in the American eel stock. Data from the survey were also used to investigate American eel hydrographic and habitat associations. Hydrographic and habitat associations were determined by the percentiles representing the largest proportion of the overall catch. Between 1988 and 1999, 82% of all yellow eel catches were observed in dissolved oxygen levels between 5 and 9 mg/L. Temperature ranges were broad, with 79% of the catch occurring from 12–28°C, with 89% of the catch at salinities below 12 ppt, and 27% in waters less than 2 ppt. Catch percentages indicate more than 75% occurred in waters between 4 and 10 m in depth. Yellow eel appear to prefer detritus (wood material, leaves, etc), hydroids, or shell bottoms, with each associated with over 30% of yellow eel catch.

Introduction

The American eel *Anguilla rostrata* is a catadromous species which ranges from Greenland to Central America along the western Atlantic and Gulf coasts, and inland to the Mississippi River and Great Lake drainages (Tesch 1977; Helfman et al. 1987; Jenkins and Burkhead 1994). The species is panmictic, supported throughout its range by a single spawning population believed to concentrate during late winter to early spring in the Sargasso Sea (Wenner 1973). All but its earliest life stages (eggs and leptocephali) are commercially exploited, supporting a small but valuable fishery. Recent declines in commercial landings and fishery independent indices have resulted in an increased interest in the management of the species (Electric Power Research Institute 1999; Haro et al. 2000). The Atlantic States Marine Fisheries Commission (ASMFC) developed a fishery management plan for its jurisdictions in 2000 (ASMFC 2000). The International Council for the Exploration of the Sea (ICES) also recently invited U.S. scientists to participate in working groups on American and European eels *A. anguilla* which resulted in recommending an international recovery plan for both stocks (International Council for the Exploration of the Seas 2002).

American eels support commercial fisheries along the Atlantic coast throughout North America from Canada to Florida as well as in

---

*Present Address: Georgia Department of Natural Resources, Coastal Resources Division, Commercial Fisheries Program, One Conservation Way, Brunswick, Georgia 31520–8687, USA; pat_geer@coastal.dnr.state.ga.us

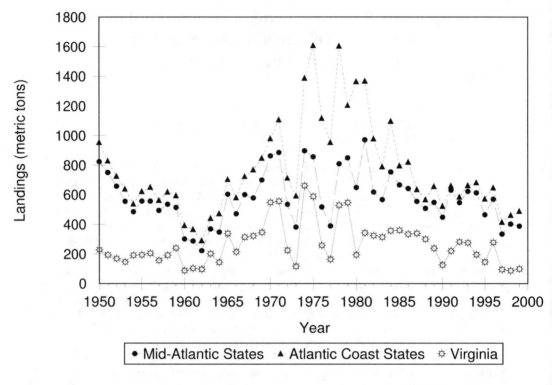

Figure 1. Commercial landings of American eel by state (Virginia), region (Mid-Atlantic) and Atlantic coast. Source: NMFS and Virginia Marine Resources Commission.

the Gulf of Mexico. In recent years, the Mid-Atlantic region (defined as the states of New York, New Jersey, Delaware, Maryland, and Virginia by the National Marine Fisheries Service [NMFS]), has comprised most of the U.S. Atlantic Coast catch with 88% of the reported landings since 1988 (NMFS 2000; Figure 1). Since 1980, the Mid-Atlantic states have taken an increasing proportion of the coast-wide landings (the Chesapeake Bay states of Virginia and Maryland alone represent nearly 36% and 27%, respectively; Figure 1). However, this increase in proportion does not correspond to an increase in landings, with Virginia's annual harvest decreasing from 1980 to 1998, while effort rose 36% during the same time period (Figure 2).

Yellow eels are primarily harvested in the upper portions of the tidal tributaries in Virginia. During the past ten years, landings have been highest on the Rappahannock River (27.7% of total state landings), followed by the James (24.9%), the Potomac (22.5%), and the York Rivers (12.0%). Total landings for Virginia the past three years (1997–1999) has shown a nearly 70% decline since the high of the mid-1980s, with similar trends seen in the Mid-Atlantic region (–42.1%) and the entire Atlantic coast (–49.9%; Figure 1).

Castonguay et al. (1994) suggested several hypotheses for the decline of the St. Lawrence River and Gulf eel fishery. Contamination by mirex and PCB were often found to be at toxic levels in some silver eel samples. Dams constructed for hydroelectricity and other purposes have limited upstream migration. Commercial fishing may be a factor since all harvest is on pre-spawning individuals. Subtle shifts in the Gulf Stream may affect recruitment of glass eels to coastal estuaries (Castonguay et al. 1994). More recently, Barse and Secor (1999) suggested the invasive nematode parasite, *Anguillicola crassus*, may infect (where it is geographically present) the American eel's swim bladder in ways that limit seaward migration.

Whatever the reason for declining harvests, commercial data and fishery independent

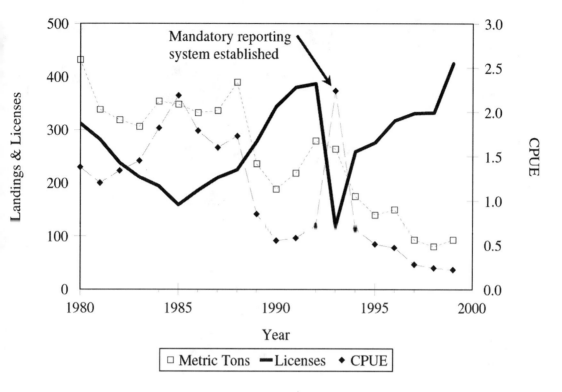

Figure 2. Virginia commercial pot landings, effort (number of licenses), and associated catch-per-unit-of-effort (CPUE) from 1980 to 1999. Source: Virginia Marine Resources Commission

surveys generally show similar trends. The Virginia Institute of Marine Science's (VIMS) bottom trawl program has monitored the Virginia portion of Chesapeake Bay and its tributaries since 1955, with yellow eels a regularly occurring component of the upriver catches. Although the reliability of landings data may not discover true trends in abundance (e.g., due to spatial coverage, reporting bias), it is hypothesized that analysis of results from this survey may determine whether there has been a true decline in abundance.

Methods

Trawl sampling was conducted monthly on the James, York, and Rappahannock Rivers as well as the Virginia portion of the main stem Chesapeake Bay, utilizing a combination of a random stratified design based on depth (Geer and Austin 1999), and a series of fixed locations which have been sampled nearly monthly since 1955. This stratification system was established for the Bay in 1988, based on four depth strata within each of four 15 nautical mile regions. The system for the tributaries was established in 1991 for the York River, and 1995 for the James and Rappahannock Rivers. Depth criteria for these tributaries was identical to the Bay, with regions designated in 10 nautical mile increments (Figure 3; Geer and Austin 1999). Fixed stations and those sampled prior to 1988 were placed in the appropriate stratum for analysis purposes. Throughout the survey, the sampling gear was always a 9.14 m semi-balloon otter trawl (38 mm stretch mesh body, 19 mm stretch mesh in the cod-end; Marinovich Net Company, Biloxi, Mississippi) with various configurations (e.g., a 6.35 mm liner, tickler chain, different doors). Since 1979, the gear has used a 9.5 mm link tickler chain and 6.35 mm liner.

These various configurations were tested extensively, indicating no significant differences in eel catches (Hata 1997). Examining this further, we were able to create a theoretical length retention size based on the length-girth relationship

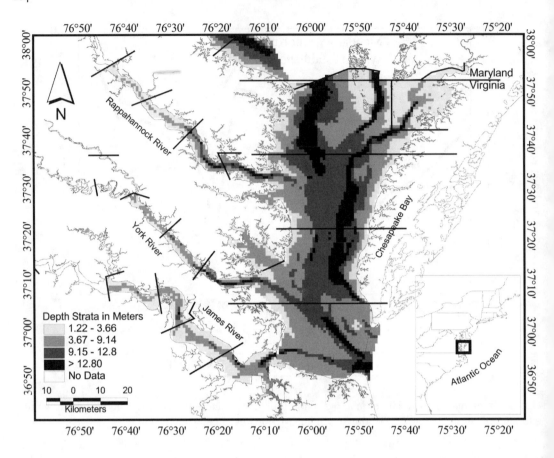

Figure 3. Depth stratification used by the VIMS trawl survey and the associated geographic regions.

and the known circumference of the gear's mesh. The length-girth relationship was determined using a linear regression model expressed as:

$$Log_{10}(Girth) - 0.6140 + 0.9439 Log_{10}\ TL$$
$$n = 1388,\ R^2 = 0.787,\ \text{size range: } 65\text{--}776\ mm$$

The theoretical mesh retention length (RTL) and 95% prediction limits (Neter et al. 1985) were calculated from the girth corresponding to mean cod end mesh circumference (Hata 1997) as expressed by:

$$Log_{10}(RTL) = CIRMESH - (-0.6140)/(0.9439)$$

where: Log (RTL) is the log_{10} of the theoretical retention length (total) and CIRMESH is the log_{10} of the mesh circumference.

Estimates of abundance were created by first separating catch information into various age/size components. Length-at-age of eels varies greatly among studies (Johnson 1974; Harrel and Loyacano1980; Hedgepeth 1983; Foster and Brody 1982; Weeder 1998; Owens and Geer, this volume), and length frequencies do not always suggest discernable year-classes (Figure 4). The trawl data suggests American eels begin to recruit to the gear at 150 mm (Figure 4), as such, eels less than 153 mm were removed from the analysis to correspond with the minimum legal harvest size in Virginia. Using these length criteria, monthly catch rates were map-plotted and strata-specific abundances and occurrence rates calculated. Numbers of individuals caught were logrithmetically transformed (ln[n + 1]) prior to abundance calculations, as this transformation has repeatedly been shown to best normalize collection data for contagiously distributed organisms such as fishes (Taylor 1953) and has been verified as the best suited transformation

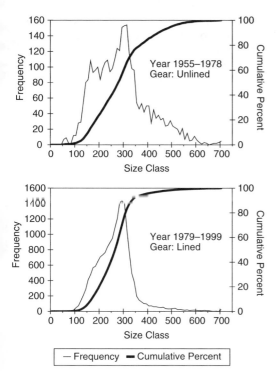

Figure 4. American eel length frequency and associated cumulative percent from the VIMS Trawl Survey, 1955–1978 with an unlined trawl, and 1979–1999 with a lined trawl.

for Chesapeake Bay trawl collections (Chittenden 1991). Resultant average catch rates (and the 95% confidence intervals as estimated by ±2 standard errors) were then back-transformed to the geometric means. Plots and data matrices were then examined for the area-time combinations that appeared to provide the best basis for calculations. Criteria applied during the selection process included identification of maximal abundance levels, uniformity of distribution, minimization of overall variance, and avoidance of periods which indicated distribution patterns suggesting migratory behavior was occurring. These spatial areas for American eels were identified as the upper two regions of each tributary (Figure 3), providing the highest and most consistent catches.

Surveys were planned on regular time intervals which might or might not coincide with periods of maximal recruitment to the sampling areas. Using a very limited portion of the overall data set would decrease sample sizes, increasing both confidence intervals, and the risk of sampling artifacts influencing results. As a result, the use of a single (maximal) month's survey results was deemed inappropriate. Conversely, a conscious effort was made not to incorporate any longer temporal series of data into abundance estimates than was necessary in order to capture the period of maximal utilization of the sampling area. Using this approach it was possible to identify a four-month window, which consistently captured the months of highest abundance for American eels. The temporal component selected was April to June. Analysis of variance indicated these months had significantly higher catch rates ($\alpha = 0.05$, $P < 0.0001$) than all other months.

Annual estimates of relative abundance were calculated as the summation of the products of stratum specific means and ratio of area to total area sampled, to produce a weighted geometric mean catch per tow (Geer and Austin 1999; Cochran 1977). Since stratum areas were quite variable, use of a weighted mean provided an estimate that more closely mirrored actual population sizes than would a simple mean. Hydrographic data were collected at both the surface and bottom at each trawl site using a HYDROLAB© unit. Bottom hydrographic variables (temperature, dissolved oxygen and salinity) were assessed as to their importance to yellow eel abundance. Catches were assigned to an interval for each variable and compared to the total number of trawls occurring at that interval. Preference was recognized as those frequencies in which the greatest percentile of catches occurred (Rubec et al. 1999; Able 1999).

Habitat samples were collected from May 1998 through December 1999 and sorted into 13 different categories (Table 1). These categories were coarse and amounts were quantified from each trawl using a standard volume container (72 liters) in an attempt to establish relationships with fish distributions. These data are limited with only 2303 stations evaluated over the entire sampling range. Given additional sampling, a more robust multi-variant statistical analysis may be applied to determine the relationships between fish abundance and habitat usage. However, for this exercise, habitat usage was expressed as percentiles of the total catch in a manner similar to the hydrographic data.

Table 1. Substrate, or habitat types, with various statistical information and observed associations with yellow eels (≥ 153 mm) in Chesapeake Bay, May 1998–1999. Values for unknown, sand, and mud are always given as 1.0. Percent (Pct) refers to the ratio of trawls or catch associated with that habitat. Habitat quantity was estimated relative to the capacity of a commercial fishing tote (internal dimensions 25.7" x 16.7" x 10", approximately 72 liters). A full tote was given a quantity of 4, a half of tote was 2, a quarter tote was 1, etc.

Substrate Description	Code	Dominant Species or Material	Habitat statistics			Trawls		Eel catch	
			Mean	Min	Max	No.	Pct	No.	Pct.
Artificial	ART	Man-made. Ballast stones, crab pots, etc.	0.58	0.1	3.0	42	1.82	0	0.00
Detritus	DET	Organic Material. Mainly land based. Leaves, twigs, marsh grasses, etc.	0.30	0.1	10.0	704	30.58	271	23.30
Dead man's fingers	DMF	*Alcyonidium* spp	0.41	0.1	5.0	193	8.38	0	0.00
Hydroid	HYD	There are 43 reported species in Chesapeake Bay, including: white hair *Sertularia cupressina*, rope grass *Garveia franciscana*	0.33	0.1	5.0	1009	43.83	290	24.94
Mud (soft) bottom	MUD	Soft bottom—always given value of 1.0	1.00	1.0	1.0	179	7.78	151	12.98
Submerged Aquatic Vegetation	SAV	eel grass *Zostera marina*, widgeon grass *Ruppia maritima*, coontail *Ceratophyllum demersum*	0.32	0.1	3.0	154	6.69	1	0.09
Shell (oyster, clam, etc)	SHL	Oyster *Crassastrea virginica* Blue Mussel *Mytilus edulis*, Bent Mussel *Ischadium recurvum*, wedge rangia *Rangia cuneata*	0.47	0.1	5.0	534	23.20	338	29.06
Sand (hard) bottom	SND	Hard bottom—always given value of 1.0.	1.00	1.0	1.0	158	6.86	1	0.09
Sponges (yellow, orange, etc)	SPG	Redbeard *Microciona prolifera* Loosanoff's Haliclona *Haliclona loosanoffi*	0.70	0.1	6.0	191	8.30	1	0.09
Sea Squirts *Mogula* spp.	SQT	*Molgula* spp.	0.79	0.1	14.0	460	19.98	1	0.09
Seaweeds (red, green, or brown)	SWD	Brown, green and red algaes. sea lettuce *Ulva lactuca*, tappered red weed *Agardhiella tenera*, coarse red weed *Gracilaria verrucosa*	0.36	0.1	10.0	392	17.03	62	5.33
Tube Worms	TUB	parchment worms *Chaetopterus* spp	0.22	0.1	1.0	151	6.56	0	0.00
Undetermined	UNK		1.00	1.0	1.0	261	11.34	114	9.80

Total number of trawls evaluated: 2303   Total number of eels captured: 1163

Results

Distribution and Abundance Estimates

Yellow eels were most abundant in the upper reaches of the tributaries (river miles 20 to 40), near or above the freshwater interface (Figure 5). Some areas of the James and Rappahannock Rivers produced very high catches. In contrast, the main-stem Bay had very few eels present. Since the inception of the random stratified surveys in 1988, Bay sampling has comprised over 44% of the total sampling effort, but less than ½ of 1% of the total yellow eels captured. The Bay

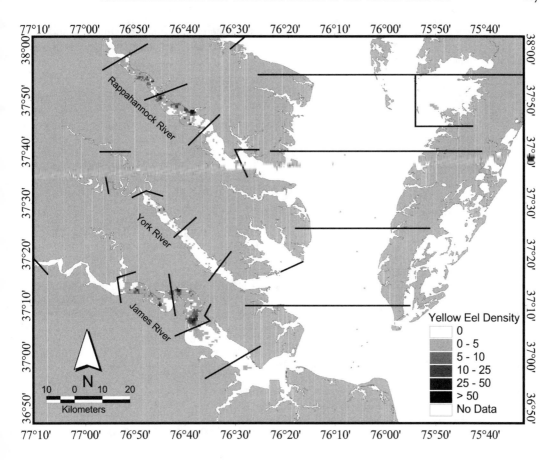

Figure 5. Yellow eel density in Virginia 1988–1999 from the VIMS Trawl Survey.

was sampled inconsistently between 1979 and 1987, but still consisted of 34% of the total sampling. From 1979 to 1999, the Rappahannock River accounted for 42% of the yellow eel catch, with the James and York Rivers making up 30% and 24%, respectively. This period provided the longest and most intense sampling period for the VIMS trawl survey, with 62% of all samples collected occurring during these 21 years.

Sampling between 1955 and 1978 was more inconsistent (Table 2). The limited sampling of only the Bay and York River until the mid-1960s makes a Virginia-wide estimate of abundance bias (Table 2; Geer and Austin 1999). Sampling during this period was 38% of the total effort (1955–1999), but only 26% of yellow eels were captured during this stanza of the survey.

The theoretical retention length ($T_L R$; Hata 1997) calculations produced a minimum value of 318 mm for the unlined gear (1955–1978), and 107 mm for the lined trawl gear (1979 to present; 95% C.I.'s = ±0.08; Figure 6). The mean length for the period from 1955 to 1978 was 294 mm (±1.97, n = 3,188). Since the introduction of a liner to the trawl gear in 1979, the mean length has declined to 269 mm (±0.57). The large $T_L R$ of the unlined gear suggests smaller individuals were not being captured efficiently by the gear. Clearly, the data since 1979 would provide more accurate results for yellow eels since, in this case, the life stage was defined as those eels greater than or equal to 153 mm TL. These earlier data (prior to 1979) will be excluded from further examination due to the generally low catch rates, sporadic sampling, high variability, and low gear efficiency as compared to the most recent data (Table 2 and Figure 6).

A fair amount of variability in catch per unit effort occurred between years as well as between aquatic systems. The Rappahannock

Table 2. Weighted geometric mean catch per trawl and associated 95% confidence intervals and sample size (N) for the James, Rappahannock, and York Rivers, as well as overall values for Chesapeake Bay tributaries for the years 1955 to 1978.

Year	James River Mean	C.I.	N	Rappahannock River Mean	C.I.	N	York River Mean	C.I.	N	Overall Mean	C.I.	N
1955	No Sampling performed during this period						6.35	0.22–43.09	2	6.35	0.22–43.09	2
1956							1.84	0.55–4.22	10	1.84	0.55–4.22	10
1957							0.16	0–0.55	5	0.16	0–0.55	5
1958							1.96	1.02–3.34	17	1.96	1.02–3.34	17
1959							0.53	0–1.57	11	0.53	0–1.57	11
1960							0.18	0.06–0.33	10	0.18	0.06–0.33	10
1961							1.77	0.44–4.3	8	1.77	0.44–4.3	8
1962							1.10	0.01–3.33	9	1.10	0.01–3.33	9
1963							0.60	0–1.68	12	0.60	0–1.68	12
1964	1.32	0–5.42	4				0.00	0.00	13	0.82	0–2.74	17
1965	0.39	0–1.66	9				0.00	0.00	12	0.26	0–1	21
1966	0.22	0–0.65	12				0.69	0.34–1.12	15	0.35	0.08–0.68	27
1967	0.35	0–1.03	12	2.38	0–12.77	5	0.34	0–0.9	15	0.72	0.11–1.66	32
1968	0.26	0–0.68	9	2.91	1.36–5.46	15	4.57	1.74–10.34	15	1.34	0.82–2.01	39
1969	0.32	0–1.06	12	0.79	0.23–1.6	15	4.42	1.98–8.84	15	0.99	0.49–1.64	42
1970	0.00	0.00	12	0.59	0–1.75	14	0.14	0–0.35	15	0.17	0–0.37	41
1971	0.41	0.1–0.8	12	5.82	3.34–9.73	15	1.71	0.71–3.29	15	1.40	0.97–1.93	42
1972	0.04	0–0.13	16	1.92	0.46–4.84	12	0.90	0.47–1.45	31	0.54	0.3–0.82	59
1973	0.03	0–0.09	8	2.25	0.89–4.58	16	0.83	0.46–1.29	65	0.78	0.47–1.17	89
1974	0.00	0.00	9	1.05	0.19–2.54	12	1.13	0–4.51	8	0.35	0.11–0.64	29
1975				1.25	0.24–3.11	12	0.15	0–0.51	8	0.82	0.2–1.76	20
1976	0.00	0.00	16	2.02	0.84–3.94	18	0.43	0–1.21	12	0.42	0.23–0.65	46
1977	0.25	0–0.69	8	3.48	2.14–5.38	12	0.30	0.01–0.67	8	1.05	0.69–1.49	28
1978	0.15	0–0.51	8	2.89	1.82–4.37	12	0.67	0–1.88	8	0.96	0.61–1.39	28
1955–1978	0.20	0.10–0.32	147	2.10	1.51–2.81	158	0.68	0.53–0.84	339	0.66	0.55–0.79	644

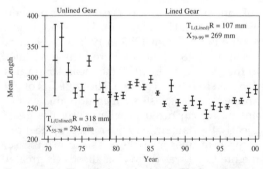

$T_{TL(GearType)}R$ = Theoretical retention length (TL) for each gear

Figure 6. Mean length and 95% confidence intervals by year for American eels captured by the VIMS trawl survey, with the theoretical retention length and mean length for the two primary gears (unlined—1955–1978 and lined (1979–1999). The length data prior to 1970 are sporadic (26 observations between 1955–1969) and, as such, were removed for clarity purposes.

River typically had the highest catch rates (overall mean$_{79-99}$ = 4.40 eels per trawl), followed by the James (2.38) and York Rivers (1.23; Table 3 and Figure 7). All three river systems indicate a decline in CPUE since the late eighties (Figure 7). Since 1985, there has been a significant decline in CPUE on both the York ($P < 0.001$) and Rappahannock Rivers ($P < 0.1$). Catch rates on the James River were significantly lower since 1996 than they were in the early 1990s. Rappahannock River catches appear to be at the lowest levels since sampling began on that system in 1967 (Tables 2 and 3, and Figure 7). The York River has CPUE's in recent years well below the long-term average (1979 to 1999; Table 3 and Figure 7), and often comparable to historical catch rates seen with the inefficient trawl gear used from 1955 to 1978 (Table 2). The overall index does not show a significant trend from 1979 to present; however, an analysis of variance indicates a downward decline since 1985

Table 3. Weighted geometric mean catch per trawl and associated 95% confidence intervals and sample size (N) for the James, Rappahannock, and York Rivers, as well as overall values for Chesapeake Bay tributaries for the years 1979 to 1999.

	James River			Rappahannock River			York River			Overall		
Year	Mean	C.I.	N	Mean	C.I.	N	Mean	C.I.	N	Mean	C.I.	N
1979	1.23	0–9.43	7	6.74	2.09–18.43	12	0.66	0.06–1.59	10	1.97	0.47–4.98	29
1980	4.64	1.92–9.87	8	2.01	0.45–5.26	12	6.66	4.49–9.7	18	4.15	2.66–6.25	38
1981	4.81	1.22–14.22	8	0.78	0.51–1.09	12	4.11	2.34–6.83	18	2.06	1.28–3.1	38
1982	1.18	0–4.2	18	4.97	2.59–8.93	18	3.16	2.07–4.63	26	2.27	1.03–4.27	62
1983	4.94	1.19–15.11	8	35.42	16.59–74.43	12	3.78	1.85–7	22	9.01	5.03–15.63	42
1984	3.81	2.59–5.44	18	24.11	16.72–34.59	18	5.18	2.12–11.1	20	5.61	4.20–7.31	56
1985	7.11	3.01–15.4	9	16.59	8.05–33.17	10	2.12	0.86–4.22	18	8.06	4.7–13.39	37
1986	5.14	3.54–7.3	16	5.82	3.98–8.35	15	1.76	0.56–3.88	16	4.60	3.46–6.04	47
1987	3.27	1.19–7.33	14	5.13	2.35–10.24	15	3.72	1.4–8.29	16	3.62	1.86–6.46	45
1988	3.24	1.16–7.34	4	0.61	0.04–1.51	6	1.90	0.16–6.24	8	1.26	0.48–2.46	18
1989	12.00	7.23–19.53	8	5.92	0.7–27.2	10	2.86	0.89–6.86	13	7.69	4.49–12.75	31
1990	5.15	3.34–7.72	8	4.74	1.15–14.32	10	1.87	0.43–4.78	12	4.50	3.04–6.49	30
1991	1.72	0.08–5.85	8	5.48	2.49–11.03	10	0.99	0.34–1.97	19	2.00	0.82–3.92	37
1992	15.88	11.29–22.19	8	2.30	0.7–5.41	10	0.83	0.35–1.48	28	7.04	5.34–9.21	46
1993	8.36	4.79–14.12	8	1.11	0.6–1.77	10	0.53	0.16–1.02	25	2.96	2.08–4.1	43
1994	2.68	0.72–6.85	8	4.21	1.55–9.63	10	0.60	0.27–1.01	25	2.13	1.06–3.75	43
1995	3.34	2.32–4.67	10	2.87	1.61–4.74	10	0.31	0.07–0.61	25	2.33	1.76–3.01	45
1996	1.62	0.71–3.02	28	6.58	4.4–9.63	29	0.48	0.23–0.77	27	2.55	1.76–3.57	84
1997	1.59	0.56–3.28	30	4.50	1.01–14.08	30	0.52	0.12–1.07	30	2.16	1.03–3.93	90
1998	1.93	1.15–3.01	30	2.67	0.29–9.45	30	0.76	0.29–1.41	30	1.98	0.98–3.47	90
1999	1.25	0.46–2.48	30	1.83	0.33–5.01	30	0.46	0.2–0.78	30	1.31	0.63–2.28	90
1979–1999	2.38	1.79–3.10	286	4.40	2.66–6.962	319	1.23	1.00–1.47	436	2.90	2.28–3.62	1041

($P < 0.05$). Overall, the grand mean for all three tributaries from 1979 to 1999 was 2.90 eels per trawl (Table 3). The overall mean length from 1979 to 1999 was 268 mm but varied significantly between tributaries (ANOVA, $P < 0.05$; James = 258 mm [±0.99, $n = 6,403$]; York = 271 [±1.27, $n = 5,112$]; Rappahannock = 275 mm [±0.75, $n = 7,156$]; Figure 8).

Hydrographic Associations

There does not appear to be a distinct association between dissolved oxygen (DO) and catch. Nearly 82% of the eel catch from 1988 and 1999 occurred at DOs between 5 and 9 mg/L, while 64% of stations sampled had values within that range (Table 4). Eels can absorb as much as 60% of the required oxygen through the skin (Sheldon 1974) such that their survival is better in air than in low oxygenated (≥ 4 mg/L) or polluted waters (Tesch 1977).

Eels have been documented to have wide temperature requirements across their large geographic range. Barila and Stauffer (1980) found the preferred temperature to be at 16.7°C (range 6–30°C), with feeding ceasing below 14°C. In Virginia, eels were often found between temperatures of 13 and 27°C. The only noticeable increase in abundance across temperatures was at 26–28°C, with 17% of the total catch (Table 4 and Figure 9). The low catches observed at temperatures below 8°C suggests periods of inactivity. At temperatures above 13°C they become more active and susceptible to the commercial fishing gear (Van Den Avyle 1982). This threshold corresponds to the period of peak abundance (April to June) when mean monthly water temperatures range from 11.7 to 20.8°C.

Salinity tolerances for glass eels, elvers, and silver eels are also broad (Van Den Avyle 1982) due to the diverse aquatic environments eels encounter during inland (juvenile) and seaward (adult) migrations. In keeping with this observation, yellow eels were usually most abundant in the fresher waters of the upper tributaries, with 89% of the catch occurring in salinities below 12 ppt, of which, nearly 27% occur in waters less than 2 ppt (Table 4 and Figure 9).

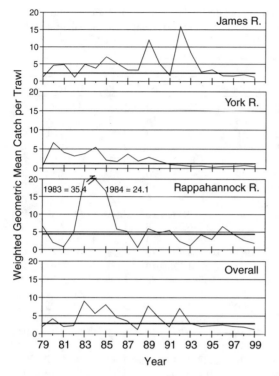
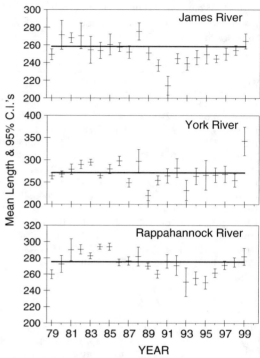

Figure 7. Estimates of relative abundance for yellow eels by tributary (James, York and Rappahannock Rivers) and overall Virginia estimate from the months of peak abundance April to June, 1979–1999.

Figure 8. Mean length (TL) and 95% confidence intervals of American eels captured by trawls from the Rappahannock, York, and James Rivers, 1979–1999. The horizontal line represents the grand mean for the river.

Depth association indicates most (75%) catches occurred at depths from 4 to 10 m, while nearly 50% of all trawl locations occurred at these depths (Table 4 and Figure 9). The upper part of the tributaries were typically shallower than the lower rivers and the main stem of Chesapeake Bay and, since eel densities are highest in these upper reaches, it's evident that these shallower depths would represent the bulk of the eel catch.

Habitat Associations

The habitat types occurring in Chesapeake Bay are not mutually exclusive and may appear in various combinations. Unfortunately, these data have been collected for only 18 months and, without further sampling and extensive multivariate statistical analyses, results will be expressed only as observed percentiles. For yellow eels, the predominant habitat types were detritus, hydroid, or shell bottoms. Each of these habitat types were associated with over 30% of the yellow eel catch (Table 1), but of these three, only the shell bottom type showed a difference between the percent of stations where it occurred versus the total catch (Table 1). Upper regions of the tributaries were characterized by soft mud bottoms with large quantities of land based detritus (e.g., leaves, tree limbs, twigs). Several areas of the James and Rappahannock Rivers had abundant shell bottoms comprised mainly of American oyster reefs *Crassostrea virginica*, wedge rangia *Rangia cuneata*, and bent mussels *Ischadium recurvum*.

Discussion

The yellow eel decline observed in the St. Lawrence River and Gulf between 1985 and 1992 (81-fold decline in catch at the Saunders Power Dam Ladder; Castonguay et al. 1994) has not been as dramatic in the Chesapeake Bay. Virginia landings show a 5.5-fold decline between 1979

Table 4. Hydrographic associations relative to eel catch and stations sampled.

	Water temperature					Dissolved oxygen					Salinity					Depth				
	Number		Percent			Number		Percent			Number		Percent			Number		Percent		
°C	Trawls	Catch	Stations	Catch	mg/L	Trawls	Total	Station	Catch	ppt	Trawls	Total	Station	Catch	Meters	Trawls	Catch	Station	Catch
1	119	0	1.05	0.00	0	49	6	0.44	0.08	1	573	1964	5.08	26.47	1	242	88	2.07	1.15
3	325	5	2.86	0.07	1	77	7	0.70	0.10	3	240	1010	2.13	13.61	3	1228	749	10.53	9.75
5	811	52	7.15	0.70	2	148	57	1.34	0.79	5	248	985	2.20	13.27	5	1720	2153	14.75	28.03
7	958	168	8.44	2.28	3	280	52	2.54	0.72	7	282	919	2.50	12.38	7	2186	1806	18.74	23.52
9	777	336	6.85	4.56	4	538	378	4.88	5.26	9	425	736	3.77	9.92	9	1910	1820	16.38	23.70
11	872	382	7.69	5.18	5	959	1173	8.69	16.33	11	571	1014	5.06	13.66	11	1587	541	13.61	7.04
13	922	614	8.13	8.32	6	1498	1227	13.58	17.08	13	734	375	6.50	5.05	13	1143	174	9.80	2.27
15	869	705	7.66	9.56	7	1570	1426	14.23	19.86	15	921	181	8.16	2.44	15	698	138	5.98	1.80
17	769	735	6.78	9.96	8	1675	1306	15.18	18.18	17	1214	120	10.76	1.62	17	451	78	3.87	1.02
19	795	570	7.01	7.73	9	1375	730	12.46	10.16	19	1445	67	12.81	0.90	19	326	88	2.79	1.15
21	895	770	7.89	10.44	10	1308	434	11.85	6.04	21	1425	28	12.63	0.38	21	83	27	0.71	0.35
23	830	478	7.32	6.48	11	965	291	8.75	4.05	23	1229	17	10.89	0.23	23	34	3	0.29	0.04
25	1270	760	11.20	10.30	12	449	73	4.07	1.02	25	872	2	7.73	0.03	25	27	3	0.23	0.04
27	891	1217	7.85	16.50	13	118	22	1.07	0.31	27	597	1	5.29	0.01	27	6	4	0.05	0.05
29	208	400	1.83	5.42	14	25	0	0.23	0.00	29	297	1	2.63	0.01	29	9	1	0.08	0.01
31	31	184	0.27	2.49						31	211	1	1.87	0.01	31	5	0	0.04	0.00
33	2	0	0.02												33	9	7	0.08	0.09
Total	11344	7376				11034	7182				11284	7421				11664	7680		

Total number of trawls evaluated: 12,310   Total number of eels captured: 7,680

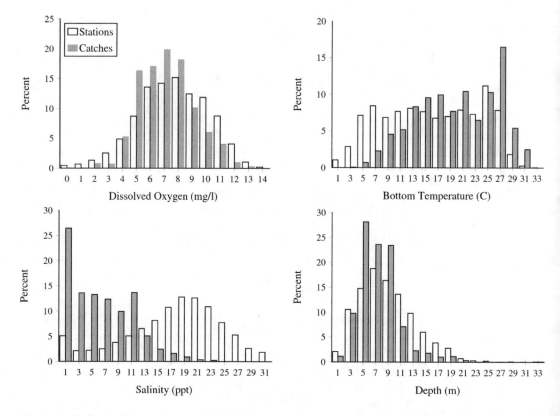

Figure 9. Hydrographic preferences (dissolved oxygen, bottom temperature, salinity, and depth) of yellow eels (≥ 153 mm) in Chesapeake Bay, 1988–1999.

and 1999, with the VIMS estimates of relative abundance observing only a 2-fold decrease for the same period. The early-to-mid 1980s were periods of peak abundance for yellow eels in both the York and Rappahannock Rivers. The VIMS Trawl data indicates very high CPUE from 1983 to 1985 on the Rappahannock, followed by a marked declined near or below the historical average (6.88 eels per trawl from 1979–1999). The York River estimates of eel CPUE remained above the historical average for ten years (1980–1990), with estimates well below average since. The James River does not show a similar pattern to that of the York and Rappahannock Rivers, with large fluctuations and declining catch rates not becoming evident until the late 1990s. This may be due to a river wide closure on commercial fishing (associated with kepone contamination) from 1975 to 1989. With no fishing pressure, the population could increase. When the ban was lifted in late 1989, there was an immediate decline observed in both average length and abundance (Table 3; Figures 7 and 8).

What can be causing this decline in Virginia? The four theories proposed by Castonguay et al. (1994) for the St. Lawrence water system may not all apply in Chesapeake Bay. Contamination from pollutants is of little concern except possibly on the James. Loss of habitat from hydroelectric dams and other man-made structures does not seem plausible, as these structures in Virginia are over 160 km upriver and many were built nearly one hundred years ago before the observed decline. In addition, in recent years there has been an active movement to remove low head dams to re-open historical spawning grounds for such species as American shad *Alosa sapidissima*. Recruitment failures due to Gulf Stream and ocean current shifts, or inadequate spawning escapement, are difficult to assess without a long-term consistent estimate of recruitment (young-of-the year). However, with over 36% of all U.S. landings occurring in Virginia waters, and increasing effort being placed on the fishery (Virginia Marine Resources Commission, unpublished data), commercial harvests may be reducing local stock density of future spawning stock.

The more recent finding of *Anguillicola crassus* in Maryland and Virginia tributaries of Chesapeake Bay provides an alternative theory. In 1995, the organism was reported in Texas (Fries et al. 1996) and first observed in the Chesapeake Bay in 1997 (Barse and Secor 1999). If infection by this nematode decreases swimming efficiency (Sprengel and Luchtenberg 1991), eels ability to migrate over long distances to their spawning grounds in the Sargasso Sea may be affected. Preliminary examination of 126 fish from the Virginia tributaries indicated a 61% infection rate (number of eels infected/number of eels examined) on the York River, 44% on the Rappahannock, and 27% on the James Rivers (Geer and Owens, unpublished data). However, these parasites may have been present for years and are perhaps just beginning to be reported.

If over-fishing is responsible for the decline in these relative abundance estimates and commercial landings then it may be evident in annual mean lengths. Recent ageing studies (1997–1999) from yellow eels captured by the VIMS trawl survey average 272 mm (±1.2), with the bulk (89%) occurring within five year-classes (ages three to seven; Owens and Geer, this volume). On average, the Rappahannock River has significantly larger (grand mean = 276.47 ± 1.30 mm) eels than the York (271.34 ± 2.40 mm) and James Rivers (252.94 ± 1.71 mm). Average length declined during the early 1990s, but appears to be above the grand mean for each system the past two years. The sinusoidal pattern observed in these annual mean length data (Figures 6 and 8) would suggest one or two year classes to be dominating the fishery. However, the estimates of abundance presented here represent multiple year classes (primarily ages 2 to 5), and the failure of one or several year classes may be masked by this very successful one. With the entire Atlantic fishery based on pre-spawning individuals, over-fishing is a possibility but available landings data are insufficient to make any conclusions. Increasing fishing effort in Virginia and other states, in association with increased natural mortality from any of the other four proposed theories, may have produced the declines seen in coastal landings.

Hydrographic associations observed from these data are consistent with those found in studies from other areas (Van Den Avyle 1982). Ford and Mercer (1986) suggested yellow eels establish a home range when they first invade freshwater, the size of which can vary with habitat type, food availability, and abundance of conspecifics. Seasonal variations in salinity (spring rains and summer drought for example) may have little impact on their distribution. However this study suggests a preference for low salinities (90% of catch below 12 ppt, 53% below 6 ppt), perhaps indicating a larger home range due to large tidal influences in the estuaries.

The information presented here can only be used to assess abundance and habitat of the entire yellow eel stock in Virginia's tidal tributaries. Without some indicator of year-class (or recruitment) success, it will be difficult to extrapolate any forecasts on future eel abundance in the Chesapeake Bay.

Acknowledgments

This study was funded in part by the Virginia Institute of Marine Science. The VIMS Trawl survey is supported by the Sportfish Restoration Project F104 through the Virginia Marine Resources Commission and the U.S. Fish and Wildlife Service. Gratitude is expressed to the entire staff of the VIMS juvenile finfish program for their dedication to collecting and managing these data for 46 years. Comments and suggestions by S. J. Owens, H. M. Austin, W. A. Lowery, J. Weeder, and two anonymous reviewers helped improve this manuscript.

References

Able, K. W. 1999. Measures of juvenile fish habitat quality: Examples from a National Estuarine Research Reserve. Pages 134–147 *in* L. Benaka, editor. Fish habitat: essential fish habitat and rehabilitation. American Fisheries Society, Symposium 22, Bethesda, Maryland.

Atlantic States Marine Fisheries Commission (ASMFC). 2000. Interstate fishery management plan for American eel. Fishery Management Report no. 36. ASMFC. April 2000.

Barila, F. Y., and J. R. Stauffer, Jr. 1980. Temperature behavioral responses of the American eel, *Anguilla rostrata* (LeSueur), from Maryland. Hydrobiologia 74:49–51.

Barse, A. M., and D. H. Secor. 1999. An exotic nematode parasite of the American eel. Fisheries 24(2):6–10

Castonguay, M., P. V. Hodson, and C. M. Couillard. 1994. Why is recruitment of American eel, *Anguilla rostrata*, declining in the St. Lawrence River and Gulf? Canadian Journal of Fisheries and Aquatic Sciences 51:479–488

Chittenden, M. E., Jr. 1991. Evaluation of spatial/temporal sources of variation in nekton catch and the efficacy of stratified sampling in the Chesapeake Bay. Final report to Chesapeake Bay Stock Assessment Committee & NOAA/NMFS. Virginia Institute of Marine Science, Gloucester Point, Virginia.

Cochran, W. G. 1977. Sampling techniques. John Wiley & Sons. New York, New York.

Electric Power Research Institute. 1999. American eel (*Anguilla rostrata*) scoping study: a literature and data review of life history, stock status, population dynamics, and hydroelectric impacts. TR–111873. Palo Alto, California.

Ford, T. E., and E. Mercer. 1986. Density, size distribution, and home range of the American eel, *Anguilla rostrata*, in a Massachusetts salt marsh. Environmental Biology of Fishes 17:309–314.

Foster, J. W. S. III, and R. W. Brody. 1982. Status report: The American eel fishery in Maryland, 1982. Maryland Department of Natural Resources, Tidal Fisheries Division, Annapolis, Maryland.

Fries, L. T., D. J. Williams, and S. K. Johnson. 1996. Occurrence of *Anguilla crassus*, an exotic parasitic swim bladder nematode of eels, in the southeastern United States. Transactions of the American Fisheries Society 125:794–797.

Geer, P. J., and H. M. Austin. 1999. Estimation of relative abundance of recreationally important finfish in the Virginia portion of Chesapeake Bay. Annual report to Virginia Marine Resources Commission and U.S. Fish and Wildlife Service. Sportfish Restoration Project F104R9. July 1998 to June 1999. Virginia Institute of Marine Science, Gloucester Point, Virginia.

Haro, A., W. Richkus, K. Whalen, A. Hoar, W. Dieter-Busch, S. Lary, T. Brush, and D. Dixon. 2000. Population Decline of the American Eel: Implications for Research and Management. Fisheries 25(9):7–16

Harrell, R. M., and H. A. Loyacano. 1980. Age, growth, and sex ratio of the American eel in the Cooper River, South Carolina. Proceedings of the Annual Conference of the Southeastern Association of Fish and Wildlife Agencies 34:349–359

Hata, D. N. 1997. Comparison of gears and vessels used in the Virginia Institute of Marine Science juvenile finfish trawl survey. Special Report in Applied Marine Science and Ocean Engineering No. 343. Virginia Institute of Marine Science, Gloucester Point, Virginia.

Hedgepeth, M. Y. 1983. Age, growth and reproduction of American eels, *Anguilla rostrata*, Lesueur, from the Chesapeake Bay area. Masters thesis, The College of William and Mary. Williamsburg, Virginia.

Helfman, G. S., D. E. Facey, L. S. Hales, Jr. and E. L. Bozeman, Jr. 1987. Reproductive ecology of the American eel. Pages 42–56 in Common Strategies of Anadromous and Catadromous Fishes. American Fisheries Society, Symposium 1, Bethesda, Maryland.

International Council for the Exploration of the Seas. 2002. Working group on eels. http://www.ices.dk/icework/wgdetailacfm.asp?wg=WGEEL. (August 30, 2002).

Jenkins, R. E., and J. M. Burkhead. 1994. Freshwater Fishes of Virginia. American Fisheries Society, Bethesda, Maryland.

Johnson, J. S. 1974. Sex distribution and age studies of, *Anguilla rostrata*, in freshwaters of the Delaware River. Masters thesis, East Stroudsburg State College, East Stroudsburg, Pennsylvania.

National Marine Fisheries Service. 2000. Marine fisheries annual landing results. http://remora.ssp.nmfs.gov/MFPUBLIC/owa/mrfss (30 June 2000).

Neter, J., W. Wasserman, and M. H. Kunter. 1985. Applied linear regression models. Second Edition. Richard D. Irwin, Inc., Homewood, Illinois.

Rubec, P. J., J. C. W. Bexley, H. Norris, M. S. Coyne, M. E. Monaco, S. G. Smith, and J. S. Ault. 1999. Suitability modeling to delineate habitat essential to sustainable fisheries. Pages 108–133 *in* L. Benaka, editor. Fish habitat: essential fish habitat and rehabilitation. American Fisheries Society, Symposium 22, Bethesda, Maryland.

Sheldon, W. M. 1974. Elvers in Maine; techniques of locating, catching, and holding. Maine Department of Marine Resources, Augusta, Maine.

Sprengel, G., and H. Luchtenberg. 1991. Infection by endoparasites reduces maximum swimming speed of European smelt *Osmerus eperlanus* and European eel *Anquilla anguilla*. Diseases of Aquatic Organisms 11:31–35

Taylor, C. C. 1953. Nature of variability in trawl catches. Fisheries Bulletin 54:142–166

Tesch, F. W. 1977. The eel. J. Greenwood, translator. Chapman and Hall, London, England.

Van Den Avyle, M. J. 1982. Species profiles: Life histories and environmental requirements of coastal fishes and invertebrates (South Atlantic)—American eel. U.S. Fish and Wildlife Service FWS/OBS-82/11.24. U.S. Army Corps of Engineers, TR EL-82-4.

Weeder, J. A. 1998. Maryland American eel population study. Completion Report - Project 3-ACA–026. Maryland Department Natural Resources, Tidewater Administration, Fisheries Division, Annapolis, Maryland.

Wenner, C. A. 1973. Occurrence of American eels, *Anguilla rostrata*, in waters overlying the eastern North American continental shelf. Journal Fisheries Research Board of Canada 30: 1752–1755

Size and Age of American Eels Collected from Tributaries of the Virginia Portion of Chesapeake Bay

STEPHEN J. OWENS

Virginia Department of Game and Inland Fisheries,
1320 Belman Road, Fredericksburg, Virginia 22401, USA

PATRICK J. GEER*

Virginia Institute of Marine Science, Department of Fisheries Science,
Post Office Box 1346, Gloucester Point, Virginia 23062, USA

Abstract.—Efforts to quantify impacts of commercial harvests on the age and size structure of American eels *Anguilla rostrata* in Virginia waters are lacking, and an overall understanding of the species status is minimal. Virginia supports a viable commercial fishery for American eels, which ranks number one in the nation. The bulk of the landings are exported to foreign markets. American eels were collected from the major tributaries (James, Rappahannock, and York rivers) of the Virginia portion of the Chesapeake Bay utilizing trawls, electrofishing, and the commercial eel pot fishery over a four year period (1997–2000). Age and size distribution of collected eels were evaluated. Ages were determined by examining transverse sections of sagittal otoliths. Individuals in the two to six year range dominated American eel populations from Virginia tributaries, though specimens were collected up to age 18. York and Rappahannock River growth rates were similar and higher than those collected from the James River, although comparisons of eel growth rates revealed no significant difference in specific growth between rivers.

Introduction

The American eel *Anguilla rostrata* is widely distributed throughout North Western Atlantic slope drainages from Greenland to South America. In Virginia, American eels occur in all Atlantic slope drainages except for the Roanoke and Peedee rivers (Jenkins and Burkhead 1994). In some mountainous Virginia streams, American eels may be the largest native predatory fish present which may significantly influence the resultant species assemblage (Jenkins and Burkhead 1994). A viable commercial fishery exists within the Chesapeake Bay and its tributaries. Commercial landings from Virginia waters during a twenty-year period (1978–1998) ranged between 86.5 and 544.6 metric tons, which consistently dominated U.S. landings (NMFS 2000).

Increasing concern for declining stocks of American eels along the U.S. Atlantic coast has sparked considerable interest in assessing the species' status. Since the late 1970s, commercial landings of American eels from Virginia waters have declined to the current low of 86.5 metric tons in 1998 (Virginia Marine Resources Commission, unpublished data; NMFS 2000). Accounts from local waterman regarding declining size and numbers of eels captured have led to concerns that the Chesapeake Bay population may be suffering from overfishing. Currently, Virginia eel's are managed with a six-inch minimum size limit that restricts harvest of glass eels. Interest in the age and growth status of Virginia's eel populations led to the initiation of this study.

Methods

American eels were collected from the major tributaries (James, Rappahannock, and York Rivers) of the Virginia portion of the Chesapeake Bay from 1997 to 2000 in order to assess age and size distribution (Figure 1). Samples were collected by the Virginia Institute of Marine Science (VIMS) juvenile finfish survey, Virginia

---

*Present Address: Georgia Department of Natural Resources, Coastal Resources Division, Commercial Fisheries Program, One Conservation Way, Brunswick, Georgia 31520-8687, USA.

Department of Game and Inland Fisheries (VDGIF) fisheries biologists, and cooperating commercial fisherman utilizing a variety of sampling techniques including bottom trawls, electrofishing gear, and commercial eel pots. The VIMS Juvenile Finfish Survey collected samples from approximately 60 fixed and stratified randomly selected stations per month in the Chesapeake Bay and tidal rivers. Samples from the tidal rivers for this study were collected from the river mouth to approximately river km 72. Eels were obtained with a 9.1-m otter trawl fitted with a liner and tickler chain towed on the bottom for five minutes per station. Five commercial eel pot fisherman collected specimens from each of the tidal rivers during the Fall/Winter of 1999 (October–December). Generally, commercially caught eels were obtained from the river mouth upstream to approximately river km 64. The sample size of commercially caught eels from the James River was less than the other rivers due to escapement of eels from a temporary holding facility used by the participating fisherman. Samples obtained from VDGIF Biologists were collected during their routine electrofishing surveys conducted above the freshwater/saltwater interface in each tidal river.

Location of capture, total length (mm), weight (g), and girth (mm) were recorded prior to the removal of the sagittal otoliths. Once removed, otoliths were stored dry in individually numbered trays for future age analysis. Otoliths from 594 American eels were then embedded in molds filled with Buehler epoxide resin and sectioned transversely with an Isomet saw (Secor et al. 1992; Fitzhugh and Rice 1995; Oliveira 1996). Each section was mounted on a glass slide, immersed in oil, and examined under a dissecting microscope at 40× using transmitted light. Ages were determined by counting annuli (Beamish 1979). Three independent readings were made and the mean age recorded. Ages represent total years, which includes one year for leptocephalus and glass eel stage (Helfman et al. 1984).

Sex determination was not made due to limited resources and time constraints that prevented the histological examination of gonadal tissue. Studies have shown that a failure to examine gonadal tissue by histological methods could result in erroneous sex determinations (Dolan and Power 1977). Consequently, sex ratios reported by morphological techniques (Böetius and Böetius 1967; Winn et al. 1975) have been questioned (Dolan and Power 1977; Facey and LaBar 1981).

Results

Distribution of American eels collected for age determination varied from freshwater tributaries to locations in the lower estuary (Figure 1). Gear selectivity was evident throughout the study. Most eels collected by the VIMS Trawl Survey were less than 35 cm while commercially obtained eels were generally greater than 40 cm. Eels collected by VDGIF biologists during electrofishing samples were more variable in size than the other two collection methods. Gear selectivity may have biased the size structure to not adequately represent larger specimens that are present in the Virginia population. It was hoped that the combination of the three capture methods adequately reduced bias to an acceptable level.

Length frequency distributions for collected eels ranged 6.0–77.6 cm, with most specimens falling in the 11–56 cm size range (Figure 2). Individuals missing from the 38–40 cm size classes provided for a bimodal distribution for all river systems (Figure 2). James River eels ranged from 10.0 to 74.6 cm, though eels greater than 36 cm were difficult to obtain (Figure 2). Weight ranged between 2.5 and 769.5 g for James River eels. Larger eels were more easily obtained for age and growth analysis from the Rappahannock and York Rivers due to a greater fishing effort by cooperating commercial fisherman (Figure 2). American eels from the Rappahannock River ranged from 8.0 to 66.3 cm total length and weighed between 2.1 and 671.8 g. York River eels ranged from 9.0 to 77.6 cm in length and from 1.3 to 1060.8 g in weight.

Eel populations from the James, Rappahannock, and York Rivers were evaluated to assess whether differential growth occurred between river systems. Weight–length relationships for American eels from the James, Rappahannock, and York Rivers exhibited a power curve with between 92 and 98% of the variation explained and b values which ranged 2.91–3.15 (Figure 3). Girth–length relationships were significant for all systems and were stronger in the York River (73%) than for specimens collected from the James (63%) and Rappahannock (61%) Rivers (Figure 4).

Examination of American eel otoliths identified specimens from age-1 to age-18. However, most eels were in the age-3 to age-6 range (Figure 5). Eels less than age-3 (< 20 cm) were infrequently collected which was probably related to gear selectivity. Mean length at age was cal-

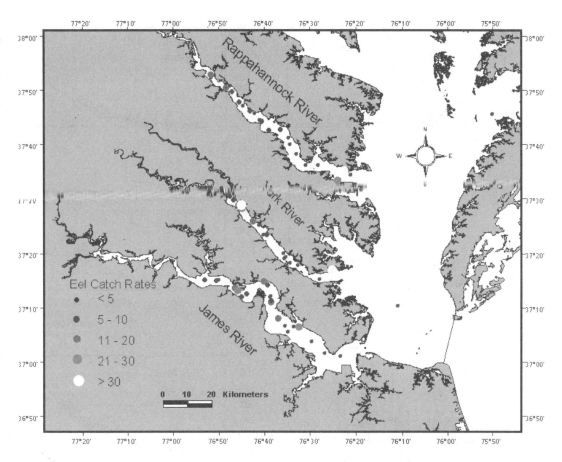

Figure 1. American eels collected for age and growth analysis from the tributaries of Chesapeake Bay, Virginia, 1997–2000.

culated for eel populations from each of the major tributaries of the Chesapeake Bay (Table 1). Individuals from age-2 to age-6 dominated the James River population of American eels (Figure 6). The population of Rappahannock River American eels was comprised mostly of individuals from age-2 to age-7, while the York River population was comprised mostly of age-3 to age-6 specimens (Figure 6). Although growth rates appeared to be faster for eels collected from the York River compared to the other river systems (Table 1), analysis of covariance (ANCOVA) found that total length at age was not significantly different between river systems ($P > 0.05$).

Age classes exhibited considerable overlap between length ranges (Figure 5), which prevented the reliable assignment of ages with an age-length key for Virginia eels. Mean length at age calculations exhibited an average standard error of 1.6 cm between all age classes (Table 2) and as a result, only ages associated with otolith examination were used.

Discussion

The absence of larger and older eels in our collections from the James River may have been related to intense commercial fishing pressure or sampling inefficiency. Previously mentioned escapement from a commercial holding facility reduced the representation of commercially caught eels from the James River which may have produced larger eels than was documented. Nevertheless, growth rates for eels from the James River system were slower (though not significantly) in comparison to the Rappahannock and York Rivers. Many of the older specimens collected were found in second or third order streams well upstream of younger eels. These larger and older eels may follow radically different emigrational cues as compared to estuarine residents.

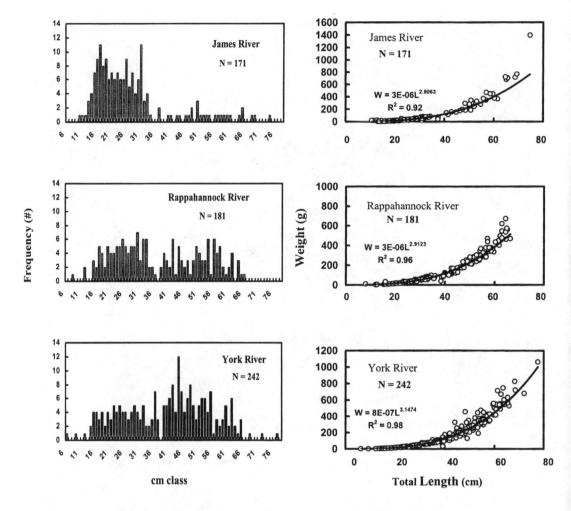

Figure 2. Length frequency distribution for American eels collected for age and growth analysis from tributaries of Chesapeake Bay, Virginia, 1997–2000.

Figure 3. Weight-length relationship for American eels collected from tributaries of Chesapeake Bay, Virginia, 1997–2000.

Investigations have shown that otoliths are the most suitable structure for aging eels (Frost 1945; Bertin 1956; Sinha and Jones 1966). Validation studies (Liew 1974; Oliveira 1996) demonstrated that American eel otoliths typically form a single complete annulus per year, which allows for reliable age determination.

American eels older than age-7 and greater than 65 cm in length were scarce from the Cooper River, South Carolina (Hansen and Eversole 1984) and the Altamaha River, Georgia (Helfman et al. 1984). This may represent an age and size when eels from these systems enter the silver phase and migrate back out to sea (Hansen and Eversole 1984; Helfman et al. 1984). Samples from Virginia's tidal rivers were also dominated by individuals younger than age-7. Eels from Virginia's three tidal rivers grew faster (4.4 cm at ages 2–9) than eels from Newfoundland, Canada but slower (8.4 cm at ages 2–9) than specimens from the Cooper River, South Carolina (Table 2). Similarities between population age and size structure for eel populations examined in this study and Cooper River populations may be explained by the transformation and subsequent migration of silver eels to the sea.

Compared to American eels in the Chesapeake Bay system, northern populations of

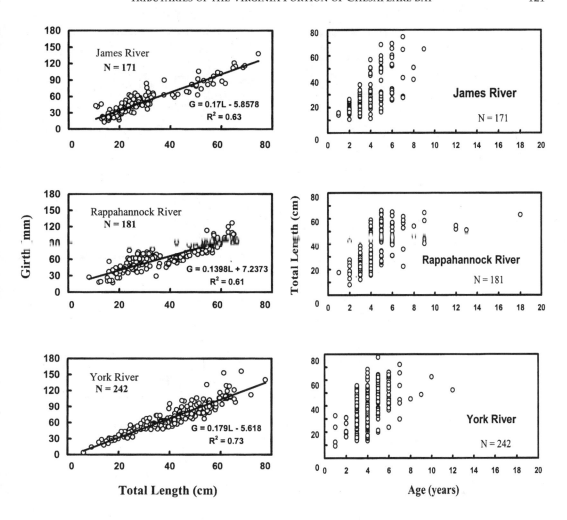

Figure 4. Girth-length relationship for American eels collected from tributaries of Chesapeake Bay, Virginia, 1997–2000.

Figure 5. Length-at-age data for American eels collected between 1997 and 2000 for age and growth analysis from Chesapeake Bay and its major tributaries.

American eels tend to grow slower and to remain longer in freshwater or estuarine systems before migration back to the sea (Gray and Andrews 1971; Hurley 1972; Facey and LaBar 1981). Differential growth rates documented for eels found in brackish and fresh Canadian waters may be tied to food availability, density dependent mechanisms, or environmental conditions (Gray and Andrews 1971). The amount of time young eels spend during the migration up freshwater systems has been thought to influence age and length attained before re-entering the sea (Hansen and Eversole 1984). Information collected from Newfoundland waters suggested that eels changed into the silver stage and migrated back to sea after 12–13 years and at a mean size of 69 cm (Gray and Andrews 1971). Similarly, populations from Lake Ontario and Lake Champlain began seaward migrations at around 13 years and at sizes greater than 80 cm (Hurley 1972; Facey and LaBar 1981). High growth rates exhibited for southern populations of eels may be due to more favorable growing conditions related to warmer more stable water temperatures, abundant forage, and shorter migration (Gunning and Shoop 1962; Böetius and Böetius 1967; Harrell and Loyacano 1982; Helfman et al. 1984). Southern populations of

Table 1. Mean total length at age for American eels collected from the tributaries of Chesapeake Bay, Virginia, 1997–2000. Standard errors denoted by SE.

System	Age	N	Mean TL (mm)	SE
James River	1	5	155	6
	2	22	200	11
	3	51	224	10
	4	37	268	18
	5	34	362	25
	6	15	433	46
	7	4	313	39
	8	3	525	68
Rappahannock River	1	1	175	–
	2	13	223	24
	3	38	269	13
	4	44	341	20
	5	44	446	20
	6	21	488	18
	7	10	487	37
	8	1	462	–
	9	5	539	49
	10	1	629	–
	12	2	513	4
	18	1	629	–
York River	1	4	194	53
	2	9	226	18
	3	61	321	15
	4	82	397	14
	5	53	468	15
	6	22	524	20
	7	6	525	63
	8	1	452	–
	10	3	544	41
	11	1	456	–

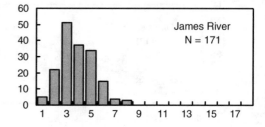
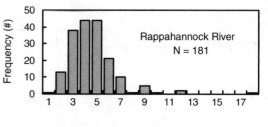
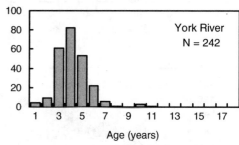

Figure 6. Age frequency distribution for American eels collected from the Chesapeake Bay tributaries, 1997–2000.

American eels tended to develop into silver eels about five years sooner than northern populations (Hansen and Eversole 1984; Helfman et al. 1984).

Consistent with previous studies at other locations, a size range that overlapped several age classes was present for American eels from Virginia waters (Böetius and Böetius 1967; Ogden 1970; Hurley 1972; Facey and LaBar 1981; Harrell and Loyacano 1982; Helfman et al. 1984). The occurrence of a wide range of lengths within an age-class could be linked to variable growth rates among yellow-phase eels and to problems with aging (Sinha and Jones 1966; Ogden 1970; Facey and LaBar 1981). Additionally, capture location (salt/fresh water) may have influenced length at age. Similar size and age structure for populations of American eels from Virginia and South Carolina waters supports the hypothesis that yellow-phase eels may change into silver-phase eels at around seven years of age and begin their seaward migration.

Larger eels may have constituted a greater proportion of the historical catch for the Virginia eel fishery, but the available data cannot be used to address growth over-fishing concerns. Mean length at age may have decreased since the 1970s; but American eel harvest information is generally lacking from Virginia waters limiting our understanding of the fishery both past and present. Furthermore, recorded landings are not classified by river system and information collected from fish buyers may be misleading due to a limited number of dealers who obtain fish from different drainages. Current Virginia regu-

Table 2. Mean total length (cm) at age comparisons for American eels collected from Atlantic slope drainages.

Age	Chesapeake Bay tributaries		Cooper River, SC (Hansen and Eversole 1984)		Newfoundland, Canada (Gray and Andrews 1971)	
	N	Mean TL ± SE	N	Mean TL ± SE	N	Mean TL ± SE
1	7	17.6 ± 2.9	2	29.2		
2	42	20.4 ± 0.9	33	36.1 ± 0.9	2	17.4
3	162	27.4 ± 0.8	97	41.1 ± 0.7	6	22.0
4	151	34.6 ± 1.0	126	45.5 ± 0.6	18	26.8
5	124	45.1 ± 1.2	108	48.2 ± 0.7	27	31.8
6	55	47.6 ± 1.6	57	51.1 ± 0.7	46	38.1
7	18	49.3 ± 3.7	9	58.0 ± 2.1	42	44.8
8	5	47.6 ± 2.2	6	51.4 ± 1.1	57	51.4
9	6	53.6 ± 4.3	?	61.1	66	50.1
10	1	62.4			35	66.3
11					12	75.5
12	3	52.8 ± 0.8	1	55.1	3	75.6
18	1	62.9				

lations restrict harvest of glass eels and elvers less than six inches in length, but impacts of such regulation are unknown. Historical landings are recorded as poundage, and no classification is made between sizes of eels harvested or yellow versus silver phase. Also, information concerning fishing effort within Virginia waters (and more importantly within different river systems) is lacking.

Acknowledgments

This study was funded in part by the Virginia Institute of Marine Science. Special thanks are extended to the members of the VIMS juvenile finfish survey, J. Odenkirk and T. Gunter of the Virginia Department of Game and Inland Fisheries, B. Green, J. Hinson, E. Inge, and L. Jenkins for assistance in the collection of specimens. Comments and suggestions by H. Austin, D.M. Seaver, and R.J. Wood helped improve this manuscript. We also recognize Wendy Morrison and Robert Graham for their independent reviews and comments on an earlier draft.

References

Beamish, R. J. 1979. Differences in age of Pacific hake (*Merluccius productus*) using whole and sections of otoliths. Journal of the Fisheries Research Board of Canada 36:141–151.

Bertin, L. 1956. Eels: a biological study, Cleaver-Hume Press Ltd., London, England.

Böetius, I., and J. Böetius. 1967. Eels, *Anguilla rostrata* (LeSueur) in Bermuda. Videnskabelige Meddelelser fra Dansk Naturhistorisk Forening i Khobenhavn 130:62–84.

Dolan, J. A., and G. Power. 1977. Sex ratio of American eels, *Anguilla rostrata*, from the Matamek River system, Quebec, with remarks on problems in sexual identification. Journal of the Fisheries Research Board of Canada 34: 294–299.

Facey, D. E., and G. W. LaBar. 1981. Biology of American eels in Lake Champlain, Vermont. Transactions of the American Fisheries Society 110:396–402.

Fitzhugh, G. R., and J. A. Rice. 1995. Error in back-calculation of lengths of juvenile southern flounder, *Paralichthys lethostigma*, and implications for analysis of size selection. Pages 227–246 *in* D. H. Secor, J. M. Dean, and S. E. Campana, editors. Recent Developments in Fish Otolith Research. University of South Carolina Press, Columbia, South Carolina.

Frost, W. E. 1945. The age and growth of eels (*Anguilla anguilla*) from the Windmere catchment area. Journal of Animal Ecology 14:26–36, 106–124.

Gray, R. W., and C. W. Andrews. 1971. Age and growth of the American eel, *Anguilla rostrata*, in Newfoundland waters. Canadian Journal of Zoology 49:121–128.

Gunning, G. E., and C. R. Shoop. 1962. Restricted movements of the American eel, *Anguilla rostrata* (LeSueur), in freshwater streams, with comments on growth rates. Tulane Studies in Zoology 9:265–272.

Hansen, R. A., and A. G. Eversole. 1984. Age, growth, and sex ratio of American eels in brackish-water portions of a South Carolina River. Transactions of the American Fisheries Society 113:744–749.

Harrell, R. M., and H. A. Loyacano, Jr. 1982. Age, growth and sex ratio of the American eel in the Cooper River, South Carolina. Proceedings of the Annual Conference Southeastern Association of Fish and Wildlife Agencies 34:349–359.

Helfman, G. S., E. L. Bozeman, and E. B. Brothers. 1984. Size, Age, and sex of American eels in a Georgia River. Transactions of the American Fisheries Society 113:132–141.

Hurley, D. A. 1972. The American eel, *Anguilla rostrata*, in Eastern Lake Ontario. Journal of the Fisheries Research Board of Canada 29:535–543.

Jenkins, R. E., and N. M. Burkhead. 1994. Freshwater fishes of Virginia. American Fisheries Society, Bethesda, Maryland.

Liew, R. K. 1974. Age determination of American eels based on the structure of their otoliths. Pages 124–136 *in* T. B. Bagenal, editor. Ageing of Fish, Proceedings of an International Symposium. Unwin Brothers Limited, Surrey, England.

NMFS (National Marine Fisheries Service). 2000. MF Annual LandingsResults:http://www.st.nmfs.gov/webplcomm/plsql/webst1.MF_ANNUAL_LANDINGS.RESULTS. (July 12, 2000).

Ogden, J. C. 1970. Relative abundance, food habits, and age of the American eel, *Anguilla rostrata* (LeSueur) in certain New Jersey streams. Transactions of the American Fisheries Society 99(1):54–59.

Oliveira, K. 1996. Field validation of annular growth rings in the American eel, *Anguilla rostrata*, using tetracycline-marked otoliths. Fishery Bulletin 94:186–189.

Secor, D. H., J. M. Dean, and E. H. Laban. 1992. Otolith removal and preparation for microstructural examination. Pages 119–127 *in* D. K. Stevenson and S. E. Compana, editors. Otolith microstructure examination and analysis. Canadian Special Publication of Fishery and Aquatic Science, Canada.

Sinha, V. R., and J. W. Jones. 1966. On the sex and distribution of the freshwater eel (*Anguilla anguilla* L.). Journal of Zoology 150:371–385.

Winn, H. E., W. A. Richkus, and L. K. Winn. 1975. Sexual dimorphism and natural movements of the American eel (*Anguilla rostrata*) in Rhode Island streams and estuaries. Helgolaender Wissenschaftliche Meeresuntersuchungen 27:156–166.

Recruitment of American Eels in the Richelieu River and Lake Champlain: Provision of Upstream Passage as a Regional-Scale Solution to a Large-Scale Problem

RICHARD VERDON

Hydro-Québec, Hydraulique et Environnement, 75 René-Lévesque Ouest,
Montréal, Québec H2Z 1A4, Canada
Email: verdon.richard@hydro.qc.ca

DENIS DESROCHERS

Milieu Inc., 188 Henrysburg, Saint-Bernard-de-Lacolle, Québec J0J 1V0, Canada
Email: denis.desrochers@MILIEUInc.com

PIERRE DUMONT

Faune et Parcs Québec, Direction de l'aménagement de la faune la Montérégie
201 Place Charles-Lemoyne, Longueuil, Québec J4K 2T5, Canada
Email: pierre.dumont@fapaq.gouv.qc.ca

Abstract.—For 150 years, the Richelieu River (Québec, Canada) supported a sizeable commercial silver American eel fishery. Between 1920 and 1980, annual landings fluctuated from 0 (World War II) to 74.8 metric tons (1938), averaging 34.6 metric tons. A sharp and constant decline since 1981 (from 72.9 to 4.7 metric tons) and a significant increase in eel size (from 890 to 1,017 mm between 1987 and 1997) pointed to a decline in recruitment in the Lake Champlain watershed. Although a decline in the eel fishery has also been recorded in the rest of Québec, the decrease in the Richelieu watershed is more dramatic. This enhanced decline has been at least partly related to the rebuilding, in the 1960s, of two old cribwork dams, Saint-Ours and Chambly. In 1997, a ladder was retrofitted on the 270-m Chambly Dam to enhance eel recruitment. This ladder was installed on a concrete cutoff wall (12.6 m) set on the crest of the dam to favor eel concentration. In the first year of operation, a total of 10,863 eels (mean TL = 368.5 mm) used the ladder. In 1998, eel migration peaked in June and, each year, maximum daily captures occurred between 23:00 and 2:00 hours. Ladder efficiency, estimated initially at 43.7%, was increased to 57.4% in 1998 after extending the operation period (51 to 133 days), improving the cutoff wall and widening the entrance. The number of eels passing over the dam has nevertheless decreased in 1999, when only 3,685 eels were electronically counted. This decrease suggests a low recruitment related in part to a possible blockage at a lower dam, at Saint-Ours, 50 km downstream.

Introduction

The American eel *Anguilla rostrata* fishery in Québec has a long history. It was practiced by native people before the arrival of the first European settlers. In New France, eel was fished for its flesh, skin and oil; it has also been used as fertilizer and even as currency (Axelsen 1997; Robitaille and Tremblay 1994). For at least 150 years, the Richelieu River (Québec, Canada) supported significant commercial American eel fisheries. Between 1920 and 1980, annual landings fluctuated from 0 (World War II) to 74.8 metric tons (in 1938), averaging 34.6 metric tons (SD = 20.7), or 6.6% of Québec total landings of 522.6 metric tons (Figure 1). Between 1981 and 1997, the catch averaged 27.9 metric tons (SD = 18.7), and a sharp and constant decline was observed, from 72.9 to 4.7 metric tons. The fishery closed in 1998.

During the same period, as is observed in many parts of the species' range in North America (Nilo and Fortin 2000; EPRI 1999; Haro et al. 2000), Québec landings also decreased, but the change was more gradual, from 603.5 to 184 tons. Local causes were suspected to explain, at

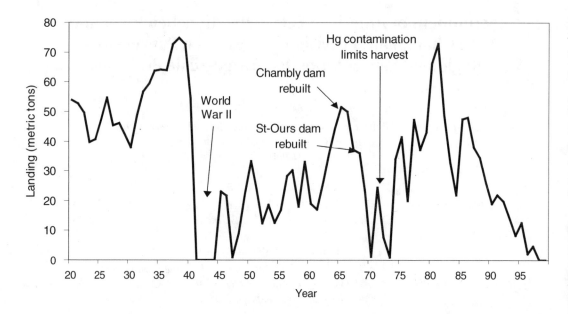

Figure 1. Annual American eel landing in the Richelieu River (Québec) during the 1920–1998 period. Data were obtained from Robitaille and Tremblay (1994) and from the commercial landings database of the Ministère de l'Agriculture, des Pêcheries et de l'Alimentation du Québec.

least partly, the more pronounced depletion of American eel in the Richelieu River watershed.

The population structure and size, growth and food habits of the American eel in Lake Champlain have been described by Facey and LaBar (1981), and LaBar (1987), while their movements in the river and lake were described by Vladykov (unpublished data), Vladykov (1955), and LaBar and Facey (1983). All eels obtained from the lake and river have been females (Facey and LaBar 1981; Dutil et al. 1985; Pierre Dumont, unpublished data). In mark and recapture experiments, all the recaptures outside the marking sectors of the lake ($n = 5$; LaBar and Facey 1983) and the upper part of the river ($n = 67$; Vladykov, unpublished data) were reported in the lower catchment, most of them ($n = 64$) in the St. Lawrence estuary, from Québec City to Rivière Ouelle, 130 km downstream. This indicates that the Richelieu catchment contributes to the silver eel migration in the St. Lawrence River. Eel was mainly a commercial species in the watershed and most of the harvest has historically been taken at a weir located on the Richelieu River at Saint-Jean, about 40 km downstream of Lake Champlain outlet. This weir was made of four V-shaped traps covering two-thirds of the river width, and has been operated at the same site for 150 years by six generations of the same family (Eales 1968). Traps were usually installed after spring flood, by the end of May, up to mid-October, when the run ends. Most of the catch was composed of emigrating silver-phase eels (Dutil et al. 1985). In 1982, the State of Vermont authorized the commercial harvest of yellow eel in Lake Champlain by electro-fishing. This experimental fishery soon closed by reason of low profitability. Prior to that, there had never been a commercial eel fishery on the lake (LaBar 1987).

The aim of this paper is to describe the approach used since 1996 to better understand the yellow eel upstream migration in the Richelieu River and the corrective passage measures that were implemented in an attempt to reduce the population decline.

Study Area

The Richelieu River drains Lake Champlain, a large (1,140 km^2), deep (mean depth, 22 m; maximum depth, 122 m) and narrow oligotrophic lake bordering New York and Vermont States and extending into Québec (Figure 2). Lake

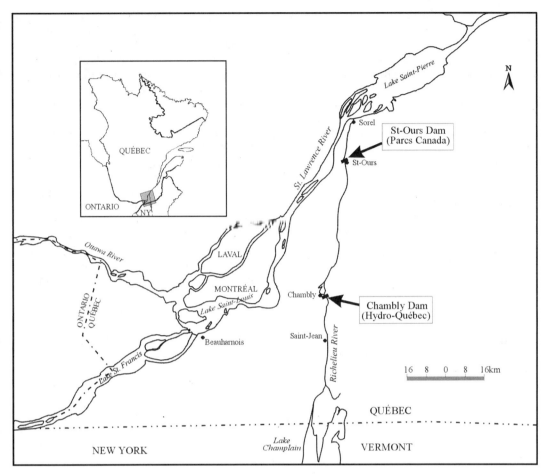

Figure 2. Location of the study area.

Champlain is connected to the Hudson River and to the Great Lakes by a network of locks and canals. The Richelieu River extends for 115 km between Lake Champlain and the St. Lawrence River. From June to September, the average discharge is 271 m^3/s (1937–1996) or about 3.5% of the St. Lawrence River flow at the confluence of the two rivers. Before the middle of the 19th century, no artificial barrier impeded fish migration in the Richelieu. Two cribwork dams were constructed along the Richelieu River, at Saint-Ours in 1846 and at Chambly, 52 km upstream, in 1896. The original structures were frequently damaged and were equipped with fishways that were not replaced when the dams were rebuilt, between 1965 and 1969. None of these dams are used for power production. The Saint-Ours Dam is equipped with radial gates, and fish passage is possible for a few weeks during high discharge periods, typically in early spring. Chambly Dam is a 270-m-long concrete overflow weir and the water head is approximately 5 m. Recreational navigation locks allow boat passage around both dams and probably provides some upstream passage for a limited number of eels.

Methods

Silver Eel Characteristics

To monitor changes in length frequency distribution of outmigrating eel from the Richelieu River, the Saint-Jean commercial harvest was sampled in 1995, 1996, and 1997 using a stratified random sampling regime similar to one used in 1987 by Dumont et al. (1997). Samples of 30–40 eels were purchased monthly, measured (TL in mm) and weighed (nearest g) from the beginning to the end of the season, which gradually decreased from five to three months of operation between 1995 and 1997. In 1997, additional length data ($n = 107$) were obtained from the New York Power Authority as part of a separate study (Normandeau Associates Inc. and Skalski 1998). Monthly size distributions were weighted by monthly landings of the fishery to obtain annual length and weight distributions that were compared by one-way analysis of variance (ANOVA), followed by year-to-year SNK tests. The Fulton condition Factor (Ricker 1975), generally used as a "well-being" index, was compared between years by Kruskal-Wallis nonparametric ANOVA. Statistical analysis was performed using SPSS/PC + for microcomputers (Norusis 1990).

Eel Ladder

In 1997, an experimental eel ladder was put in place on the Chambly dam. For this purpose, a 12.6-m cutoff wall made of seven concrete blocks, each weighing 12 tons, was first installed on the weir crest, near the left riverbank. This wall was necessary to stop the flow and create a relatively still area, allowing eels to swim to the foot of the dam. An eel ladder made of a wood and plastic climbing ramp, covered with staggered studs, was then fixed on the concrete wall and the down face of the dam (Figure 3). The ladder was 8.8-m long, 0.55 m wide, and had a slope of 51°. A submersible pump provided a flow of 0.6 L/s on the climbing ramp, while an additional attraction flow of 14.4 L/s was provided by gravity at the entrance. In 1998 and 1999, the design was improved by widening the entrance to 1.1 m and eliminating the leaks through the cutoff wall. In 1997 and 1998, a live trap at the ladder outlet captured the migrating eels, while a photoelectric counter (Smith-Root model PEC-92) was installed in 1999. At the outset, the electronic counts were validated with manual counts, and thereafter processed to eliminate false signals. Multiple signals within 200 ms were treated as a single individual, so that the discrepancies between manual and electronic counts were less than 2%. Eel were counted at 30-minute intervals during the night for the 1997 and 1998 seasons, and all day long in 1999. The ladder operated from 14 July to 5 September in 1997, 11 June until 21 October in 1998, and 21 June until 14 September in 1999.

Local Movements of Migrant Eels

In order to study the local movements of migrant eels downstream from the dam, most of the eels captured in the ladder during the 1997

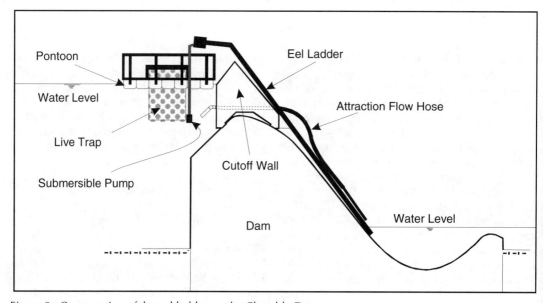

Figure 3. Cross section of the eel ladder on the Chambly Dam.

migration season were tagged and/or marked, and then released at different sites downstream. Prior to handling, eels were anesthetized with a 0.1% solution of 2-phenoxyethanol. A first group of 1,990 individuals were tagged with PIT tags (AVID Canada #2101, 14 × 2.1 mm, and BIOMARK USA #TX 1400 L, 12 × 2.1 mm) inserted with a syringe under the skin, on the back of the fish, about 1 cm in front of the dorsal fin. All tagged fish were also fin-clipped to evaluate the tag retention rate. A second group of 8,873 eels were hot branded on their right side, using six different mark patterns. All tagged and marked eels were released evenly at six locations downstream from the dam: along each shoreline at 400 m, 200 m and a few meters from the dam. All recaptures were made in the ladder, and recaptured eels were released downstream. Including recaptures, 15,196 eels were released, each site receiving between 2,511 and 2,553 eels (Figure 4). The proportion of eels recaptured from these six sites were compared by Chi-Square tests. In 1998, 298 eels were tagged with PIT tags inserted under the skin, just behind the top of the head. They were released evenly at the same sites as in 1997 (Figure 4).

In order to evaluate fallback of eels after they had climbed the ladder, in 1998, 996 recaptured PIT-tagged eels were released at two locations in the forebay. They were released alternately, on a weekly basis, immediately upstream from the ladder and along the riverbank, some 38 m from the waterfall (Figure 5).

Population Estimate and Efficiency of the Ladder

The migrant population was estimated in 1997 using Schumacher and Eschmeyer's method (Ricker 1975). Since all individuals were released downstream from the dam, no correction was made for stratification. The efficiency of the ladder was estimated in 1997 and 1998, using the recovery rate of the PIT-tagged eels released downstream. Efficiency is defined as the number of eels recaptured at the top end of the ladder over the number of eels tagged and released downstream.

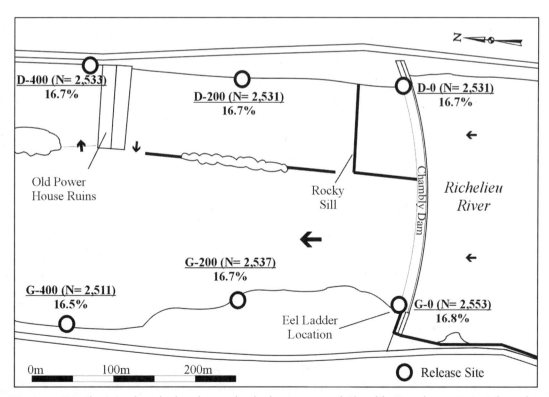

Figure 4. Distribution of marked and tagged eels downstream of Chambly Dam between 14 July and 4 September 1997.

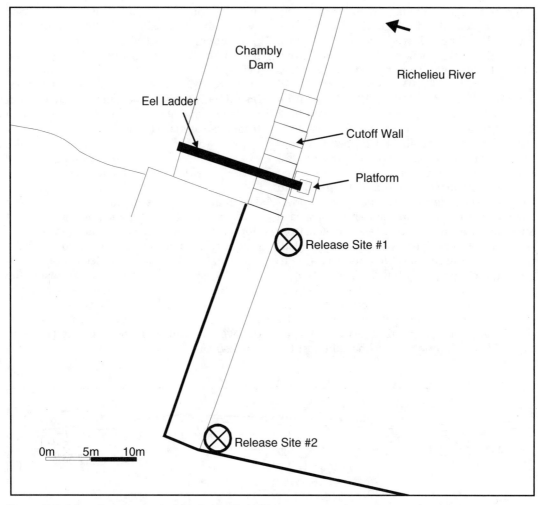

Figure 5. Release sites upstream of the ladder in 1998.

Results

Silver Eel Characteristics

Eels from the commercial fishery during the 1990s were considerably larger (Figure 6) than those collected in 1987 ($P < 0.001$). Eels in the 1997 sample are also longer than those measured in 1995 and 1996 ($P < 0.01$). From 1987 ($n = 181$) to 1997 ($n = 494$), average length increased from 890 mm (range 688–1086; SD = 75.6) to 1,017 mm (range 820–1200; SD = 69.5) and weight, from 1,497 g (range 520–2790; SD = 406.7) to 2,244 g (range 1192–3536; SD = 457). Extremely large eels (up to 1,255 mm and 4 kg) were caught in the 1990s, but size variability decreased: for example, the coefficient of variation for weight decreased from 27.2% in 1987, to 20.4% 10 years later. No change was observed for the Fulton condition factor, which averaged 0.21.

Chronology of Yellow Eel Migration

Table 1 summarizes, for the years 1997 through 1999, the counts in the ladder, excluding recaptures. Counts decreased from 10,863 in 1997 to 3,685 in 1999. Because of late installation of the ladder, the onset of upstream migration cannot be accurately determined. However, some eels were caught as early as 11 June in 1998 when

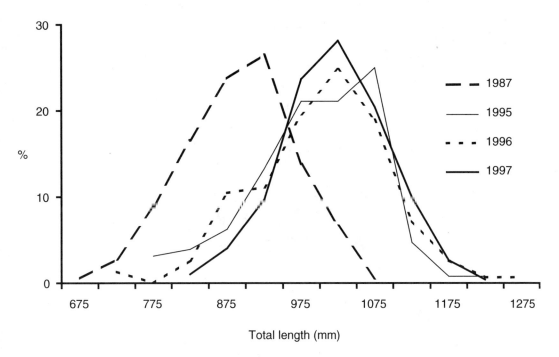

Figure 6. Evolution of the size distribution of eels sampled in the catch of the St-Jean fishery, Richelieu River, Québec (1987–1997).

water temperature was 18°C (Figure 7). Movements were practically over by the end of September. Daily counts were very irregular during the three years, with peaks close to or over 1,000 individuals per day, interspersed with near-zero lows. No direct correlation between eel catches and temperature, flow or lunar phase was found. However, most of the peaks coincided with increased water temperature, which is most obvious in 1998.

Diel Movement

Upstream movement was very similar for the three seasons studied (Figure 8). Most of the migratory behavior occurs at night, starting at about 21:00 hours, peaking near 1:00 hours, and declining at dawn. For the three years, more than 50% of passages occurred between 23:00 and 2:00 hours. Occasional migration might occur during daylight hours.

Size Distribution

The average size of eels captured in 1997 and 1998 in the ladder was 379.7 mm (SD = 73.2) and 386.2 mm (SD = 79.2) respectively (Figure 9), and the difference is significant ($P < 0.001$). Length ranged between 192 and 687 mm.

Local Movements

Out of the 10,863 different marked (8,873) and tagged (1,990) eels released in 1997, 4,759 were recaptured that same year, including multiple recaptures. From the tagged eels, 122 (6.1%) had lost their PIT tags, so that the release location of recaptured eels could be traced back on 4,637 individuals (Table 2). Individuals released at the foot of the dam, on either banks of the river, were recaptured in equal proportion (18.7% versus 17.9%; $P > 0.05$) in the ladder which is located on the left bank. It is therefore obvious that they are actively moving along the dam, seeking an upstream passage. In the same way, eels released along the left bank 200 m and 400 m downstream from the dam were recaptured in the same ($P > 0.05$) proportion (17.9% and 17.6%) as the ones released at the foot of the dam (17.9%), implying an active upstream movement. Eels released along the right bank of the river, 200 m and 400 m downstream from the dam were recaptured

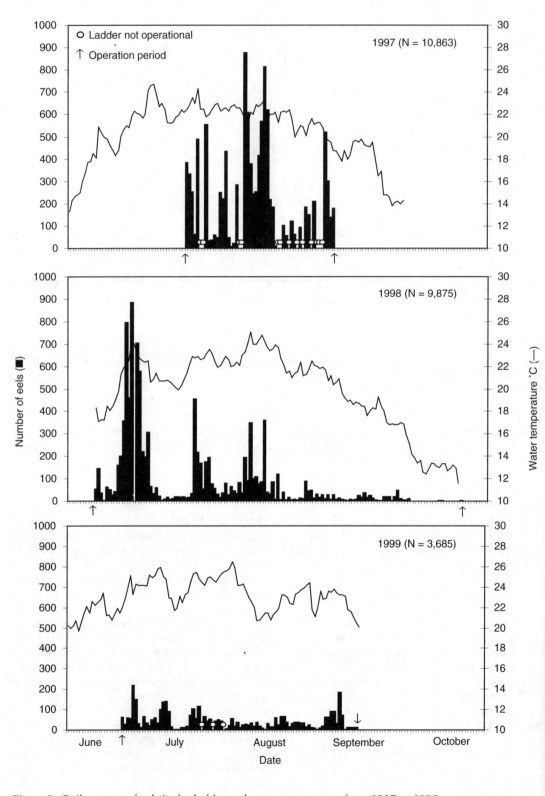

Figure 7. Daily counts of eels in the ladder and water temperature from 1997 to 1999.

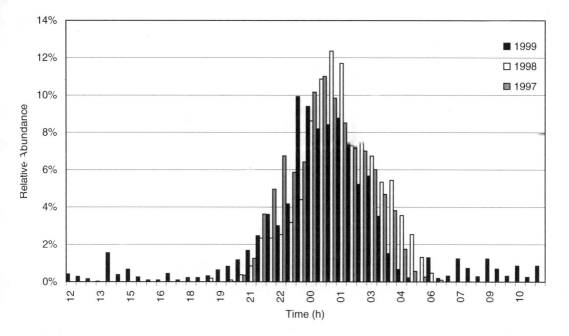

Figure 8. Diel movement of upmigrating eels in the ladder from 1997 to 1999.

Table 1. Number of eels counted in the eel ladder at the Chambly Dam from 1997 to 1999.

			Total	June	July	August	September	October
Operating Days	1997		45	-	16	25	4	-
	1998		133	20	31	31	30	21
	1999		86	10	31	31	14	-
Number of Eels	1997	N	10,863		3,249	6,471	1,143	
		%	100	-	29.9	59.6	10.5	-
	1998	N	9,875	5,257	2,012	2,126	455	25
		%	100	53.2	20.4	21.5	4.6	0.3
	1999	N	3,685	721	1,440	848	676	
		%	100	19.6	39.1	23.0	18.3	-
Daily Average	1997		241	-	203	259	286	-
	1998		74	263	65	69	15	1.2
	1999		43	72	46	27	48	-

\- Eel ladder not operating

Table 2. Origin of eels recaptured in the ladder in 1997 and average time between release and recapture for the six release sites. The location of these sites is shown on Figure 4.

	Left shoreline			Right shoreline		
	G-0	G-200	G-400	D-0	D-200	D-400
Number released	2,553	2,537	2,511	2,531	2,531	2,533
Recaptures (%)	17.9	17.9	17.6	18.7	15.4	12.5
Time between release and recapture (d)	9.8	11.2	13.3	13.1	19.7	21.3

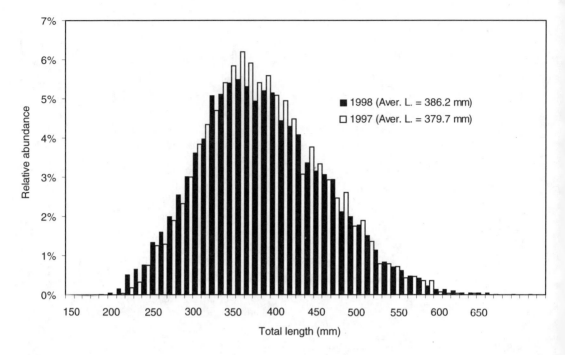

Figure 9. Length distribution of eels in the ladder in 1997 and 1998.

in significantly ($P < 0.001$) lower proportions (15.4% and 12.5% of the total respectively).

From PIT tag recaptures ($n = 1,137$), it was possible to assess, for the six release locations, the time between release and recapture (Table 2). The average time varied between 9.8 d (station G-0) and 21.3 d (station D-400) and is a function of distance and velocity barriers. Eels released at stations D-0 and G-400 would take, on average, about three more days to enter the ladder than the ones released at station G-0, near the ladder entrance. Eels released at station D-400 would take about 11 more days than individuals released at G-0.

Of the 996 tagged eels which were released upstream from the dam in 1998, only six (0.6%) were recaptured in the ladder. They had fallen back downstream from the dam before resuming their upstream migration. Of these six, one had been released near the ladder exit, while the five others had been released near the riverbank, where the water velocity is very low. It is very unlikely that the current at the release location had entrained them. Instead, they had probably actively swum to such an area before being entrained.

Population Estimate

In 1997, the total population at the foot of the dam was estimated at 19,650 individuals, with a confidence interval ($P = 0.05$) lying between 18,746 and 20,646.

Efficiency of the Ladder

In 1997, 870 of the 1,990 tagged eels were recaptured at the top end of the ladder, corresponding to 43.7%. In 1998, after improvement of the ladder and cutoff wall, 171 of the 298 tagged eels were recovered, corresponding to 57.4%, significantly higher than in 1997 ($P < 0.001$; chi-square test). In 1998 and 1999, tagged eels from previous years were also detected (Table 3). For eels released in 1997, the cumulative recovery rate at the end of the 1999 season goes up to 55.1%. For eels released in 1998, the cumulative recovery rate in the ladder, after two seasons, is 68.5%.

Discussion

The length and weight of out-migrating eels caught commercially at Saint-Jean increased sig-

Table 3. Cumulative recaptures from 1997 to 1999 of eels PIT tagged and released in 1997 and 1998.

Year of release	Number released	Number of cumulative recaptures (%)		
		1997	1998	1999
1997	1,990	870[1] (43.7%)	1,018 (51.2%)	1,097 (55.1%)
1998	298	-	171 (57.4%)	204 (68.5%)

[1] In 1997, recaptured eels were released downstream from the dam.

nificantly between 1987 and 1995–1997. During the same period, landings showed a 88% decrease, from 20 to 4.7 metric tons. Earlier data collected by LaBar and Facey (1983) and LaBar (1987) in three bays of Lake Champlain, between 1979 and 1985, also indicated that mean length increased by between 12.2 and 16.5%, while population estimates decreased by between 13 and 50%. These observations suggest a sharp decline in eel recruitment in the Lake Champlain watershed that may be related, at least in part, to the rebuilding, between 1965 and 1969, of the two cribwork dams.

The onset of upstream migration in the Richelieu River occurs earlier than in the upper St. Lawrence. At Beauharnois (Figure 2), it generally begins in late June and peaks at the end of July (Verdon and Desrochers, this volume). The same is observed at the Moses-Saunders Power Dam eel ladder, some 80-km upstream. Several authors linked yellow eel movements to water temperature for American eel (Haro and Krueger 1991; Hartley 1992; Thibault and Verreault 1995; EPRI 1999) or European Eel *Anguilla anguilla* (Westin and Nyman 1979; Moriarty 1986; White and Knights 1997a, 1997b; Knights and White 1998). For the latter species, active migration of juveniles is stimulated by temperatures greater than 14–16°C, increasing to maxima at greater than 20°C (Knights and White 1998). Our results lead to the same conclusion for *A. rostrata*. In larger lake-fed river systems, such as the St. Lawrence, the increase in temperature is more gradual and less variable, generally involving a later migration. In smaller systems, such as the Richelieu River, movements occur earlier and eels seem more responsive to temperature fluctuations. In 1998, the beginning and peak of migration occurred about a month earlier at Chambly Dam than at Beauharnois (Verdon and Desrochers, this volume), although the distance to the sea is comparable.

Yellow eels are more active at night than during the day (Mack 1983; Helfman 1986; Facey and Van Den Avyle 1987). They typically move from dusk to dawn, although day movements occur during peak migration. Thibault and Verreault (1995) showed that yellow eel migration on the Sud-Ouest River occurred mainly between 18:00 and 24:00 hours, although daylight migration was also observed. Our observation indicated similar diurnal patterns at the Richelieu Dam eel ladder. We also noted that artificial light disrupted this night time migration. For optimal eel ladder efficiency, it is therefore essential that there should be no artificial lighting near eel ladders.

Average size of yellow eel generally increases with distance from the sea (Haro and Krueger 1991; see also White and Knights 1997b for *A. anguilla*). The modal size observed in the Petite Trinité River, for eels captured with small-mesh dip-nets, was 80–90 mm, and 225–234 mm for the Sud-Ouest and Rimouski rivers (Thibault and Verreault 1995), three tributaries of the lower St. Lawrence River. The average length at the Chambly Dam in 1997 and 1998 (371.7 and 386.2 mm, respectively) was higher. On the other hand, the average length of eels captured at Beauharnois in 1998 with a similar device (average = 471.7 mm, n = 5,442, SD = 69.9 mm; Verdon and Desrochers, this volume) is much larger than at Chambly, despite the site being at a similar distance from the sea. It is likely that large shallow lakes along the St. Lawrence provide much better feeding opportunities than the Richelieu River, and that eel could travel more slowly through such productive habitats. Eel accumulating downstream from the Chambly Dam over the past years might also have suffered from a lack of food availability. Age and growth studies are needed to confirm these postulates.

Results from the tagging and recapture study undertaken below the Richelieu Dam clearly indicate an active movement of eels along the dam face. Results also suggest that sills or velocity barriers hinder upstream movement.

Conversely, downstream movement of yellow eels during the migrating season is rather unusual. This was shown by the small number of eels (0.6%) which fell back from the forebay and were recaptured in the Chambly eel ladder in 1998. This is consistent with the marking and recapture experiments carried out in the Beauharnois region; these studies have shown that, when facing a barrier, migrants do not generally move downstream for a great distance during the migration season, but might do so at the end of the season (Desrochers 1995; 1996; Verdon and Desrochers, this volume). White and Knights (1997a) also demonstrated some downstream movement of juvenile European eels after tagging before resumption of their upstream migration.

The installation of a cutoff wall on the weir crest is an important component of ladder efficiency. Preliminary attempts to capture eels in 1996 some distance from the dam with a climbing ramp failed, as eels passed by the entrance and concentrated near white waters at the foot of the dam, as could be observed during electrofishing. In 1997, after installation of the wall, leaks between concrete blocks seemed to reduce ladder efficiency as several eels were observed trying to climb unsuccessfully on the wetted face of the dam. After adjustments in 1998, the efficiency of the ladder improved noticeably. Knights and White (1998) found that European eel tend to swim in quieter water near the bottom and sidewalls, and therefore, the entrance to eel passes should be located near the base of walls and/or extended around corners into calmer water. They also conclude that strong attraction flow close to the entrance will help attract eels. This is consistent with our observations on American eel at Chambly Dam.

Ladder efficiency, estimated as the number of detected eels over the number of tagged and released eels (57.4% after one season and 68.5% after two seasons), is considered minimum, since it does not take into account deaths, lost tags, eels that emigrated downstream, or eels that may have made their way through the locks. Very few studies aimed at assessing fish ladders efficiencies for eels have been made, although most authors acknowledge that pool and weir fishways are generally inefficient at passing eels. Thus, according to Knights and White (1998), fishways designed for salmonids are usually unsuitable for eels, based on their swimming ability. Similarly, Moriarty (1986) reports that elvers fail to use an overflow pool-type fish pass, while Baras et al. (1996) achieved only a 2.1% recapture from 3,724 yellow eels released below a Denil pass on the River Meuse, Belgium. On the other hand, at the Arzal Dam, on the Vilaine River, France, the daily efficiency of the eel ladder was estimated at 4% for the actively migrating elvers, allowing the passage of about 30% of late arrivals (C. Briand, Instit. d'Aménag. de la Vilaine, France, personal communication).

The installation of an eel ladder at Chambly allowed the upstream passage in 1998 of eels that had been accumulating downstream from the dam over the past years. As a result, the number of eels ascending the ladder dropped from 9,875 in 1998, to 3,685 in 1999. From the estimated population, and assuming that the ladder efficiency was the same in 1999 as that in 1998, the recruitment between both seasons would have been of low extent, resulting from a blockage of upstream migrants at the lower dam, at Saint-Ours, 50 km downstream. A similar eel ladder has been retrofitted on this dam in 2001.

Barriers that impede or restrict upstream passage of juvenile eels can cause a significant decline in recruits in the upstream habitat, leading to a reduction in the output of outmigrating females from that upstream habitat. Elvers are able to colonize upstream areas, be it at a lower rate, even when barriers are present (EPRI 1999). However, vertical falls, even those a few centimeters high, and moderate to high water velocity can prevent upstream migration of yellow eel (Porcher 1992). Lary et al. (1998) estimated at 16,000 the number of potential barriers for American eel along the Atlantic Coast. At Chambly, an eel ladder with a 1.1-m wide entrance, corresponding to 0.4% of the dam length, fed with a minute portion of the river flow, allowed passage of more than 60% of the eels present downstream from the dam. Thus, cost-effective passage facilities for upstream migrating eels are feasible.

Acknowledgments

The authors would like to thank M. Binet, N. Fournier, M. LaHaye, J. Leclerc, K. McGrath and R. Thuot for providing data on the commercial harvest, and R. Beauchemin, S. Beaudin, S. Desloges, C. Fleury, J. F. Gaudreault, P. Gosselin and S. Labarre for field work assistance. We also acknowledge the referee comments by Paul Jacobson and Jacques Boubée on an earlier draft of this paper.

References

Axelsen, F. 1997. The status of the American eel (*Anguilla rostrata*) stock in Québec. P. 121-131 *in* R. H. Peterson, editor. The American eel in Eastern Canada: stock status and management strategies. Proceeding of Eel Workshop, January 13–14, 1997, Québec City, Québec. Canadian Technical Report of Fisheries and Aquatic Sciences No. 2196.

Baras, E., J. C. Philippart, and B. Salmon. 1996. Estimation of migrant yellow eel stock in large rivers through the survey of fish passes: a preliminary investigation in the River Meuse (Belgium). *in* I.G. Cowx, editor. Stock Assessment in Inland Fisheries, Oxford: Fishing News Books: 315–325.

Desrochers, D. 1995. Suivi de la migration de l'anguille d'Amérique (*Anguilla rostrata*) au complexe Beauharnois 1994. MILIEU & Associés pour le service Milieu naturel, vice-présidence Environnement, Hydro-Québec.

Desrochers, D. 1996. Étude de faisabilité d'une passe migratoire à anguilles (*Anguilla rostrata*) à la centrale de Beauharnois. MILIEU Inc. pour le service Milieu naturel, vice-présidence Environnement et Collectivités, Hydro-Québec.

Dumont, P., M. LaHaye, J. Leclerc, and N. Fournier. 1997. Caractérisation des captures d'anguille d'Amérique dans des pêcheries commerciales de la rivière Richelieu et du lac Saint-François. P. 25-34 *in* M. Bernard et C. Groleau, editors. Compte rendu du deuxième atelier sur les pêches commerciales, Duchesnay (Qué.), 10–12 décembre 1996.

Dutil, J. D., B. Légaré, and C. Desjardins. 1985. Discrimination d'un stock de poisson, l'anguille (*Anguilla rostrata*), basée sur la présence d'un produit chimique de synthèse, le mirex. Canadian Journal of Fisheries and Aquatic Sciences 42:455–458.

Eales, J. G. 1968. The eel fisheries of eastern Canada. Fisheries Research Board of Canada Bulletin 166.

EPRI (Electric Power Research Institute). 1999. American Eel (*Anguilla rostrata*) Scoping Study: A literature and data review of life history, stock status, population dynamics, and hydroelectric impacts. Technical Report TR-111873. Palo Alto, California.

Facey, D. E., and G. W. LaBar. 1981. Biology of American eels in Lake Champlain, Vermont. Transaction of the American Fisheries Society 110:396–402.

Facey, D. E., and M. J. Van Den Avyle. 1987. Species profile: life histories and environmental requirements of coastal fishes and invertebrates (North Atlantic). American Eel. U.S. Fish and Wildlife Service Biological Report 82 (11.74). U.S. Army Corp of Engineers, TR EL-82-4.

Haro, A. J., and W. H. Krueger. 1991. Pigmentation, otolith rings, and upstream migration of juvenile American eels, *Anguilla rostrata*, in Coastal Rhode Island stream. Canadian Journal of Zoology 69:812–814.

Haro, A., W. Richkus, K. Whalen, A. Hoar, W. D. Busch, S. Lary, T. Brush, and D. Dixon. 2000. Population Decline of the American Eel. Fisheries 25 (9):7–16.

Hartley, K. A. 1992. 1991 operation of the eel ladder at the R.H. Saunders generating station, Cornwall, Ontario, pages 2.1-2.3 *in*: S.J. Kerr and A. Schiavone editors 1992. Annual report of the St. Lawrence River subcommittee to the Lake Ontario Committee and the Great Lakes Fishery Commission.

Helfman, G. S. 1986. Diel distribution and activity of American eels, *Anguilla rostrata*, in a cave-spring. Canadian Journal of Fisheries and Aquatic Sciences 43:1595–1605.

Knights, B., and M. White. 1998. Enhancing immigration and recruitment of eels: the use of passes and associated trapping systems. Fisheries Management and Ecology 5:459–471.

LaBar, G. W. 1987. Changes in population structure of American eels, *Anguillla rostrata*, in Lake Champlain, Vermont, U.S.A., after initiation of a commercial fishery. Presented at the 1987 meeting of the European Inland Fisheries Advisory Council Working Party on Eel, Bristol, England, 12–16 April, 1987.

LaBar, G. W., and D. E. Facey. 1983. Local movements and inshore population sizes of American eels in Lake Champlain, Vermont. Transaction of the American Fisheries Society 112:114–116.

Lary, S. J., W. -D. N. Busch, and C. M. Castiglione. 1998. Distribution and availability of Atlantic coast freshwater habitat for American eel (*Anguilla rostrata*). American Fisheries Society 128th Annual Meeting, 23-27 August, Hartford, Connecticut.

Mack, R. 1983. Operating of the R.H. Saunders generating station eel ladder - 1981. Ontario Ministry of Natural Resources, Cornwall.

Moriarty, C. 1986. Riverine migration of young eels *Anguilla anguilla* (L.) Fisheries Research, 4:43–58.

Nilo, P., and R. Fortin. 2000. Synthèse des connaissances et établissement d'une programmation de recherche sur l'anguille d'Amérique (*Anguilla rostrata*). Société de la Faune et des Parcs du Québec, Direction du développement de la faune, Québec, Canada.

Normandeau Associates Inc., and J. R. Skalski. 1998. Draft final report. Estimation of survival of American eel after passage through a turbine at the St. Lawrence-FDR power project, New York. Prepared for New York Power Authority, White Plains, New York.

Norusis, M. J. 1990. SPSS/PC+ Statistics 4.0 for IBM PC/XT/AT and PS/2. SPSS Inc., Chicago, Illinois.

Porcher, J. P. 1992. Les passes à anguilles. Bulletin Français de Pêche et Pisciculture 326-327:5–14.

Ricker, W. E. 1975. Computation and interpretation of biological statistics of fish populations. Fisheries Research Board of Canada Bulletin 191.

Robitaille, J., and S. Tremblay. 1994. Problématique de l'anguille d'Amérique (*Anguilla rostrata*) dans le réseau du Saint-Laurent. Québec, Ministère de l'Environnement et de la Faune, Direction de la faune et des habitats. Rapport technique.

Thibault, J., and G. Verreault. 1995. Décompte et capture des anguillettes des rivières Verte, du Bic, du Sud-Ouest et Rimouski, été 1984. Ministère de l'Environnement et de la Faune.

Vladykov, V. D. unpublished. Contribution à la biologie de l'anguille d'après les données des étiquetages de. 1945. à 1957 dans la rivière Richelieu et le fleuve Saint-Laurent. Département des Pêcheries du Québec. MS Report.

Vladykov, V. D. 1955. Fishes of Québec. Eels. Québec Department of Fisheries, Album 6.

Westin, L., and L. Nyman. 1979. Activity, orientation, and migration of Baltic eel (*Anguilla anguilla* L.). ICES Report 174:115–123.

White, E. M., and B. Knights. 1997a. Dynamics of the European Eel, *Anguilla anguilla* (L.), in the Rivers Severn and Avon, England, with special reference to the effects of man-made barriers. Fisheries Management and Ecology 4:311–324.

White, E. M., and B. Knights. 1997.b. Environmental factors affecting migration of the European eel in the Rivers Severn ans Avon, England. Journal of Fish Biology 50:1104–1116.

Upstream Migratory Movements of American Eel *Anguilla rostrata* between the Beauharnois and Moses-Saunders Power Dams on the St. Lawrence River

RICHARD VERDON

Hydro-Québec, Hydraulique et Environnement, 75 René-Lévesque Ouest, Montréal, Québec H2Z 1A4, Canada
Email: verdon.richard@hydro.qc.ca

DENIS DESROCHERS

Milieu inc., 1435 Chemin de Saint-Jean, La Praire, Quebec J5R 2L8, Canada
Email: denis.desrochers@MILIEUinc.com

Abstract.—The Beauharnois Generating Station, west of Montreal, is the first barrier encountered by the American eel during their upstream migration in the St. Lawrence River. The next power dam, Moses-Saunders, near Cornwall (Ontario)/Massena (New York), is located 80 km upstream. To study eel movements, 3,980 eels were marked with PIT tags in 1998 and released upstream (1,546) and downstream (2,434) of the Beauharnois Dam. In 1998 and 1999, 353 and 146 tagged eels, respectively, were recaptured at Beauharnois. In 1998, the movement of eels released downstream from the generating station was rather limited, since a great proportion (329/353 or 93%) of eels recaptured in the tailrace had been released at this site. However, results from 1999 indicate extensive interannual movements of a limited number of eels. At Moses-Saunders, in 1998 and 1999, 23 and 71 tagged eels, respectively, from Beauharnois were detected; most of these were released in the Beauharnois Canal. The time between release and recapture in 1998 indicates that the average migration speed during summer was approximately 1 km per day, although some eels can travel at a speed of 2.3 km per day. It is believed that most migrating juvenile eels would not have enough time to cross both dams in less than three seasons. A comparison of approximately 20 years of count data from the eel ladder at the Moses-Saunders Power Dam and lock usage in the St. Lawrence Seaway suggests that decreasing lockages are associated with the decreased number of eels at the ladder. Lastly, a change in eel commercial harvest between the two dams might have had an impact on yellow eel movements between the two sites.

Introduction

Over the last two decades, American eel *Anguilla rostrata* landings have declined substantially in the United States and Canada (EPRI 1999; Haro et al. 2000). During the same period, the recruitment of yellow eels in the upper St. Lawrence River, as shown by the number of eels ascending the fish ladder at Cornwall, Ontario, underwent a dramatic decrease (Marcogliese et al. 1997). Castonguay et al. (1994) examined the possible causes of the species' decline in the St. Lawrence, namely toxicity from chemical contamination, habitat modification, commercial fishing, and oceanic changes. They concluded that it is not possible to identify a leading factor, but the long delay between decline onset and habitat modification (St. Lawrence Seaway and hydroelectric dams) argues against this disruption being the primary cause of the decline. Haro et al. (2000) also state that causes for this decline are unknown, but a number of potential factors may contribute, including barriers to migration.

Recently, questions have been raised about upstream migratory movements of eels between the two major dams on the St. Lawrence, the Beauharnois (Québec) and the Moses-Saunders (Ontario/New York) Power Dams. Because eels likely pass the Beauharnois hydroelectric complex through navigation locks, we looked at the relationship between lock usage and the number of upstream migrants. Lastly, we also examined if a change in eel commercial harvest between the two dams may have had an impact on yellow eel movements between the two sites.

Study Area

The study area covers the St. Lawrence River between Lake St. Louis, an enlargement of the St. Lawrence River near Montreal, and the Moses-Saunders Power Dam near Massena (New York) and Cornwall, Ontario (Figure 1). The Beauharnois Power Dam consists of a 36-unit generating station and is the first barrier encountered by upstream migrating eels in the St. Lawrence River. It is built at the outlet of the 24.5-km-long man-made Beauharnois Canal, where 85% of the total river flow was diverted (Figure 2). Water diversion through the Beauharnois Canal increased gradually between 1932 and 1961, following commissioning of the different units of the power station. During the study period, from June to September of 1998 and 1999, the flow at the power station varied between 7,367 and 8,325 m^3/s during the first year and between 5,527 and 6,128 m^3/s during the second. In order to control the flow and water levels in the original St. Lawrence riverbed, a series of smaller dams were built between 1933 and 1965. Water head at these dams is between 2 and 7 m. Due to these smaller dams and the Beauharnois Project, it is likely that upstream eel migration in the river has been blocked or seriously impeded since the early 1940s.

On the north side of the Beauharnois Power Dam, the St. Lawrence Seaway allows commercial ship navigation through two locks between Lake St. Louis and the Beauharnois Canal, a 24-m water head (Figure 2). The lower lock entrance is located about one kilometer downstream from the power station. The Beauharnois locks were commissioned in 1959, replacing a smaller navigation canal located on the north St. Lawrence riverbank. Eels migrating upstream are believed to use the St. Lawrence Seaway at Beauharnois.

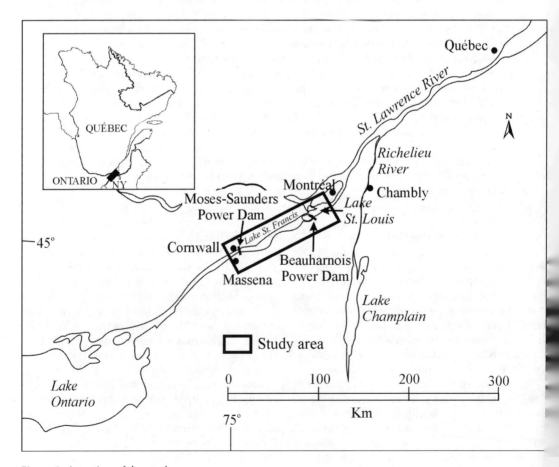

Figure 1. Location of the study area.

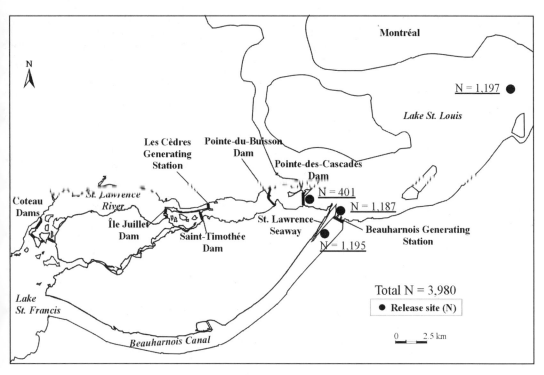

Figure 2. Number of PIT tagged eels released in the Beauharnois vicinity in 1998.

Vertical rock walls in the Beauharnois Power Station tailwater prevent eel climbing. In the original St. Lawrence riverbed, the configuration of dams, with access roads on both sides, would have made upstream passage via this route virtually impossible.

The Moses-Saunders Power Dam is located about 80 km upstream from Beauharnois; its main constituent is a 32-unit generating station. It was built between 1954 and 1959, and the average flow is 7,394 m³/s at the dam site (New York Power Authority 1996). In 1974 a prototype eel ladder was constructed in one of the ice sluices on the Ontario (north) side of the dam. This ladder operated from 1974 to 1980 and was then replaced by a permanent structure that has operated each summer since 1980.

Methods

Eel Capture and Tagging

From 19 June to 6 August 1998, 3,980 eels were PIT-tagged and released in the vicinity of Beauharnois Power Dam. Eels were captured in the Beauharnois Generating Station tailwater using a trap consisting of a climbing ramp and a net. Before tagging, eels were anesthetized using 2-phenoxyethanol with a dilution of 1.15 mL per liter of water. They were then measured, and a PIT tag (14 × 2.1-mm AVID Canada) was inserted under the skin behind the head, using a syringe. Tagged eels were also fin-clipped to facilitate identification and evaluate tag loss.

Tagged eels were released at 4 different sites (Figure 2): immediately downstream from the Pointe-des-Cascades Dam ($n = 401$), in the Beauharnois Power Dam tailrace ($n = 1,187$), 18 km downstream, in Lake St. Louis ($n = 1,197$), and in the Beauharnois Canal, 1.5 km upstream from the Power Dam ($n = 1,195$). Individuals released in Lake St. Louis were distributed along a transect perpendicular to the main lake axis.

Eel Recapture

Eels were recovered and/or detected at the Beauharnois and Moses-Saunders Power Dams. At Beauharnois, the recapture site was the same as the capture site, in the Generating Station tailwater. The trap operated between 10 June and 14 September 1998, and from 27 June to 16 September 1999. Eels were placed in ice water for a few minutes to slow their movement and to facilitate subsequent examination. During both years, recaptured eels were released upstream from the Beauharnois Power Dam. At Moses-Saunders, from 26 June to 26 October 1998, tagged eels were electronically detected and identified in the eel ladder. From 1 June to 2 November 1999 eels were electronically detected at the Moses-Saunders eel ladder, and also from captures made with five traps located in ice sluices in the Moses-Saunders Power Dam, adjacent to its tailwaters.

St. Lawrence Seaway Traffic

Since it is likely that eels bypass the Beauharnois Dam mainly through the St. Lawrence Seaway, eel counts in the Moses-Saunders ladder were plotted against lockage data for the corresponding years. Eel counts were also plotted against previous year's lockages, because of the time for eel to travel the 80-km distance between the Beauharnois locks and the Moses-Saunders ladder. Since upstream migratory eel movement at Beauharnois occurs mainly from June to August (Desrochers 1995, 1996, and 2000; Desrochers and Fleury 1999) ship movements in the locks were examined for these three months. The Beauharnois lockage data were not available, therefore, data from the St. Lambert locks were used. The St. Lambert locks are located 35 km downstream from Beauharnois, and since there is no harbor between the two, ship movements in both locks correspond, with a few hours difference.

Results

Chronology of Migration

In 1998, 5,798 eels were captured in the Beauharnois Generating Station tailwater from 19 June to 14 September and in 1999, 17,285 from 29 June to 16 September (Figure 3). In both years, daily captures were irregular and peaked after water temperature reached 22–23°C. In 1999, a bimodal distribution was observed, one mode at the end of July and one at the end of August; the decrease between the two modes coincides with a decrease in water temperature from 24 to 21°C. Average daily counts were highest in July of 1998 (113.4 eel/d) and in August of 1999 (280.2 eel/d). In 1998, numbers rarely exceeded 20 eels per day when water temperature was below 21°C.

Size Distribution

The average size of eels captured in 1999 at Beauharnois (468.7 mm, $n = 10,353$, SD = 79.2) was significantly ($P < 0.01$) smaller than in 1998 (471.7 mm, $n = 5,442$, SD = 69.9). The range was between 196 and 726 mm (Figure 4). The average length in both those years was larger ($P < 0.001$) than in 1994 (430.0 mm, $n = 8,756$, SD = 70.7) and 1995 (449.6 mm, $n = 6,086$, SD = 70.1) (Desrochers 1996). The average size at Beauharnois for all years was also much larger than at Chambly Dam, on the Richelieu River, where the average size for eels captured with a similar device was 379.7 mm ($n = 7,609$, SD = 73.2) in 1997 and 386.2 mm ($n = 6,536$, SD = 79.2) in 1998 (Verdon et al., this volume).

For the 186 tagged eels that were recaptured in 1999 after a period of 10–14 months, both at Beauharnois and Moses-Saunders, the average length increase was 10.7 mm or 2.3%. Half of them exhibited an annual growth rate of less than 1.4%.

Migratory Movements in the Beauharnois Region

In 1998 and 1999, 353 and 146 tagged eels, respectively, were recaptured at Beauharnois. An additional four eels in 1998 and three eels in 1999 were recaptured, but had missing tags. The overall recovery rate was 9.0% in 1998 and 3.7% in 1999.

In 1998, within a three-month summer period, the movement of eels released downstream from the generating station was rather limited, since a great proportion (329/353 or 93%) of individuals recaptured in the tailwater had been released at this site (Figure 5). For these, the average time between release and recapture was 7.7 days. Most of the other recaptures ($N = 22$) were eels which had been released 18 km downstream, in Lake St. Louis. On average, they were recaptured 19.8 d after release, corresponding to a mean migrating speed of 0.9 km/d, excluding the time for the eels to find the trap entrance. The minimum time was four days, corresponding to a migration speed of 4.5 km per day. Only 2 of the 1,195 eels

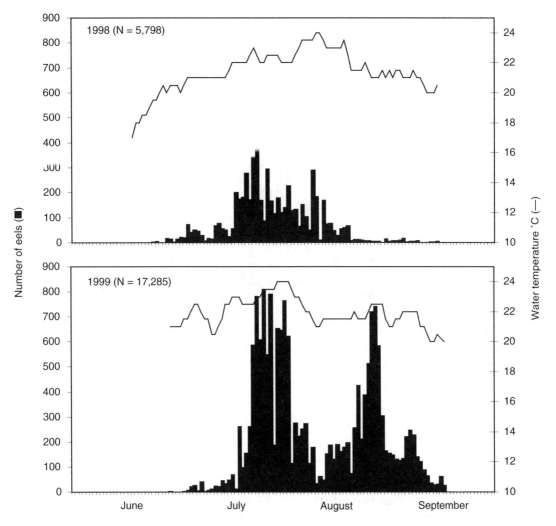

Figure 3. Daily captures of eels at Beauharnois and water temperature in 1998 and 1999.

released 1.5 km upstream from the generating station fell back, likely through one of the unit intakes, before resuming their upstream migration. Eels released at the Pointe-des-Cascades Dam were not recaptured at the Beauharnois Power Dam during the 1998 season.

In 1999, the eels exhibited more extensive movements. Of 146 recaptures, 90 came from Lake St. Louis, 45 from the Beauharnois tailrace, 8 from the Pointe-des-Cascades Dam and 3 from the Beauharnois Canal (Figure 6). In addition, one eel, which was tagged and released in the Moses-Saunders tailrace in 1999, was recaptured in the Beauharnois tailwater, after having moved about 80 km downstream.

Migratory Movements between Beauharnois and Moses-Saunders

In 1998, 23 tagged eels from Beauharnois were detected in the Moses-Saunders eel ladder. All had been released in the power dam forebay of the Beauharnois Canal, and they correspond to 1.5% of the eels that were released at this site (Figure 7). On average, the 80-km distance between the two dams was covered in 72.8 days, corresponding to a speed of 1.1 km per day, approximately the same as for the eels released in Lake St. Louis (0.9 km/d). The minimum-recorded time to cover the distance was 35 days or 2.3 km per day.

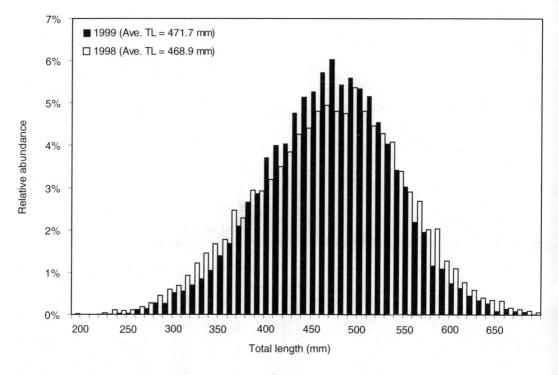

Figure 4. Length frequencies of eels captured at Beauharnois in 1998 and 1999.

In 1999, 71 eels which had been tagged at Beauharnois were detected at Moses-Saunders, 31 (44%) along the south (U.S.) side, 20 (28%) in the middle, and 20 (28%) along the north (Canadian) side of the dam (Figure 8). Among the latter, 19 were identified in the eel ladder. Out of the 71 detected eels, 66 originated from the Beauharnois Canal and 5 downstream of the Beauharnois and Pointe-des-Cascades Dams (Figure 7). They represent 4.3% (66 of 1,546) and 0.2% (5 of 2,434) of the fish released upstream and downstream, respectively, from the Beauharnois Power Dam.

Influence of St. Lawrence Seaway Traffic

The number of eels ascending the Moses-Saunders eel ladder is available from 1975 until 1999, except for the years 1980 and 1996. There is a strong relationship ($r = 0.83$, $P < 0.001$) between the average daily count at the Moses-Saunders eel ladder and average monthly ship passages in the Seaway from June to August of the previous year (Figure 9). This relationship is stronger than with ship passages in the corresponding year. Average eel numbers were generally high from 1975 to 1985, between 6,000 and 14,000 per day. From 1974–1984, average monthly ship passages during the summer in the Beauharnois locks varied between 450 and 750. After 1985, the average daily number of eels at Moses-Saunders dropped below 4,000, while average ship passages after 1984 were below 400.

Discussion

The chronology of migration of yellow eel at Beauharnois follows a similar pattern that is observed at the upstream Moses-Saunders eel ladder, where peak migrations have typically occurred between 16 July and 15 August. These peak migrations tend to correspond to the warmest seasonal temperature, above 20°C (Hartley 1992). On the Richelieu River, a tributary of the St. Lawrence, migration starts much earlier and appears to peak in June (Verdon et al., this volume). Earlier migration in the Richelieu is probably due to an earlier increase

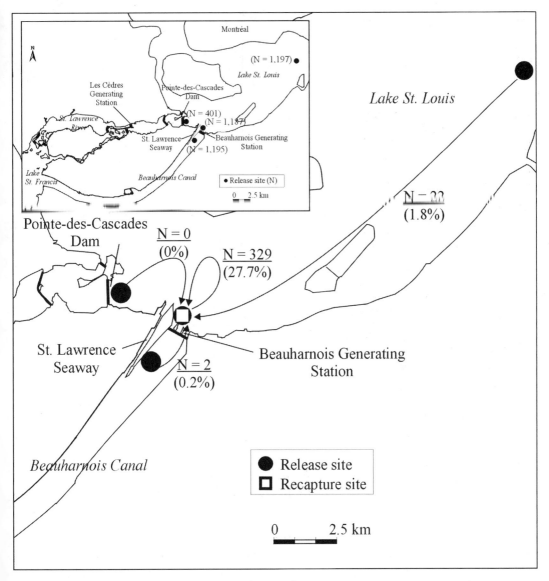

Figure 5. Number of PIT tagged eels recaptured at Beauharnois in 1998.

in water temperature, in comparison to the St. Lawrence, where discharge is much higher. Other authors have found relationships between the migration of yellow eel and water temperature. In the Sud-Ouest and Rimouski rivers, two tributaries of the lower St. Lawrence, Thibault and Verreault (1995) found that water temperature triggers upstream migration, which peaks during the second half of July, when water temperature reaches 23°C. On the Susquehanna River (Maryland), the first catches coincide with increases in water temperature from 12 to 15°C (EPRI 1999). For European Eel *A. anguilla*, White and Knights (1997a) found that speed of migration increased with temperatures above 15–16°C, with peaks being reached above 20°C. In the Shannon River, Ireland, season, temperature and age or size of eels influences the upstream migration of pigmented eels (Moriarty 1986). The migratory season will be advanced by exceptionally warm conditions and retarded by cold. In the migratory models generated by

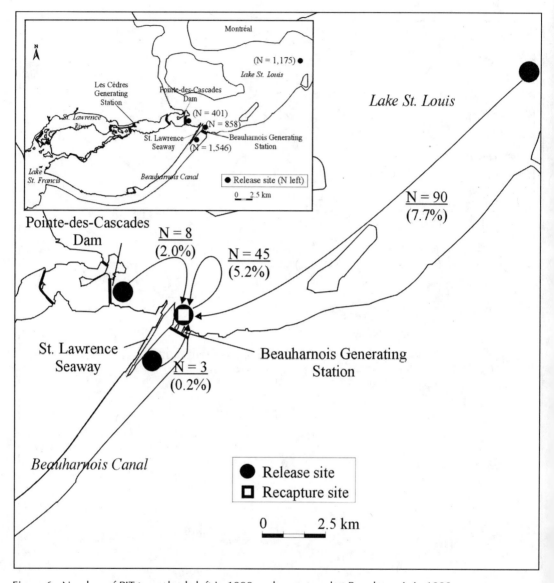

Figure 6. Number of PIT tagged eels left in 1999 and recaptured at Beauharnois in 1999.

White and Knights (1997b) for juvenile eels, water temperature accounted for the greatest amount of the variation in the catch.

The movement of tagged eels during the season they were released (1998) was rather limited, since more than 90% of the recaptured eels were caught near the release site. This conclusion is similar to results obtained in 1994 at the same site, where 91.3% ($n = 403$) of recaptured eels did not show movement during the first season (Desrochers 1995). Swimming between Beauharnois and Pointe-des-Cascades Dams, which involves downstream swimming for some distance was very rare. Few recaptures from the Beauharnois Canal in 1998 (2 out of 1,195) also supports minimal downstream movement during summer. Interannual movement was much more pronounced, suggesting fallback between two consecutive migrating seasons. For European eel, Bartel and Kosior (1991) also found that individuals smaller than 60 cm did not migrate far, moving within the release place during the first two months after release.

The large number of eels counted in the Moses-Saunders ladder, more than one million annually, at the beginning of the 1980s suggests

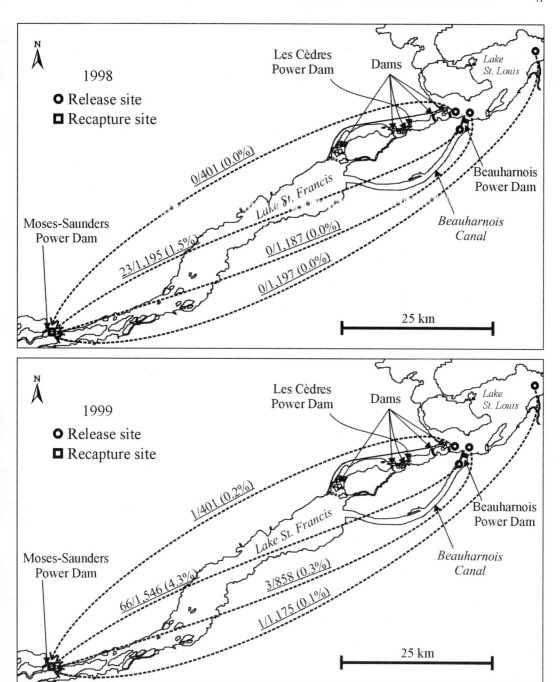

Figure 7. Number of PIT tagged eels released at Beauharnois in 1998 and recaptured at the Moses-Saunders Power Dam in 1998 and 1999.

substantial passage through the navigation locks since complete damming of the St. Lawrence occurred at Beauharnois in the early 1940s. On the other hand, eels released upstream from the Beauharnois Power Dam showed, during the following two summers, a greater proportion of recapture at Moses-Saunders than for those released downstream. This implies that the Beauharnois Dam reduces and/or slows down migration, and that passage

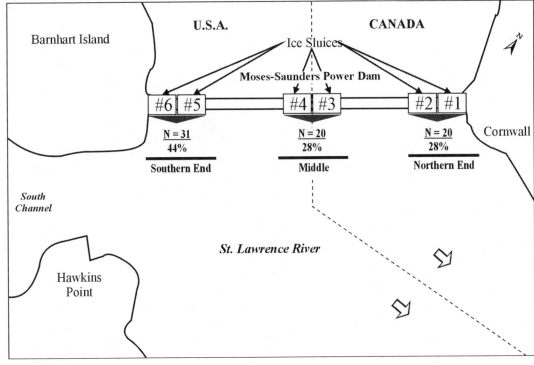

Figure 8. Spatial distribution of recaptured eels at Moses-Saunders Power Dam in 1999.

through the locks is not an easy process for upstream migrating eels.

Eel migration through locks is not well documented, although eels were caught in trap nets set in the Beauharnois locks from May to July of 1989 (CSSA et al. 1990). Tesch (1977) mentions that some migration may occur via sluices and locks during normal operation and leakage. From harvest data in Lake Ontario, Kolenosky and Hendry (1982) also suggested that eel migration was blocked, or severely reduced, during the construction of the Seaway. This suggests that locks provide passage around Beauharnois Power Dam. Average flow in the Beauharnois locks is low, less than 0.5% of the Beauharnois Generating Station, but being just a short distance away, it still might attract migrating eels. Discharge water during the emptying of the locks, and/or water flow generated by large moving ships may produce attraction flows. The Seaway traffic was reduced almost two-fold between the 1974–1984 and 1985–1998 periods. This alone cannot explain the decrease in the ladder, where numbers of eels were reduced by more than one order of magnitude during the same period. However, the decreasing traffic in the locks of the St. Lawrence Seaway at Beauharnois could have reduced the opportunities for juvenile eels to move upstream.

Results show that most eels do not cross the Beauharnois Power Dam and Moses-Saunders eel ladder in less than three migrating seasons. The travel distance between both power dams, about 80 km, would probably be covered in more than one season for the majority of the migrants. Lake St. Francis, a large, shallow water body, represents a good feeding area and a potential stopover for juveniles. In Lake Champlain, LaBar and Facey (1983) studied movements of radio-tagged eels in 1978 and 1979 and found that total displacement distances ranged from 0.6 to 4.9 km. For European eel, mean migration rate in the nontidal rivers was 0.64 ± 0.6 km/d and migration speeds of up to 1.61 km/d have been reported (White and Knights 1997a). From the low recapture rate, the authors also conclude that only a very small proportion of the migratory population moved a long distance or continued migration for an extended period. It might, therefore, be too soon

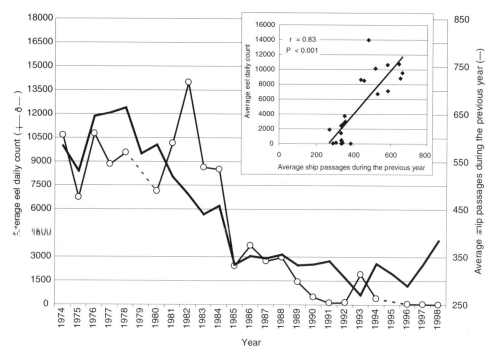

Figure 9. Relationship between average daily counts at the Moses-Saunders eel ladder (1975–1999, except 1980 and 1996) and average ship passages for June to August of the previous year in the St. Lawrence Seaway.

to draw a conclusion on the low numbers recaptured at the Moses-Saunders Dam, since most of the tagged eels might not have had enough time to reach it.

Commercial harvesting of eels in Lake St. Francis has increased substantially since 1986, both in Québec and Ontario, after the sturgeon fishery closure. Between 1975 and 1985, total landings averaged 11,000 kg annually, compared to almost 27,000 kg annually for the following years, an increase of almost 2.5-fold (Figure 10). With an average weight of 678 g on the Québec side (P. Dumont, personal communication), the average annual number of eels harvested in the last 15 years is approximately 40,000 individuals. This increased harvest has likely contributed to the decrease in the number of eels ascending the Moses-Saunders ladder, directly by removal of eels, but also because of the increase in available habitat and food to the remaining eels in Lake St. Francis. From 1982 on, there is a negative relationship ($r = 0.75$) between the average daily number of eels at the Moses-Saunders ladder and Lake St. Francis commercial eel landings.

Acknowledgments

P. Dumont, for providing commercial harvest data of Lake St. Francis; K. McGrath, for providing recapture data at Moses-Saunders Power Dam; S. Beaudin, Y. Chagnon, C. Côté, S. Desloges, S. Fleury, J. F. Gaudreault, R. Misson, M. Morin, A. Normand, R. Perreault, Y. Poiré, F. Poirier, B. Sansregret, and J. Talbot for field assistance. We also acknowledge the referee comments of Cédric Briand and an anonymous reviewer on an earlier draft of this paper.

References

Bartel, R., and M. Kosior. 1991. Migrations of tagged eel released into the lower Vistula and the Gulf of Gdansk. Pol. Arch. Hydrobiol. 38(1):105–113.

Castonguay, M., P. V. Hodson, C. M. Couillard, M. J. Eckersley, J. D. Dutil, and G. Verreault. 1994. Why is recruitment of the American Eel, *Anguilla rostrata*, declining in the St. Lawrence River and Gulf? Canadian Journal of Fisheries and Aquatic Science 51:479–488.

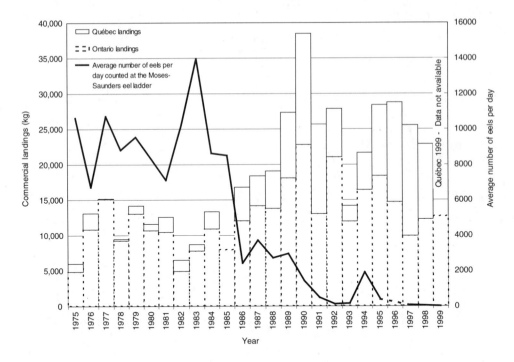

Figure 10. Average number of eels per day counted at the Moses-Saunders eel ladder for the period 1975–1999 and Lake St. Francis commercial eel landings (Québec and Ontario, Canada).

CSSA, Environnement Illimité, and Pierre Mousseau. 1990. Centrale hydroélectrique Les Cèdres. Études complémentaires de la faune ichtyenne, de la sauvagine et du milieu support—1989. Report prepared for Hydro-Québec, Montreal, Québec, Canada.

Desrochers, D. 1995. Suivi de la migration de l'anguille d'Amérique (Anguilla rostrata) au complexe Beauharnois, 1994. Report prepared for Hydro-Québec, Montreal, Québec, Canada.

Desrochers, D. 1996. Étude de faisabilité d'une passe migratoire à anguilles (Anguilla rostrata) à la centrale de Beauharnois. Report prepared for Hydro-Québec, Montreal, Québec, Canada.

Desrochers, D. 2000. Passe migratoire à anguille (Anguilla rostrata) au barrage de Chambly et étude de la migration des anguilles juvéniles du Saint-Laurent–1999. Report prepared for Hydro-Québec, Montreal, Québec, Canada.

Desrochers, D., and C. Fleury. 1999. Passe migratoire à anguille (Anguilla rostrata) au barrage de Chambly et étude de la migration des anguilles juvéniles du Saint-Laurent. Report prepared for Hydro-Québec, Montreal, Québec, Canada.

EPRI (Electric Power Research Institute). 1999. American Eel (Anguilla rostrata) Scoping Study: A literature and data review of life history, stock status, population dynamics, and hydroelectric impacts. Report TR-111873, Palo Alto, California.

Haro, A., W. Richkus, K. Whalen, A. Hoar, W. D. Busch, S. Lary, T. Brush, and D. Dixon. 2000. Population Decline of the American Eel. Fisheries 25(9):7–16.

Hartley, K. A. 1992. 1991 Operation of the eel ladder at the R.H. Saunders generating station, Cornwall, Ontario In S. J. Kerr and A. Schiavone (editors). 1992 Annual report of the St. Lawrence River subcommittee to the Lake Ontario Committee and the Great Lakes Fishery Commission, Kingston, Ontario, Canada.

Kolenosky, D. P., and M. J. Hendry. 1982. The Canadian Lake Ontario Fishery for American Eel (Anguilla rostrata). In K. M. Loftus (Editor), Proceedings of the 1980 North American Eel conference, Toronto, Ontario, Canada. Ontario Ministry of Natural Resources, Toronto, Ontario, Canada. Technical Series No 4:8–16.

LaBar, G. W., and D. E. Facey. 1983. Local movements and inshore population sizes of American eels in Lake Champlain, Vermont. Transaction of the American Fisheries Society 112:114–116.

Marcogliese, L. A., J. M. Casselman, and P. V. Hodson. 1997. Dramatic declines in recruitment of American Eel (*Anguilla rostrata*) entering Lake Ontario –Long term trends, causes and effects. Plenary presentation at the 3rd National EMAN Meeting, Saskatoon, Saskatchewan, Canada.

Moriarty, C. 1986. Riverine migration of young eels *Anguilla anguilla* (L.) Fisheries Research 4:43–58.

New York Power Authority. 1996. St. Lawrence-F.D.R. Power Project FERC No. 2000. Initial consultation package for relicensing. New York, New York.

Tesch, F. W. 1977. The Eel: Biology and Management of Anguillid Eels. London: Chapman and Hall.

Thibault, J., and G. Verreault. 1995. Décompte et capture des anguillettes des rivières Verte, du Bic, du Sud-Ouest et Rimouski, été 1984. Ministère de l'Environnement et de la Faune, Direction régionale du Bas-St-Laurent, Québec, Canada.

White, E. M., and B. Knights. 1997a. Dynamics of upstream migration of European eel, *Anguilla anguilla* (L.), in the Rivers Severn and Avon, England with special reference to the effects of man-made barriers. Fisheries Management and Ecology 4(4):311–324.

White, E. M., and B. Knights. 1997.b. Environmental factors affecting migration of the European eel in the Rivers Severn ans Avon, England. Journal of Fish Biology 50:1104–1116.

Studies of Upstream Migrant American Eels at the Moses-Saunders Power Dam on the St. Lawrence River near Massena, New York

KEVIN J. MCGRATH*

New York Power Authority, 123 Main Street, White Plains, New York 10601, USA
Corresponding author: mcgrath.k@nypa.gov

DENIS DESROCHERS AND CAROLE FLEURY

Milieu inc., 1435 Chemin de Saint-Jean, La Praire, Quebec J5R 2L8, Canada

JOSEPH W. DEMBECK IV

Kleinschmidt Associates, 75 Main Street, Pittsfield, Maine 04967, USA

Abstract.—A number of environmental studies have been conducted in association with the relicensing of the St. Lawrence-FDR Power Project located on the St. Lawrence River near Massena, New York. Some of these studies have been directed towards determining if there is a need for additional passage, beyond the existing ladder at the northern end of the Moses Saunders Power Dam, for upstream migrating American eels *Anguilla rostrata*. The studies conducted in 1997–1999 were designed to assess the efficiency of the existing eel ladder and to learn how and where eels approach the Power Dam. To assess the efficiency of the ladder, 485 eels were captured and PIT tagged in 1997, then released immediately downstream of the Power Dam. As of 1999, 34% had entered the ladder and 23% had exited the ladder and passed upstream. Of eels that successfully exited the ladder into Lake St. Lawrence (impoundment), between 6% and 12%, depending upon the year, were recaptured in the tailwaters, indicating that they probably returned through the Power Dam turbines. In 1999, eel traps were located along the face of the Power Dam and at an adjacent water control structure, Long Sault Dam, to assess where and how eels approach the Power Dam. Eels approached the Power Dam unevenly, with substantially more eels at the southern end of the Power Dam (43%) than at the middle (23%), northern end (33%), or Long Sault Dam (1%). Recapture of eels following their initial capture and release 0.5 km downstream from their capture location indicated that the eels exhibited variable levels of fidelity for their original capture location. Overall, these studies suggest that another eel ladder located on the southern end of the Power Dam would provide additional passage opportunities for upstream migrating eels. Additional studies are being conducted to assess the optimum ladder entrance and release location.

Introduction

The American eel *Anguilla rostrata* is a catadromous panmictic species that spawns in the Sargasso Sea (Schmidt 1923; Avise et al. 1986). The leptocephali larvae drift with the ocean currents and as they approach the east coast of North America they undergo metamorphosis into glass eels (Tesch 1977). Elvers, which are pigmented glass eels, migrate into estuaries, rivers, ponds and lakes (Dutil et al. 1989; Haro and Krueger 1991). In the St. Lawrence River, elvers develop into yellow eels that migrate up the River and into Lake Ontario for 5–12 years. They reside in the River or Lake for 8–15 years and migrate back to the Sea as maturing yellow or silver eels (Liew 1982; Casselman 2001).

To reach Lake Ontario, upstream migrant eels must pass the Beauharnois Generating Station (Verdon and Desrochers, this volume) which was completed in the early 1940s and then the Moses-Saunders Power Dam (Power Dam), 80 km upstream, which was constructed in 1958. No formal passage exists at the Beauharnois Generating Station and eels are thought to pass the Station via the nearby navigation locks. No passage existed at the Moses-Saunders Power Dam between 1958 and 1974 other than through the adjacent navigation locks. In 1974, an eel ladder was installed on the north side of the Power Dam by Ontario Power Generation (Lannin and Liew 1979; Eckersley 1982). The Ontario Ministry of Natural Resources operates the ladder jointly with Ontario Power Generation and counts upstream migrants and collects data on length, weight, and age of the migrants. Average annual counts at the ladder were about 890,000 eels through 1985, but have declined markedly since that time (Castonguay et al. 1994; Casselman et al. 1997a) and have been less than 4,000 over each of the past three years (Casselman 2001). This marked decline coincides with a general decline along the eastern seaboard (Electric Power Research 1999; Haro et al. 2000).

The studies described in this paper are part of larger number of studies that have been conducted in association with the U.S. Federal Energy Regulatory Commission relicensing of the St. Lawrence-FDR Power Project. Part of these efforts have been devoted to understanding the behavior and movement patterns of upstream migrant eels near the Power Dam, with the ultimate goal of learning how best to pass migrants upstream of the Power Dam, should the need be demonstrated. The specific purpose of the studies described in this paper was to assess the efficiency of the existing eel ladder and to assess how and where eels approach the Power Dam. This paper describes the main findings of these studies, which were conducted in 1997, 1998, and 1999 (with a focus on the more comprehensive 1999 studies).

Methods

Study Area

The Moses Saunders Power Dam is located on the St. Lawrence River approximately 950 km upstream from the Gulf of St. Lawrence (Figure 1). The Power Dam, which is jointly owned by the New York Power Authority and Ontario Power Generation, is 1 km wide, has a head of 25 m, contains 32 fixed blade propeller turbines, and has a total generating capacity of approximately 1,800 MW.

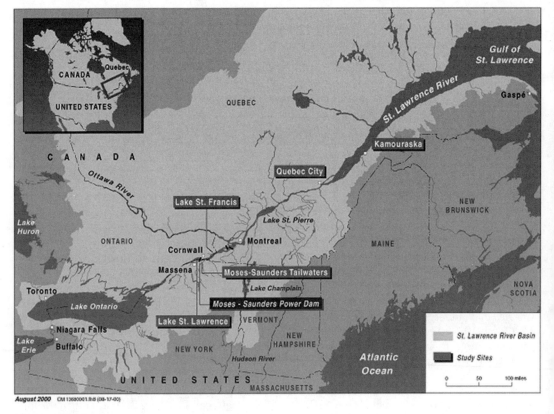

Figure 1. Study area.

The Power Dam has six ice sluices, located in pairs, two at its northern end, two in the middle and two at the southern end (Figure 2). The ice sluices were intended to pass ice in the winter but have never been used. The ice sluices at the ends of the Power Dam are each approximately 23 m wide, while those in the middle of the Dam are approximately 15 m wide. All sluices open into the tailwater and provide a calm backwater area that extends approximately 36 m under the Power Dam. Water depth in the ice sluices is approximately 2 m. The eel ladder is located in ice sluice #1 on the northern end of the Power Dam.

Eel Ladder and Eel Traps

The eel ladder is constructed of aluminum and has a vertical rise of approximately 27 m and a total length of approximately 150 m. The ladder rises in a series of eight and a half double (parallel) troughs, which zigzag across the face of the ice sluice at a 12-degree slope. The entrance chute at the base of the ladder is widened into a funnel shape to promote access. Aluminum bars are set in a herringbone pattern on the bottom of the troughs at 45-degree angles to reduce water velocities and provide climbing substrate. Artificial plastic vegetation is loosely laid in the bottom of the troughs to provide cover (light protection) and provide additional substrate to facilitate eel climbing. Conveyance flow in the ladder (measured in 1999 only) was approximately 2.5–6.0 L/s and attraction flow was approximately 7.5–11 L/s. A net was set below the exit chute from the ladder and was used to count and capture eels exiting the ladder.

Eels were also captured in eel traps that were set in the ice sluices in 1997 and 1999. The traps consisted of a climbing ramp made of ABS plastic with stud substrate positioned in a staggered pattern (0.6 cm on center), a water trough integral to the climbing ramp that passed an attraction flow of approximately 5.6 L/s, electric pumps to water the surface of the ramp and provide the attraction flow, and a floating platform that supported a submerged collection net that was capable of holding several days' catch (Figure 3). Floats supported the traps and climbing ramps in 1997, resulting in some movement and swaying of the climbing ramp. The traps and ramps were rigidly supported in 1999, which precluded movement.

PIT Tagging and Monitoring

Eels were PIT (Passive Integrated Transponder) tagged in 1997 and 1999 in order to monitor their movements. In 1997 eels were anesthetized with 2-phenoxyethanol (1.2 mL/L) while in 1999 eels

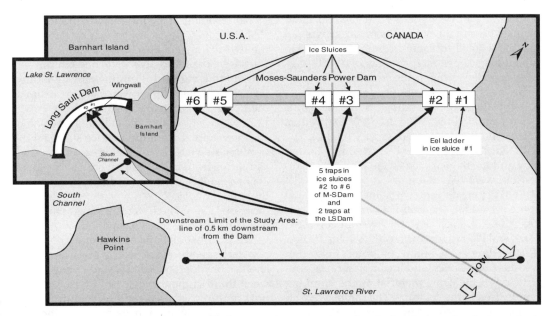

Figure 2. Location of the ice sluices and eel traps at the Moses-Saunders Power Dam and at the Long Sault Dam in 1999.

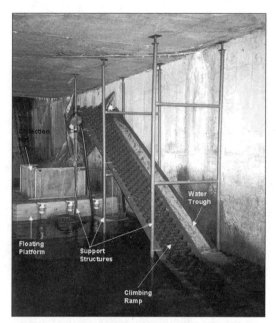

Figure 3. Eel trap installed in Ice Sluice #5 of the Moses-Saunders Power Dam in 1999.

were anesthetized in a clove oil solution (0.1 mL/L). All captured eels were scanned for the presence of a PIT tag, measured to the nearest mm (total length), and weighed (±1 g). If the eel was not a recapture (PIT tag present), then it was tagged. In 1997, PIT tags were inserted into the body cavity; in 1999, eels were tagged subcutaneously behind the head. The process involved making a small incision through the eel's skin utilizing a scalpel. Sterilized (alcohol) PIT tags were inserted through the incision with a syringe injector. The incision was sealed with cyanoacrylate glue. Depending upon availability, two types of PIT tags were used, AVID (product #2101) and Biomark (product #TX 1400 L). The AVID tags were 2 mm in diameter and 13 mm long and the Biomark tags were the same diameter but were 10 mm long.

Two PIT tag detection systems were installed at the eel ladder, one near the entrance and one near the exit. Each detection system consisted of several PIT tag readers (AVID, Power Tracker II) for redundancy, laptop computers housed in weatherproof boxes, and a climbing ramp constructed of the same material as the climbing ramps used in the eel traps. The ramps were elevated and constructed of plastic in order to isolate the PIT tag readers from electromagnetic interference associated with the aluminum structure of the existing ladder. A continuous flow (0.4 L/s) of water was supplied to each ramp. The PIT tag readers were set in series to provide direction of travel and data on the number of eels that turned back. Eels captured in the traps were monitored for PIT tags using hand held PIT tag readers.

Annual Studies: 1997–1999

Eel traps were deployed in 1997 and 1999, while the eel ladder was monitored all three years, 1997, 1998, and 1999. The deployment of eel traps in 1997 was largely experimental in nature, testing deployment techniques over the course of the migration season. Traps were deployed intermittently from July 1 through August 25. Few eels (63) were captured in the traps, although the experience gained in these efforts was valuable in the design and the more extensive deployment in 1999. Since the number of eels captured in the traps in 1997 was limited, the results on the pattern of eels approaching the Power Dam and the information on movements was of limited value and is not discussed in this paper. However, all eels captured in the traps and a portion (422/3,273) of the eels captured at the eel ladder were tagged with PIT tags, then released either 0.1 or 0.5 km downstream of the Dam. Their success rate in finding and ascending the eel ladder was monitored in 1997, 1998, and 1999.

No trapping or tagging of eels occurred in 1998; however, the eel ladder was monitored for the 1997 PIT tagged eels. In 1999, a total of seven traps were deployed, five at the Power Dam and two at the Long Sault Dam (a spillway located on a side channel of the River, 6 km upstream of the Power Dam; Figure 2). Traps at the Power Dam were deployed in the tailwater opening of each ice sluice, except the first ice sluice on the northern side where the existing eel ladder acted as a surrogate trap. Two traps were also deployed at Long Sault Dam along the downstream face of the spillway adjacent to minor seepages through the closed gates. There were no spills at the Long Sault Dam during the study period.

During 1999, all traps were operated 24 hours per day, seven days a week. The traps were checked on Monday, Tuesday, and Friday of each week from 1 June through 29 October. All first time captures were tagged except during a period of high catches in the later part of the 1999 season when one out of every three eels was PIT tagged. All eels (1997 and 1999 eels) were scanned for PIT tags. All eels (including recaptures) were released in the tailrace approximately 0.5 km directly downstream from their capture locations. The eel ladder was monitored continuously from 1 June to 2 November.

The distribution of upstream migrants as they approach the Power Dam in 1999 was determined by comparing relative count data of initial capture eels from the traps and the ladder. Initial captures, and not recaptures, were used to eliminate the possibility of bias due to handling and tagging and, because it is thought that initial capture eels approach the Power Dam freely based upon their natural upstream migratory tendencies. In order to make the comparison between the traps and ladder it was necessary to derive a ladder estimate that was comparable to the traps. We assumed that numbers at the entrance of ladder, approximately 10 m up the ladder, would be comparable to numbers at traps, which were approximately 6 m long. An estimate of initial captures at the entrance to the ladder was calculated by adjusting the number that exited the ladder, the only initial capture ladder data available, by two adjustment factors. The first adjustment factor accounted for eels that exit the ladder, are entrained, return and reascend the ladder, thus being counted at the exit more than once. Tagging studies in 1999 showed that 5.3% of the eels that exited the ladder ascended it a second time. Therefore, the total count at the exit was reduced by this proportion to provide an estimate of initial counts at the exit. The second adjustment factor accounted for the longer length of the ladder (150 m) and the inefficiencies associated with eels climbing this distance. Since only a portion of the eels that enter the ladder reach the exit, the ladder efficiency was applied to the reduced count. The efficiency rate was applied on a seasonal basis because eels were more successful at ascending the ladder earlier in the season than later in the season. The 1999 rates were 70.8% from 1 June to 31 August and 28.5% from 1 September to 31 October. This final adjustment provided an estimate of eels at the entrance of the ladder comparable to the trap counts.

Results

Average daily flow during June through October through the Power Dam from 1960 to 1993 has been approximately 7,700 m^3/s. Flows during June through October were above average in 1997 and 1998; 8,700 m^3/s and 8,300 m^3/s respectively, and below average in 1999 at 6,200 m^3/s.

The upstream migration of eels approaching the Power Dam was distinctly bimodal in 1999, with peaks in early July and early October (Figure 4). A similar bimodal pattern was observed for eels that exited the eel ladder in 1997, 1998, and 1999 (Figure 5). Eels exhibited a distinct nocturnal pattern as they entered the eel ladder (Figure 6). Approximately 94% entered the ladder between 19:30 and 05:30 hours. The average total length of eels utilizing the ladder in 1999 was 464.8 mm. This average length is approximately 100 mm greater than the length of eels observed at the eel ladder 25 years ago (Table 1).

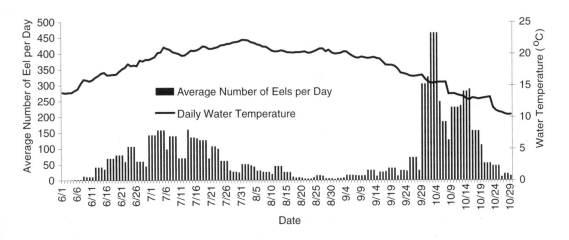

Figure 4. Average number of eels counted per day in the traps and at the ladder of the Moses-Saunders Power Dam from 1 June to 29 October 1999.

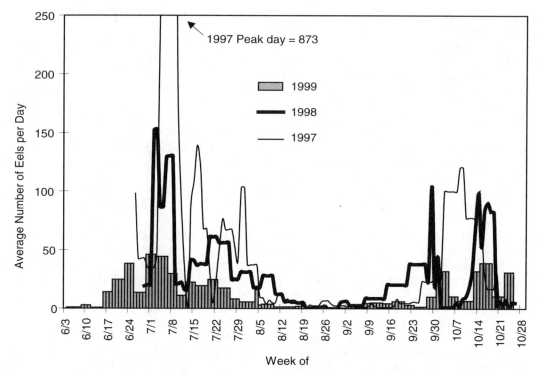

Figure 5. Average number of eels counted per day at the exit to the eel ladder at the Moses-Saunders Power Dam for the years 1997, 1998, and 1999.

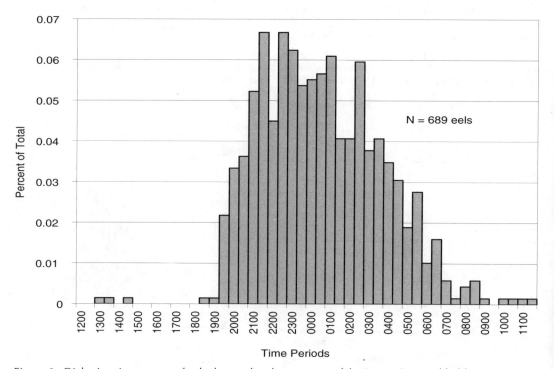

Figure 6. Diel migration pattern of eels detected at the entrance of the Power Dam eel ladder in 1999.

Table 1. Size of Upstream Migrating Juvenile Eels Captured at the OMNR Eel Ladder Installed at the Moses-Saunders Power Dam, 1967–1999[1]

		Average (mm)	Standard deviation (mm)	Minimum (mm)	Maximum (mm)	N
Seine (ice sluice #1[2])	1967	346.3	64.4	178	538	300
Eel ladder	1975–1977[3]	352.8	84.1	138	788	3,201
	1980–1984	368.0	77.1	126	785	2,322
	1985–1989	408.0	83.0	201	832	969
	1990–1994	439.9	77.2	238	790	347
	1997	449.6	75.6	285	829	557
	1999	464.8	80.5	210	834	289

1. Sources 1967–1994, Ontario Ministry of Natural Resources (1967–1994) and New York Power Authority (1997–1999).
2. Eels collected at the location of the eel ladder prior to its construction.
3. Eels frozen for at least one month before processing and selected smaller eels are excluded.

A total of 11,954 eels (9,602 captures; 2,352 recaptures, some of which were multiple recaptures) were collected in the eel traps and at the ladder in 1999. Greater numbers of eels were captured at the southern end (43%) of the Power Dam than at the middle (23%) and at the northern end (33%; Figure 7). These differences were statistically significant (chi-square, p < 0.01). Only 1% of the eels were captured at the Long Sault Dam. Of the 21% of eels captured at the northern end of the Power Dam, most (17.5%) were from the eel ladder. Attraction flow at the ladder was slightly higher than at the traps.

Fidelity of recaptured eels for the original capture location was relatively high at the northern end (55%) and the southern end (61%) of the Power Dam, but considerably lower in the middle (19%; Figures 8, 9, and 10). The recapture rate of eels released in the middle of the Power Dam was highest at the southern end of the Dam (61%) and approximately equal at the

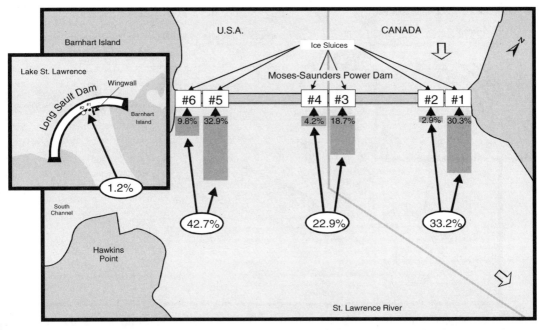

Figure 7. Spatial distribution of eels captured at the Moses-Saunders Power Dam and at the Long Sault Dam from 1 June to 29 October 1999.

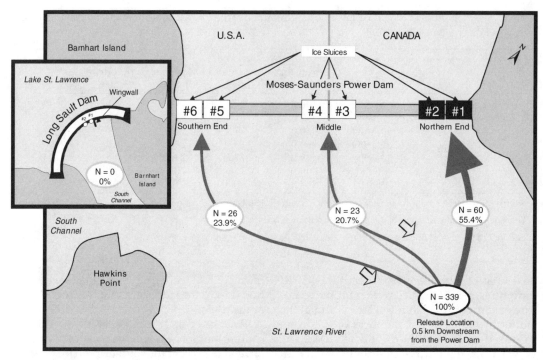

Figure 8. Movement pattern of the eels captured and PIT tagged at the northern end of the Moses-Saunders Power Dam in 1999 and released 0.5 km downstream from the capture location.

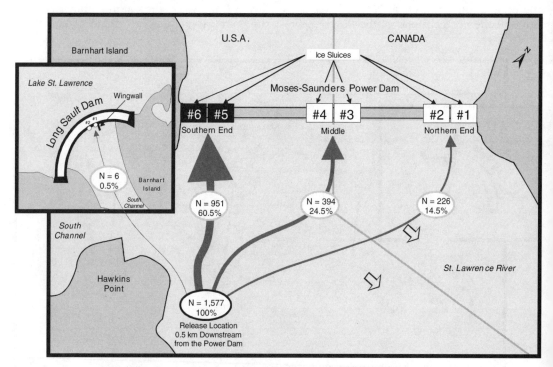

Figure 9. Movement pattern of the eels captured and PIT tagged at the southern end of the Moses-Saunders Power Dam in 1999 and released 0.5 km downstream from the capture location.

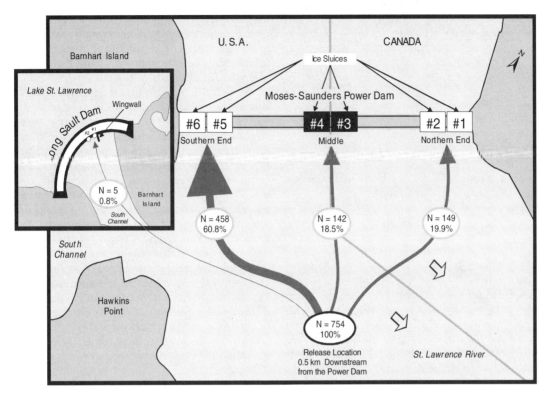

Figure 10. Movement pattern of the eels captured and PIT tagged in the middle of the Moses-Saunders Power Dam in 1999 and released 0.5 km downstream from the capture location.

other locations (19% in the middle and 20% at the northern end; Figure 10).

By the end of the 1999 field season, 34.4% of the 485 eels that had been released in 1997 had entered the ladder and 23.5% had successfully exited the ladder (Table 2). The success rate in ascending the ladder (percentage of eels that exited the ladder after entering the ladder), depending upon the year, varied between 47.4% (1999) and 78.8% (1998). These data have not been adjusted for mortality, tag loss, missed tags (tag readers not detecting a tag) or exodus from the study area since we have no estimates for any of these parameters.

Of the PIT tagged eels that successfully exited the ladder into the impoundment in 1997, 1998 and 1999, 6.0%, 8.5%, and 12.1% respectively, were detected a second time in the ladder or in a trap in the tailrace (Table 3). These eels presumably returned to the tailwater by

Table 2. Number of PIT tagged eels released downstream of the Moses-Saunders Power Dam in 1997 that entered or exited the eel ladder in 1997 and subsequent years (n = 485).

Year of capture	Entered				Exited				Ladder efficiency
	Number[1]	Percentage[2]	Cumulative number	Cumulative percentage[2]	Number[1]	Percentage[2]	Cumulative number	Cumulative percentage[2]	
1997	63	13.0%	63	13.0%	44	9.1%	44	9.1%	69.8%
1998	66	14.9%	129	26.6%	52	11.1%	96	19.8%	78.8%
1999	38	9.7%	167	34.4%	18	4.6%	114	23.5%	47.4%

1. Includes eels that exited the ladder then returned to the tailrace through the turbines.
2. Does not incorporate mortality, tag loss, or exodus from study area—no estimates available.

Table 3. Return rates of eels after successful passage through the eel ladder.

Year of study	Successful passages number	Returns	
		Number	Percentage
1997	50	3	6.0%
1998	59	5	8.5%
1999	206	25	12.1%

passing through the Power Dam turbines. We refer to this as the return rate.

Discussion

Seasonal Movement

Historically, 1974–1996, the migration of eels at the eel ladder showed only one peak, between mid-July and mid-August (Hartley 1992; Casselman et al. 1997a, b). However, during the past three years of study, 1997 through 1999, we have noted a distinctly bimodal migration pattern, with a second peak in late September/early October. During the 1974–1996 period, the ladder was operated past the end of September to mid-October in only three years (1979, 1981, and 1987), so there is little comparative data during the later part of the migration season. At the downstream Beauharnois Generating Station, Verdon and Desrochers (this volume) noted a unimodal pattern in 1998 (peak in July) and a bimodal pattern in 1999 (peaks in July and August). In studies conducted in 1997 through 1999 at the Chambly Dam on the Richelieu River (a tributary of the St. Lawrence approximately 100 km downstream from the Power Dam) eel use of a ladder was spread over a four month period from June through September, however, there was no late season peak (Desrochers 2000). Autumnal upstream migrations have been noted in some locations for the European eel (Penaz and Tesch 1970 as reported in Tesch 1977). Moriarty (1986) noted unimodal upstream migration extending from the end of May until mid-September (and in one year into October) over seven years (1977–1983) of study on the River Shannon in Ireland.

The use of the Power Dam eel ladder was markedly reduced in the autumn when water temperatures declined towards 10°C. Tesch (1977) also noted that European eels showed little movement once temperatures reached 10°C. Moriarty (1986) noted, on the River Shannon, a relationship with temperature with the initiation of the migration in the spring but no relationship with temperature with the decline in upstream migrants in the fall.

We initially thought that the second pulse of eels at the Power Dam might be related to the early season migrant passage around the downstream Beauharnois Project and then the migrants reaching the Moses-Saunders Power Dam late in the season. However, the tagging studies conducted by Verdon and Desrochers (this volume) show substantial variability in the rate of travel from Beauharnois to the Power Dam, which does not support this hypothesis. The reason for the change from the historical unimodal peak in mid-July through mid-August to a bimodal pattern with peaks in June and October at the Power Dam remains unknown. We speculate that it may be related to the lower density of eels now present in the River (Casselman 2001) which may effect dispersal and upstream migration patterns (White and Knights 1997a; Moriarty 1986).

Nocturnal Movement

The movement of eels to the ladder was distinctly nocturnal. Tesch (1977) citing Mann (1961) and Rosengarten (1954) indicates that ascent of eel ladders by the European eel takes place mostly at night and that as young eels get larger they become more nocturnal. Almost all movement of upstream migrant American eels occurred at night during three years of study at an eel ladder on the Richelieu River (Desrochers 2000). Dutil et al. (1988) noted that most upstream movement of American eels in a tidal tributary of the St. Lawrence River occurred at night.

Length

Eels utilizing the ladder have shown a trend of increasing average length over time (1974–1999). In the years immediately following the installation of the ladder in 1974, there was a decrease in the size of the eels (Liew 1982). This was attributed to larger and older eels initially being blocked from passing the Power Dam, however, with the installation of the ladder these eels moved upstream of the Power Dam and younger and smaller eels were thought to have moved into the tailrace region. Following this initial equilibration, there has been a distinct trend in increasing size first noted by Hartley (1992) in 1991. Eels collected in the 1980–1984 period averaged 368.0 mm; in the early 1990s they were close to 440 mm; in 1999 the average length was 464.8 mm, an approximate 100 mm increase in size over 20 years (Table 1). Hartley (1992) associated the increase in the eel average length at the Power Dam through time with the decline in the number of migrating eels at the ladder. A decrease in the abundance of migrants downstream from the Power Dam would provide more available downstream habitat for the eels, which would increase the time spent in downstream areas and could result in older, larger eels arriving at the Power Dam. Casselman et al. (1997a) and Casselman (2001) have reported an increase in age of the eels at the ladder. White and Knights (1997b) noted an increase in length and age of European eels as the distance upstream increased and as the number of upstream barriers increased. Tesch (1977) cites several examples where growth of European eels increases at lower densities.

The recent length data, however, presents a confusing picture. Eels at the Power Dam have had to migrate a considerable distance up the St. Lawrence River, approximately 950 km. This migration includes passage through a navigation lock in order to bypass the downstream (80 km) Beauharnois Power Project (Verdon and Desrochers, this volume). Given the distance and obstruction, one would expect the eels at the Power Dam to be larger and older relative to most upstream migrating populations in smaller drainages with no obstructions (White and Knights 1997b; Haro and Krueger 1991). There is little comparative data at northern latitudes and it is very difficult to compare across water bodies with potentially different growth rates. The size (Table 1) and age of eels at the Power Dam (Liew 1982; Casselman 2001), however, is roughly comparable to upstream migrants collected in a brook immediately adjacent to an estuary in Nova Scotia (Jessop 1987), while at the same time these eels are substantially larger than most upstream migrants (< 250 mm) in a small tidal river in Rhode Island (Oliveira 1997). Furthermore, one would expect eels arriving at Beauharnois to be larger than those in Nova Scotia and, given the obstruction at Beauharnois, one would expect them to be smaller than at the Power Dam. However, the average size of eels collected below the Beauharnois Power Project in 1999 (468.7 mm; Verdon and Desrochers, this volume) is very similar to the average size of eels at the Power Dam in 1999 (464.8 mm) observed in this study and roughly comparable to the population sampled in Nova Scotia.

Eel Approach to the Power Dam and Fidelity

The spatial distribution of eels captured in the traps and at the ladder in 1999 showed that the eels arrived in greater numbers (43%) at the southern end of the Power Dam than at the middle (23%) or the northern end (33%). It is unknown what accounts for these differences, but we speculate it may be related to the axis of the Dam relative to the pattern of water as it exits the Dam (see Figures 7, 8, or 9) and to the substantially higher flows in the southern channels downstream of the Project (Nettleton 1989). The upstream migrating eels may key on the higher flows along these southern channels.

Eels exhibited variable levels of fidelity to their original capture location with this characteristic being more pronounced for eels originally captured at the southern (61%) and northern ends (55%) of the Power Dam. However, eels captured in the middle of the Power Dam exhibited a relatively low fidelity (19%) for this location, with most moving towards the southern end of the Power Dam and a substantial number moving towards the northern end of the Power Dam. The lack of fidelity from the middle of Power Dam may, in part, be attributable to the turbulent nature of the tailrace waters.

Efficiency of the Ladder

The data indicate that after three years, approximately 35% of the eels in the tailrace region below the Power Dam find the eel ladder and of these, between 48% and 79% exit the ladder into the forebay, depending on the year. Cumulatively, after three years at large, approximately 23% of the eels reached the forebay of the Power Dam. At the Chambly Dam, a water control dam on the Richelieu River, 55% of the eels PIT tagged and released in the "tailrace" in 1997 had reached the impoundment of the Dam via an eel ladder after three seasons at large (Desrochers 2000). The actual percentage exiting the ladder at the Chambly Dam is likely greater than 55% since counts at the exit to the ladder were only made in 1998 and 1999, thus an additional number of uncounted eels probably exited in 1997, the first year of their release. The large difference between Moses-Saunders Power Dam and Chambly Dam results is likely attributable to the larger size of the Power Dam (1 km wide), the larger extent of the tailrace area, and the height and length of the ladder. At the Chambly Dam, the Richelieu River is only 270 m wide and the eel ladder is more than 16 times shorter than that at the Power Dam.

Some of the variability in ladder efficiency rates displayed by the eels at the Power Dam may be related to colder water and air temperatures late in the migration season. Eels ascending the ladder are exposed to cooler air temperatures, particularly at night, as they ascend the approximately 150 m long ladder and these cooler temperatures are thought to reduce the ability of eels to successfully ascend the ladder.

Entrainment

The recapture rates (eels that exited the ladder into the impoundment then were recaptured in the tailrace in the traps or at the ladder; presumably having passed through the turbines) ranged from 6 to 12% for the three years, 1997–1999. This represents only a portion of eels that returned to the tailwaters, since the recapture rate does not account for the effectiveness of the ladder or traps in recapturing eels in the tailrace. If one assumes that the capture rate of these gear is roughly comparable to the annual percentage of the 1997 pool of PIT tagged eels recaptured at the entrance to the ladder (approximately 10–15%, see Table 2) then the actual return rates, the proportion of eels that return to the tailrace (recaptured or not), would be proportionately much higher than the recapture rates.

The recapture rates of eels that exit the ladder at the Moses-Saunders Power Dam is higher than seen at the Beauharnois Power Dam (1.3%) and at the Chambly Dam (0.6%; Desrochers 1996; Desrochers and Fleury 1999). The position of the ladder exit relative to the turbine intakes at the Power Dam is probably partially responsible for these higher rates; they are separated by less than 100 m. Additionally, the forebay waters are turbulent at the Power Dam while at Beauharnois the eels were released in a relatively calm area adjacent to a side wall of the dam. The turbulence in the Power Dam forebay may contribute to higher entrainment and return rates.

Future Studies

Future studies are planned to monitor the movement patterns of the pool of PIT tagged eels that remain in the tailwater, the effects of water and air temperature on movement behavior, and to further address the entrainment of eels.

Acknowledgments

We would like to thank the New York Power Authority for support and funding; Daniel Parker of the New York Power Authority and André Bourbonnais and Del Potts of Ontario Power Generation for substantial logistical support; Tom Tatham of the New York Power Authority for sound scientific guidance; and Scott Ault of Kleinschmidt Associates for substantial support in study planning, design and coordination. Special thanks to Alastair Mathers from the Ontario Ministry of Natural Resources and Richard Verdon from Hydro-Québec who allowed the use of their data. Additionally we thank Richard Verdon for his helpful technical comments on an earlier draft of this paper. Lastly we want to thank Doug Dixon of EPRI for his never ending patience in allowing us to finalize this paper and his very helpful review.

References

Avice, J. C., G. S. Helfman, N. C. Saunders, and L. S. Hales. 1986. Mitochondrial DNA differentiation in North Atlantic eels: population genetic consequences of an unusual life history pattern. Proceedings of the National Academy of Science 83:4350–4354.

Casselman, J. M., L. A. Marcogliese, T. Stewart, and P. V. Hodson. 1997a. Status of the Upper St. Lawrence River and Lake Ontario American Eel Stock-1996. In R. H. Peterson editors. The American Eel in Eastern Canada: Stock Status and Management Strategies, Proceedings of Eel Management Workshop, 13–14 January 1997, Quebec City, Quebec, Canada. Technical Report of Fisheries and Aquatic Sciences 2196.

Casselman, J. M., L. A. Marcogliese, and P. V. Hodson. 1997b. Recruitment index for the Upper St. Lawrence River and Lake Ontario eel stock: a re-examination of eel passage at the R. H. Saunders Hydroelectric Generating Station at Cornwall, Ontario, 1974–1995. In R. H. Peterson editors. The

American Eel in Eastern Canada: Stock Status and Management Strategies, Proceedings of Eel Management Workshop, January 13–14, 1997, Quebec City, Quebec. Canada. Technical Report of Fisheries and Aquatic Sciences 2196.

Casselman, J. M. 2001. Dynamics of American Eel, *Anguilla rostrata*, Resources: Declining Abundance in the 1990's. International Symposium, Advances in Eel Biology, University of Tokyo, Tokyo, Japan, September 2001.

Castonguay, M., P. V. Hodson, C. M. Couillard, M. J. Eckersley, J. -D. Dutil, and G. Verreault. 1994. Why is recruitment of the American eel, Anguilla rostrata, declining in the St Lawrence River and Gulf? Canadian Journal of Fisheries and Aquatic Sciences 51:479–488.

Desrochers, D. 1996. Étude de Faisabilité d'une Passe Migratoire à Anguilles (*Anguilla rostrata*) à la Centrale de Beauharnois, [by] Milieu inc. [for] le service Milieu naturel, vice-présidence Environnement et Collectivité, Hydro-Québec, Montreal, Canada.

Desrochers, D. 2000. Passe migratoire a anguille (*Anguilla rostrata*) au barrage de Chambly et etude de la migration des anguilles juveniles du Saint-Laurent—1999. by Milieu, inc., for Hydraulic et Environnement, Groupe Production, Hydro-Quebec, Montreal, Canada.

Desrochers, D., and C. Fleury. 1999. Passe migratoire à anguille (Anguilla rostrata) au barrage de Chambly et étude de la migration des anguilles juvéniles du Saint-Laurent—1999, [by] Milieu inc., [for] Hydraulique et Environnement, Groupe Production, Hydro-Québec.

Dutil, J.-D., A. Giroux, A. Kemp, G. Lavoie, and J.-P. Dallaire. 1988. Tidal influence on movements and on daily cycle of activity of American eels. Transactions of the American Fisheries Society 117:488–494.

Dutil, J.-D., M. Michaud, and A. Giroux. 1989. Seasonal and diel patterns of stream invasion by American eels (*Anguilla rostrata*) in the northern Gulf of St. Lawrence Canadian Journal of Zoology 67:182–188.

Eckersley, M. 1982. Operation of the Eel Ladder at the Moses-Saunders Dam, Cornwall 1974–1979, Pages 4–7 *In* K. H. Loftus editors Proceedings of the 1980 North American Eel Conference. Ontario Fisheries Technical Report Serial Number 4.

Electric Power Research Institute. 1999. American eel (*Anguilla rostrata*) scoping study: a literature and data review of life history, stock status, population dynamics, and hydroelectric impacts. TR-111873. Palo Alto, California.

Haro, A. J., and W. H. Krueger. 1991. Pigmentation, otolith rings, and upstream migration of juvenile American eels, Anguilla rostrata, in a coastal Rhode Island Stream Canadian Journal of Zoology 69:812–814.

Haro, A., W. Richkus, K. Whalen, A. Hoar, W. D. Busch, S. Lary, T. Brush, and D. Dixon. 2000. Population Decline of the American Eel. Fisheries 25(9):7–16.

Hartley, K. A. 1992. 1991 Operation of the eel ladder at the R. H. Saunders generating station, Cornwall, Ontario. *In* S. J. Kerr and A. Schiavone editors, 1992 Annual Report of the St. Lawrence River Subcommittee to the Lake Ontario Committee and the Great Lakes Fishery Commission Ontario Ministry of Natural Resources, Kingston, Ontario, Canada.

Jessop, B. M. 1987. Migrating American Eels in Nova Scotia. Transactions of the American Fisheries Society 116:161–170.

Lannin, W. R., and P. Liew. 1979. The eel ladder. Ontario Fish and Wildlife Review 18(1).

Liew, P. K. L. 1982. Impact of the eel ladder on the upstream migrating eel (*Anguilla rostrata*) population in the St. Lawrence River at Cornwall: 1974–1978. *In* K. H. Loftus editors, Proceedings of the 1980 North American Eel conference, Ontario Fisheries Technical Report Series No 4: 17–22.

Mann, H. 1961. Der Aalaufstieg in der Aalleiter an der Staustufe Geesthacht. Fischwirt 11:69–74.

Moriarty, C. 1986. Riverine migration of young eels *Anguilla anguilla* (L.). Fisheries Research 4:43–58.

Nettleton, P. 1989. St. Lawrence River Environmental Investigations Volume 5: Hydrodynamic and Dispersion Characteristics of the St. Lawrence River in the Vicinity of Cornwall-Massena. Unpublished Manuscript, Environment Ontario, Toronto, Canada.

Oliveira, K. 1997. Movements and growth rates of yellow-phase American eels in the Annaquatucket River, Rhode Island Transactions of the American Fisheries Society 126:638–646.

Penaz, M., and Tesch, F.-W. 1970. Geschlechtsverhaltnis und Wachstum beim Aal (*Anguilla Anguilla*) an verschiedenen Lokalitaten von Nordsee und Elbe. Ber Dt Wiss Kommn Meeresforsch 21:290–310.

Rosengarten, J. 1954. Der aufstieg der Fische im MoselfischpaB Koblenz im Fruhjahr 1952 und 1953. Z. Fisch 3:489–552.

Schmidt, J. 1923. The breeding places of the eel. Philosophical Transactions of the Royal Society of London, Series B, Biological Sciences 211: 179–208.

Tesch, F.W. 1977. The eel. Chapman and Hall, London, England.

White, E. M., and B. Knights. 1997a. Environmental factors affecting migration of the European eel in the Rivers Severn and Avon, England Journal of Fish Biology 50:1104–1116.

White, E. M., and B. Knights. 1997b. Dynamics of upstream migration of the European eel, Anguilla anguilla (L.), in the Rivers Severn and Avon, England, with special reference to the effects of man-made barriers. Fisheries Management and Ecology 4:311–324.

PART II
Eel Fisheries

Effect of Changes in Growth and Eel Pot Mesh Size on American Eel Yield per Recruit Estimates in Upper Chesapeake Bay

JULIE A. WEEDER

*Maryland Department of Natural Resources Fisheries Service, Matapeake Work Center,
301 Marine Academy Drive, Stevensville, Maryland 21666, USA
410-643-6785, ext. 106, Fax: 410-643-4136, Email: jweeder@dnr.state.md.us*

JAMES H. UPHOFF, JR.

*Maryland Department of Natural Resources Fisheries Service, Tawes State Office
Building, 580 Taylor Avenue C-2, Annapolis, Maryland 21401, USA
410-260-8304, Fax: 410-260-8278, Email: juphoff@dnr.state.md.us*

Abstract.—A Thompson-Bell yield-per-recruit model (YPR) was used to model the effect of changes in selectivity associated with an eel pot fishery that used a mixture of mesh sizes (a description of the current fishery) and one based on large mesh sizes only (the ideal fishery) at current (1999) and historic (1981) growth patterns. Perceived changes in growth patterns between 1981 (linear growth with large eels present) and 1999 (asymptotic growth with large eels absent) had far more influence on YPR than modeled changes in selectivity. Yield-per-recruit was far less with the current growth pattern than with the historic pattern. Eels in upper Chesapeake Bay would have been considered overfished ($F_{0.1}$ and F_{max} exceeded) at the current level of F when historic growth was simulated. Reference points were not exceeded when current growth estimates were used. Simulated mixed mesh and large mesh pot fisheries had similar YPR at current F when current growth estimates were used; mixed pot YPR was about 14% less when historic growth was modeled. The existence of demographic data from 1981 has left us skeptical of growth estimates derived in 1999 and we believe that strong size selective fishing mortality biased our 1999 growth estimates and YPR calculations. We recommend evaluating the status of Maryland's eels using growth estimates based on areas or time periods (such as 1981) that have not been subjected to intense fisheries.

Introduction

American eels *Anguilla rostrata* have been harvested from the Chesapeake Bay since colonial times for food and bait (Jones et al. 1988). Commercial harvest in Maryland has been recorded since 1920 and was up to 1 million pounds during the 1940s (Hildebrand and Schroeder 1972; Jones et al. 1988). Maryland eel landings have periodically risen and fallen through the last four decades (Figure 1). In the late 1960s, an international live eel market developed with the advent of airfreight transportation (Foster 1981). After a harvest of over 700,000 lbs was recorded during 1981, landings during the remainder of the 1980s were much lower than in previous decades (Weeder and Uphoff 1999). During 1989–1999, landings were stable, with median landings greater than those of any other period since 1962 (Figure 1). Sexually maturing adult eels are the focus of Maryland's fishery because they are more valuable; however, small eels also have considerable commercial value. Weeder and Uphoff (1999) suggested that smaller eels may be heavily exploited because large eels are rare and the live market for eels of all sizes has grown. While some eels are kept locally for use as bait, many more are shipped live overseas to markets throughout Europe and Asia for food, aquaculture and stocking of streams depleted of native eels (Weeder and Uphoff 1999).

The eel fishery in Maryland is regulated with a minimum length limit and pot mesh size. A six inch minimum size limit for commercially captured eels and a minimum mesh size of 1.27 cm × 1.27 cm on pots ("large mesh" for convenience) were required in 1994. However, the

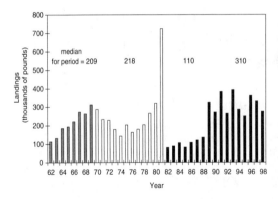

Figure 1. Commercial American eel landings in Maryland, 1962–1998. Data are separated into periods based on observed trends.

fishery did not fully implement these mesh size changes (Weeder and Uphoff 1999) and from 1996 to January 1998, mesh of sizes 0.64 cm × 1.27 cm ("small mesh" for convenience) and 0.85 cm × 0.85 cm ("medium mesh" for convenience) were allowed to accommodate current practices of the fishery. In February 1998, an escape panel of large mesh was required on pots made of mesh smaller than 1.27 cm × 1.27 cm to assist eelers in the transition from smaller mesh pots to legal gear. Currently, eelers often fish a mix of large mesh pots and small mesh pots with escape panels ("current fishery" for convenience).

Foster (1980) and Foster and Brody (1982) examined the fishery and described catch per unit effort, size, and age distribution of eels in two of Maryland's upper Chesapeake Bay tributaries in the early 1980s. In 1997, we began sampling the fishery and initiated fishery-independent research. We have used these data to estimate yield per recruit (YPR) of the current mixed pot fishery and an "ideal fishery" in which only large mesh pots are used. We estimated growth, the relationship of weight to length, fishing and natural mortality, and mesh selectivity of eel pots. We contrasted YPR based on 1981 and 1999 growth estimates.

Methods

Eels were collected from five tidal rivers during spring 1999 with commercial eel pots of various mesh sizes (Figure 2). Some eels were purchased from eelers and others were collected. Each river was sampled at least two days. Total length (mm) and weight (g) were recorded for each fresh eel. Five eels from each 2-cm length increment were subsampled for aging from each river. Both sagittal otoliths were removed, affixed to glass slides, and polished (Secor et al. 1992). Otoliths were read on a dissecting microscope using both transmitted and reflected light. Each set of otoliths was read at least twice by the same biologist. One year in freshwater was defined on the otolith as one opaque zone and one translucent zone (Sinha and Jones 1967; Gray and Andrews 1971; Liew 1974; Chisnall and Kalish 1993). Aging began at the first opaque zone after the central nucleus, to include only bands produced in freshwater (Tesch 1977). Any otoliths with suspected supernumary zones were excluded from analysis to avoid overaging (Deelder 1976; Berg 1985; Lecomte-Finiger 1992; Oliveira 1996). Although growth differed significantly between the five rivers sampled in spring 1999 (Weeder and Uphoff 1999), we pooled the data to estimate growth in order to encompass the range of variation observed in Maryland eel populations. Readable otoliths from eels captured in spring 1981 during Foster and Brody's (1982) fishery-independent survey of the Sassafras and Susquehanna Rivers were recovered from storage and aged using the same methods. Because the age interpretation of otoliths in 1981

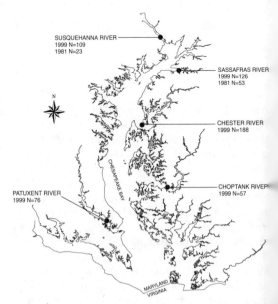

Figure 2. American eel sampling locations and sample sizes in Maryland's Chesapeake Bay tributaries, spring 1981 and 1999.

differed from current guidelines, we could not use the ages originally assigned to them for estimating mortality rates or YPR. We examined the original ages for growth trends and aged a subsample using current methods.

We evaluated eel growth in 1981 and 1999 with the von Bertalanffy growth function (von Bertalanffy 1938) and linear regression. The length-weight relationship for each year was described using the allometry equation: $W = \alpha L^\beta$; where W = weight (g), L = length (mm) and α and β are model parameters. Instantaneous natural mortality (M) was assumed to be $3/T_{max}$ (Anthony 1982), where T_{max} was the oldest age in the population. The total instantaneous mortality rate (Z) for spring 1999 was calculated using a catch curve for ages 2–12 based on numbers at age derived from an age-length key (Ricker 1975; Allen 1997). We used a modified Vartot optimum allocation method (Lai 1987) to minimize error in the age-length key. The instantaneous fishing mortality rate (F) was derived $F = Z - M$.

We used a Thompson-Bell YPR analysis as described by Gabriel et al. (1989) to estimate YPR, determine F-based biological reference points and determine the status of the stock at the current level of F for combinations of selectivity and growth. This method uses partial recruitment vectors (proportion of each age-class vulnerable) to scale F and define YPR more precisely. It also requires assigning the proportion of F and M that occurred prior to spawning. These proportions were both set to one because eels migrate out from the Chesapeake Bay to spawn and then die. We used an arbitrary initial cohort of 100,000 at age 0 to run the model through age 12. We simulated YPR for values of F from 0 to 1 at 0.05 increments. The selectivity patterns for large mesh pots, the 'ideal fishery', and a mix of large mesh pots and small mesh pots with panels, the 'current fishery', were determined from spring 2000 length-frequencies.

Two biological reference points were calculated, F_{max} and $F_{0.1}$. The fishing mortality rate that maximizes YPR is denoted F_{max} and occurs where the slope of the YPR function is zero (Hilborn and Walters 1992). This reference point is relevant to short-term yield, but ignores the effect of fishing on future generations (Sissenwine and Shepherd 1987). $F_{0.1}$ occurs where the slope of the YPR function is 0.1 (10%) of the initial slope. $F_{0.1}$ is a somewhat arbitrary, conservative choice of F (Hilborn and Walters 1992).

During spring 2000, we set eel pots of different mesh sizes together in the Wye River. The lengths of eels caught by us in the Wye River with large mesh pots, small mesh pots with escape panels, and medium mesh pots with and without escape panels were examined, as were similar data from commercially caught eels from the Nanticoke River and Wye River. Maryland eelers often use pots made of more than one size mesh at one time, and when commercially caught eels were purchased at the dock, they were already mixed. Fifty percent of the pots used on the Nanticoke River by eelers were made of the smaller mesh paneled pots, while 30% of the Wye River eel pots were of this type. We looked at length frequencies of eels caught in mixed mesh size pots and noted the first abundant size-class in the distribution for each mesh size. An age-length key for spring 1999 was then used to set the recruitment vectors as the proportion of eels at each age in our sample that exceeded the first abundant size-class in each length-frequency. These proportions could not be determined for 1981 because data were insufficient for creating an age-length key. We did not use Maryland's six inch minimum size for commercial harvest in our vector calculation because none of the mesh sizes used in the Bay retain eels that are less than six inches long (Foster and Brody 1982; Weeder 1998; Weeder and Uphoff 1999).

Results

We aged 523 eels of the 1592 eels weighed and measured in 1999. We only aged 76 of the 578 eels weighed and measured in 1981 due to the poor condition of many otoliths. The growth of eels collected in spring 1999 was best described by the von Bertalanffy growth function ($L_\infty = 330$, SE = 5.43; $K = 1.53$, SE = 0.37; $t_0 = 0.11$, SE = 0.23; Figure 3), while linear regression provided the best fit to spring 1981 eel size at age (slope = 82.8, SE = 4.35; intercept = 73.18, SE = 20.03; $r^2 = 0.83$; Figure 3). The length-weight relationship $W = \alpha L^\beta$ was best described: 1981 slope = 2.68, SE = 0.042, intercept = 1.4×10^{-5}, SE = 3.64×10^{-6}; 1999 slope = 3.18, SE = 0.01, intercept = 5.96×10^{-7}. Most eels were less than 400 mm in 1999, while most were above this size in 1981 (Figure 3). Length at age in 1981 and 1999 was similar for about the first four years. After age 4, increasingly larger eels appeared in the older ages in 1981 data; eels older than four years were

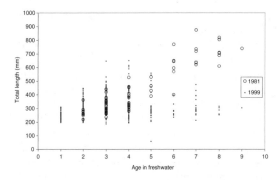

Figure 3. Length at age of eels collected from Maryland's Chesapeake Bay tributaries during 1981 and 1999.

present in the 1999 data, but they were not larger than many of the younger eels. Most eels were age 4 or younger in 1999, while there was a wider range of ages observed in 1981. Our analysis of the original ages from 1981 also showed that growth was linear and consistent with our reduced otolith sample. This agreement indicated that growth based on our reduced sample reflected the larger, original sample.

The smallest eels effectively captured by the 'current' fishing gear array were 29 cm long, so partial recruitment vectors for this pot mixture were 0.32 for age 1, 0.71 for age 2, 0.81 for age 3 eels, and 1.00 for older eels. The smallest eels effectively captured by the 'ideal' fishing gear array were 31 cm long, the same size as those captured in 1980. The partial recruitment vectors for these large mesh pots alone were 0.09 for age 1, 0.51 for age 2, 0.62 for age 3 eels, and 1.00 for older ages. The large mesh proportions used for 1999 were applicable to 1981 data, as the size ranges at ages 1–3 during 1981 were similar to those in 1999 (Figure 3).

Using 1999 growth parameters, reference points were higher for the ideal, large mesh pot fishery (F_{max} = 1.28 and $F_{0.1}$ = 0.49) than for the current, mixed pot fishery (F_{max} = 0.63 and $F_{0.1}$ = 0.36). Differences between gear arrays were not as apparent when 1981 growth was considered (current fishery $F_{0.1}$ = 0.14, F_{max} = 0.19; ideal fishery $F_{0.1}$ = 0.17, F_{max} = 0.23 (Figure 4).

Total instantaneous mortality rate (Z) estimated by the catch curve was 0.52 in 1999 (95% CI 0.40–0.64). The oldest observed age was 12, and M = 0.25. Current F was 0.27 (95% CI 0.15–0.39). At current F and 1999 growth, YPR was slightly higher for large mesh pots than for the current, mixed pot fishery. With 1981 growth, yield was 14% lower for the current fishery selection pattern at current F (Figure 4). Current F was lower than all reference points with 1999 growth. However, it was consistently higher than either reference point with 1981 growth estimates (Figure 4). Yield per recruit of the current, mixed pot fishery fell below that of large mesh pots at F as low as 0.1 when 1981 growth was simulated (Figure 4). When 1999 growth was simulated, YPR of the current fishery did not become notably less than that of the large pot fishery until F reached 0.45. Yield per recruit of large mesh pots based on 1981 growth was 19–84% greater than when 1999 growth was applied (Figure 4).

Discussion

Perceived changes in growth patterns between 1981 (linear growth with large eels present) and 1999 (asymptotic growth with large eels absent) had far more influence on YPR than our modeled, modest changes in selectivity. At current F, YPR based on 1999 growth was about half of YPR based on 1981 growth. Both $F_{0.1}$ and F_{max} were exceeded at the current level of F when historic growth was simulated, indicating overfishing. When current growth estimates were used, reference points were not exceeded and

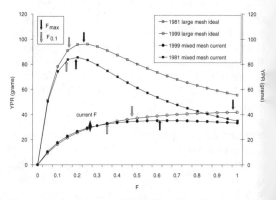

Figure 4. American eel Yield Per Recruit, current F, F_{max} and $F_{0.1}$, spring 1981 and 1999—'ideal' fishing scenario of only 1.27 cm × 1.27 cm 'large mesh' eel pots and 'actual' fishing scenario of a mixture of 'large mesh eel pots and 0.64 cm × 1.27 cm 'small mesh' eel pots fitted with 1.27 × 1.27 cm 'large mesh' escape panels.

overfishing was not indicated. Simulated mixed pot and large pot fisheries had similar yields at current F when 1999 growth estimates were used; the mixed pot fishery exhibited a modest reduction in YPR when 1981 growth was modeled.

Two tributaries of Maryland's Chesapeake Bay were intensively studied in 1981 and again in 1998–1999. Large eels (> 700 mm) were present in 1981 samples but almost not at three collected during 1999. Eels were much less abundant during 1999, and the size and age range in the Sassafras River was truncated compared to 1981 (Foster and Brody 1982; Weeder and Uphoff 1999). In 1998 and 1999, eels longer than 50 cm were virtually absent in the Sassafras River, while they made up 16% of the sample in 1981 (Foster and Brody 1981). Despite intensive sampling in a variety of locations, we recovered far less eels larger than 50 cm during spring 1999. These eels may have been removed from the system by fishing, they may have matured and left at a smaller size or they may no longer reach this size due to changes in biotic or abiotic factors.

Intense size selectivity could have increased mortality of large fish and changed the depiction of growth from linear in the early 1980s to asymptotic in the late 1990s. A shift in population makeup toward younger, smaller eels is consistent with responses to increased size selective fishing pressure (Trippel 1995). Maryland's eel fishery is currently dominated by eels less than four years old, while a much greater range of sizes and ages was observed during 1981 (Foster and Brody 1981; Weeder and Uphoff 1999). Anguillid species in other areas have shown both linear and asymptotic growth patterns. Growth in unfished areas of Virginia's Shenandoah River during the late 1990s was best described as linear (Goodwin 1999), while European eels *Anguilla anguilla* caught in a commercially fished Italian lagoon grew asymptotically (De Leo and Gatto 1995). Fully recruited fish remaining after size-selective fishing pressure would be slower growers. We believe that experimental bias during fish collection was unlikely, and that if we caught slower growing individuals it was because the faster growing fish had been harvested.

Maryland's eels could be maturing and leaving to spawn at a smaller size than in the past in response to increased exploitation. Many exploited fish stocks decrease in size at maturity as a compensatory response (Trippel 1995). Fish can reach sexual maturity at a small size if they can obtain the necessary nutrition for gonadal maturation (Trippel 1995). Sexually maturing eels leave during fall to spawn (Helfman et al. 1987). Some of the eels we collected in spring 1999 were large enough ($\approx$ 35 mm) to be approaching sexual maturity. Wenner and Musick (1974) found American eels with silver coloring (a sign of sexual maturity) in the ocean southeast of the Chesapeake Bay (habitat of migrating silver eels) that were less than 40 cm (15.7 in) long, but mature female American eels on the Atlantic coast generally average 50–80 cm in length (Helfman et al. 1987). Barbin and McCleave (1997) and Wenner and Musick (1974) were able to obtain individual eggs from silver American eel females as small as 45 cm, which indicated these eels were sexually mature. Sexually mature male American eels caught along the Atlantic coast from Newfoundland to Georgia averaged 30–40 cm long (Helfman et al. 1987). Weeder and Uphoff (1999) did not find eels longer than 37 cm in the Sassafras River between fall 1997 and spring 1998. Eels longer than 37 cm were present in spring 1981 in the Sassafras River but were not present in fall 1997 (Weeder and Uphoff 1999). Eels may have left this river at a smaller size in fall 1997 than in fall 1980. Whether they immediately journeyed to the ocean to spawn or migrated farther down the Bay and remained there to grow further is unknown. Males generally mature at a smaller size than females (Helfman et al. 1987; De Leo and Gatto 1995) and a general reduction in the size of eels captured could be associated with an increased percentage of males in the population. It is unknown whether a reduced size would cause, versus result from, such a sex ratio shift. We histologically determined the sex of 483 yellow eels captured during 1999 and 2000 (Weeder and Uphoff 2000, 2001; methods followed Todd 1981 and Beullens et al. 1997). All eels longer than 40 cm were female, consistent with findings in Maine and Rhode Island (Oliveira and McCleave 2000; Krueger and Oliveira 1997). Among eels measuring 40 cm or less, males made up only 14% of the total observed, whereas 47% were female and 39% were not differentiated enough to sex. We have no comparable historic sex information from this region with which to detect any change in sex ratio over time. However, the percentage of males observed in Maryland's Chesapeake Bay estuary was lower than we expected, given that males are thought to be more abundant in estuarine habitats and made up 25–55% of samples from Georgia estuaries (Helfman et al. 1987).

We derived our estimate of current F from catch curve analysis. Catch curves assume recruitment and mortality rates (F and M) are uniform and unchanging with time over age-groups in question (Ricker 1975; Allen 1997). Neither moderate fluctuations in recruitment (Ricker 1975) nor highly variable recruitment (Allen 1997) preclude catch curves from being useful in some situations. Allen (1997) simulated the effect of highly variable recruitment on catch curve mortality rate estimates for crappie *Pomoxis* spp. When true mortality was high ($Z = 1.21$) it was determined that these estimates of Z would be within 10% of those meeting the constant recruitment assumption. Relative recruitment levels of American eels to the Chesapeake Bay region were unknown to us. Castonguay et al. (1994) described declining recruitment in the St. Lawrence River. If this trend is occurring in Chesapeake Bay, it would have a tendency to negatively bias the catch curve because younger age classes would have proportionally less representation in the stock than the older age classes with higher initial recruitment. This would decrease the slope of the descending limb of the catch curve and the estimate of total mortality.

Fall data from 1997 to 1999 were not included in the current analysis due to seasonal differences in size of fish of the same age and uncertainty about age determination of eels collected during the autumn months (September through December). A tentative estimate of Z of all American eels collected during fall 1997, 1998 and 1999 was 0.99, twice that of the spring 1999 estimate, and American eels older than eight years were absent. This large variation in estimates of F between seasons indicated our single catch curve estimate of current F in spring 1999 should be viewed cautiously.

Foster and Brody (1982) studied American eels during the single largest year of harvest on record between 1962 and the present; this harvest was nearly twice as high as any other harvest during this period. Landings were depressed for six years following 1981. Market conditions have a large influence on American eel landings; however, it is possible that this nadir in harvest reflected severe overharvest in 1981. Large American eels were present and would have buoyed landings considerably in 1981. If many of these large American eels were later removed from the population by high exploitation, yield would be depressed and probably dependent on new recruits.

Foster (1980) interviewed a number of eelers in 1980 that were concerned that the average American eel size was getting smaller and that too many small American eels were being taken. During 1997–2000, we heard many of the same concerns about declining eel size and increasing pressure on small American eels (Weeder and Uphoff 2000). Restoration of large American eels to the population would probably require very low F at a higher age at entry than provided for by current minimum length limits and American eel pot mesh sizes.

Changes in Maryland's American eel stock could impact East Coast American eel recruitment. Adult American eels from the entire species range migrate to the Sargasso Sea to mate randomly and die, and their offspring drift back to places in North America independent of their parental origin (Helfman et al. 1987). Any changes in fecundity or abundance of spawners could impact larval production and overall recruitment to the Atlantic coast. The amount of spawner biomass coming from Maryland could be lower than in previous years, due to decreases in the number and size of spawning adults produced. A reduction in the number of spawning adults would decrease reproductive output, as would a reduction in spawner size. The combined reproductive potential of many small spawners may be less than an equivalent biomass of large ones (Trippel 1995). Smaller eels are less fecund than large ones (Boëtius and Boëtius 1980; Wenner and Musick 1974; Barbin and McCleave 1997). Smaller fish may be less reproductively successful due to reduced egg size and size-specific behaviors that influence egg success (Trippel 1995). If there were a sufficient reduction in the reproductive contribution from particular areas, overall egg production would likely be impacted. Because larval dispersal is random, a decline in larval production would impact the entire species range, including those areas from which the reproductive contribution of spawners was high.

The existence of demographic data from 1981 altered our perspective of eel population status in 1999. We have strong suspicions that size selective fishing mortality biased our 1999 growth estimates and YPR calculations. We recommend evaluating status of Maryland's eels using growth estimates based on areas or time periods (1981 in this case) not subjected to intense fisheries. Surveys of additional Chesapeake Bay estuaries may yield areas where size selective mortality is absent or less intense, and these should be used to estimate YPR once they become available. Until such data becomes available, we will use the 1981 data to model current YPR of lightly exploited populations. Heavily exploited populations, which are possibly the most biased, are some of those most in need of evaluation. The actual yield of these areas is less accurately described using 1981 or equivalent data. The knowledge that Maryland's tidal American eel population has changed so substantially is critical for assessing the impact of exploitation and the direction of future American eel fishery management.

Acknowledgments

We thank M. Baugh, S. Hammond, R. Lukacovia, C. Walstrum, and K. Whiteford for assistance in the field and laboratory. We also appreciate the comments by S. Owens and an anonymous reviewer on an earlier draft of this manuscript. This work would not have been possible if not for the investigations of J. Foster and R. Brody in the early 1980s. This research was supported by funds from Atlantic Coastal Fisheries Cooperative Management Act Grant Numbers NA76FG0306 and NA86FG0395.

References

Allen, M. S. 1997. Effects of variable recruitment on catch-curve analysis for crappie populations. North American Journal of Fisheries Management 17:202–205.

Anthony, V. 1982. The calculation of $F_{0.1}$: a plea for standardization. Northwest Atlantic Fisheries Organization, Serial Document SCR 82/VI/64, Halifax, Nova Scotia, Canada.

Barbin, G. P., and J. D. McCleave. 1997. Fecundity of the American eel Anguilla rostrata at 45E N in Maine, USA. Journal of Fish Biology 51:840–847.

Berg, R. 1985. Age determination of eels, Anguilla anguilla (L): Comparison of field data with otolith ring patterns Journal of Fish Biology 26:537–544.

Boëtius, I., and J. Boëtius. 1980. Experimental maturation of female silver eels, Anguilla anguilla. Estimates of fecundity and energy reserved for migration and spawning Dana 1:1–29.

Beullens, K., E. H. Eding, P. Gilson, F. Ollevier, J. Komen, and C. J. J. Richter. 1997. Gonadal differentiation, intersexuality and sex ratios of European eel (Anguilla anguilla L.) maintained in captivity. Aquaculture 153:135–150.

Castonguay, M., P. V. Hodson, C. M. Couillard, M. J. Eckersley, J.-D., and G. Verreault. 1994. Why is recruitment of the American eel, Anguilla rostrata, declining in the St Lawrence River and Gulf? Canadian Journal of Fisheries and Aquatic Sciences 51:479–488.

Chisnall, B. L., and J. M. Kalish. 1993. Age validation and movement of freshwater eels (Anguilla diffenbachii and A. australis) in a New Zealand pastoral stream. New Zealand Journal of Marine and Freshwater Research 27:333–338.

Deelder, C. L. 1976. The problem of the supernumary zones in otoliths of the European eel (Anguilla anguilla (Linnaeus, 1958)): A suggestion to cope with it. Aquaculture 9:373–379.

De Leo, G. A., and M. Gatto. 1995. A size and age-structured model of the European eel (Anguilla anguilla L.). Canadian Journal of Fisheries and Aquatic Sciences 52:1351–1367.

Gabriel, W. L., M. P. Sissenwine, and W. J. Overholtz. 1989. Analysis of spawning stock biomass per recruit: An example for Georges Bank haddock. North American Journal of Fisheries Management 9:383–391.

Foster, J. W. S. 1980. Investigations of the Maryland eel fishery, spring summer 1980. Maryland Department of Natural Resources Fisheries Service, Annapolis, Maryland.

Foster, J. W. S. 1981. The American eel in Maryland—A situation paper. Maryland Department of Natural Resources, Tidewater Administration, Tidal Fisheries Division, Annapolis, Maryland.

Foster, J. W. S., and R. W. Brody. 1981. Status Report: The American Eel Fishery in Maryland, 1981. Maryland Department of Natural Resources, Tidal Fisheries Division, Annapolis, Maryland.

Foster, J. W. S., and R. W. Brody. 1982. Status Report: The American Eel Fishery in Maryland, 1982. Maryland Department of Natural Resources, Tidal Fisheries Division, Annapolis, Maryland.

Goodwin, K. R. 1999. American eel subpopulation characteristics in the Potomac River drainage, Virginia. Master's thesis. Virginia Polytechnic University, Blacksburg, Virginia.

Gray, R., and C. W. Andrews. 1971. Age and growth of the American eel in Newfoundland waters. Canadian Journal of Zoology 49:121–128.

Helfman, G. S., D. E. Facey, L. S. Hales Jr., and E. L. Bozeman Jr. 1987. Reproductive ecology of the American eel. Pages 42–56 in Common Strategies of Anadromous and Catadromous Fishes. American Fisheries Society, Symposium 1, Bethesda, Maryland.

Hilborn, R., and C. J. Walters. 1992. Quantitative Fisheries Stock Assessment: Choice, Dynamics & Uncertainty. Chapman and Hall, New York, New York.

Hildebrand, S. F., and W. C. Schroeder. 1972. Fishes of Chesapeake Bay. Reprint of 1929 edition. T. F. H. Publications Inc. for the Smithsonian Institution, Neptune, New Jersey.

Jones, P. W., H. J. Speir, N. H. Butowski, R. O'Reilly, L. Gillingham, and E. Smoller. 1988. Chesapeake Bay Fisheries: Status, Trends, Priorities & Data Needs. Maryland Department of Natural Resources Tidewater Administration and Virginia Marine Resources Commission.

Krueger, W. H., and K. Oliveira. 1997. Sex, size, and gonad morphology of silver American eels. Anguilla rostrata. Copeia 2:415–420.

Lai, H.-L. 1987. Optimum allocation for estimating age composition using age-length key. Fishery Bulletin 85(2):179–183.

Lecomte-Finiger, R. 1992. The crystalline ultrastructure of otoliths of the eel (*A. anguilla* L. 1758). Journal of Fish Biology 40:181–190.

Liew, P. K. L. 1974. Age determination of American eels based on the structure of their otoliths. Pages 124–136 in T. B. Bagenal, editor. Proceedings of an international symposium on the aging of fish. Unwin Brothers, Surrey, England.

Oliveira, K. 1996. Field validation of annular growth rings in the American eel, *Anguilla rostrata*, using tetracycline-marked otoliths. Fishery Bulletin 94:186–189.

Oliveira, K., and J. D. McCleave. 2000. Variation in population and life history traits of the American eel, Anguilla rostrata, in four rivers in Maine Environmental Biology of Fishes 59:141–151.

Ricker, W. E. 1975. Computation and interpretation of biological statistics of fish populations. Fisheries Research Board of Canada Bulletin 191.

Secor, D. H., J. M. Dean, and E. H. Laban. 1992. Manual for otolith removal and preparation for microstructural examination. Electric Power Research Institute and Belle W. Baruch Institute for Marine Biology and Coastal Research, Belle W. Baruch Institute for Marine Biology and Coastal Research Technical Publication Number 1991–01.

Sinha, V. R. P., and J. W. Jones. 1967. On the age and growth of the freshwater eel (*Anguilla anguilla*). Journal of Zoology 153:99–117.

Sissenwine, M. P., and J. G. Shepherd. 1987. An alternative perspective on recruitment overfishing and biological reference point. Canadian Journal of Fisheries and Aquatic Sciences 44:913–918.

Tesch, F. W. 1977. The Eel: biology and management of Anguillid eels. Chapman and Hall, London.

Todd, P. R. 1981. Morphometric changes, gonad histology, and fecundity estimates in migrating New Zealand freshwater eels (*Anguilla* spp.). New Zealand Journal of Marine and Freshwater Research 15:155–170.

Trippel, E. A. 1995. Age at maturity as a stress indicator in fisheries. Bioscience 45(11):759–771.

von Bertalanffy, L. 1938. A quantitative theory of organic growth. Human Biology 10:181–213.

Weeder, J. W. 1998. Completion Report—Maryland Eel Population Study Project 3-ACA-026. Maryland Department of Natural Resources Fisheries Service, Annapolis, Maryland.

Weeder, J. W., and J. H. Uphoff, Jr. 1999. Completion Report—Maryland Eel Population Study Project 3-ACA-041. Maryland Department of Natural Resources Fisheries Service, Annapolis, Maryland.

Weeder, J. W., and J. H. Uphoff, Jr. 2000. Completion Report—Maryland Eel Population Study Project 3-ACA-055. Maryland Department of Natural Resources Fisheries Service, Annapolis, Maryland.

Weeder, J. W., and J. H. Uphoff, Jr. 2001. Completion Report—Maryland Eel Population Study Project 3-ACA-065. Maryland Department of Natural Resources Fisheries Service, Annapolis, Maryland.

Wenner, C. A., and J. A. Musick. 1974. Fecundity and gonad observations of the American eel, Anguilla rostrata, migrating from Chesapeake Bay, Virginia Journal of the Fisheries Research Board of Canada 31:1387–1391.

Effect of Harvest on Size, Abundance, and Production of Freshwater Eels *Anguilla australis* and *A. dieffenbachii* in a New Zealand Stream

BENJAMIN L. CHISNALL AND M. L. MARTIN

*National Institute of Water and Atmospheric Research Ltd,
Post Office Box 11-115, Hamilton, New Zealand*

BRENDAN J. HICKS

*Centre for Biodiversity and Ecology Research, Department of Biological Sciences,
the University of Waikato, Private Bag 3105, Hamilton, New Zealand*

Abstract.—Sympatric longfinned eels *Anguilla dieffenbachii* and shortfinned eels *A. australis* in a pastoral stream that had not been fished for more than 20 years, in the Waikato region, New Zealand, were studied over a 10-year period. Species dominance changed over the period, as the size, density, and biomass of the two species responded differently to four years of simulated commercial harvest. Geometric mean weight of longfinned eels greater than or equal to 200 mm was reduced from 597 g before harvest to 246 g after harvest, and the geometric mean weight of shortfinned eels declined from 62 to 44 g. The density of small shortfinned eels increased threefold following the removal of 26 large longfinned eels from the 600-m² reach. The density of longfinned eels was 4.4 fish/100 m² (38 g/m²) before harvest compared to 1.2 fish/100 m² (5 g/m²) one year after harvest. The density of shortfinned eels was 17 fish/100 m² (12 g/m²) before harvest compared to a maximum of 56 fish/100 m² (15 g/m²) two years after harvest. Following five years without harvest, the estimated total eel biomass had returned to preharvest levels, but in contrast to the situation before harvest when the biomass of longfinned eels was three times that of shortfinned eels, the biomass of each species in the postharvest period was about equal. Large eels, particularly longfinned eels, appeared to regulate density and structure of the resident population. Harvest appeared to cause increased recruitment of juvenile shortfinned eels to the study reach, but did not increase the total eel production. This study highlights the vulnerability of eel populations in small streams to overfishing. The large size at maturity of female shortfinned eels and longfinned eels of both sexes, and low minimum harvestable size (220 g) make these species susceptible to growth overfishing.

Introduction

Eels support the most important commercial freshwater fishery in New Zealand. Commercial exploitation of wild longfinned eels *Anguilla dieffenbachii* and shortfinned eels *Anguilla australis* takes place alongside an indigenous Maori customary fishery. The national commercial catch is approximately 1200 metric tons annually (Jellyman 1993; Annala et al. 2000), but there are no estimates of the customary catch. The commercial fishery is managed by limited entry, restrictions on equipment, and minimum size of eels (220 g).

Streams are a significant source of wild eels for the commercial fishery, but little is known about the impact of repeated fishing on eel populations. In the North Island, eel catches from streams comprised 40% of the landed eel weight between 1995 and 1998 (22% out of 116 landings sampled; Chisnall and Kemp 1998). Eel populations in streams and rivers have been investigated in New Zealand (Burnet 1952; 1969; Hopkins 1971; Hicks and McCaughan 1997) and elsewhere (Rasmussen and Therkildsen 1979; Naismith and Knights 1993; Lobón-Cerviá et al. 1990), but most studies have been of short duration, conducted in larger waters, and often in sites exposed to commercial fishing.

Our study was carried out in a pastoral tributary of the Ahirau Stream in lowland Waikato over a 10-year period. Age and growth have been previously estimated and validated for this population (Chisnall and Hayes 1991; Chisnall and Hicks 1993; Chisnall and Kalish 1993). Resident tagged eels were shown to have limited home ranges over one to three years (Chisnall and Kalish 1993).

The objectives of our study were to investigate the abundance, size structure, and biomass of shortfinned and longfinned eels in a stream that had not been fished for more than 20 years, and to monitor changes during four years of harvest, followed by five years with no harvest. We compared the populations before and after harvest to test the hypotheses that: removal of commercial-sized eels will increase the recruitment of juveniles of both shortfinned and longfinned eels; and removal of commercial-sized eels will increase eel production because of the increased resources available to small eels.

Methods

Study Area

The study reach was located in a tributary of the Ahirau Stream in the Waikato region, New Zealand. The Ahirau Stream drains incised pastoral hill country, and is a tributary of the Waipa River and the Waikato River (Figure 1). The basin area upstream of the study reach is 1.44 km².

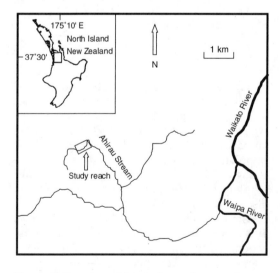

Figure 1. Location of the study reach on a tributary of the Ahirau Stream, Waikato region, North Island, New Zealand.

The study reach was 469 m long, with a mean width of 1.38 m, a mean depth of 0.26 m, and surface area of 597 m². The channel structure was predominantly pools and runs, with riffles comprising only 10% of the total reach length (Table 1). The downstream end of the reach was 5 m upstream from the confluence with another minor tributary and about 50 km upstream from the ocean. The study reach had about 240 m of low-gradient, meandering channel downstream, with about 230 m of steeper gradient channel upstream, culminating at a 1.5-m waterfall at its upstream end. The study stream had not been commercially fished for at least 20 years before the study began (G. McLoughlin, landowner, personal communication).

The channel gradient measured from a 1:50,000 NZMS 260 topographical (map number S14; Department of Lands and Survey 1979) was 0.025 m/m. Stream discharge was 49 L/s during settled weather in spring (25 September 1991), and was typical of the sampling occasions. Temperature was not measured during the study, but similar-sized pastoral streams in the region had a mean temperature of about 15°C, with maxima of greater than 22°C between November and early December (Quinn et al. 1997).

Sampling Plan

The study reach was divided into 21 sites, each approximately 20 m in length. Most downstream was site 1. A combination of fyke netting and electroshocking was used to catch eels throughout the study period. Two unbaited fyke nets (one 6-mm and one 12-mm stretched-mesh), each with a single 3-m long leader, were set at either end of each 20-m site. The fyke nets were set with their leaders angled upstream towards the bank so that the fyke net blocked the channel, and eels could enter only from upstream of the net. The nets were left to fish overnight, and then left in place while the stream was electroshocked in a single pass to catch eels that had not been caught by the fyke nets. Electroshocking was carried out using a 300–W generator-powered bank-mounted pulse unit connected to the single anode with a 200-m cable. Captured eels were anesthetized with benzocaine, weighed, and measured for total length.

The study reach was fished in late August or in early September from 1988 to 1992, and in 1995, 1997, and 1998. In 1988, only eels greater than or equal to 220 g were measured,

Table 1. Physical attributes of the study reach in a tributary of the Ahirau Stream, Waikato region, measured at a discharge of 49 L/s on 25 September 1991.

Channel unit type	Number of channel units	Total length (m)	Total area (m²)	Mean length (m)	Mean depth (m)	Mean width (m)	Mean area (m²)
Pools	22	157	253	7.1	0.36	1.62	11.5
Runs	19	273	289	14.4	0.20	1.06	15.2
Riffles	7	39	55	5.6	0.23	1.36	7.8
Sum	48	469	597				
Grand mean					0.30	1.36	

but all eels caught were replaced. In 1989, all eels caught were removed to fully assess the size and age structure of the population before simulated commercial harvest began. In 1990–1992, eels less than 220 g (the minimum commercial size) were replaced but eels greater than or equal to 220 g were removed, and in 1995, 1997, and 1998, all eels caught were replaced. During years of simulated harvest (1990–1992), eels less than 200 mm had only species and total counts recorded. In 1998, only eels greater than or equal to 220 g were measured and replaced, to assess the recovery of the harvestable biomass in the study reach. Eels that were replaced were released back to their original site of capture (Table 2).

The biomass of shortfinned eels less than 200 mm (juveniles) in 1990–1992 was calculated by multiplying the number of individuals by the mean weight of the same size-class in 1995 and 1997 (6.42 g). Production was estimated by the methods of Hicks and McCaughan (1997), using the measured size frequency to estimate biomass increment, and the growth models obtained from eels removed from the study reach in 1989 (Chisnall and Hayes 1991; Chisnall and Hicks 1993). These growth models apply only to the length and weight ranges analyzed. Weights of eels were calculated from measured lengths using weight-length regressions from data for 1988–1990 and 1998 (Table 3), except for eels measured during years of harvest. To confirm growth estimates, eels greater than or equal to 220 g were tagged with individually coded steel tags clipped to the base of the right pectoral fin in 1997; seven of these eels were recaptured in 1998.

Geometric mean weights of eels were calculated from log-transformed data (Elliot 1983). Geometric means were used to compare weights because of the strongly lognormal distributions of the individual weights.

Table 2. Biomass and density of commercially harvestable shortfinned and longfinned eels (≤ 220 g) caught by fyke netting and electroshocking between 1988 and 1998 in the 470-m study reach of a tributary of the Ahirau Stream, Waikato region.

Year	Biomass (g/m²)			Density (fish/100 m²)			Harvest condition	Treatment
	Short-finned	Long-finned	Total	Short-finned	Long-finned	Total		
1988	2.30	26.8	29.1	0.67	3.35	4.02	Pre-harvest	All eels replaced
1989	2.48	37.1	39.5	0.67	3.69	4.36	Pre-harvest	All eels removed
1990	1.88	5.1	7.0	0.67	0.84	1.51	Harvest	Eels = 220 g removed
1991	2.54	14.6	17.1	0.84	1.84	2.68	Harvest	Eels = 220 g removed
1992	1.00	6.4	7.4	0.34	1.01	1.34	Harvest	Eels = 220 g removed
1995	4.87	22.7	27.6	1.68	2.01	3.69	Post-harvest	All eels replaced
1997	2.61	22.9	25.5	0.67	2.35	3.02	Post-harvest	All eels replaced
1998	4.01	20.3	24.3	0.67	1.84	2.51	Post-harvest	All eels replaced

Table 3. Weight-length and length-age regressions for shortfinned and longfinned eels captured from the 470-m study reach of a tributary of the Ahirau Stream, Waikato region. n = sample size, a = Y intercept, b = line slope for the linear regression models. $P < 0.001$ for all regression models.

Species	n	a	b	r^2	Data ranges	Source
Ln(weight)-ln(length)						
Shortfinned	1191	−14.83	3.28	0.98	102–694 mm, 1.4–814 g	This study
Longfinned	132	−14.60	3.28	0.98	163–1095 mm, 8.3–4191 g	This study
Length-age						
Shortfinned	85	153.6	17.1	0.50	209–666 mm, 5–18 years	Chisnall and Hayes (1991)
Longfinned	29	58.7	35.8	0.79	231–1095 mm, 4–25 years	Chisnall and Hicks (1993)

Results

Biomass and Density

From weight-length regressions, the length at the minimum commercially harvestable weight (220 g) was 475 mm for shortfinned eels, and 443 mm for longfinned eels. Longfinned eels dominated the commercially harvestable catch by weight and number throughout the study period, and their biomass recovered rapidly from intensive harvest through recolonisation by nonjuveniles (eels ≥ 200 mm; Table 2, 1991). In the preharvest period, the biomass of commercially harvestable eels (≥ 220 g) was 29–40 g/m², of which greater than or equal to 90% were longfinned eels (Table 2). During the harvest period, the commercially harvestable eel biomass was reduced to 7–17 g/m², 73–86% of which were longfinned eels. In the postharvest period, the biomass of harvestable eels had increased almost to preharvest levels (24–28 g/m²), 82–90% of which were again longfinned eels. There was no difference between the biomass of shortfinned eels in the harvest, and postharvest periods (P = 0.064; analysis of variance [ANOVA] $F_{1,4}$ = 6.45), but biomass of longfinned eels was significantly reduced during the harvest period (P = 0.013; ANOVA $F_{1,4}$ = 18.60). In the postharvest period, biomass was similar to preharvest levels.

Densities of commercially harvestable eels were always greater for longfinned eels than for shortfinned eels, and there were never more than 26 eels of both species greater than or equal to 220 g in the entire 600-m² reach (Table 2). Before harvest, total density was 4.0–4.4 fish/100 m², compared to 1.3–2.7 fish/100 m² during harvest, and 2.5–3.7 fish/100 m² more than three years after harvest (Table 2). There was no difference in density between the harvest and postharvest periods for either shortfinned eels (P = 0.346; ANOVA $F_{1,4}$ = 1.14), or for longfinned eels (P = 0.072; ANOVA $F_{1,4}$ = 5.92).

Comparing eels of all sizes before harvest, longfinned eels dominated the total biomass of 50 g/m², and formed 76% in 1989 (Table 4). However, species dominance changed over the 10-

Table 4. Biomass and density of all shortfinned and longfinned eels caught by fyke netting and electroshocking between 1989 and 1997 in the 470-m study reach of a tributary of the Ahirau Stream, Waikato region.

Year	Biomass (g/m²)			Density (fish/100 m²)		
	Shortfinned	Longfinned	Total	Shortfinned	Longfinned	Total
1989	12	38	50	17	4.4	22
1990	15	5	20	25	1.2	26
1991	22	15	36	56	2.5	59
1992	18	7	25	49	1.5	50
1995	27	23	50	51	2.5	53
1997	21	23	44	45	3.0	48

year period, as the density and biomass of the two species responded differently to harvest. The density of small shortfinned eels increased threefold following the removal of 26 large longfinned eels from the 600-m² reach (Figure 2, 1989). The density of longfinned eels of all sizes was 4.4 fish/100 m² (38 g/m²) before harvest compared to 1.2 fish/100 m² (5 g/m²) one year after harvest (Table 4). The density of shortfinned eels was 17 fish/100 m² (12 g/m²) before harvest compared to a maximum of 56 fish/100 m² (15 g/m²) two years after harvest. Following five years without harvest, the total biomass had returned to preharvest levels, but in contrast to the situation before harvest, the biomass of both species was about equal. The longfinned eel density had increased to 3.0 fish/100 m² (23 g/m²) and the shortfinned eel density had declined to 45 fish/100 m² (21 g/m²).

Size Structure

The size structure of the eel populations before harvest in 1989 (when all eels were measured) showed that longfinned eels were much larger than shortfinned eels (Figure 2). Correspondingly, the mean weights of longfinned eels were always much heavier than shortfinned eels (Figure 3). For eels greater than 200 mm, the geometric mean weight of longfinned eels (246–597 g) was on average about six times greater than geometric mean weights for shortfinned eels (44–62 g). For the three years of response to harvest the geometric mean weights of longfinned eels was 246–365 g, compared to 597 g before harvest, and 512–543 g three to five years after the cessation of harvest. The mean weights increased back to preharvest levels because of recolonisation of the reach by nonjuvenile longfinned eels. However, there were no statistically significant differences among years ($P = 0.553$; ANOVA $F_{6,107} = 0.85$ for natural-log transformed weights; Figure 2), probably because of the small sample sizes. The mean weights of shortfinned eels declined over the study period; means in 1992 and 1997 were significantly lower than in 1989 ($P < 0.001$; ANOVA $F_{6,954} = 7.72$ for natural-log transformed weights).

Before harvest (shown by the 1989 catch), there were few juvenile shortfinned eels less than 200 mm and no juvenile longfinned eels. After removal of all eels in 1989, the number of shortfinned eels of all sizes increased, but in 1991 there was a dramatic increase of juvenile

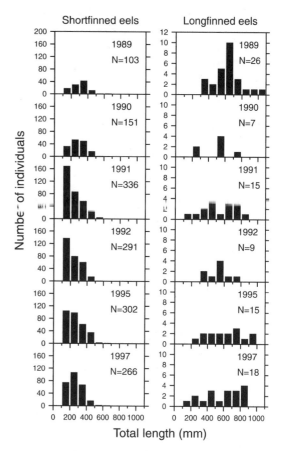

Figure 2. Length-frequency distribution of shortfinned and longfinned eels, captured in a 470-m (600-m²) reach of a tributary of Ahirau Stream, between 1989 and 1997. Eels were harvested from 1989 to 1992, but not in 1995 or 1997. Note difference in vertical scales between species.

shortfinned eels less than 200 mm (Figure 2). This pattern persisted in 1992. There was no corresponding increase in the number of juvenile longfinned eels; just one longfinned eel less than 200 mm was found in the study reach between 1989 and 1997 (Figure 2). In 1995 and 1997 the size structure of shortfinned eels had nearly returned to its preharvest form, and the size of longfinned eels had increased. During these changes to the shortfinned eel population, the longfinned eel population fluctuated at low numbers, only showing reasonable numbers of large eels in 1997 (> 600 mm, $n = 10$, compared with $n = 17$ in 1989).

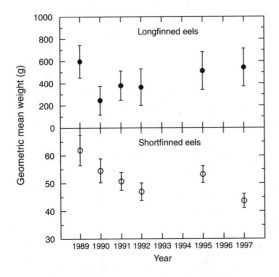

Figure 3. Geometric mean weights of shortfinned and longfinned eels greater than or equal to 200 mm in a 470-m (600-m²) reach of a tributary of Ahirau Stream between 1989 and 1997. Results for 1990–1992 show the response of weight to harvest between 1989 and 1992. Vertical lines are 1 standard error (Elliot 1983).

Table 5. Area of pools, runs, and riffles in four site groups of the 470-m study reach of a tributary of the Ahirau Stream, Waikato region, measured at a discharge of 49 L/s on 25 September 1991.

Habitat type	Area (m²)	Percent of area
Sites 1–5		
Pool	80	54
Run	62	43
Riffle	4	3
Sites 6–10		
Pool	64	47
Run	71	53
Riffle	0	0
Sites 11–15		
Pool	34	26
Run	97	74
Riffle	0	0
Sites 16–21		
Pool	76	39
Run	59	30
Riffle	62	31

Distribution of Eels Within the Reach

Habitat structure differed between downstream sites (1–5) and upstream sites (16–21) within the study reach (Table 5). There was more pool habitat and less riffle habitat downstream compared to upstream. Eel biomass and density was greater at downstream sites than at upstream sites apparently in response to these differences in habitat (Figure 4). At sites 1–5, total eel biomass before harvest in 1989 was almost 100 g/m². Following harvest (shown by the catches from 1990 to 1992), there was a trend of decreasing density of shortfinned eels greater than 200 mm with increasing distance upstream. In 1991 and 1992, the smaller shortfinned eels showed the same pattern, reaching a maximum density of 51 fish/100 m² in sites 1–5 in 1991. For the shortfinned eels greater than 200 mm, this pattern was still apparent in 1997, long after the pulse of juvenile shortfinned eels (< 200 mm) first seen in 1991 had declined. Shortfinned eels recruited to the study reach primarily as juveniles, so they presumably came as migrants from downstream.

Before harvest (1989 catch), the biomass of longfinned eels was correlated with percentage pool habitat at each site (Spearman's $r = 0.63$, $P = 0.002$, $n = 21$). The largest biomass of longfinned eels was at sites 1–5, which had the greatest area of pools. During harvest (1990–1992), the biomass of longfinned eels among the sites was unpredictable because of the low numbers of individuals. Most longfinned eels entered the reach as nonjuveniles greater than 200 mm long, and could have come from upstream or downstream of the study reach.

Production and Growth

Production of shortfinned eels increased after the removal of large longfinned eels (Table 6A). Before harvest, total production for both species was 16 g × m⁻² × year⁻¹, with longfinned eels contributing 77% of this production. During harvest (1990–1992), total production fell to 6.8–12.2 g × m⁻² × year⁻¹, of which 31–43% was attributable to longfinned eels. After harvest (1995 and 1997), production had returned to preharvest levels (15–16 g × m⁻² × year⁻¹), but only 45–52% was attributable to longfinned eels.

The production (P) to total biomass (B) ratio was fairly constant throughout the preharvest

Table 6. Production estimates and ratio of production to total eel biomass (P/B) in the 470-m study reach of a tributary of Ahirau Stream, Waikato region.

A. Production of all sizes combined.

	Production ($g \times m^{-2} \times year^{-1}$)			
Year	Short-finned	Long-finned	Total	P/B
1989	3.6	12.3	15.9	0.32
1990	4.7	2.1	6.8	0.33
1991	6.9	5.3	12.2	0.34
1992	5.0	2.8	8.7	0.35
1995	8.7	7.2	15.9	0.32
1997	7.1	7.7	14.8	0.34

B. Production of eels below and above the commercial minimum size (220 g).

	Production ($g \times m^{-2} \times year^{-1}$)			
	Shortfinned		Longfinned	
Year	< 220g	= 220g	< 220g	= 220g
1989	3.1	0.5	0.5	11.7
1990	4.3	0.4	0.1	1.9
1991	6.4	0.5	0.3	5.1
1992	5.6	0.2	0.4	2.4
1995	7.6	1.1	0.3	6.9
1997	6.6	0.5	0.2	7.4

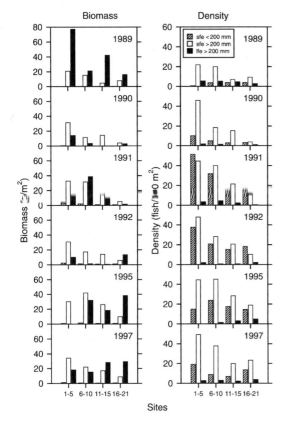

Figure 4. Biomass and density of shortfinned (sfe) and longfinned (lfe) eels in four site groups captured in a 470-m reach of a tributary of Ahirau Stream, Waikato region, between 1989 and 1995. Sites were numbered from downstream (site 1) to upstream (site 21).

harvest, and postharvest phases (P/B = 0.32 – 0.35; Table 6A), and the relationship of production to total biomass was curvilinear, described by the quadratic

$$P = -2.75 + 0.535B - 0.00323B^2.$$

For this relationship, r^2 = 0.998, n = 6, and P is less than 0.001. For longfinned eels, fish above the legal size limit (220 g) contributed most ($\geq$ 95%) to the production, whereas for shortfinned eels, fish below the legal size limit contributed most (86–96%) to the production (Table 6B). Harvest had little effect on the distribution of production between these size classes.

For production estimates, we assumed that growth remained linear with increasing eel size, which has been confirmed from mark–recapture study of this population (Chisnall and Kalish 1993). This assumption is further supported by the rapid growth of tagged large eels in the stream from 1997 to 1998 (Table 7).

Although sex was not actively determined in our study, there were two visually apparent shortfinned males caught in 1992 (122 g and 133 g in weight, and 400 mm each in length). These two eels exhibited mature morphological features.

Discussion

Response to Harvest

We made two assumptions in our study that require examination before we can fairly interpret the changes that we saw as responses to harvest. Firstly, we assumed that the combination of fyke netting and electroshocking that we

Table 7. Growth of individually tagged eels in the 470-m study reach of a tributary of Ahirau Stream, Waikato region.

Size in 1997		Size in 1998		Annual increment	
Length (mm)	Weight (g)	Length (mm)	Weight (g)	Length (mm)	Weight (g)
Longfinned					
743	1305	762	1421	19	116
659	672	720	1380	38	271
732	1109	770	1380	38	271
800	1538	839	1913	39	375
730	1099	768	1415	38	316
865	1765	880	1870	15	105
Shortfinned					
592	475	630	838	38	363

used resulted in accurate information about the density, biomass, and size distribution. Fyke nets selectively capture larger eels (Berg 1990), which can be difficult to extract from deep cover by electroshocking. However, on the whole eels are very susceptible to electroshocking, and this was the method by which most of the eels in this study were caught. The first electroshocking pass in pasture streams can catch 83% of the total eel biomass present (Hicks, unpublished data from Hicks and McCaughan 1997), and electroshocking has been widely used to assess eel populations overseas (Rasmussen 1983; Lobón-Cerviá et al. 1990). The fyke nets also limited movement between sites during electroshocking, and trapped any eels that tried to escape downstream. The use of a combination of electroshocking and fyke netting is a recommended capture technique for eels (Naismith and Knights 1990). The combined methods we used were likely to be more efficient than using fyke nets alone as are typically used by commercial harvesters. Thus our removals represent maximum effects of harvest rather than true simulation of commercial harvest.

Secondly, we did not have a control to evaluate the parallel changes in an unharvested stream. This was due partly to limited resources, and partly to the variable habitat in pastoral streams in the area. Because each stream is unique in its physical characteristics, the concept of a control has limited use in stream ecology. To assess the effectiveness of our eel capture by netting combined with single-pass electroshocking, we can compare our results to other studies of unharvested streams in the area. Our preharvest estimates in 1989 of 50 g/m^2 and 22 fish/100 m^2 (Table 4) are within the range of eel biomass estimates for other Waikato pasture sites (22–168 g/m^2), but less than the eel density estimates in the same streams (107–156 fish/100 m^2; Hicks and McCaughan 1997). An earlier estimate of eel biomass in the nearby pastoral Kaniwhaniwha Stream was 58 g/m^2 (Burnet 1952). In Hawkes Bay rivers, Rowe et al. (1999) estimated that the mean densities of both eels species at pasture sites was 271 fish/100 m^2, with a mean biomass of 88 g/m^2. The length-frequency distribution found by Hicks and McCaughan (1997), who sampled by removal electroshocking in short reaches (46–94 m), suggests that while our biomass estimates are probably accurate, juvenile eels may have been underrepresented in our samples (Figure 2). In addition, the first electroshocking pass in pasture streams caught 74% of the total number of eels (Hicks, unpublished data from Hicks and McCaughan 1997). The use of identical capture methods in all years means that our results can be compared between years, and can, we believe, be interpreted as a response to harvest. However, we cannot exclude the possibility that in part the changes that we observed over the study period were due to variable recruitment. Jellyman et al. (1999) have shown that young-of-the-year-eels caught by electroshocking in streams can show 10-fold variations in abundance between successive years.

We found evidence that partly supports our first hypothesis: The density of juvenile shortfinned eels increased dramatically following the removal of the larger eels, and three to five years after the cessation of fishing, the preharvest size frequency distributions were becoming reestablished. However, contrary to our first hypothesis, recruitment of juvenile longfinned eels did not increase following harvest. The increased abundance of small shortfinned eels following harvest of commercially harvestable eels (primarily longfinned eels) implies that the recruitment of juveniles was limited by the presence of larger individuals. Increased abundance of small fish following harvest of larger individuals is consistent with some previous studies (*A. anguilla*, Moriarty and Nixon 1990; Lobón-Cerviá et al. 1990, 1995; shortfinned eels in Lake Ellesmere, Jellyman 1992; white suckers, Johnson 1994).

Eels greater than 500 mm are primarily piscivorous (Moriarty 1972; Ryan 1986; Jellyman 1989; Barak and Mason 1992). Survival of eels can also be density-dependent (e.g., *A. anguilla*, Vollestad and Jonsson 1988). Eels become more aggressive and territorial as they grow (Knights 1987), and from our study, large eels appear to regulate population density and structure. The mechanism of regulation is probably a combination of habitat segregation, as seen in *A. australis* and *A. reinhardtii* in Australia (Sloane 1984), and cannibalism, e.g., in *A. anguilla* (Sinha and Jones 1967). Cannibalism was observed in longfinned eels greater than or equal to 220 g that were sacrificed for aging in 1989 (Chisnall, unpublished data from Chisnall and Hicks 1993). Four out of 22 eels had eaten 1–3 smaller eels. Because of the slow recruitment rate of longfinned eels, re-establishment of the species ratios and size frequency distributions observed in 1989 might take a few more years.

Production and Growth

Contrary to our second hypothesis, total eel production did not increase during the harvest period (1990–1992; Table 6A). Despite the large influx of small shortfinned eels, total production was lower during the harvest period than the preharvest period (1989) and for the postharvest period (1995 and 1997). However, production of shortfinned eels less than 220 g was increased following harvest. Because eels in our study continued to increase in length at the same rate regardless of size, the largest eels are the most productive, and their removal by harvest lowers total eel production.

Our preharvest estimate of total production (16 g × m^{-2} × year^{-1}) was very similar to other pastoral Waikato streams (mean 18 g × m^{-2} × year^{-1}; Hicks and McCaughan 1997), and within a previous range of estimates for the lower North Island, New Zealand (10–59 g × m^{-2} × year^{-1}; Hopkins 1971). Estimates of eel production in New Zealand streams are greater than estimates for the English Severn River catchment (Aprahamian 1986) and a small Danish stream (Rasmussen and Therkildsen 1979).

Production to biomass ratios in our study were virtually unchanged by harvest, but our mean ratio (0.33) was greater than the mean ratio estimated by Hicks and McCaughan (1997; $P/B = 0.23$). Our ratio was also larger than Norwegian estimates for *A. anguilla* (e.g., 0.23, Vollestad and Jonsson 1986), but similar to English estimates (e.g., 0.37, Aprahamian 1986).

Longfinned eels in this stream grew relatively quickly, and at twice the rate of sympatric shortfinned eels (Table 3). Our estimates of annual length increment for longfinned eels (36 mm/year from regression; Table 3) were greater than the 26 ± 5 mm/year (mean ± 95% confidence interval) estimated by Hicks and McCaughan (1997) for other pastoral streams in the same area, but the estimates for shortfinned eels (17 mm/year; Table 3) we used were half their estimates (30 ± 3 mm/year). These differences are probably attributable to variability of the growth conditions among streams (Jellyman 1997).

Sexually dimorphic growth may have influenced our findings. Many studies have discussed apparent sexual dimorphic growth being variously attributed to variation in sex ratios, population density and geographic occurrence (Sinha and Jones 1967; Helfman et al. 1987; Rossi et al. 1987, 1988; De Leo and Gatto 1996; Pool and Reynolds 1996; Holmgren et al. 1997). Either sex can be cited as growing faster than the other, but most often females faster than males. Admittedly, having more potential females than males included in the longfinned eel data set, and vice versa in the shortfinned eel data set, may have biased the age models used in our study based on eels removed from the study reach in 1989.

As sex was not determined, we did not examine the influence of sex ratios in our study. However, given that large (characteristically female) longfinned eels tended to dominate the biomass throughout the study, we contend that successful competition for food and space by large females in streams will tend to define the population density and structure, and thereby influence sex ratios too. The latter may account for the so-called influence of river habitat on sex determination concluded by Oliveira et al. (2001).

The minimum size of shortfinned eels in January in 10 Waikato streams was 101 ± 5 mm (mean ± 95% confidence interval; unpublished data from Hicks and McCaughan 1997). Assuming that juveniles of both species recruit to the study stream at 100 mm in length, then longfinned eels would take about 10 years to reach harvestable size (220 g, 443 mm) compared to 22 years for shortfinned eels to reach 220 g, or 475 mm.

Male shortfinned eels emigrate at a mean length of 432–465 mm (Todd 1980), which equates to a weight range of 160–204 g using our weight-length regressions (Table 2). Thus, most male shortfinned eels will have left the stream before they reach harvestable size. The only two mature male shortfinned eels caught in this study confirm this, as they weighed 122–133 g, and were about 400 mm long. Female shortfinned eels and both sexes of longfinned eels are much more vulnerable to harvest because they migrate at greater sizes and ages than male shortfinned eels. Using Todd's (1980) estimates of mean length at migration, female shortfinned eels entering the fishery at 220 g and growing at 17 mm/year would take would take 8–17 years to grow from 475 mm to 609–764 mm. Female longfinned eels entering the fishery at 220 g and growing at 36 mm/year would be vulnerable to harvest for 17–20 years as they grew from 475 mm to 1063–1156 mm.

Recolonisation by Nonjuveniles

Large eels of both species were present in each year of harvest, indicating that all sizes of eels re-invaded the study reach. Removal of the population in 1989 probably exaggerated the recolonising observed in subsequent years. Although several earlier eel studies have indicated that upstream migration ceases after about 300 mm in length (e.g., *A. anguilla*, Naismith and Knights 1988), other work has shown upstream colonising movement of nonjuveniles (> 350 mm e.g., Naismith and Knights 1993). Other sources of nonjuveniles in our study include the downstream movement of sexually mature eels, and the downstream displacement of upstream residents (feeding eels) through floods. However, it would seem that unharvested populations remain generally immobile from mark–recapture studies undertaken in this reach that showed localized movement of tagged eels (343–828 mm, Chisnall and Kalish 1993).

Management Implications

In small New Zealand streams, wild stocks of eels are distinctly vulnerable to the effects of overfishing because of the large sizes of individuals in unharvested populations at equilibrium and the long time taken to reach maturity. Longfinned eels are more vulnerable to overfishing than shortfinned eels, and females are more vulnerable than males. Simulated commercial harvest of eels greater than or equal to 220 g, the minimum legal size, exceeded production estimates in two out of four years of harvest. There were only 8–26 commercially harvestable individuals, equaling 4–24 kg of biomass, in the entire 470-m reach in any one year.

Because of the size-selective nature of the fishery (commercial and customary) and the smaller size at maturity for shortfinned eels, harvest reduces the density of longfinned eels more than shortfinned eels. Longfinned eels dominated the preharvest biomass in our pastoral stream (Table 4), confirming previous findings that longfinned eels predominated the biomass of eels in unfished streams in the Waikato (Chisnall and Hicks 1993; Hicks and McCaughan 1997). We conclude that longfinned eels probably once dominated eel populations throughout the Waikato, but that a combination of eel fishing, land-use conversion from native forest to pasture, accompanied by wetland drainage, has increased shortfinned eel abundance at the expense of longfinned eels.

Commercial fishing has been shown to selectively remove longfinned eels more than shortfinned eels, and females more than males, e.g., in Lake Ellesmere, South Island, New Zealand (Jellyman 1992; Jellyman et al. 1995) and in the Waikato River hydro-electric lakes (Chisnall et al. 1998). Unsubstantiated evidence from Waikato Maori and staff at North Island eel processing plants is consistent with the widespread replacement of longfinned eels by shortfinned eels. In the 1960s, eel populations in the Waikato River basin had substantial numbers of longfinned eels, but are now predominated by shortfinned eels.

The increase in production of shortfinned eels less than 220 g in response to harvest of commercial-sized eels in pastoral streams, is of little benefit to the fishery, because most shortfinned eels will become males and migrate before attaining harvestable size. With continued commercial fishing pressure on wild eel populations, longfinned eels are mostly unable to attain sufficient size to dominate the population, profoundly affecting the eel population ecology in favor of shortfinned eels. Without better protection of broodstocks, such as through substantial increase to the minimum size, or the introduction of maximum size limits, the long-term future of the endemic longfinned eel is uncertain.

Acknowledgments

We especially thank the landowners G. McLaughlin and family, for access to the study stream and for as far as practical, preventing fishing of the study reach. We are also grateful to Eric Graynoth and Gordon Glova, for their thoughtful review of the manuscript before submission, as well as the technical comments of Bruce Pease and two anonymous reviewers on an earlier draft of this manuscript. Thanks also go to those who assisted with sampling over the 10 years of study, especially John Hayes, Bob Van Boven, Theo Stephens, Dave West, and Ralph Morse. Recent sampling and the preparation of this manuscript, was funded by Foundation for Research, Science and Technology (New Zealand) Contract CO1605.

References

Annala, J. H., K. J. Sullivan, and C. J. O'Brien. 2000. Report from the Fishing Assessment Plenary, May 2000: Stock assessments, and yield estimates. Unpublished report held in the National Institute of Water, and Atmosphere library, Wellington, New Zealand.

Aprahamian, M. W. 1986. Eel (*Anguilla anguilla* L.) production in the River Severn, England. Polskie Archiwum. Hydrobiologii 33:373–389.

Barak, N. A.-E., and C. F. Mason. 1992. Population density, growth and diet of eels, *Anguilla anguilla* L., in two rivers in eastern England. Aquaculture and Fisheries Management 23:59–70.

Berg, R. 1990. The assessment of size-class proportions and fisheries mortality of eel using various catching equipment. Internationale Revue der Gesamten Hydrobiologie 75:775–780.

Burnet, A. M. R. 1952. Studies on the ecology of the New Zealand longfinned eel, *Anguilla dieffenbachii* Gray. Australian Journal of Marine and Freshwater Research 3:32–63.

Burnet, A. M. R. 1969. The growth of New Zealand freshwater eels in three Canterbury streams. New Zealand Journal of Marine and Freshwater Research 3:376–384.

Chisnall, B. L., and J. W. Hayes. 1991. Age and growth of shortfinned eels (*Anguilla australis*) in the lower Waikato Basin, North Island, New Zealand. New Zealand Journal of Marine and Freshwater Research. 25:71–80.

Chisnall, B. L., and B. J. Hicks. 1993. Age and growth of longfinned eels (*Anguilla dieffenbachii*) in pastoral and forested streams in the Waikato River basin, and in two hydro-electric lakes in the North Island, New Zealand. New Zealand Journal of Marine and Freshwater Research. 27:317–332.

Chisnall, B. L., and J. M. Kalish. 1993. Age validation and movement of freshwater eels (*Anguilla diffenbachii* and *A. australis*) in a New Zealand pastoral stream. New Zealand Journal of Marine and Freshwater Research. 27:333–338.

Chisnall, B. L., and C. Kemp. 1998. Size, age, and species composition of commercial eel catches from North Island market sampling (1997-1998). NIWA Research Report for the Ministry of Fisheries. EEL9701. National Institute of Water and Atmospheric Research, Hamilton, New Zealand.

Chisnall, B. L., M. P. Beentjes, J. A. T. Boubée, and D. W. West. 1998. Enhancement of the New Zealand eel fishery by elver transfers. New Zealand Fisheries Technical Report 37. National Institute of Water and Atmospheric Research, Hamilton, New Zealand.

Department of Lands and Survey. 1979. New Zealand topographical map 1:50000 NZMS260 S14. Department of Lands and Survey, Wellington, New Zealand.

De Leo, G. A., and M. Gatto. 1996. Trends in vital rates of the European eel: Evidence for density dependence? Ecological Applications. 6:1281–1294.

Elliot, J. M. 1983. Some methods for the statistical analysis of samples of benthic invertebrates. Freshwater Biological Association, Scientific Publication No. 25. Windermere Laboratory, The Ferry House, Ambleside, Cumbria, United Kingdom.

Helfman, G. S., D. E. Facey, L. S. Hales, JR., E. L. Bozeman, JR. 1987. Reproductive ecology of the American Eel. Pages 42–56 in M. J. Dadswell, R. J. Klauda, C. M. Moffitt, R. L. Saunders, R. A. Rulifson, and J. E. Cooper, editors. Common strategies of anadromous and catadromous fishes. American Fisheries Society, Symposium 1, Bethesda, Maryland.

Hicks, B. J., and H. M. C. McCaughan. 1997. Land use, associated eel production, and abundance of fish and crayfish in streams in Waikato, New Zealand. New Zealand Journal of Marine and Freshwater Research. 31:635–650.

Holmgren, K., H. Wickstrom, and P. Clevestam. 1997. Sex-related growth of European eel, *Anguilla anguilla*, with focus on median silver eel age. Canadian Journal of Fisheries and Aquatic Sciences 54:2775–2781.

Hopkins, C. L. 1971. Production of fish in two small streams in the North Island of New Zealand. New Zealand Journal of Marine and Freshwater Research 5:280–290.

Jellyman, D. J. 1989. Diet of two species of freshwater eel (*Anguilla* spp.) in Lake Pounui, New Zealand. New Zealand Journal of Marine and Freshwater Research 23:1–10.

Jellyman, D. J. 1992. Lake Ellesmere - an important fishery with an uncertain future. Freshwater Catch 48:3–5.

Jellyman, D. J. 1993. A review of the fishery for freshwater eels in New Zealand. New Zealand Freshwater Research Report Number 10. National Institute of Water and Atmospheric Research, Christchurch, New Zealand.

Jellyman, D. J. 1997. Variability in growth rates of eels (*Anguilla*spp.) in New Zealand. Ecology of Freshwater Fish 6:108–115.

Jellyman, D. J., B. L. Chisnall, B. L. Bonnett, and J. R. E. Sykes. 1999. Seasonal arrival patterns of juvenile freshwater eels (*Anguilla* spp.) in New Zealand. New Zealand Journal of Marine and Freshwater Research. 33:233–248.

Jellyman, D. L., B. L. Chisnall, and P. R. Todd. 1995. The status of eel stocks of Lake Ellesmere. NIWA Science and Technology Series 26. National Institute of Water and Atmospheric Research, Christchurch, New Zealand.

Johnson, L. 1994. Pattern and process in ecological systems: a step in the development of a general ecological theory. Canadian Journal of Fisheries and Aquatic Sciences 51:226–246.

Knights, B. 1987. Agonistic behaviour and growth in the European eel *Anguilla anguilla* L., in relation to warm water-aquaculture. Journal of Fish Biology 31:265–276.

Lobón-Cerviá, J., Y. Bernat, and P. A. Rincon. 1990. Effects of eel (*Anguilla anguilla* L.) removals from selected sites of a stream on its subsequent densities. Hydrobiologia 206:207–216.

Lobón-Cerviá, J., C. G. Utrilla, and P. A. Rincon. 1995. Variations in the population dynamics of the European eel *Anguilla anguilla* (L.) along the course of a Cantabrian river. Ecology of Freshwater Fish 4:17–27.

Moriarty, C. 1972. Studies of the eel *Anguilla anguilla* in Ireland 1. In the lakes of the Corrib System. Irish Fisheries Investigation Series A 10:3–40.

Moriarty, C., and E. Nixon. 1990. Eel biomass in a small lowland stream. Internationale Revue der Gesamten Hydrobiologie 75:781–784.

Naismith, I. A., and B. Knights. 1988. Migrations of elvers and European eels, *Anguilla anguilla* L., in the River Thames. Journal of Fish Biology 33 (supplement A):161–175.

Naismith, I. A., and B. Knights. 1990. Studies of sampling methods and of techniques for estimating populations of eels, *Anguilla anguilla* L. Aquaculture and Fisheries Management 21:357–367.

Naismith, I. A., and B. Knights. 1993. The distribution, density and growth of the European eel, *Anguilla anguilla*, in the freshwater catchment of the River Thames. Journal of Fish Biology 42:217–226.

Oliveira. K., J. D. McCleave, and G. S. Wippelhauser. 2001. Regional variation, and the effect of lake: river area on sex distribution of American eel. Journal of Fish Biology. 58: 943–952.

Pool, W. R., and J. D. Reynolds. 1996. Growth rate and age at migration of *Anguilla anguilla*. Journal of Fish Biology. 48:633–642.

Quinn, J. M., A. B. Cooper, R. J. Davies-Colley, and R. B. Williamson. 1997. Land use effects on habitat, water quality, periphyton, and benthic invertebrates in Waikato, New Zealand. NZ Journal of Marine and Freshwater Research 31:579–597.

Rasmussen, G., and B. Therkildsen. 1979. Food, growth, and production of *Anguilla anguilla* L. in a small Danish stream. Rapports et Proces-verbaux des Reunions, Conseil International pour l'Exploration de la Mer 174:32–40.

Rasmussen, G. 1983. Recent investigations on the population dynamics of eels (*Anguilla anguilla*) in some Danish streams. Pages 71–77 *in* Proceedings of the 3rd British Freshwater Fisheries Conference.

Rowe, D. K., B. L. Chisnall, T. L. Dean, and J. Richardson. 1999. Effects of land use on native fish communities in East Coast streams of the North Island, New Zealand. New Zealand Journal of Marine and Freshwater Research 33: 141–151.

Rossi, R., A. Carrieri, P. Franzoi, G. Cavallini, and A. Gnes. 1987. -1988. A study of eel (*Anguilla anguilla* L.) population dynamics in the Comacchio lagoons (Italy) by mark-recapture method. Oebalia. 14:1–14.

Ryan, P. A. 1986. Seasonal and size related changes in the food of the shortfinned eel, *Anguilla australis* in Lake Ellesmere, Canterbury, New Zealand. Environmental Biology of Fishes 15:47–58.

Sinha, V. R. P., and J. W. Jones. 1967. On the food of the freshwater eels and their feeding relationship with the salmonids. Journal of Zoology 153:119–137.

Sloane, R. D. 1984. Distribution, abundance, growth and food of freshwater eels (*Anguilla* spp.) in the Douglas River, Tasmania. Australian Journal of Marine and Freshwater Research 35:325–339.

Todd, P. R. 1980. Size and age of migrating New Zealand freshwater eels. New Zealand Journal of Marine and Freshwater Research 14:283–293.

Vollestad, L. A., and B. Jonsson. 1986. Life history characteristics of the European eel *Anguilla anguilla* in the Imsa River, Norway. Transactions of the American Fisheries Society 115:864–871.

Vollestad, L. A., and B. Jonsson. 1988. A 13-year study of the population dynamics and growth of the European eel *Anguilla anguilla* in a Norwegian river: Evidence for density-dependant mortality, and development of a model for predicting yield. Journal of Animal Ecology 57:983–997.

Enhancement and Management of Eel Fisheries Affected by Hydroelectric Dams in New Zealand

JACQUES BOUBÉE*, BEN CHISNALL, ERINA WATENE, AND ERICA WILLIAMS

*National Institute of Water and Atmospheric Research Ltd,
Post Office Box 11–115, Hamilton, New Zealand
Corresponding author: j.boubee@niwa.cri.nz

DAVID ROPER

Mighty River Power, Box 445, Hamilton, New Zealand

ALEX HARO

S. O. Conte Anadromous Fish Research Center, Biological Resources Division, US Geological Survey, Post Office Box 796, Turners Falls, Massachusetts 01376, USA

Abstract.—Two freshwater anguillid eel species, *Anguilla australis* and *A. dieffenbachia*, form the basis of important traditional, recreational, and commercial fisheries in New Zealand. These fisheries have been affected by the damming of many of the major waterways for hydroelectric generation. To create fisheries in reservoirs that would be otherwise inaccessible, elvers have been transferred from the base of dams into habitats upstream. Operations in three catchments: the Patea River (Lake Rotorangi), Waikato River (eight reservoirs notably the two lowermost, lakes Karapiro and Arapuni), and Rangitaiki River (lakes Matahina and Aniwhenua) are discussed. In all reservoirs, the transfers have successfully established fishable populations within six years of the first transfers and, in Lake Arapuni eels have reached the marketable size of 220 g in less than four years. In comparison, it typically takes from 13 to 17 years before eel populations are fishable in the lower Waikato River where direct access to the sea is available. Telemetry and monitoring at the screens and tailraces of several power stations have been used to determine migration timing, triggers, and pathways of mature eels. Successful downstream transfer of mature migrating adults has been achieved by spillway opening and netting in headraces during rain events in autumn, but means of preventing eels from impinging and entraining at the intakes are still required. An integrated, catchment-wide management system will be required to ensure sustainability of the fisheries.

Introduction

The New Zealand freshwater wild eel fishery is based on the two indigenous anguillid species, the shortfinned eel *Anguilla australis* and the longfinned eel *A. dieffenbachii*. Very small numbers of the self-introduced Australian longfinned eel *Anguilla reinhardtii* are also occasionally captured in the northern part of the country. The fishery began in the 1960s, peaked at 2,434 metric tons per annum in 1975, and declined to a roughly stable 1,400 metric tons per annum over recent years (Jellyman 1994). The wild stock fishery is now considered fully developed and the number of fishing permits has been limited by regulation since the 1980s. As fishing pressure increased, the proportion of longfinned eels in the catch has declined (Beentjes and Chisnall 1997). To maintain returns, eel fishermen have progressively pushed into more difficult terrain and have increased their netting effort. The declining resource, an impending Quota Management System, and a greater recognition of the cultural importance of eels for Maori has led to efforts to increase the wild eel resource through translocation of elvers into areas with poor recruitment. The eel industry, and other resource users (Maori tribes, the Ministry of Fisheries, hydroelectric power companies, and the Department of Conservation) have focused their enhancement efforts on a number of hydroelectric reservoirs (Figure 1; Table 1).

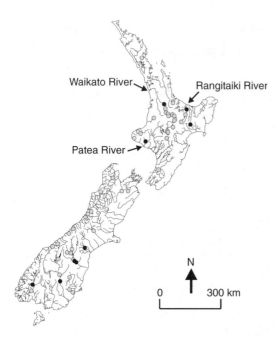

Figure 1. Location of major hydroelectric projects New Zealand. A black marker identifies projects where permanent upstream passage facilities (not necessarily functional) for elvers and eels were in place during the summer of 1999/2000.

Surveys of both remnant and enhanced eel populations have been undertaken at several sites (Beentjes et al. 1997; Chisnall et al. 1998). These studies concluded that: manual transfer was the most effective means of re-stocking hydro reservoirs; the fastest growth was obtained in reservoirs with low eel densities and high food availability; water/land ecotones provided nursery habitat for juvenile eels and habitat for prey species; and there was a need to regulate stocking rates to achieve optimum production.

With the stocking of eel populations in hydro reservoirs, the issue of downstream passage for mature adults has become more of an issue, especially in reservoirs where eels were previously nonexistent or where, through lack of recruitment, the populations have virtually disappeared through natural attrition. Renewed effort has, therefore, been made to understand migrant behavior, and to apply tentative solutions to the problem of turbine mortality (Boubée et al. 2001). This paper discusses the results of studies that aimed to balance eel fishery management goals with those of hydro-power generators. In particular, we provide examples of elver catch, transfer and monitoring procedures, results of the enhancement, and discuss alternative means of maintaining sustainable fisheries.

Study Area

For the purposes of this paper, we have focused on three sites in the central region of the North Island, where provisions for upstream and downstream passage have been implemented (or are currently being considered), and where some information on the existing eel populations is available. Study sites from west to east were on the Patea, Waikato, and Rangitaiki Rivers (Figure 2). There is one hydro reservoir (Lake Rotorangi) on the Patea River (Figure 2), two on the Rangitaiki (lakes Matahina and Aniwhenua, Figure 2) and eight on the Waikato (Figures 2 and 3), although only lowermost three (Karapiro, Arapuni and Maraetai) will be discussed herein. The six reservoirs studied vary with respect to size, trophic status, depth, temperature, and surrounding land use and topography (Table 2; Beentjes et al. 1997).

Patea River

The Patea River is part of the Taranaki commercial eel fishery (Figure 2). In 1982, extensive sampling of Taranaki rivers revealed that eel densities, distributions and species composition in the Patea River and in headwater tributaries were similar to other parts of New Zealand (Taranaki Catchment Board 1984). An earth-core dam was completed in the lower reaches of the river in 1983. Lake Rotorangi formed behind the dam is narrow and deep. At 46 km, it is the longest man-made lake in New Zealand. Land use is largely pastoral but there are significant areas of native bush cover remaining. The catchment is flood prone and these events often require the opening of the spillway gates. The lake is mildly eutrophic and exhibits thermal stratification during the summer (Burns 1995).

There is an elver pass on the dam but all other indigenous diadromous fish and the migratory shrimps *Paratya curvirostris* have been excluded from the upper catchment. However, three exotic species (rainbow trout *Oncorhynchus mykiss*; brown trout *Salmo trutta;* and perch *Perca fluviatilis*) are now present in good numbers (Taranaki Catchment Board 1988).

Table 1. Weight (kg) of elvers transferred to upstream reservoirs from the base of hydroelectric structures through authorized programs in New Zealand

Region	Location	1992–1993	1993–1994	1994–1995	1995–1996	1996–1997	1997–1998	1998–1999	1999–2000
Waikato	Lake Waikare	–	150	120	–	2	–	–	–
	Lake Karapiro	110	320	316	564	208	137	561	346
	Lake Arapuni	–	300	10	442	550	400	175	135
	Lake Waipapa	–	–	5	132	100	103	52	63
	Lake Maraetai	–	–	7	79	149	251	–	–
	Lake Whakamaru	–	–	–	–	253	427	119	15
	Lake Atiamuri	–	–	–	–	136	62	39	131
	Lake Ohakuri	–	–	–	–	159	551	272	184
	Wairere Falls	–	–	–	?	?	?	?	160
Hawkes Bay	Waikaretaheke River	–	–	–	–	2	7	3	2.5
	Northland Wairua River	–	–	1	?	?	?	?	?
	Bay of Plenty								
	Lake Matahina	–	–	–	–	–	339	–	?
	Lake Aniwhenua	–	–	–	–	59	627	1,002	?
	Tarawera River	–	–	–	–	–	63	–	–
Taranaki	Lake Rotorangi							12	182
Otago	Lake Dunstan	–	–	–	–	1	4	7	?
	Lake Wanaka	–	–	–	–	–	3	15	?
	Lake Te Anau	–	–	–	–	–	–	90	85
	Mataura Falls	–	–	–	–	–	250	190	?
Total transferred		110	770	459	1,217+	1,917+	3,224+	2,537+	1,303+

–, no transfer made; ?, no records but transfers made.
From Annala et al. 2000 and NIWA records.

Waikato River

The Waikato River basin in the central North Island supports New Zealand's single most productive eel fishery (> 25% of the national total). Since the 1960s, most of the lower basin has been fished intensively, and this, together with an encroachment on aquatic habitats by agriculture has decreased the number of larger marketable eels.

Eight hydro dams have been constructed in the upper reaches of the Waikato River (Figures 1–3). Extensive tracts of dairying pastoral land surround the two lowermost reservoirs (lakes Karapiro and Arapuni) but exotic pine forestry is the dominant land use upstream of Lake Arapuni. Associated landforms are mostly rolling hills, but aquatic margins are often steeply incised. Extensive beds of exotic aquatic macrophytes, dominated by *Egeria densa* and *Ceratophyllum demersum*, are present throughout the littorals. The three lower reservoirs are eutrophic (Livingston et al. 1986).

Historically, eels were rare in the upper catchment because of limited access caused by numerous rapids, notably the Maungatautari Falls which were submerged when Arapuni Dam was constructed in 1929 (Hobbs 1940; Cairns 1941). After the completion of Arapuni Dam, and the construction of the lowermost dam at Karapiro in 1947, very few new recruits reached upstream habitats. Once commercial fishing began in the mid 1980s, the existing stocks were soon depleted and by 1992 very few eels remained above Arapuni Dam (Chisnall and Hicks 1993).

Rangitaiki River

The Rangitaiki River drains the Urewera mountain range on the east of the central North Island (Figure 2). The steady flow and predominantly pumiceous sediments of the main river and western tributaries differ markedly from the more flashy eastern catchments. About 75% of the catchment is covered either by indigenous

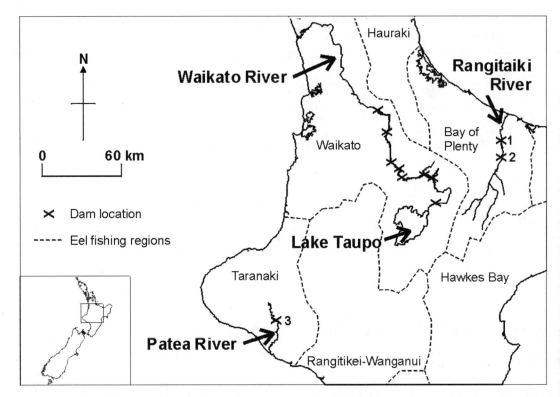

Figure 2. Map of the Central North Island showing commercial eel fishing regions and the location of the Waikato, Patea and Rangitaiki rivers. 1) Matahina Dam, 2) Aniwhenua Dam, 3) Patea Dam. The names of the Waikato River hydroelectric dams are given in Figure 3.

forest (Urewera National Park and Whirinaki State Forest) or by exotic pine plantations (Kaingaroa Forest).

Lake Matahina, the lowermost reservoir on the Rangitaiki River, has very little shallow littoral margin, and aquatic macrophytes *Egeria densa* occur only in very small (< 5 m^2) stands. In contrast, the upper reaches of Lake Aniwhenua have extensive shallow areas infested with *Egeria densa* and *Ceratophyllum demersum*. Most of the margins are lined with extensive stands of raupo *Typha orientalis*.

Upstream Passage and Population Enhancement

Patea

A bottlebrush lined PVC pipe elver pass was installed on the Patea Dam face in January 1985. This was upgraded in 1991 to gravel-lined piping (Taranaki Regional Council 1992). Elvers have climbed the pass since its installation with peak migrations being recorded consistently over a 20-day period in mid January. The pass has improved immigration to the upper catchment, but density of eels in the lake remains low (Beentjes et al. 1997). The pass is also selective for the smaller shortfinned elvers and few of the larger longfinned elvers reach the top of the 75 m high dam (Taranaki Catchment Board 1984). Although more recent upgrading of the pass has further improved its efficiency, monitoring during the 1999/2000 migration indicated that the pass only allowed the passage of about 27,000 elvers compared to 455,000 transferred manually (H. McWilliams, Taranaki Regional Council, personal communication). It is, therefore, only with the implementation of a manual transfer program in 1998 (Table 1) that recruitment to the lake has improved substantially.

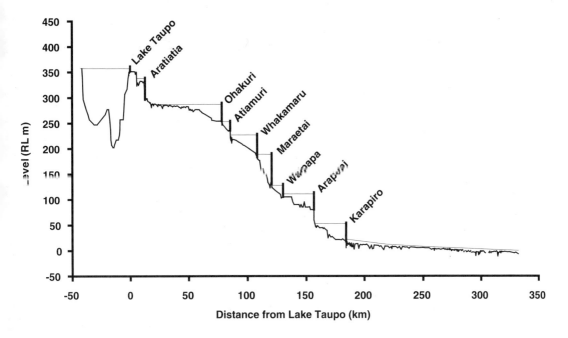

Figure 3. Longitudinal distribution of hydroelectric dams along the Waikato River, North Island, New Zealand. Relative Level (RL) shown is the height above sea level using Moturiki datum.

Karapiro

An elver trap and transfer system has been operating at Karapiro Dam since 1992/93 (Tables 1 and 3). Initially the trapping systems were very rudimentary, but after a number of trials with a variety of designs, the two current traps were constructed. The first is a static trap with an inclined ramp lined with nylon brushes supplied by Fish-Pass (Legault 2000). This trap is located in the stoplog gallery at the outlet of the turbines and the attraction flow is provided by tapping into the station's transformer cooling water outlet (Boubée and Barrier 1996). The second trap is a floating trap and lift system installed in the tailrace near the transformer cooling water outlet (Williams et al. 1999). Two inclined ramps lined with nylon brushes allow elvers and juvenile eels to reach the trap. Aerated filtered lake water is piped into both traps to keep the catch healthy.

The traps are currently in operation from November to March, and are monitored daily at the peak of the season, but only at two to three day intervals at other times. When more than 5 kg of elvers has been collected, the elvers are removed, passed through a 4 mm grate to separate the larger juveniles from the rest of the catch and weighed. The catch is then transferred to a 2 m^3 holding tank supplied with aerated lake water.

At no more than seven day intervals, the content of the holding bin is transferred to one of three 175 L transport tanks on the back of a vehicle. At this stage the catch is weighed again and a sub-sample retained to determine the species composition by number and weight. The weight of elvers in the tanks is restricted to a maximum of 40 kg and the tanks are aerated with an air compressor during transport.

The elvers are transported upriver and released into seven of the eight upper impoundments (Table 3). Prior to stocking, a few large eels had been recorded as far upstream as Ohakuri, the seventh hydro lake in the Waikato chain (Coulter 1977). These eels presumably reached the site by natural means, but human transfer is another possibility as it was a common practice of Maori in earlier times to transfer elvers above natural barriers (Hutton and Hector 1872; Mead 1966).

Table 2. Lake characteristics.

River system	Lake	Construction period[1]	Inland distance[1] (km)	Catchment area (km^2)	Lake area (km^2)	Length of lake shore[1] (km)	Max depth (m)	Max width (km)
Patea River	Rotorangi	1982–1983	45	871[2]	6.2[2]	92	58[2]	0.5[2]
Rangitaiki	Aniwhenua	1977–1980	67	2,413[3]	2.5[3]	12	–	0.8[1]
	Matahina	1961–1967	38	431[3]	2.6[3]	14	52[5]	1.1[5]
Waikato River	Karapiro	1940–1947	166	839[4]	7.6[4]	53	31[5]	0.9[5]
	Arapuni	1923–1929	193	249[4]	9.4[4]	60	64[5]	0.9[5]
	Waipapa	1955–1961	221	254[4]	1.5[4]	25	17[5]	0.3[5]
	Maraetai	1947–1953	231	588[4]	4.1[4]	36	61[5]	1.2[5]
	Whakamaru	1949–1956	244	575[4]	7.8[4]	32	38[4]	1.0[4]
	Atiamuri	1953–1958	267	285[4]	2.4[4]	19	29[5]	0.4[5]
	Ohakuri	1956–1961	275	1,528[4]	13.7[4]	47	38[5]	2.0[5]
	Aratiatia	1959–1964	344	245[4]	0.6[4]	12	12[4]	0.2[4]

Note the catchment above Aratiatia, on the Waikato River, is 3,289 km^2, the area below Lake Karapiro is 6,406 km^2).
From: [1] NIWA unpublished data; [2] Taranaki Catchment Board 1988; [3] NZSOLD 1989; [4] Mighty River Power unpublished data; [5] Viner 1987)

Rangitaiki

Following concerns expressed in 1983 by traditional Maori fishermen, the Department of Internal Affairs, resource managers at that time, began a program of manual transfers of elvers from the tailrace of the Matahina Dam to the upper catchment, including Lake Aniwhenua. Station staff and other interested parties also made transfers, but few records of the numbers of juvenile eels transferred or the timing of transfers were kept (Beentjes et al. 1997). A gravel-lined fish pass was installed on the Matahina Dam in 1991 and a resistance counter used to monitor elver movement during the three following summers (Boubée and Mitchell 1994). No further monitoring of the pass was made, but observations indicated that elver numbers using the pass remained substantially smaller than those accumulating at the base of the dam. The pass was upgraded and a trap and road transfer operation implemented in 1996 (Table 1). Unfortunately, the importance of maintaining long-term transfer records has not been recognized in this catchment and no regular catch data were collected in the 1999/2000 season.

Table 3. Estimated total number of elvers (in 1000s) transferred from the Karapiro Dam tailrace to upstream reservoirs from 1992 to 2000 (Boubée et al. 2000). The percentage of longfinned elvers is given in brackets. In addition to elvers, a very small number of juvenile eels were also transferred. Permanent trapping facilities were installed in 1995/1996.

Reservoir	1992–1993	1993–1994	1994–1995	1995–1996	1996–1997	1997–1998	1998–1999	1999–2000	Total
Karapiro	92 (34)	267 (34)	264 (34)	504 (37)	130 (21)	272 (45)	488 (22)	397 (6)	2,414 (28)
Arapuni	0	251 (34)	8 (34)	459 (19)	342 (21)	375 (43)	108 (56)	119 (16)	1,662 (29)
Waipapa	0	0	4 (34)	126 (22)	62 (21)	66 (42)	60 (10)	63 (10)	381 (22)
Maraetai	0	0	6 (34)	70 (19)	92 (30)	161 (50)	0	0	329 (37)
Whakamaru	0	0	0	0	173 (16)	247 (69)	109 (35)	14 (4)	543 (44)
Atiamuri	0	0	0	0	80 (32)	53 (45)	37 (18)	123 (10)	293 (23)
Ohakuri	0	0	0	0	327 (15)	483 (55)	214 (38)	153 (20)	1,177 (36)
Total	92 (34)	518 (34)	282 (34)	1,159 (27)	1,206 (20)	1,657 (52)	1,016 (29)	869 (11)	6,799 (31)

Timing, Size, and Species Composition of the Elver Migration

Karapiro

Previous studies reported that elvers arrived at Karapiro Dam, on the Waikato River, in mid to late January (Cairns 1941; Jellyman 1977). The more extensive records obtained as part of this study indicate that the migration can start at least one month earlier (Table 4). However, peak runs have mostly occurred in January. Similar results have been recorded at other sites around the country (Beentjes et al. 1997; Chisnall et al. 1999).

Although the timing of the migration at Karapiro Dam has varied little, the number of elvers captured and the species composition has changed between months and between years (Table 4). Studies of glass eel recruitment to New Zealand have associated this pattern of variability with the El Niño Southern Oscillation (ENSO) cycle and its effect on recruitment (Chisnall et al. 2002). Because of this variation in the species composition of the catch, it is not possible to transfer a consistent proportion of each species to all the sites to be enhanced. This does, however, present an opportunity to distribute the catch according to the habitat preferences of the two species.

In the first five seasons of the program, as trapping efficiency increased, the catch of elvers at Karapiro Dam also increased. However, since 1997/98, numbers captured have been decreasing. This could reflect either a change in the size of the elver run, or a gradual removal of elvers that may have accumulated over several years. Indeed, elvers have often been found to remain for long periods below dams. An extreme example has been found at Roxburgh Dam in the South Island, where elvers and juvenile eels smaller than 200 mm have been aged at between 4 and 22 years (Beentjes et al. 1997). Therefore, until effective upstream passage systems have been in operation for several seasons, it is probably premature to try to estimate annual recruitment into New Zealand waters from catches made below barriers.

Although a variety of pass designs have been tried to overcome the loss of recruitment to upper catchments, manual stocking through trapping and transfer has been the most effective. However, some of the elvers transferred continue to migrate upriver to the next dam where they accumulate and attempt to climb the structure. The possibility of modifying this behavior, or of reducing its consequence through changes in the current transfer techniques, warrants further investigation.

Monitoring of the Stocked Populations in the Waikato Hydro Reservoirs

Although elver transfer operations are now operating at several sites around the country, detailed information of the success of the operations is only available for the Waikato River, which was one of the first catchments where enhancement procedures were implemented. In this system, the most complete information has been collected in the lowermost reservoirs where remnant eel populations were present when the enhancement program started.

Enhancement programs in New Zealand are operated under a permit system, administered by the Ministry of Fisheries. These permits allow the holder to collect juvenile eels below an

Table 4. Seasonal catches of elvers at the Karapiro Dam.

	Season											
	1994–1995		1995–1996		1996–1997		1997–1998		1998–1999		1999–2000	
Month	N	LF (%)	N	LF (%)	N	LF (%)	N	LF (%)	N	LF (%)	N	LF (%)
November	–	–	–	–	1	2	1	5	–	–	–	–
December	–	–	179	30	139	14	614	66	144	13	122	–
January	–	–	630	40	578	27	642	61	657	39	249	17
February	358	34	317	6	346	19	359	37	133	16	358	6
March	12	–	110	2	143	2	57	30	82	1	139	<1
Total	370	33	1,236	26	1,207	20	1,673	52	1,016	29	868	11

N, total number (in 1000's) of shortfinned and longfinned elvers; LF, longfinned eels; –, no collection made.

Note: not all the elvers captured were transferred to the upstream reservoirs.

obstruction and to transfer a maximum weight to stipulated areas within the catchment. As no local information was available on the most appropriate seeding density, based largely on European information, a criterion of about 60 kg per square kilometer of lake area was used (D. Allen, Ministry of Fisheries, Auckland, personal communication). In reality, the limits set by the Ministry were rarely reached in any of the reservoirs as catches have always been below the total permitted.

Information on the size, age structure and species composition of the stocked population in lakes Karapiro and Arapuni was obtained during February (summer) 1996, 1997, and 1999 (Beentjes et al. 1997; Chisnall et al. 1998; Boubée et al. 2000). The lakes were sampled using a fleet of fine meshed fykes (D-opening, winged hoods, 0.75 mm square mesh; Chisnall and West 1996), in conjunction with standard fyke nets (D-opening, double funneled, 20 mm stretched mesh).

Net sites were chosen to represent the range of lake margin and open water habitats available in each location. Nets set on the margins were generally positioned perpendicularly or obliquely to the shoreline with the cod-ends outermost. Deep water was fished using trains of standard fyke nets strung together in pairs (leader to leader) and anchored at each end of the train. All nets were set overnight and none were baited.

As the nets were lifted, the eel numbers and species composition per net were recorded. Catches were then pooled and the catch measured. A length-stratified sub-sample (where available, 10–20 eels per 100 mm size interval) was retained for aging using the otolith crack and burn method (Hu and Todd 1981).

Following stocking of Lake Arapuni in 1993/1994, catch rates rapidly increased from a mean catch of 0.64 eels (both species combined) per standard fyke net in 1995, 6.13 eels per fyke net in 1999 (Table 5). In 1992, the size distribution of eels collected indicated that only a remnant population of large eels remained in the lake (Figure 4). However, surveys conducted in 1996 and 1997 showed that a fishable eel population had developed in the lake (Beentjes et al. 1997; Chisnall et al. 1998). The size distribution was skewed towards smaller eels in 1996, but by 1997, a larger proportion of the catch had reached the commercial size limit of 220 g (Figure 5).

The large proportion of longfinned eels present in the catches in 1997 had declined by 1999 (Figure 5), probably as a result of intensive commercial fishing. Other investigations have also indicated that fishing selectivity for longfinned eels occurs. This is based on the more territorial behavior of this species, along with its greater weight to length ratio (girth), making it more vulnerable to netting than shortfinned eels (Chisnall 1994; Chisnall et al. 1998; Chisnall and Kemp 2000).

By 1999, the population structure of eels in Lake Arapuni was typical of fished populations elsewhere in the Waikato (Beentjes and Chisnall 1997, 1998). Very few large eels were left, and the population mean length of 450 mm was about the same as the legal market size. However, the size range of the sample obtained in 1999 indicates that, in time, some eels will reach sexual maturity. The present level of commercial harvesting does not, therefore, completely eliminate the need to provide downstream passage.

Table 5. Eel catch per unit effort (CPUE) obtained with fine-meshed fykes (Fm fyke) and standard fykes (Std fyke) in Lake Arapuni between 1995 and 1999.

Year	Net type	No. nets	CPUE SF	CPUE LF	CPUE LF + SF
1995[1]	Fm fyke	4	0.25	0.50	0.75
	Std fyke	14	0.50	0.14	0.64
1996[2]	Fm fyke	21	10.47	2.90	13.37
	Std fyke	114	1.35	0.25	1.60
1997[3]	Fm fyke	20	4.83	6.00	10.83
	Std fyke	66	1.89	1.24	3.04
1999[4]	Fm fyke	7	6.86	1.43	8.29
	Std fyke	8	5.25	0.88	6.13

SF = shortfinned eel, and LF = longfinned eel.
From [1]Boubée et al. 1995; [2]Beentjes et al. 1997; [3]Chisnall et al. 1998; [4]Boubée et al. 2000.

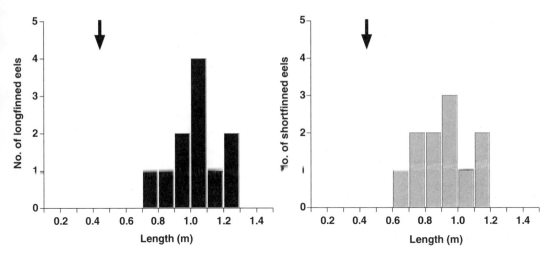

Figure 4. Length distributions of eels sampled from Lake Arapuni in 1992. Arrows show the approximate length at the current 220 g commercial harvest size limit. (From Chisnall 1993.)

Growth rates estimated from otoliths indicated that eels in Lake Arapuni reached the market size of 220 g in two to three years (Chisnall et al. 1998). This was particularly rapid compared with the typical 13–17 years in the lower Waikato (Chisnall 1989; Chisnall and Hayes 1991). However, slower growth could result if stocking density exceeds food resources. As vegetated land/water ecotones play an important role in the food resource of eels (Chisnall 1987), there is a potential to increase eel production by manipulating water levels in these zones (e.g., Duncan and Kubecka 1995).

Manual stocking, which began in summer 1992/1993, has been highly effective in establishing/re-establishing eel stocks throughout the upper Waikato River hydro project system. Since then, close to six million elvers have been collected and transferred upstream. Stocks released at the start of the program began to be harvested in lakes Karapiro and Arapuni about six years later. Lake Maraetai (upstream of Lake Arapuni), was first stocked in the summers of 1995 and 1996, and by 1999 it had around 30% of its eel population above the 220 g harvest weight limit (Boubée et al. 2000). Further upstream, the proportion of eels weighing more than 220 g is below 20%, but even based on a relatively slow growth estimate of 40 g per year (Chisnall et al. 1998), eels from the first elver releases to these upper reservoirs will be harvestable in 2001–2002. Thus, within six years of being transferred, most stocked elvers have grown sufficiently to be commercially harvested. Whether the existing populations are large enough to sustain a profitable commercial fishery, and whether growth rates in the less productive upstream reservoirs will be maintained if the population size is increased, are unknown. If, however, effective harvest does not take place, it is likely that within a decade large numbers of the transferred eels will mature, begin to migrate downstream, and will be killed during passage through the turbines.

Downstream Passage

With the successful implementation of trap and transfer operations for elvers and the development or redevelopment of eel populations above hydro dams, attention has subsequently shifted to the issue of downstream passage. A review of downstream eel passage issues in New Zealand concluded that turbine survival of the large New Zealand migrant eels was likely to be poor (Mitchell and Boubée 1992). Ongoing studies were undertaken in a number of catchments, with a view of understanding the behavior of downstream migrating adults so that means of allowing safe passage over the dams could be developed. As information on the behavior of downstream migrant eels was obtained, protection concepts were formulated and potential mitigation measures tested.

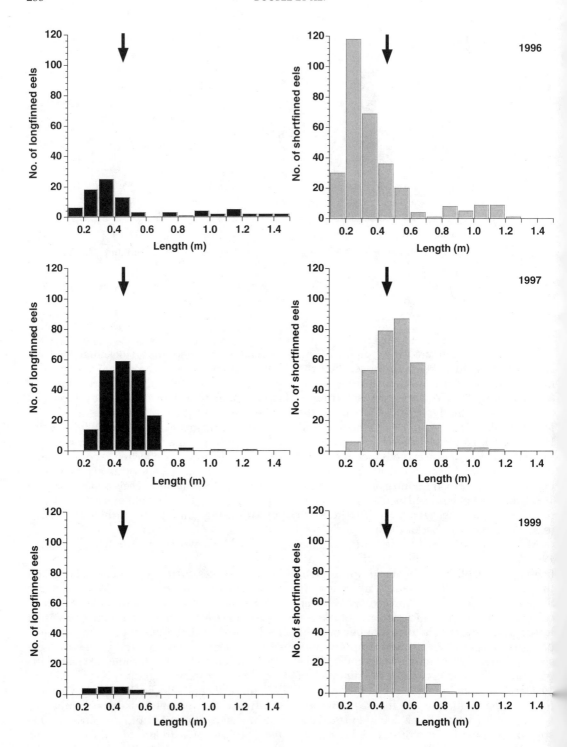

Figure 5. Length distributions of eels sampled from Lake Arapuni in 1996, 1997, and 1999. Arrows show the approximate length at the current 220 g commercial harvest size limit.

Patea River—A Single Hydro Station, In a Flood Prone System

An 83 m high, 190 m long earth filled dam forms Lake Rotorangi on the Patea River. The intake and spillway, on the left abutment, are at the end of a short 8–9 m deep canal. The intake screens for the three main turbines are 1.6 m wide and 6.1 m high with bar spacing at 45 mm center. The two bottom opening radial spillway gates are hydraulically operated. Annual mean river flow is 20 m^3/s and the estimated 100-year flood 1030 m^3/s

One option for protecting migrant eels at the flood prone Patea Dam is to allow spill events to coincide with downstream migration peaks. This provides a compromise between conservation issues and maximizing power production, but can only be advocated if migration can be accurately predicted. Since downstream migration of mature eels tends to occur during rain events and at night (Boubée et al. 2001), it was considered that safe passage could be achieved by opening the spillway and limiting generation for a few hours. To assess the success of such an operation, a spill event was monitored by deploying a large double winged funnel net (25 mm square mesh) across the river channel 1 km below the dam. The spill event occurred between 1930 and 2200 hours, after heavy rain had fallen in the headwaters for about 24 hours. The turbines were closed during the spill, but were operating before and after the spill. The net was set in position between about 1900 hours and 0700 hours and emptied at 2230 hours and at the end of the trial. Captured eels were counted, identified, measured, examined for external signs of damage, and released downstream.

A total of 119 eels were caught with 98% exhibiting the distinguishing features of migrating eels (enlarged eyes, elongated head and black pectoral fins) described by Hobbs (1947) and Todd (1980). Only 5% showed signs of damage (bruising, skin abrasion and cuts). However, very few large female migrant eels (the target of the exercise), passed during the 2.5 hours that the spillway was opened (Figure 6). Thus, the fact that the timing of the migration can be extended over several hours if not days, and an inability to target large females would preclude spillway opening as a mitigation measure unless a means of preventing eels from entering the intake between spills can be devised.

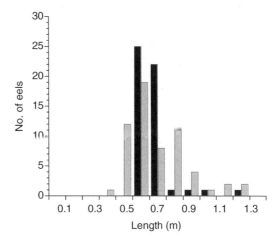

Figure 6. Length frequency distribution of shortfinned and longfinned eels captured below the spillway on the Patea Dam. Dark shading, longfinned eels. Light shading, shortfinned eels. All the eels captured were mature migrants. (Shortfinned eels > 0.5 m and longfinned eel > 0.8 m are females; Todd 1980)

Waikato River—A Large Multi Hydro, Flow Controlled System

The mean flow of the Waikato River at Karapiro is about 247 m^3/s, making this system the North Island's largest river. It is also New Zealand's longest river, rising from the slopes of the central North Island volcanoes, flowing through Lake Taupo and out into the Tasman Sea, a distance of approximately 425 km. There are control gates at the outlet of Lake Taupo and eight hydro dams downstream. Thus, except for large floods, flows are largely modified for power generation.

Karapiro, the last hydroelectric dam on the Waikato River, is a 30 m high by 330 m long arch dam structure. The right abutment incorporates the intake, spillway, and station building which houses three 32 MW turbines. There are four vertical lift spillway gates, each 7 m high and 6.4 m wide. The intakes are 5.4 m high and 6.3 m wide and protected by screens with about 100 mm bar spacing.

To determine the timing of the downstream migration of eels in the Waikato River, and to get an estimate of the number of eels damaged during passage through Karapiro Dam, a commercial fisherman was hired to collect and record eel bodies from the riverbank in the 16 km reach below the dam. Over two years, visits were made at no more than 15 day intervals in summer, and at monthly intervals in winter. Data collected included time of sighting, species, size, and state of the eels.

Between November 1998 and May 2000 (two migration periods), 436 eels were recovered from the 16 km reach below the Karapiro Dam. All but a few were migrants. The majority of bodies recovered were severed (mean length to cut 450 mm) and the few recovered whole were on average over 950 mm long. As migrant females are generally larger than 500 mm for shortfinned and 800 mm for longfinned eels (Todd 1980), most of the eels killed at Karapiro were mature females. Furthermore, based on our observations, and an estimate of mortality based on the Larinier and Travade (1999) formula which accounts for fish size and turbine characteristics, it can probably be assumed that the majority of eels smaller than about 600 mm pass safely through the station's turbines. However, mortality estimates are close to 70% for a 950 mm eel, and are considerably higher for larger eels.

Most of the migrant eel bodies were found between January and May, although a small number were found as early as November (Figure 7). The largest number of shortfinned eels were recovered in March/April, and longfinned eels in April/May. This timing is consistent with migration at other sites (Cairns 1941; Hobbs 1947; Burnett 1969; Todd 1981; Palmer et al. 1987). Peaks in the number of eels recovered did not follow periods of heavy rain as has been found in the Patea and Rangitaiki catchments. Therefore, attempting to predict migration peaks based on rainfall events, and implementing brief protective measures such as spillway openings, does not appear to be possible in large regulated systems such as the Waikato River. In such large multi-station systems, and until effective protective and bypass devices can be devised, we instead advocate intensive harvesting as a mean of reducing losses caused by turbine mortality. However, commercial harvesting does not completely eliminate the need to provide downstream passage at hydro dams, and every effort should be made to devise an effective solution to the problem of turbine mortality. Furthermore, to ensure the sustainability of the populations, sufficient mature eels must be allowed to reach spawning grounds, and fisheries downstream of barriers must be managed to ensure that this occurs.

Rangitaiki River—A Multi-Station, Flood Prone System

Aniwhenua Dam on the Rangitaiki River presented a near ideal situation for experimenting with a capture net in a catchment where large downstream migrating eels were still present. The reservoir is formed by a 200 m long, 10 m high, earth filled dam and a 2.2 km long canal directs flows to a 38 m high head pond dam. Two 3.4 m penstocks deliver water through the turbines and back into the Rangitaiki. Flows in autumn months are usually around 25–40 m^3/s, with freshets peaking at 55–75 m^3/s (Mitchell 1996). The Aniwhenua intake has a 30 mm screen rack which retains all female and most longfinned male migrant eels.

A "net and transfer" operation for downstream adult migrating eels has been in place at Aniwhenua Dam since 1994 (Mitchell 1996; Boubée et al. 2001). The 25 mm mesh net stretches at an angle across the entire 25–30 m wide and 7 m deep intake canal. The net is deployed in autumn when more than 40 mm of rain falls in the catchment over three days. When in place, the net is emptied at intervals throughout the night to collect the eels and clear the net of debris. Eels captured are transferred to fine

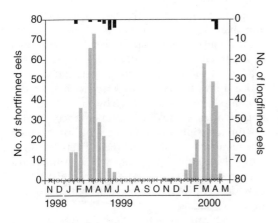

Figure 7. Number of dead migrant eels recovered downstream of the Karapiro Dam, Waikato River, November 1999 to May 2000. Dark shading, number of longfinned eels; gray shading, number of shortfinned eels; X axis, dates of surveys.

mesh holding bags, and then transported and released in the tailrace of Matahina Dam 33 km from the sea.

Monitoring of the netting operations has been rather ad hoc, however, during the two year period when the number of eels impinging on the power station's intake screens were obtained (Boubée et al. 2001), the capture net was deployed on three occasions and catches noted. Of the 282 migrant eels recorded, 72 were collected live in the net, with 40 eels captured during one single, well timed, deployment.

Once released below Matahina Dam, the eels appear to have successfully completed their journey to the sea, although possibly not immediately. This was shown by a telemetry study of five of the released eels implanted with radio transmitters. The tagged eels were monitored with automatic recorders at two sites located 10 and 21 km downstream of the release site. Tracking by boat was also used to locate the eels before and after they passed the recorders. Only localized movement was recorded in the first month after release, but four of the five eels migrated out of the system on the first rain event that occurred after they were released. This rain event only caused the lower river to rise slightly, but also resulted in an increase in the number of migrant eels impinging on the Aniwhenua Dam intake screen further upstream. Although the migration delay recorded could have been a response to surgery, there is a potential risk that the released eels could be harvested before conditions induce them to finally move out to sea. Therefore, catch and transfer practices for mature eels should only be considered viable if an integrated approach to fisheries management, including the protection of downstream migrants, is applied throughout the catchment.

Conclusions

Upper catchment populations of European eels *Anguilla anguilla* are known to contain a predominance of females (e.g., Aprahamian 1988), and the same differential distribution of sexes seems to occur for longfinned and shortfinned eels in New Zealand reservoirs (Chisnall and Hicks 1993). The reservoirs could therefore be used as reserves to favor the production of female eels. However, because of the as yet unresolved problems associated with downstream passage, we currently advocate the intensive harvest of the new populations, and protection of downstream habitats where free passage is available.

There is a need to either harvest eels before they reach maturity, or to deter/bypass them to prevent entry into the intakes in order to minimize losses to the fishery. Rainfall triggers have been used at Aniwhenua Dam to predict migrations and to deploy a capture net across the intake (Boubée et al. 2001). Rainfall triggers could probably also be used at Patea Dam (Watene et al., this volume), where mature eels could be safely transferred over the dam by controlled openings of the spillway.

The Aniwhenua canal was netted effectively on several occasions in the past, however, increasing problems with floating swards of the introduced macrophyte *Ceratophyllum demersum*, have severely limited its use in the last two seasons. Furthermore, most dams are large, and deploying and retrieving a net across large intakes is impractical. Assembling resources is also difficult, given the unpredictability of rainfall events. We conclude that netting is generally not feasible on medium to large-scale operations, and only possible where drift of plant material is not a problem.

Ultimately, there needs to be a carefully balanced approach taken to achieve the three main goals of creating/re-creating a fishery, allowing escapement of some mature eels, and maximizing power generation. These goals can only be achieved if all users participate in integrated management of the resource. Under appropriate stocking regimes, hydro lakes have the potential to support highly productive eel fisheries. Given present control of the fishery by limitations to the number of fishers, this could play an important role in reducing the fishing pressure on lower river areas and in catchments with free access to the sea. In some systems, intensive commercial harvesting of the hydro lake habitats could overcome the need for downstream eel passage. However, this would only be possible if the reservoirs are treated as an integral part of a fishery managed for sustainability.

Acknowledgments

A large number of people contributed to this study, but special thanks go to Mike Holmes and Grant Williams not only for their assistance in the field, but also for sharing their long experience as commercial fishermen. Joshua Smith and

Eddy Bowman provided field support. Jody Richardson, Peter Todd and Jean Robitaille provided useful comments and suggestions on the draft. TrustPower Ltd. granted access to Patea Power Station. These studies were funded by Mighty River Power Ltd. and the New Zealand Foundation for Research, Science and Technology, contract number 1611.

References

Annala, J. H., K. J. Sullivan, and C. J. O'Brien. 2000. Report from the fishery assessment plenary, May 2000: Stock assessments, and yield estimates. Ministry of Fisheries, Wellington, New Zealand.

Aprahamian, M. W. 1988. Age structure of eel, Anguilla (L.), populations in the River Severn, England, and the River Dee, Wales. Aquaculture and Fisheries Management 19:365–376.

Beentjes, M. P., and B. L. Chisnall. 1997. Trends in size and species composition and distribution of commercial catches. New Zealand Fisheries Data Report No. 89.

Beentjes, M. P., B. L. Chisnall, J. A. T. Boubée, and D. J. Jellyman. 1997. Enhancement of the New Zealand eel fishery by elver transfers. New Zealand Fisheries Technical Report No. 45.

Beentjes, M. P., and B. L. Chisnall. 1998. Size, age and species composition of commercial eel catches from market sampling, 1996—97. New Zealand Fisheries Technical Report No. 29.

Boubée, J. A. T., and C. Mitchell. 1994. The eels that climb over a dam. Water and Wastes in New Zealand 82:23–26.

Boubée, J. A. T., D. West, D. Kane, R. Barrier, and J. Richardson. 1995. Fish population survey of the Waikato River upstream and downstream of the Kinleith Mill discharge. NIWA Consultancy Report PPL301/2.1.

Boubée, J. A. T., and R. Barrier. 1996. Elver collection and transfer programme —Karapiro Dam, 1995-96. NIWA Consultancy Report ELE602.11.

Boubée, J. A. T., B. Chisnall, E. Williams, and J. Smith. 2000. Fish survey of the Waikato hydro reservoirs—1999. NIWA Client Report ELE90502/9.

Boubée, J. A. T., C. P. Mitchell, B. L. Chisnall, and E. J. Bowman. 2001. Factors regulating the downstream migration of mature eel (Anguilla spp.) at Aniwhenua Dam, Bay of Plenty, New Zealand. New Zealand Journal of Marine, and Freshwater Research 35:121–134.

Burnett, A. M. R. 1969. Migrating eels in a Canterbury river, New Zealand. New Zealand Journal of Marine and Freshwater Research 3:230–244.

Burns, N. M. 1995. Using hypolimnetic dissolved oxygen depletion rates for monitoring lakes. New Zealand Journal of Marine and Freshwater Research 29:1–11.

Cairns, D. 1941. Life history of the two species of New Zealand freshwater eel. Part I. Taxonomy, age and growth, migration and distribution. New Zealand Journal of Science and Technology, Series B 23:53–72.

Chisnall, B. L. 1987. Juvenile eel biology in the backwaters of the Waikato River. Unpublished MSc thesis, University of Waikato, New Zealand.

Chisnall, B. L. 1989. Age, growth and condition of freshwater eels (Anguilla spp.) in backwaters of the lower Waikato River, New Zealand. New Zealand Journal of Marine and Freshwater Research 23:459–465.

Chisnall, B. L., and J. W. Hayes. 1991. Age and growth of shortfinned eels (Anguilla australis) in the lower Waikato basin, North Island, New Zealand. New Zealand Journal of Marine and Freshwater Research 25:71–80.

Chisnall, B. L. 1993. Age, and growth of freshwater eels in the Waikato River. NIWA Consultancy Report ELE113/1.

Chisnall, B. L., and B. J. Hicks. 1993. Age and growth of longfinned eels (Anguilla dieffenbachii) in pastoral and forested streams in the Waikato River basin, and in two hydro-electric lakes in the North Island, New Zealand. New Zealand Journal of Marine and Freshwater Research 27:317–332.

Chisnall, B. L. 1994. An unexploited mixed species eel stock (Anguilla australis and A. dieffenbachii) in a Waikato pastoral stream, and its modification by fishing pressure. Conservation Advisory Science Notes ISSN 1171-9834, Wellington, New Zealand.

Chisnall, B. L., and D. West. 1996. Design and trials of a fine meshed fyke net for eel capture, and factors affecting size distribution of catches. New Zealand Journal of Marine and Freshwater Research 30:355–364.

Chisnall, B. L., M. P. Beentjes, J. A. T Boubée, and D. W. West. 1998. Enhancement of the New Zealand eel fishery by elver transfers. New Zealand Fisheries Technical Report No. 37.

Chisnall, B. L., and C. Kemp. 2000. Size, age, and species composition of commercial eel catches from North Island market sampling. 1997-1998). New Zealand Fisheries Technical Report No. 87.

Chisnall, B. L., D. J. Jellyman, M. L Bonnett, and J. R. E Sykes. 2002. Spatial, and temporal variability in length of glass eels (Anguilla spp.) in New

Zealand. New Zealand Journal of Marine, and Freshwater Research 36:89–104.

Coulter, G. W. 1977. The ecological impact on the Waikato River of untreated effluent from the proposed Broadlands Geothermal Power Station. New Zealand Energy Research and Development Committee, Report 26.

Duncan, A., and J. Kubecka. 1995. Land/water ecotone effects in reservoirs on the fish fauna. Hydrobiologia 303:11–30.

Hobbs, D. F. 1940. Natural reproduction of trout in New Zealand and its relation to density of populations. New Zealand Marine Department, Fisheries Bulletin No. 8.

Hobbs, D. F. 1947. Migrating eels in Lake Ellesmere. Transactions of the Royal Society of New Zealand 77:228–232.

Hu, L. C., and P. R. Todd. 1981. An improved technique for preparing eel otoliths for ageing. New Zealand Journal of Marine and Freshwater Research 15:445–446.

Hutton, F. W., and J. Hector. 1872. Fishes of New Zealand–Catalogue with diagnoses of the species. Notes on the edible species. Wellington, Colonial Museum, and Geological Survey Department.

Jellyman, D. J. 1977. Summer upstream migration of juvenile freshwater eels (*Anguilla*) in New Zealand. New Zealand Journal of Marine and Freshwater Research 11:61–71.

Jellyman, D. J. 1994. The fishery for freshwater eels (*Anguilla* spp.) in New Zealand. New Zealand Fisheries Assessment Research Document 94 (14).

Legault, A. 2000. Sélectivité des substrats a anguilles. Note de travail Fish-Pass 1997 mise à jour 2000. Fish-Pass, Renne, France.

Larinier, M., and F. Travade. 1999. La dévalaison des migrateurs: problèmes et dispositifs. Bulletin Français de la Pêche et de la Pisciculture 353/354:181–210.

Livingston, M. E., B. J. Biggs, and J. S. Gifford. 1986. Inventory of New Zealand lakes. Part 1, North Island. Water and Soils Miscellaneous Publication No. 80.

Mead, A. D. 1966. Richard Taylor missionary tramper. Reed, Wellington, New Zealand.

Mitchell, C. P., and J. A. T Boubée. 1992. Impacts of turbine passage on downstream migrating eels. New Zealand Freshwater Fisheries Miscellaneous Report No. 112.

Mitchell, C. P. 1996. Trapping the adult eel migration at Aniwhenua Power Station. New Zealand Department of Conservation. Science for Conservation Report 37.

NZSOLD. 1989. Dams in New Zealand, The New Zealand Society on Large Dams, Wellington, New Zealand.

Palmer, D., J. A. T. Boubée, and C. Mitchell. 1987. Impingement of fish and crustacea at Huntly Thermal Power Station. New Zealand Freshwater Fisheries Report No. 91.

Taranaki Catchment Board. 1984. The Ecology of eels in Taranaki. Taranaki Catchment Board and Regional Water Board, Stratford, New Zealand.

Taranaki Catchment Board. 1988. Lake Rotorangi – Monitoring a new hydro lake. Taranaki Catchment Board and Regional Water Board, Stratford, New Zealand.

Taranaki Regional Council. 1992. Egmont electricity Lake Rotorangi 1991/92 monitoring annual report. Water quality and biological programmes. Taranaki Regional Council, Stratford, New Zealand.

Todd, P. R. 1980. Size and age of migrating New Zealand freshwater eels (*Anguilla* spp.). New Zealand Journal of Marine and Freshwater Research 14:283–293.

Todd, P. R. 1981. The timing and periodicity of migrating New Zealand freshwater eels (*Anguilla* spp.). New Zealand Journal of Marine and Freshwater Research 15:225–235.

Viner, A. B. 1987. Inland waters of New Zealand. DSIR Science Information Publishing, Wellington, New Zealand.

Williams, E., J. A. T. Boubée, J. Smith, and C. Wirihana. 1999. Karapiro elver transfer programme 1998/99. NIWA Client Report ELE90211/1.

Age, Growth, and Catch-Related Data of Yellow Eel *Anguilla anguilla* (L.) from Lakes of the Erne Catchment, Ireland

MILTON A. MATTHEWS

Northern Regional Fisheries Board, Ballyshannon, CO Donegal, Ireland

DEREK W. EVANS

*Queen's University Belfast, Faculty of Agriculture and Food Science,
Newforge Lane, CO Antrim, Northern Ireland*

CHARLES A. MCCLINTOCK

Northern Regional Fisheries Board, Ballyshannon, CO Donegal, Ireland

CHRISTOPHER MORIARTY

University of Dublin, Zoology Department, Trinity College, Dublin 2, Ireland

Abstract.—Results are presented from an extensive fyke net study of the Erne lakes undertaken in 1998 and 1999 with a sampling effort of over 56,000 net-nights. Highest catches were obtained from Upper and Lower Lough Erne in the lower reaches of the catchment where the greater part of commercial eel fishing takes place. CPUE ranged from 30 to 45 eels weighing 4.5–8.5 kg per 10 fyke nets per night. Mean length ranged from 42 to 44 cm with growth rates of 2.5–3.5 cm per year. In contrast, numbers from the upper reaches of the catchment were lower (4–15 eels, 1.3–4.2 kg) and were characterized by the preponderance of large 46–64 cm, fast growing females (4.1–4.8 cm per year). The results indicate that: the effect of stocking by overland transport of elver can be discerned using commercial gear after as little as five years; and that eel production can be greatly enhanced by stocking, thanks to the existence of extensive areas of excellent habitat which are seriously under-populated. The data provide essential baseline information for the assessment of the progress of a major enhancement program initiated in 1997. The results clearly support the initial assumption, made on the basis of small samples taken in the 1970s, that a stocking program would be effective.

Introduction

Ireland is one of the largest exporters of wild eel in Europe. Output is dominated by Lough Neagh, Northern Ireland, which has sustained a yield in the order of 700 metric tons per annum. By comparison the eel industry in the Republic of Ireland remains underdeveloped with total annual production estimated at 250 metric tons (Moriarty 1999).

Under the European Union's Special Support Program for Peace and Reconciliation, a major cross border endeavor, the Erne Eel Enhancement Program, was initiated in 1997 to monitor and develop the commercial eel fishery of the Erne, which currently produces an estimated annual yield of 50–100 metric tons (Rosell 1997). The first requirement of the Program was the establishment of a base line against which the effects of enhancement work may be measured. The aim of the present study was, by observing the operation of a fyke net fishery and sampling the catch, to provide basic data on the population structure, growth and food of the eel throughout the catchment. This study, conducted from 1998 to 1999, draws conclusions relative to the future management of the fishery.

The Erne Catchment

Straddling the border between Northern Ireland and the Republic of Ireland, the Erne catchment with an area of 4,375 km^2, is the fourth largest in Ireland. The lakes have a total area of 33,000 ha, which drain in a northwesterly direction through Loughs Gowna, Oughter, Upper Lough Erne, and Lower Lough Erne before entering the Atlantic Ocean at Ballyshannon (Figure 1). A summary of physical and chemical parameters is given in Table 1. The catchment is dominated by Lough Erne, situated mainly in Northern Ireland. Lower Lough Erne is divided into the Broad Lough to the north, with a maximum depth of 62 m, and the Narrows to the south, which with numerous shallow bays and inlets contains some of the richest eel fishing grounds. The flooded drumlin terrain of the Narrows extends into Upper Lough Erne (maximum depth 21 m) before radiating out into a multitude of smaller lakes, channels, bays and islands contained in the upper catchment.

Base rich geology throughout much of the catchment, together with leached acid soils and blanket peats, result in a complex water chemistry that is moderately hard and mildly eutrophic, reflecting geology and agricultural land use, which is predominantly pasture with additional intensive production of cattle and pigs. Sediment deposition in Upper and Lower Lough Erne increased dramatically from the 1950s onwards (Battarbee 1976). This increase in sediment deposition is derived from nutrient input from agriculture and sewage (Gibson 1998). Intensive livestock rearing has resulted in eutrophication of many of the lakes in the upper catchment, with Lough Oughter classified as hypertrophic or seriously polluted due to agricultural enrichment (Lucey et al. 1997). Nevertheless, the lakes of the upper catchment support an extensive and renowned sport fishery attracting some 16,000 anglers annually.

Hydroelectric Generation on the Erne

A major feature of the Erne system is the presence of two hydroelectric power stations upstream of Ballyshannon, which were commissioned in the early 1950s. Erection of two dams across the lower reaches of the River Erne at Cathaleen's Fall (height 27 m) and at cliff (height 18 m) included provision for upstream movement of migratory fish including salmon, sea trout and eel through the construction of salmon passes and elver ladders at both dams. Water levels are tightly regulated within a band of approximately 1 m to meet the differing requirements of agriculture, navigation and electricity. The average flow rate through the stations is 92 m^3/s, rising to over 400 m^3/s during major flooding.

Elver Recruitment to the Erne

Records of elver catches to Cathaleen's Fall date to the early 1960s when partial trapping was carried out, with an unknown proportion of elver ascending through elver ladders at each of the dams. Total trapping commenced in 1970 providing a long-term dataset of recruitment to the Erne. As both dams span the entire width of the River Erne, additional recruitment into the system is likely to be negligible. A steady decline in the numbers of glass eel to the coasts of mainland Europe has been reflected in generally declining elver catches to the neighboring catchments of the Shannon and the Bann since the early 1980s (Moriarty and Reynolds 1997; Rosell 1997). In contrast, the elver run to the Erne, while fluctuating, has shown no overlying downward trend. Mean catch from 1970 to 1999 was 1.3 t per annum, with peaks of 4.6 and 4.4 metric tons in 1982 and 1994.

Prior to 1993, all elver captured at Ballyshannon were released in the lower reaches of Lower L. Erne. Therefore, dispersal of stocks was entirely due to natural tendencies of elver to move upstream. In 1993, more distant transfer of elver commenced with the majority (approximately 90%) being released at points around Lower Lough Erne and the northern end of Upper Lough Erne, where the greater part of the commercial eel fishery is located. Limited annual stocking of the upper lakes (90–225 kg elver) was carried out from 1993 to 1995. Following the commencement of the Erne Eel Enhancement Program in 1997, an extensive program of glass eel and elver stocking to all parts of the Erne has been established. Total stocking to the upper lakes (upstream of Upper Lough Erne) was 365 kg in 1998, 847 kg in 1999 and 757 kg in 2000.

The Commercial Eel Fishery

Total reported catch from the Erne is 30–35 metric tons of yellow eel and 10–15 metric tons silver eel, although the true Figure for commercial landings is estimated at 50–100 metric tons per annum (Rosell 1997). The fishery supports about 60 longline and 10 fyke net license holders divided approximately equally between Upper

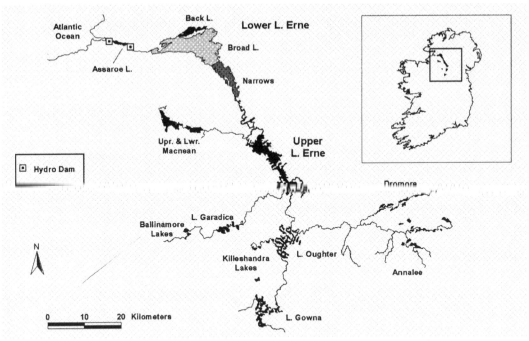

Figure 1. The Erne catchment, showing areas surveyed by fyke net crews in 1998 and 1999.

and Lower L. Erne in Northern Ireland and the upper lakes in the Republic of Ireland. Longline fishing is carried out most intensively in Upper Lough Erne and the Narrows of Lower Lough Erne, which are characterized by shallow (1–5 m depth), productive bays. Fishing in the upper lakes is much less intense, with the majority of fishermen fishing on a part-time basis. The relatively small size of many of the lakes (< 100 ha) in conjunction with lower stock densities means that they are unable to support intensive commercial fishing continuously throughout the season. As a result many are fished intensively for two to three weeks each season and left fallow for the intervening period. Small, remote lakes may not be fished for several seasons.

Methods

Extensive fyke net surveys of the Erne lakes were conducted from mid-July to September 1998 and from May to early September 1999. A total of six 2-man commercial crews in 1998 and four crews in 1999 were assigned to specified zones

Table 1. Physical and chemical parameters of the main lakes in the study.

Lake	Height (m)	Area (ha)	Transparency minutes (m)	Conductivity (mS/cm at 20°C)	Tropic status
Lower Lough Erne	45.7	10,950	1.0–2.0	221–245	Eutrophic
Upper Lough Erne	46.0	3,450	1.0	245	Eutrophic
Lough Oughter	44.9	1,300	0.7	181.3–255.0	Hypertrophic
Lough Gowna	60.5	1,290	0.9	119.4–208.7	Highly Eutrophic

within the catchment (Figure 1). Fishing was carried out using paired summer fyke nets of similar specifications to those described by Moriarty (1972) and in common use in Ireland. Each fyke net had a total length of 15 m comprising two traps (each 3.5 m in length) joined by an 8 m leader net. Each trap consisted of two chambers and a cod end with knot-to-knot sizes of 17 mm, 14 mm and 11 mm, respectively. The diameter of the trap entrance was 55 cm.

Crews typically fished 40–60 fyke nets each, set in multiples of five, with normally 5 or 10 nets per train. Nets were lifted at least five times per week, weather permitting. Number and weight of eels from each chain of five nets were recorded during each lift. Depth and water temperature were recorded by means of portable 'Fishing Buddy' echosounders. Data sheets were completed for each day's fishing, marking the location of each train of nets on maps of scale 1:50,000. Nets were relocated every three to four days to provide comprehensive coverage of the available fishing grounds. Crews were accompanied by a scientific observer at least once per week. Each crew's total catch for the week was examined and weighed prior to sale. Full commercial prices were paid to ensure full catches were presented for sale. Commercial dealers carried out size grading of the catch. Undersize (typically < 42 cm length) and poor quality eels, which were discarded by commercial dealers, were batch weighed for each crew and returned to the lakes at point of sale.

Random samples of 150–200 eels were taken (generally within 24 hours of capture) from each crew each week and anesthetized in chlorobutol. Total lengths were measured to nearest 0.1 cm and weights to nearest 5 g for eels less than 1 kg and 10 g for eels over 1 kg, respectively. A subsample of 5–10 eels per crew per week representative of all sizes was killed by anesthetic overdose, measured, sexed by examination of gonads and the sagittal otoliths extracted. Otoliths were prepared using the burning and cracking method (Christensen 1964; Moriarty 1983), mounted against a glass slide in silicone rubber (Hu and Todd 1981) and measured under 100X magnification using an eyepiece graticule. Measurements were taken from the first clearly marked annulus or the 'transition zone' representing the beginning of freshwater growth (Michaud et al. 1988). The mean length of elvers entering the Erne was previously determined to be 7.2 cm.

Assuming a linear relationship between otolith radius and total length of eel, values for the annual increments were calculated according to the formula:

$$a = i(L - 7.2)/l$$

where:
i = distance between the rings measured on otoliths
$l = \Sigma\, i$ for the individual
L = total length of the individual
a = the calculated annual length increment.

Growth was described using the von Bertanlaffy (1957) growth equation:

$$l_t = L_\infty (1 - e^{-K[t - t_o]})$$

where:
l_t = length at time t
L_∞ = maximum size towards which the length of the fish is tending
K = a measure of the rate at which length approaches L_∞
t_o = parameter indicating the (hypothetical) time at which the fish would have been zero size if it had always grown according to the von Bertanlaffy equation.

Results

Numbers

A summary of the catches is given in Table 2. A total of 6.4 metric tons was taken by six crews over the 12 week season for 1998. Greatest numbers were taken in Upper Lough Erne and the Narrows of Lower Lough Erne. Both these areas traditionally support the most intensively fished areas of the Erne system. Numbers caught declined with increasing distance from the downstream lakes and markedly fewer eels were recorded from the Annalee/Dromore system and Lough Gowna. The average weight was very much greater in the upper lakes, increasing through the system from 139 to 325 g. Examples of daily variations of weight per unit effort, from waters of high and low yield, are given in Figure 2.

Monthly CPUEs for each of the six zones were compared to determine whether catches varied significantly over the duration of the season. There was some indication of lower numbers in the second half of September, but only one zone,

Table 2. Summary fyke-net catch data from each zone of the Erne fished in 1998 and 1999. Zones as in Figure 1; CPUE as catch per 10 paired fyke nets per night.

Crew	Zone	Total effort	CPUE (10 net nights^{-1}) number	kg	Length (cm) n	mean	SD	% in length groups <40cm	>50cm
1998									
1	Broad Lough	3,074	26.6	3.7	1,453	42.8	7.8	40	16
2	Narrows	3,657	35.2	4.9	1,185	44.3	8.3	28	21
3	Upper Lough Erne	3,045	33.6	5.9	1,450	45.8	8.9	22	25
4	Lough Oughter	3,331	15.6	3.9	1,219	50.6	12.4	9	42
5	Lough Gowna	1,930	4.0	1.3	498	49.0	12.7	7	33
6	Annalee & Dromore	1,054	4.0	1.3	950	56.5	14.4	0	61
1999									
1	Broad Lough	2,470	32.0	5.1	804	43.1	7.0	39	16
	Narrows	2,300	33.7	4.5	1,050	42.6	7.0	49	15
	Back Lough	1,030	57.1	8.9	450	46.3	8.4	37	24
2	Upper Lough Erne	2,445	46.9	8.5	1,313	43.7	7.9	39	17
	Lough MacNean upper	620	4.7	1.8	107	56.8	11.8	0–26	71
	Lough MacNean lower	730	18.9	5.0	321	50.4	8.7	0	46
	R. Erne & Assaroe Loughs	640	31.3	6.4	295	48.1	8.8	0	40
3	Lough Oughter	1,760	13.1	3.4	470	47.3	10.4	0–12	29
	Killeshandra Lakes	1,845	8.8	3.0	798	55.0	13.0	0	35
	Lough Gowna	1,950	19.1	4.1	877	46.2	11.0	17–30	23
	Ballinamore Lakes	650	6.3	4.2	300	64.4	14.1	0–31	80
4	Annalee & Dromore	4,650	10.9	2.2	1,665	48.6	12.6	23	35

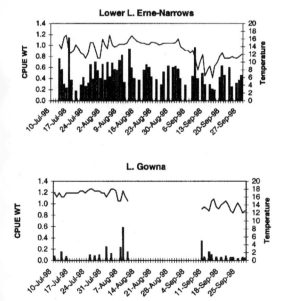

Figure 2. Daily CPUE as weight (kg) per fyke net pair per night with water temperature (°C) from the Narrows and Lough Gowna in 1998.

Upper Lough Erne, showed a significant ($P < 0.05$) decline from July to September 1998.

The 1999 season lasted for 19 weeks and a total of nine metric tons was captured. Fishing in the six 1998 zones was repeated, though with a lower effort, and six additional zones were sampled (Table 2). In two cases, Upper Lough Erne and Lough Gowna, CPUE increased substantially between years. The Back Lough, an area of Lower Lough Erne seldom fished previously, yielded nearly double the catch of the adjacent intensively fished zones. In Upper Lough Erne numbers and weight increased, with a decrease in length. A comparable, but greater, increase took place in Lough Gowna. Low catches were recorded from Upper and Lower MacNean, the Annalee-Dromore system and the Killeshandra and Ballinamore lakes.

Size Composition

A summary of the size composition of eel catches from the Erne Loughs in 1998 and 1999 is shown in Table 2. Length distributions from

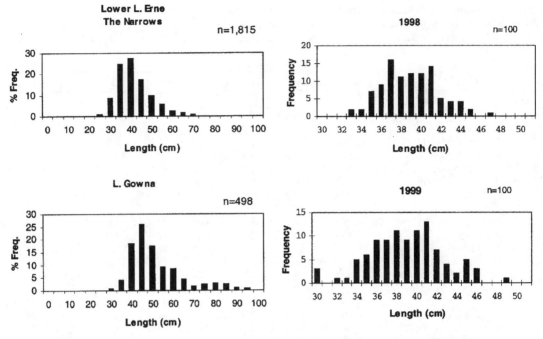

Figure 3. Lengths (cm) of yellow eel from the Narrows and Lough Gowna in 1998.

Figure 4. Representative samples of lengths (cm) of yellow eel from Lower Lough Erne discarded by commercial dealers as undersize or damaged in 1998 and 1999.

the Narrows and Lough Gowna are given in Figure 3 as representative of length distributions of catches recorded from lakes in the upper and lower areas of the catchment in 1998 and 1999.

Undersize eels, as rejected by commercial dealers on weekly sale days, were measured periodically to assess the size range of eel rejected. The majority of discarded eel were found to be less than 42 cm with a small number of larger specimens also rejected as being excessively thin, damaged (e.g., tail bites) or dead. Examples of the lengths of eels from Lower L. Erne discarded by dealers on weekly sale days are shown in Figure 4.

In 1998, length varied significantly (Kruskal-Wallis, $P < 0.01$) between each of the six zones surveyed with increasing means associated with greater numbers of large eel coinciding with distance from the river mouth. Undersized eels made up 40% of the catch by number from the Broad Lough and were absent from the Annalee and Dromore (Table 2). Similarly, the percentage of large eel (> 50 cm) increased from 16% to 61% respectively.

In 1999, length frequencies from Upper and Lower Lough Erne were very similar to those recorded in 1998. On the Back Lough, higher mean length and higher frequency of large eel were observed. In the upper catchment, the areas of greatest commercial eel activity, Loughs Oughter and Gowna, yielded smaller eels than those from the smaller lakes situated west of Killeshandra and Ballinamore, which are fished less frequently. The MacNean system, and in particular Upper MacNean, was characterized by low numbers and a preponderance of large, broad-headed females. Mean lengths were greater in the Upper MacNean.

Undersize eel were much more prevalent in catches from Lough Gowna in 1999 than recorded in 1998. Increased levels of undersized eel were also recorded from the Annalee-Dromore system, which may have been influenced by the inclusion of some of the larger, more regularly fished lakes (e.g., Lough Sillan) on this system during the 1999 season. Very few undersize eel were recorded in catches from Upper and Lower MacNean during 1999 with commercial dealers accepting the catches on most weeks without grading for size.

Age and Growth

Age distributions of female eels collected in 1998 and 1999 from Lower Lough Erne (Narrows and Broad Lough combined) and from Lough Gowna are presented in Figure 5. The majority ranged from 5 to 10 years for each zone with small numbers from 20 to a maximum of 30 years. Lower Lough Erne, the most intensively fished water on the system, and which had the longest history of stocking, showed the nearest approach to a normal curve. Age distribution in the upper lakes was more irregular, and those farthest upstream showed a high degree of bimodality, with a marked preponderance of eel of five to seven years.

Growth rates varied widely between different parts of the Erne with eels from the upper lakes increasing at 4.1–4.8 cm per year. Lower rates, 2.6 and 3.5 cm per year were observed from Upper and Lower Lough Erne. Plots of mean length at age, derived from back-calculated lengths, show that differences in growth rate between zones were evident within the first three to five years. Eels from the upstream lakes, Loughs Oughter, Gowna and the Annalee-Dromore, showed near linear growth to maturation before 16 years of age (Figure 6), while the growth of those from Upper and Lower Lough Erne was clearly asymptotic. The values for L_∞ (Table 3) from the upper lakes reflect this in being very much greater than the length of the largest eel sampled. Values for the lower lakes are closer to the observed maxima.

Discussion

Previous study of yellow eel stocks in the Erne (Moriarty 1973) was confined to lakes in the upper catchment. Eels from seven lakes in the Killeshandra and the Annalee-Dromore areas showed the highest growth rates observed in Ireland with a mean size of 51 cm at 10 years, at least 10 cm longer than the means for eels of the same age in other Irish lakes. Few eels older than 12 and none more than 16 years were captured, indicating that early maturation was a characteristic of the population.

Direct comparison of Moriarty's (1973) results with those derived from the current survey, some 25 years later, showed a slightly lower mean length of 48 cm at 10 years estimated from back-calculated eels. Maximum ages recorded from the Killeshandra and the Annalee-Dromore

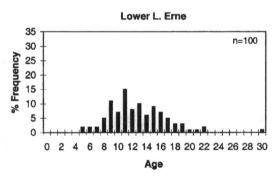

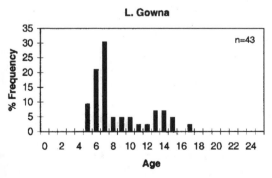

Figure 5. Age (years) of female yellow eels collected in 1998 and 1999 from Lower Lough Erne and Lough Gowna.

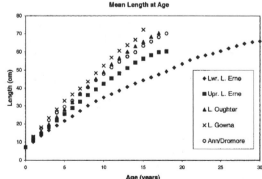

Figure 6. Back-calculated mean length (cm) at age of female yellow eel from the Erne catchment in 1998 and 1999.

Table 3. Von Bertalanffy parameters from back calculations.

Zone	L_∞	K	t_0
Lower Lough Erne	93.9	0.96	−2.19
Upper Lough Erne	104.1	0.96	−1.78
Lough Oughter	164.9	0.97	−0.58
Lough Gowna	146.7	0.96	−5.58
Annalee & Dromore	121.0	0.96	−1.38

lakes were 17 and 24 years, respectively, with 8% of the Annalee-Dromore sample older than the 1973 value of 16. The observation of older eels in the current study may well reflect the greater fishing effort and increased number of lakes sampled in each area rather than any underlying change in eel growth or maturity over the intervening 25 year period.

Stocking with elver, in progress for some 30 years, is the only means whereby eels can enter the lakes of the Erne catchment. The fact that the lakes hold abundant stocks of eel aged less than 15 years clearly indicates that small eels were present in the sampling areas. The youngest eel captured in the fyke nets were aged five years and it is assumed that younger individuals, if captured, were small enough to escape through the cod end mesh. Substantial numbers of eel of five to seven years were present in the Annalee-Dromore lakes, having been scarce or absent in the 1970s (Moriarty 1973). This coincided with the introduction of a stocking program in 1993.

Results from the current extensive fyke net survey of the Erne system conducted by teams of commercial fishermen indicates stock abundance, size composition and growth rates to vary widely between different areas of the catchment. Upper and Lower Lough Erne, which support an extensive commercial longline fishery, showed the greatest CPUE at 30–45 eels per 10 nets per night, falling to 13–15 in Lough Oughter and 4–11 in the upper reaches of the catchment. Similarly, the proportion of undersize eels declined progressively from the lower lakes to those in the upper system.

Impact of Commercial Longlining

Although every effort was made to avoid fishing in areas recently covered by commercial longlining crews, the survey results indicated depletion of the stock as the commercial season progressed. Marked differences were observed between the size and age frequencies of samples from adjacent areas known to have been recently longlined or left fallow. This was particularly evident in Lower Lough Erne in 1999 where the lengths of eel were significantly larger in the Back Lough (an area rarely longlined due to submerged trees and underwater obstructions) than elsewhere in the lake, which is heavily longlined through the commercial season.

Similar contrasts in catches were noted between Loughs Oughter and Gowna which are annually fished by commercial longline and fyke net crews, and the Ballinamore and Killeshandra lakes, which are sporadically fished due to their smaller size (< 60 ha) and difficult access.

The results indicate that there is potential for increased exploitation levels, particularly if recruitment can be enhanced. Zonation of the catchment, as implemented with fyke net survey crews during this study, would increase exploitation efficiency, create a sense of ownership among fishermen and improve catch return data for future monitoring of the fishery.

Enhancement Stocking

The current distribution and abundance of yellow eel, and in particular undersize eel, throughout the Erne reflects previous elver stocking policies in which elver were released immediately upstream of Ballyshannon or stocked out into Lower Lough Erne and the northern end of Upper Lough Erne (Table 2). The preponderance of relatively large (> 50 cm) and old eels in Upper and Lower MacNean and the Killeshandra lakes indicates recruitment to have been limited and largely confined to eels which have taken a number of years to make their way upstream through the system.

In contrast, the presence of strong five to seven year age classes in Loughs Oughter and Gowna and the larger lakes on the Annalee-Dromore system (Figure 5) suggests that overland transportation and stocking of elvers to these lakes which commenced in 1993 were recruiting to the fishery by 1998. These results give a dramatic demonstration of the effectiveness of a stocking program.

Currently, total annual recruitment of elver to the Erne averages 1.3 metric tons. Based on the European recommended elver stocking rates of 0.1 kg/ha, and a total lake area of 33,000 ha, an annual target of 3.3 metric tons of glass eel and elver would be required to fully utilize the available habitat on the Erne. Whether similarly rapid growth rates would be recorded in eels from lakes stocked at higher densities remains unknown, with the availability of additional glass eel and elver for stocking remaining the main barrier to future enhancement measures.

The extensive sampling by standard (fyke net) gear has provided an assessment of the status in 1998 and 1999 of the sub-population of eel, which are available to the commercial fishery. The data on numbers, size and age will enable reliable comparisons with this baseline data to be made as the development of the fishery progresses. The surveys have also provided evidence that measurable effects of stocking become apparent after as little as five years, using standard commercial fishing gear. Information on size and variability of catches made by commercial fishing crews will contribute to the accuracy of estimates of total catch for the entire catchment.

Acknowledgments

This study forms part of the Erne Eel Enhancement Program, which is funded under the EU Special Support Program for Peace and Reconciliation. We gratefully acknowledge the assistance of the survey crews and Emer Kelly in the preparation and reading of otoliths. Russell Poole corroborated the otolith readings. The comments of Brian Knights and an anonymous reviewer on an earlier draft of this manuscript are also greatly appreciated.

References

Battarbee, R. W. 1976. The eutrophication of Lough Erne inferred from changes in the diatom assemblages of ^{210}Pb- and ^{137}Cs-dated sediment cores. Proceedings of the Royal Irish Academy 86 (B): 141–168.

Christensen, J. M. 1964. Burning of otoliths, a technique for age determination of soles and other fish. Journal du Conseil Permanent International pour l'Exploration de la Mer 29:73–81.

Gibson, C. E. 1998. The Erne. Pages 237–256. in C. Moriarty, editor. Studies of Irish Rivers and Lakes Dublin: Marine Institute.

Hu, L. C., and P. Todd. 1981. An improved technique for eel otoliths for ageing. Journal du Conseil Permanent International pour l'Exploration de la Mer 15:115–116.

Lucey, J., J. J. Bowman, K. J. Clabby, P. Cunningham, M. Lehane, M. MacCarthaigh, M. McGarrigle, and P. F. Toner. 1997. Water Quality in Ireland 1995–1997. Environmental Protection Agency Report.

Michaud, M., J. D. Dutil, and J. J. Dodson. 1988. Determination of the age of young American eels, Anguilla rostrata, in freshwater based on otolith surface area and microstructure Journal of Fish Biology 32:179–189.

Moriarty, C. 1972. Studies of the eel *Anguilla anguilla* in Ireland. 1. In the lakes of the Corrib system. *Irish Fisheries Investigations* A. 10.

Moriarty, C. 1973. Studies of the eel *Anguilla anguilla* in Ireland. 2. In L. Conn, L. Gill and north Cavan lakes. *Irish Fisheries Investigations* A. 13.

Moriarty, C. 1983. Age determination and growth rate of eels, Anguilla anguilla (L.). Journal of Fish Biology 23:257–264.

Moriarty, C. 1999. Strategy for the development of the eel fishery in Ireland. Fisheries Bulletin (Dublin) 19, 46 pages.

Moriarty, C., and J. D. Reynolds. 1997. Ireland (Republic) in C. Moriarty and W. Dekker, editors. Management of the European eel. Fisheries Bulletin (Dublin) 15: 64–68.

Rosell, R. 1997. Ireland (Northern Ireland) in C. Moriarty and W. Dekker, editors. Management of the European eel. Fisheries Bulletin (Dublin) 15: 60–63.

von Bertanlaffy, L. 1957. Quantitative laws in metabolism and growth. Quarterly Review of Biology 32:217–231.

A Review of Eel Fisheries in Ireland and Strategies for Future Development

CHRISTOPHER MORIARTY

University of Dublin, Zoology Department, Trinity College, Dublin, Ireland
Email: cm@iol.ie

Abstract.—Key elements of the biology and current status of European eel *Anguilla anguilla* fisheries in Ireland are reviewed. This leads to the formulation of a stocking strategy based on techniques known to be successful in Northern Ireland and elsewhere in Europe. Its aim is to increase the yield of wild-caught eel in the Republic of Ireland by a factor of four, from 250 to 1,000 metric tons per annum. A co-operative fishery operated in Lough Neagh in Northern Ireland has an annual yield, sustained since 1962, of 700 metric tons, providing seasonal employment for 370 fishers. The lake is stocked with glass eel at a rate of 250 elver (0.1 kg) per hectare and yields 20 kg per hectare. The fishery is strictly regulated, with a fixed daily quota per fisher. Extensive sampling of eel throughout the Republic of Ireland has indicated the existence of 500 km^2 of surface area of lake in which eel growth is equal to or greater than that observed in Lough Neagh. The development theory is that, if stocked at the same rate as Lough Neagh, the same yield could be obtained. Management and investment problems will be considerable and entail the commitment of substantial state funding over an initial period of ten years. Although such factors as currently falling prices for eel make it impossible to forecast the economic return, the fact remains that no development of the wild eel fishery is possible in the absence of a stocking program. The best advice to the authorities, therefore, is to develop such a scheme without delay. It is concluded that a similar strategic approach could be applied to eel-bearing catchments in North America.

Introduction

Archaeological studies show that European eel *Anguilla anguilla* fishing has been in progress in Ireland for at least nine thousand years. The mesolithic site used by these early fishers is on the River Bann, downstream of Lough Neagh which today is one of the most effectively managed eel fisheries in Europe and the model for the strategic plan outlined in this paper (Figure 1). Historical records of eel fishing elsewhere in Ireland (O'Flaherty 1684), extend back to the 17th century and a large number of 'eel weirs' are marked on the first Ordnance Survey maps of Ireland, published in 1837. Other than the legal requirement that not more than 90% of the width of a river may be obstructed by eel catching gear, no management measures were taken until 1929. In that year a minimum size limit of half a pound (227 g) was imposed in some of the major fisheries, together with a minimum gape of the hook (9.5 mm) used in a long-line. Minor scientific studies on the silver eel of the Lough Neagh catchment were made by visiting consultants in the 1930s, but no systematic work was undertaken in Ireland until 1965.

Active steps in management of the eel in Ireland began in the Lough Neagh catchment in 1932, when ascending elvers were intercepted at the head of the tide and transported overland to the lake. The same practice began on the River Shannon catchment in 1959. By that time the Lough Neagh fishery was recognized as one of the most productive in Europe and its output was very much greater than the entire production of other lakes, rivers and estuaries in Ireland.

Also in 1959, this author was instructed to 'do eels', with a view to finding out why Lough Neagh was so much more successful as a fishery than the other waters and whether any measures might be taken to improve the situation. The approach taken, after reviewing the fishing methods and finding them similar to those in operation in Lough Neagh, was to embark on a wide-ranging survey of the biology of eel populations throughout Ireland. Extensive data were collected on catch per unit effort, length, diet and age. This led to a conclusion that an increase in stocking, followed in due course by an increase in fishing effort, would result in production of the same order of magnitude as prevailed in Lough Neagh.

In 1998 a national workshop on eel was held (Watson et al. 1999) and the following year this author, on behalf of the Irish Marine Institute, published a strategy plan (Moriarty 1999). The aim of this paper is to summarize these two publications and provide an outline of the cost of a development scheme that could be used as a model for eel fishery enhancement in north temperate regions.

The Lough Neagh Fishery

Lough Neagh, a freshwater lake in Northern Ireland, has long been known as one of the world's most productive eel fisheries. In 1963, an improved management regime was established and by 1982 it was evident that an annual yield in the order of 700 metric tons, or 17 kg per hectare, had been achieved and sustained. Over the following decades, this sustainability has been maintained.

The lake, with a surface area of 40,000 hectares, has several features, unique in Ireland, which have rendered it ideal for rational fishery development. It is approximately square in shape and there are very few islands or indentations on the shoreline. This makes for exceptionally easy navigation and also for patrolling for regulatory purposes. The depth is less than 30 m, with a large proportion of the bed being roughly level at a depth of 10 m. This means that the entire bed is habitable by eel—although the eel population of extensive areas of sandy substrate is so small that nobody fishes over them. The most important feature is the ownership of the fishing rights. Legislation from the 17th century has made it possible for a single individual or company to be the sole owner of the fishing. In 1971, these rights were acquired in full by the eel fishermen themselves, acting as a registered Co-operative Society. The lake is base-rich, eutrophic and highly productive (Wood and Smith 1993).

The aim of the Lough Neagh Fishermen's Co-operative is to maximize the number of fishers, while ensuring the stability of the stocks. The management system is to stock the lake at an average annual rate of 0.1 kg elver per hectare, to limit the number of licensed boats and to impose a daily catch quota on the fishers. The number of licenses issued is 185, employing some 370 fishers since two fishers usually man the boats. The quota is 51 kg per day. The catch is collected daily from the fishers and delivered to the riverside packing station where undersized eel are returned to the water. Marketable eel are graded according to varying demands in the receiving countries and dispatched alive by airfreight. Silver eel are caught at two stations downstream of the lake and exported in the same manner.

Key Aspects of the Biology of the Eel in Ireland

Glass eel begin to enter river estuaries in Ireland in December, but are not present in large enough numbers for substantial catches to be made until February. Entry into fresh water seldom begins before April and reaches a peak in May. From May until July, and sometimes August, bootlace eel may be captured on migration upstream as they ascend through eel ladders. The majority of these are less than 12 cm in length, with decreasing numbers of larger individuals up to at least 50 cm. A remarkable point about the upstream migration of these relatively large eel is that they are bigger than many in the silver phase. There is thus a degree of overlap in size of eel migrating both upstream and downstream (Moriarty 1986)

One-off surveys by fyke net of the eel in lakes and rivers throughout Ireland were undertaken between 1965 and 1975. Standard gear with 12-mm cod end mesh was used in a large number of lakes and rivers. This allowed sampling of eel down to a length of 27 cm, capturing a representative sample of 37 cm and larger. Length, weight, age and stomach contents were determined. This allowed comparisons to be made between watercourses of abundance, size, food and growth and the following conclusions were drawn (Moriarty 1999).

European eel in its growth stage is virtually absent from the fully marine coastal waters of Ireland. Substantial populations of yellow eel are present in a number of river estuaries. Commercial fisheries are based in two of these, Wexford Harbor and Waterford Harbor. But other large, and apparently suitable, estuaries including those of the Shannon and the Lee do not have large eel populations and are not exploited. In addition to males, female eel of all sizes are found in the estuaries. There is strong circumstantial evidence that these eel spend their entire lives in the estuaries, become silver and migrate to sea without ever entering fresh water.

Nearly all watercourses in Ireland have a population of eel. The exceptions are the extremely rare land-locked lakes and a small proportion of riverine habitat made inaccessible by high waterfalls. Numerous lakes are present on the river systems of the northern two-thirds of Ireland, in contrast to the rivers of the south, which have few lakes or none. The four major lake and river systems are the Bann, the Erne, the Corrib and the Shannon (Figure 1). The Bann is exceptional in that its sole major lake, Lough Neagh, was formed by tectonic subsidence and lies at a relatively high point in the river system, with no substantial inflowing streams or upstream lakes.

The other lake systems result from glacial action in geologically recent times and all are less than 10,000 years old. On each of these systems, the biggest and deepest lake lies less than 20 km from the head of the tide. The survey (Moriarty 1988) showed that, in a given catchment with increasing distance from tidal water, eel became fewer, larger and older.

Growth in all cases was relatively slow. Values for theoretical maximum length ($L\infty$) on the order of 90–100 cm, were determined using back calculations but, for practical purposes, the growth curve for eel of the sizes sampled was close to a straight line and varied between 2 and 3 cm per year. The majority of eel of market size were aged between 10 and 15 years.

Considerable variation between waters was observed in food consumed and in age and size at silvering. From the point of view of management of the stocks, however, the most important observations were that, in base-rich waters, growth rate equal to or greater than the rate in Lough Neagh was the rule. Furthermore, population density in the downstream lakes was similar to that in Lough Neagh and there was no evidence of over-population in any case. Although population density in acid waters was sometimes similar to that in base-rich ones, growth was slower, silvering took place later and it was concluded that development of the fishery should be given a low priority.

Current Fishery and Management Practice

Glass eel and elver: Fishing for young eel is prohibited by the 1959 Fisheries Act, but may be

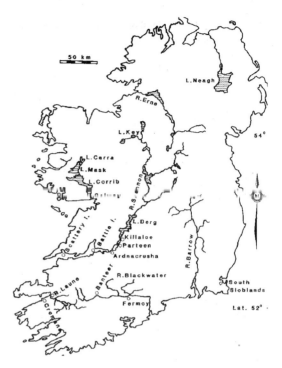

Figure 1. Major Rivers and Lakes in Ireland (from Moriarty 1987).

authorised under its Section 14 in the interests of developing the fishery, both for stocking material and as 'seed' for eel culture, since artificial propagation is not possible. Because of high demand from the Far East, prices reached extremely high levels in 1997, but declined in subsequent years. The elver is still a very valuable commodity, worth between US$50 and $100 per kg alive and in good condition. Section 18 (1a) of the 1980 Fisheries Act empowers Regional Fisheries Boards to capture glass eel or elver for research or for stocking.

Capture of glass eel for stocking did not take place in Ireland until the 1990s when anchored tidal (tela) nets and dip-nets (Matthews et al. 1999) began to be operated successfully. Tidal nets may be large and set by boat in mid-estuary or small and attached to bridges. Dip nets are about 1-m diameter and operated manually from the bank.

Elvers, migrating from salt to fresh water, have been captured systematically for transfer upstream or between catchments on the River Bann since the 1930s, on the Shannon since 1959 and later on other rivers. The usual gear is an elver trap, in the form of a catching box installed at the top of an elver pass.

Yellow eel: The feeding and growing stage is caught by baited long line or eel pot or by unbaited fyke net. In addition to the 185 licenses on Lough Neagh, about 200 fishermen held licenses in 1999 elsewhere in Ireland. It appears that all waters, in which viable eel catches are likely to be made, are exploited. Some may prove to be under-fished and the stocks in all may be enhanced by the application of suitable management measures. Many proposals for the development of unfished waters have been made and some followed up. Few succeeded. It seems reasonable to believe that, in the course of hundreds of years, eel fishers had established which watercourses could yield good catches and which were not worth the effort.

The average annual catch per fisher on Lough Neagh is on the order of 1.5 metric tons. Returns collected by the Department of the Marine and Natural Resources between 1988 and 1996 showed that 19 independent eel fishers in the Republic of Ireland regularly made catches in excess of 1 metric ton per annum. Two of these operated weirs for silver eel, one baited estuarine eel pots, four fyke nets and the remaining 12 long-lines in fresh water. The top return, 4.3 metric tons by fyke net, was nearly equaled by a long-line catch of 4.2 metric tons. At a price to the fisher of US$4 per kg, the annual income of the best workers was therefore $16,000 and, on average, $4,000 to $6,000.

The majority of fishers reported catch considerably less than 1 metric ton. The size of the catch depends to a great extent on the skill of the fisher, but also on the effort expended. It is clear that a reasonable income can be made by an exceptionally hard-working fisher, while helpful supplements to other sources of income are made by the average worker who approaches eel fishing part-time.

Silver eel: Downstream migrants are caught at 'eel weirs', in most cases, barriers of conical *coghill* nets attached to stakes in rivers. Best catches are made in stormy weather about the time of the new moon from September to December. Except in the case of the biggest river systems, Bann, Erne, Corrib and Shannon, operation of silver eel fisheries has declined greatly in the course of the 20th century. High cost of the gear has combined with low catches and increasing alternative forms of income to make silver eel fishing unattractive.

In most cases a 'free gap' of 10% of the width of the stream is required at an eel weir. This allows the escape of considerably more than one eel in every ten. The yield in a number of river systems is greatly increased by the installation of additional weirs. The installation of a weir upstream of an existing fishery does not necessarily reduce the catch of the latter. However, the development of new weirs in such situations may be highly contentious.

Management measures: Although angling for eel is increasingly popular, the greater part of the catch is commercial. All commercial fishermen must be licensed. Licenses are issued by the seven Regional Fisheries Boards which aim to control the numbers of fishers and the gear used in the interests of conserving the stocks. Following indications of reduced recruitment in recent years, attempts are being made to hold fishing effort at its 1997 level.

Size limits are technically in force on two of the major river catchments (i.e., Shannon and Corrib), having been introduced by bye-law in 1929 and 1930. The size limit is half a pound (227 g) and would, if enforced, virtually prohibit the capture of male eel, since the great majority of males in Ireland normally migrate at lower body weight. These restrictions were introduced in the absence of any relevant scientific understanding of the situation and are not enforced.

Closed seasons are also in force in a number of catchments. These were imposed largely in the interests of simplifying control of illegal salmon fishing rather than as a measure to manage the eel fishery. In combination with strict control of fishing effort, a closed season is an effective means of limiting total catch. It is, however, a very crude measure which takes no account of such important factors as the age distribution of the stock. Furthermore, it may adversely affect marketing by rendering eel unobtainable at times of high demand.

Development Strategy

One of few indisputable facts in the eel fishery of Ireland is that, for 37 years, Lough Neagh has sustained an output of eel in the order of 17 kg per hectare, providing a livelihood for nearly 400 individuals.

Lough Neagh is stocked at a rate of 0.1 kg per hectare. The elvers for this purpose are caught at a tidal barrier, which is believed to block the natural ascent. Based on the assumption that no significant numbers of elver enter the lake system by natural means, the stocking rate is taken to be equal to the total recruitment. The Lough Neagh catchment is the only one in Europe for which such detailed information on stock and crop is available. The proposed strategy for development of the eel fishery in the rest of Ireland is therefore based on the assumption that, if productive waters were to be stocked at a rate of 0.1 kg elver per hectare, they could be managed to yield 17 kg per hectare.

There is a simplistic element in this approach—but, in the absence of much basic biological data on the species and the remote chance that this can be made good in short- or mid-term, it is argued that it is the only possible model to apply.

The negative view is that Lough Neagh is an exceptionally good eel habitat and that it is already at its full carrying capacity. Therefore, it would be wrong to assume that other Irish waters can be developed to the same extent. The positive alternative, that the other waters would constitute a better habitat and that Lough Neagh itself could be managed to yield a greater catch is, at least, equally valid. While there is virtually no other information available on yield per recruit in Europe, there are cases, such as the Imsa lake system in Norway and the Ayrolles-Gruissan lagoon in France (Moriarty 1996) of more than double Lough Neagh's yield per hectare.

Since it takes between 10 and 15 years for an eel in Ireland to attain marketable size, there is a long lead-in time before any results of enhanced stocking are likely to be observed. The recommended development strategy was to initiate a stocking program for all suitably productive and exploitable lakes with elver at the observed Lough Neagh rate. Further studies aimed at confirming the appropriateness of the Lough Neagh rate would require a number of years and might not be conclusive. Any delay in developing the stocking program would undoubtedly delay the development of the fishery.

The prognosis for ultimate yield assumes that these lakes will have the same production as Lough Neagh. The following proposals for development have been made to the Irish government, on a basis of two successive five-year plans.

First five-year plan: Enhancement by elver capture and transport is likely to require a period of ten years before any substantial improvement in the stocks may take place. In the meantime, it is proposed that the recommendations of the EIFAC/ICES Working Party on Eel (Anonymous 1997) be accepted. These include the view that fishing effort should not be increased in the absence of enhancement measures.

An approach which agrees with this is to impose a moratorium at present levels. In the first place, no new licenses should be issued for yellow eel fishing in the established fisheries, nor should any licensees be permitted to increase the number of nets or other gear currently allotted. An exception to this might be made so that one licensee could operate the nets of another who wished to cease fishing.

Monitoring, by sampling the commercial catch, should take place in all the existing fisheries. If necessary, legislation should be introduced requiring all licensed fishers to co-operate with designated officials in providing samples and permitting observation. No steps to alter the intensity of fishing should be taken until the end of the five-year period so that a proper analysis of the results may be achieved. This procedure entails a risk of allowing particular fisheries to collapse, but restrictions enforced over any shorter period would make it almost impossible to draw firm conclusions about the performance of the fishery.

Second five-year plan: Stocking at the established rate will continue throughout the second period. The plan will be based on the analysis of the results of monitoring in the course of the first five years. These results are not predictable and could equally well provide evidence of under- or over-fishing. The former, under-fishing, should be addressed where necessary by increasing permitted effort. Over-fishing is likely to result in reduced effort and no active steps should be taken to impose additional restrictions. It is considered that, in the long term, more will be gained in the form of valuable information on population dynamics than would be lost in allowing a particular fishery to decline.

The most important principle to adopt is to avoid interference with the progress of the fishery as established for the first five years. Because of the slow growth-rate of the eel and the fact that the fishery captures a number of year classes, it is not possible to base predictions on observations made over a period of less than five years. The longer-term welfare of the fishery will depend on the ability to make sound predictions.

Designation of waters for fishery development and for spawning stock: The results of the nation-wide studies carried out in the 1960s and 1970s have indicated that eel productivity is high in the base-rich lakes; however, growth is slow and productivity low in oligotrophic lakes. With the exception of rare cases, such as the Burrishoole river and lake system, in which efficient downstream traps are installed and used for research purposes, the potential for developing any acid lakes for eel fishing must be questioned. These lakes, however, yield substantial numbers of spawners and, therefore, potentially contribute to the eel resource as a whole. In view of current concern among international authorities that spawner escapement should be enhanced (Anon. 1993), serious consideration should be given to declaring all such lakes eel sanctuaries—except for angling by single rod and line which is highly unlikely to have any negative impact on the eel population.

River-based eel fishing, in particular for silver eel at weirs, has been widespread but is now confined to a small number of long-established installations. Fyke-net fishing in rivers is effective, but, because of the very small area of water involved, is likely to be destructive in the sense of removing within a single season five or more year classes from any stretch of water. Because of the small size of nearly all Irish rivers, it is unlikely that an annual sustainable fishery could be developed in any of them other than the Shannon. Existing fisheries, such as those on the Barrow, should be supported but serious consideration should be given to maintaining rivers as eel sanctuaries, again with the exception of angling.

The primary argument against developing eel fishing in rivers and in acid lakes is economic. An important positive aspect of the ideal of eel sanctuaries is that, although production is low, the extent of the unfished rivers and lakes is so large that they make a substantial contribution to the spawning escapement.

Although there have been many anecdotal accounts of a negative impact of commercial eel fishing on other species, especially salmon and trout, widespread observations by scientists have failed to come up with any evidence of risk to other fish populations caused by eel fishing. The relatively high value of the eel compared to most other species and the absence of a legitimate market for them is an added inducement to the eel fisher to confine himself to his chosen species. In spite of these facts, strong objections are made in places to any use of nets to catch fish. In the absence of any sound evidence of incompatibility between sport fishers and commercial eel fishing, objections by local interests should be met by an information campaign rather than by restriction of eel netting.

Research and monitoring requirements: Maximizing the sustainable catch requires sophisticated monitoring throughout the eel's habitat. It is envisioned that the first three years of the development scheme will require an intensive scientific input to establish base-line data. Following this, monitoring may be carried out effectively by a small team. This must be continued as a permanent exercise throughout and after the initial ten-year lead-in period.

Costs and benefits: The cost of implementing the ten-year program has been estimated on the basis of employing a research and development team of 19 workers for the first three years to locate and monitor suitable glass-eel catching sites, to establish the routine stocking procedure and to carry out the base-line research (Watson et al. 1999). Thereafter the team would be reduced to six development, and two scientific workers. The total cost, including equipment but assuming that sufficient laboratory and office space already exists, was estimated at less than or equal to $5.6 million. The value of the increase in annual yield of 750 metric tons would be between $3 and $3.5 million. The research and development costs would therefore be covered by the return from the first two years of operation of the enhanced fishery.

This estimate is made simply in terms of increased sale of unprocessed eel. Eel are rarely retailed as fresh fish, but are sold in various prepared forms, particularly smoked (in northern Europe) or marinated (in Mediterranean regions). These products are sold at high prices and very substantial value added is obtained in the processing industry. It is assumed that an increased catch in Ireland would lead to the establishment of processing plants so that the gain to national income would be very much greater. The contribution to the national economy could be between three and four times the value of the catch. By this calculation, the development costs would be met in the course of the first year's return.

In recommending funding of the proposed development, two key factors are considered. First, the food of the eel is largely small aquatic invertebrates which are otherwise unexploited. The market eel, therefore, is produced without any input other than the element of manpower in the stocking and fishing operations. Second, eel fishing involves very small capital investment but is highly labor intensive. Besides maintaining the supply of a highly prized food, the fishery provides income and employment in more or less remote rural communities and makes an extremely important sociological contribution

Because of the long lead-in time for development and the problems in policing the fishery, it is considered highly unlikely that any form of venture capital could be attracted. On the other hand, the gain to the state of creating employment and a marketable product will be considerable and permanent. The recommendation, therefore, is that the project be funded entirely by the State, at least for the first ten years of development.

Conclusions

The principal factor limiting the development of the eel fishery in Ireland is the behavior pattern of the eel, which leads to under-population of large areas of potentially productive waters. The application of established technology in capture and re-distribution of the elver may solve this problem. Following a lead-in period of ten years, a highly profitable fishery, with low annual maintenance costs, would be brought into existence.

Costs and benefits were calculated on the basis of prices in operation in Ireland at this time. The price per kg paid to the fishers was declining and, should the decline continue, the ultimate value of the enhanced catch would be less than forecast. Moreover, the current trend in employment in Ireland is towards office and factory work and away from commercial fishing in inland waters. These factors, if unchanged, could lead to the production of an underexploited stock, without the benefit of increased employment and income. However, such predictions could prove false in the long term. The price of eel could increase, as could the numbers of people seeking outdoor employment. All available evidence points to the impossibility of increasing the eel population and developing the fishery without establishing a stocking program. In such a situation, the best advice possible to government is to proceed with a development scheme on the lines presented herein. Given that the same limiting factor of migratory behavior is likely to apply to the American eel, it would appear that a similar strategy could be applied to suitable catchments in North America.

Acknowledgments

This strategy was developed while the author was employed in the Irish Government's Fisheries Research Center. The Erne Eel Enhancement Program generously provided support for participation in the AFS-EPRI Symposium. Thanks to Brian Knights, Milton Matthews, and Hakan Wickström for their very helpful comments on an earlier draft of this manuscript.

References

Anonymous. 1993. Report of the eighth session of the Working Party on Eel. EIFAC Occasional Paper 27:8.

Anonymous. 1997. Report of the eleventh session of the Joint EIFAC/ICES Working Group on Eel. EIFAC Occasional Paper 33:11.

O'Flaherty, R. 1684. *West or H-Iar Connaught*. Galway: Kenny's Bookshop. 1978. (Reprint with Introduction by W J Hogan of J Hardiman's edition of 1846).

Matthews, M. A., R. S. Rosell, D. W. Evans, and C. Moriarty. 1999. The Erne eel fishery. *In* L. Watson, C. Moriarty, and P. Gargan editors 1999. Development of the Irish eel fishery. Fisheries Bulletin (Dublin) 16.

Moriarty, C. 1986. Riverine migration of young eels *Anguilla anguilla* (L.) Fisheries Research 4:43–58.

Moriarty, C. 1987. Factors influencing recruitment of the Atlantic species of anguillid eels. Pages 483–491 *in* M. J. Dadswell, R. J. Klauda, C. M. Moffit, R. L. Saunders, R. A. Rulifson and J. E. Cooper, editors. Common strategies of anadromous and catadromous fishes. American Fisheries Society, Symposium 1, Bethesda, Maryland.

Moriarty, C. 1988. The eel in Ireland. Royal Dublin Society, Occasional Papers in Irish Science and Technology 4:1–9.

Moriarty, C., editor. 1996. The European eel fishery in 1993 and 1994. Fisheries Bulletin (Dublin) 14:1–52.

Moriarty, C. 1999. Strategy for the development of the eel fishery in Ireland. Fisheries Bulletin (Dublin) 19:59–65.

Watson, L., C. Moriarty, and P. Gargan editors. 1999. Development of the Irish eel fishery. Fisheries Bulletin (Dublin) 16.

Wood, R. B., and R. V. Smith editors. 1993. Lough Neagh: The Ecology of a Multipurpose Water Resource. Kluwer, The Netherlands.

The Exploitation of the Migrating Silver American Eel in the St. Lawrence River Estuary, Québec, Canada

GUY VERREAULT*, PIERRE PETTIGREW,
RÉMI TARDIF, AND GONTRAND POULIOT

*Faune et Parcs Québec, Direction de l'aménagement de la faune du Bas Saint-Laurent,
506 rue Lafontaine, Rivière-du-Loup, Quebec G5R 3C4, Canada.
Corresponding author: guy.verreault@fapaq.gouv.qc.ca

Abstract.—Over the past decade, the St. Lawrence River Estuary's American eel *Anguilla rostrata* commercial fishery has experienced a dramatic decline (300 metric tons in 1990 to 72 metric tons in 2000). Eels have been harvested in large fixed traps on the muddy flats of the Estuary's south shore since early European colonization. Field sampling in 1996–1998 showed that catches were made up of female silver eels, much larger than reported anywhere in North America. Mean length and weight was 85.3 cm (SD = 9.31) and 1,264 g (SD = 498.5). Eye index varied from 2.7 to 10.2 and was not suitable to adequately classify immature and maturing eels. Daily catch estimates were correlated to migration chronology. Eels migrate through the area from mid-September to early November, however, 90% of all migrants are usually caught during 30 days centered on mid-October. Fishing effort decreased by 39.4% during this period and mean annual CPUE fluctuated between 3.27 and 4.27 kg/m of leader net. Therefore, CPUE seems no longer adequate to assess eel abundance and biomass in the fishery area.

Introduction

The American eel *Anguilla rostrata* is a common species in the St. Lawrence River system. Native people were already fishing eels when European settlers arrived on the shore of the St. Lawrence (Sagard 1636), and their exploitation targeted mainly seaward migrating eels. In the early 1600s, the newcomers installed hundreds of weirs on the tidal shore as eels were considered a valuable resource for the young colony. From those years to 1958, there were no specific regulations applied to eel fisheries and people were permitted to catch as much as they could. In the early 1960s, fishing regulations were introduced and the number and variety of gear types permitted were restricted.

The commercial eel fishery in the St. Lawrence watershed extends from Lake Ontario to the estuary, 450 km downstream. The most important fishing zone, both in terms of landings and value, is located in the estuary. In this area, large weirs are deployed targeting maturing eels migrating from freshwater habitat to the spawning grounds in the Sargasso Sea (Axelsen 1997). Eels coming from Lake Ontario, the St. Lawrence River, and its tributaries have to pass through the estuarine fishery en route to the Sargasso Sea. The fishery can be considered as a productivity index for the St. Lawrence above this location. Since 1986, a dramatic decline has been observed in the landings and elver recruitment in Lake Ontario (Casselman et al. 1997), and this downward trend is still being observed. Castonguay et al. (1994) critically reviewed four potential causes for the decline: anthropogenic chemical contamination, anthropogenic habitat modifications, oceanic changes, and commercial fishing. In their review, the authors noted that the major difficulty in examining the fishing hypothesis was the absence of fishing effort information. This situation was exacerbated by a lack of basic knowledge on fishing exploitation and on the eel's biological parameters. Consequently, we began research in 1996 on various aspects of the life cycle of eels and on the fishery itself.

The major objectives of our study were: to define morphometric parameters of the exploited eel stock in the commercial fishery of the St. Lawrence Estuary; and to describe the fishery exploitation parameters including fishing effort and spatial and temporal distribution of landings.

Study Area

The study area is on the south side of the St. Lawrence Estuary, beginning 80 km downstream from Quebec City and extending 120 km further downstream (Figure 1). Mean width is 17 km in a transition zone between fresh and marine waters. Water salinity ranges from 6 ppt in the upstream section to up to 30 ppt downstream (Ghaniné et al. 1990). This dynamic frontal region shows high turbidity (Ouellet and Cerceau 1976) with high suspended sediment concentration and strong currents. This zone has strong semidiurnal tides with a mean amplitude of four meters (d'Anglejean 1990). Tidal activity and sediment deposit contribute to the development of the wide tidal flats, which can extend up to 2000 m from shore. Freshwater from upstream generally flows downstream along the south shore, in contrast to the north shore, where there is a movement of marine water heading upstream.

Methods

Weirs are the only type of gear authorized for eel fishing in the estuary. As weirs are installed before the beginning of the season (1 September) and remain unchanged at the same location until the season is completed (beginning of November), we used the total length of gear (leader) in conjunction with deployment time as a measure of effort. The study area was divided into three sectors with similar physical and geographical characteristics. To obtain accurate distribution and measurements of weir lengths, we conducted an aerial photographic survey of the area in 1996. Calibration was performed by field measurements on selected weirs to the nearest meter. Photo-interpretation allowed us to measure each weir individually. Effort was then summed for the three sectors.

We organized a network of index fishermen (13 in 1996 and 15 for the following years) who collected daily catch (± 1 kg). During those years (1996–2000), index fishermen represented more than 40% of the total fishing effort for the entire area. Daily biomass measurements by fishermen were randomly checked for accuracy during field visits. For each three sectors, catch per unit of effort (CPUE) was obtained by dividing total catch by total length of the weirs they used.

Daily eel sampling was performed during the 1996, 1997, and 1998 fishing seasons. Three or more fishermen were visited each day, and for each zone, 30 eels were randomly collected for body measurements. Measurements were made on live eels: weight to the nearest 10 g and length to the nearest cm. In 1996, eyes were measured with electronic calipers (accurate to 0.1 mm), and eye index was calculated according to the method of Pankhurst (1982).

CPUE was measured from index fishermen results, and then CPUE was applied to the total effort to obtain daily catch and total landing estimates. Estimation for total catch was performed in accordance with Cochran (1977) for each sector using the following formula:

Total catch estimate:

$$Y_R = \sum_{h=1}^{3} Y_{Rh} = \sum_{h=1}^{3} R_h \cdot X_h = \sum_{h=1}^{3} \frac{\sum_{i=1}^{nh} y_{ih}}{\sum_{i=1}^{nh} x_{ih}} \cdot X_h$$

Variance estimate:

$$V(Y_R) = \sum_{h=1}^{3} V(Y_{Rh}) =$$

$$\sum_{h=1}^{3} \frac{N_h^2 (1-f_h) \cdot \sum_{i=1}^{nh} (y_{ih} - R_h x_{ih})^2}{n_h \cdot (N_h - 1)}$$

Confidence limits for estimation:

$$Y_R \pm t_{\alpha(2),(n-1)} \cdot \sqrt{V(Y_{Rs})}$$

where:
- YR_h catch (kg) per stratum
- R_h CPUE (kg/m) per stratum
- X_h total soaked leader length of weirs (m) per stratum
- y_{ih} catch per sampled weir (kg) per stratum
- x_{ih} soaked leader length per sampled weir (m) per stratum
- N_h number of weirs per stratum
- n_h number of weirs sampled per stratum
- f_h sampling fraction in each stratum

$$f_h = \frac{\sum_{i=1}^{nh} x_{ih}}{X_h}$$

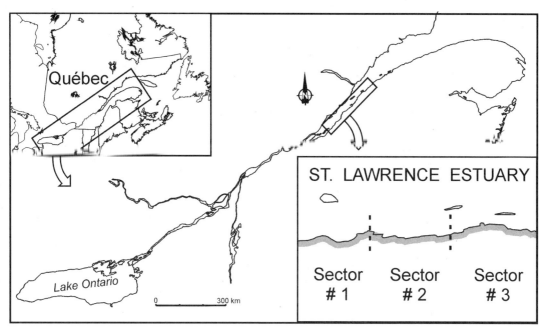

Figure 1. Study area and fishing sectors for silver eel in the St. Lawrence River Estuary.

Eel numbers were then calculated by integrating mean weight of sampled eels with total catch.

In order to be comparable to Axelsen's (1997) study of this fishery, CPUE and total fishing effort were recalculated for the county of Kamouraska (sector 1 and 2).

Results

Morphological data were collected on 4,529 eels in 1996, 1997 and 1998 (Table 1). Length frequencies show, for all three years, positively skewed ($x^2, P < 0.0001$) distributions ranging from 60 to 120 cm and did not differ significantly among years ($x^2, P \geq 0.56$). Weight frequency distributions were also positively skewed ($x^2, P < 0.0001$) and also did not differ among years ($x^2, P \geq 0.834$). All distributions were unimodal and the mode was found in the 80–85 cm class for length frequency distributions (Figure 2) and in the 900–1200 g class for weight frequency distributions (Figure 3). Body length and weight did not vary significantly over the time of capture (Scheffé's, $P \leq 0.0053$). The length-weight relationships were strong and highly significant for each year, with a coefficient of determination (r^2) always higher than 0.85 ($P < 0.0001$). The slope and elevation of these regressions did not differ among years (analysis of covariance [ANCOVA], $P \geq 0.0748$). This result suggests that the Fulton's condition factor (K) values, which varies from a mean value of 0.192–0.199, are not significantly different. Eye index was highly variable, ranging from 2.7 to 10.2 with a mean value of 6.3. This index did not increase with body length (Figure 4).

Table 1. Biological data collected during eel fishery sampling from 1996 to 1998 in the St. Lawrence Estuary.

Year	n	Length (cm)	(SD)	Weight (g)	(SD)	Fulton's K	(SD)
1996	1,081	86.1	(9.76)	1,334	(543.8)	0.199	(0.0327)
1997	1,492	84.5	(9.10)	1,203	(467.7)	0.192	(0.0308)
1998	1,956	85.4	(9.17)	1,273	(489.7)	0.196	(0.0252)
Pooled data	4,529	85.3	(9.31)	1,264	(498.5)	0.195	(0.0291)

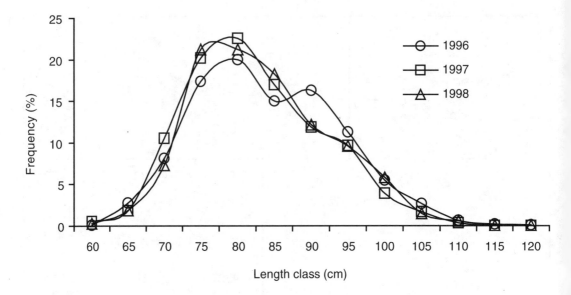

Figure 2. Length-frequency distribution of migrating eel from the St. Lawrence Estuary fishery:1996–1998.

Fishing effort, expressed as soaked leader length of weirs decreased from 34,258 m in 1996–20,905 m in 2000 (Table 2). Fishing effort for index fishermen fluctuated with the same trend and amplitude as the total area. The lowest annual CPUE was observed in 1997 and the highest in 1998 (Table 2). CPUE for index fishermen were always higher in the upstream sector #1 (Table 3). The CPUE for each individual fisherman was strongly and inversely correlated ($r = -0.590$, $P < 0.0001$) to the cumulative length of leader soaked upstream (Figure 5).

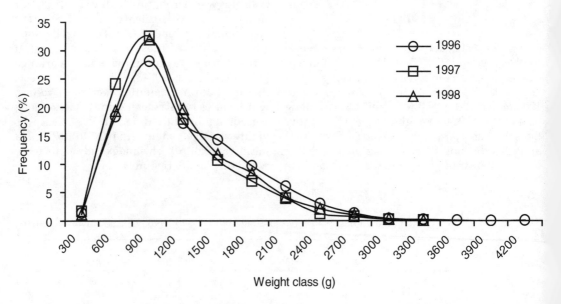

Figure 3. Weight-frequency distribution of migrating eel from the St. Lawrence Estuary fishery: 1996–1998.

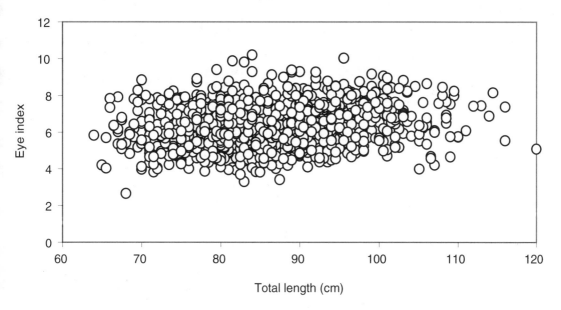

Figure 4. Eye index for silver eel sampled in the St. Lawrence River Estuary fishery in 1996.

Daily catch estimates fluctuated during the season. Fluctuations differed from year to year and peak capture was observed on one occasion in 1996 and 1998 and on two occasions in 1997. In 1999 and 2000, there was virtually no peak capture (Figure 6). In any year, more than 90% of the total harvest came within a 30-day period. Total annual harvest estimates fluctuated from a maximum of 133.5 metric tons in 1998 to a minimum of 72.8 metric tons in 2000.

When merging our catch and effort data with those of Axelsen (1997), we observed three distinct periods from 1979 to 2000 (Figure 7). In the first period (1979–1985), CPUE variations were inversely related to effort, which dropped markedly in 1982 to regain its level back in 1984. In the second period (1986–1993), effort was relatively stable or slightly increasing and greater than in period 1, while CPUE variations showed a general negative trend inversely related to effort. The third and last period (1994–2000) showed that effort was constantly dropping, while CPUE was relatively stable with no detectable trend. Due to a near constant negative

Table 2. Exploitation parameters for the St. Lawrence Estuary eel fishery: 1996–2000.

Exploitation parameters	Year				
	1996	1997	1998	1999	2000
Total fishing effort (leader length [m])	34,258	34,233	31,282	27,052	20,905
Fishing effort for index fishermen (m)	13,668	16,283	14,802	13,173	12,803
Number of index fishermen	13	15	15	14	12
Harvest from index fishermen (metric tons)	48.9	52.5	62.9	52.9	43.9
Estimated total harvest (metric tons)	119.5	111.9	133.5	109.2	72.8
Confidence interval (95%)	7.79	5.56	9.13	6.61	4.97
Standard deviation (metric tons)	3.81	2.73	4.47	3.23	2.44
Annual CPUE (kg/m)	3.49	3.27	4.27	4.04	3.48

Table 3. CPUE for the three study zones in the St. Lawrence Estuary eel fishery: 1996–2000.

Sector	Annual CPUE (kg/m)				
	1996	1997	1998	1999	2000
# 1 (upstream)	3.97	3.95	5.25	4.73	4.54
# 2 (middle)	3.96	3.05	4.28	4.16	3.09
# 3 (downstream)	1.41	2.64	2.21	2.69	2.91

relation between effort and CPUE, as demonstrated by these results and those of Figure 5, we must compare CPUE among years or group of years with comparable values of effort. For example, the same fishing effort measured between 1979, 1980, and 1996, then between 1983 and 1999–2000 leads to a CPUE three times lower.

Discussion

Biological Parameters

Three years of biological sampling data on eels harvested during their spawning migration enable us to characterize this stock. Mean length and weight, for pooled data, were 85.3 cm (SD = 9.31 g) and 1,264 g (SD = 498.5 g). These characteristics represent the highest values reported for silver eel in Canada (Table 4). As Couillard et al. (1997) reported for samples taken at the same location in 1990, body length and weight did not vary significantly over time of capture.

These eels are coming from Lake Ontario and other upstream tributaries (Castonguay et al. 1989) of the St. Lawrence-Great Lakes watershed. For the species' distribution range, these locations are among the farthest from their spawning grounds in the Sargasso Sea. All of these eels are migrating maturing females (Dutil et al. 1985; Couillard et al. 1997). Furthermore, since these eels are the largest eels observed in Canada, this sub-population could represent a major contribution to the reproductive potential of this panmictic species (Barbin and McCleave 1997). Vøllestad (1992) found a significant positive relation between mean length at migration for European eels *Anguilla anguilla* and distance to the spawning area. It seems to be the same relationship for the American eel. There was no trend in condition factor over the sampling period and this could reflect the same stage of maturation for all migrating females.

Since all eels trapped in the fishery are migrating maturing females (Couillard et al. 1997), eye indices greater than 6.5 may have been expected. Pankhurst (1982) reported that female European eel with an eye index greater than 6.5

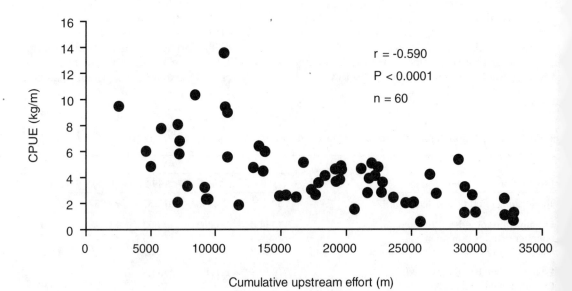

Figure 5. Annual CPUE and cumulative effort (soaked leader length) upstream of each individual fisherman (pooled data from 1996 to 2000).

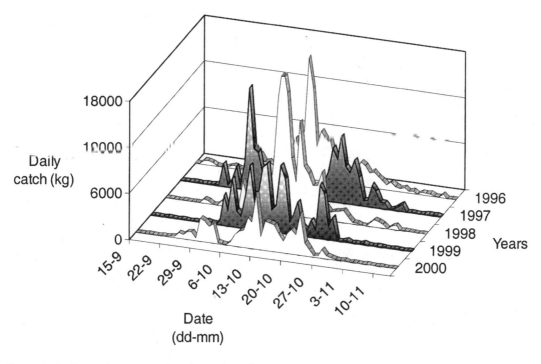

Figure 6. Daily catch estimates for silver eels in the St. Lawrence River Estuary fishery: 1996–2000.

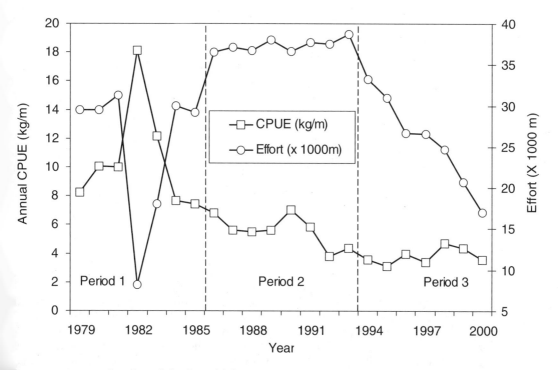

Figure 7. CPUE and total catch for the eel fishery: 1979–2000 (data for 1979–1995 are from Axelsen 1997).

Table 4. Mean total length and weight of silver American eels sampled in Canada.

River	N	Length (cm)	Weight (g)	Source
Medway (Nova Scotia)	644	49.1	247	Jessop (1987)
LaHavre (Nova Scotia)	730	56.3	372	Jessop (1987)
Topsail Pond (Newfoundland)	82	69.4	592	Gray and Andrews (1971)
Dog Bay (Newfoundland)	94	77.8	971	Bouillon and Haedrich (1985)
Holyrood Bay (Newfoundland)	90	72.2	730	Bouillon and Haedrich (1985)
Sud Ouest (Newfoundland)	51	61.4	429	Brennan (1976)
St. Lawrence (Quebec) 1996	1,081	86.1	1334	This study
St. Lawrence (Quebec) 1997	1,492	84.5	1203	This study
St. Lawrence (Quebec) 1998	1,956	85.4	1273	This study

are sexually maturing adults and Jessop (1987) found that eye indices increased with length. We did not observe such results from the 1996 sampling. Assuming that our sample had the same maturity characteristics as those reported by Couillard et al. (1997), eye index is not a useful indicator of sexual maturity for this subpopulation.

Exploitation Parameters

Fishing for eels is done yearly at the same time with the same gear at the same exact location. Individual weir length does not change during the season but may vary from year to year. Capture of eels depends on the displacement of these fish through nearly 100 fixed traps in the area, thus informing us about the chronology and the extent of the migration. Migrating eels arrive in the area at the end of September and the run is over some 30 days after. By comparing CPUE among sectors we think that the fishery is based on interception: CPUE being lower downstream (Table 3). Comparing individual annual CPUE against cumulative effort upstream (Figure 5), which is a measure of relative geographical position, enabled us to observe that the more downstream a weir is positioned along the study area, the lower is the weir's CPUE. This suggests the presence of an intercept fishery, where the initial group of migrant eels is gradually less abundant each time a fraction of it gets trapped in its downstream migration. However, the cumulative effort upstream of each trap does not alone explain the variation in CPUE as shown by the correlation coefficient ($r = -0.590$, $P < 0.0001$; Figure 5). The individual characteristics of each trap and their exact location could explain some of the variation in CPUE as some traps always have, year after year, a greater CPUE. Other variations in CPUE, as reported by fishermen, could include current conditions, weather, and wind.

The exploited eels migrate close to the shore in shallow waters, so weirs are located in water depth less than five meters during high tide. What phenomenon could drive these migrating fish so close to the shore in shallow water when mean width in this area is 17 km with the presence of channels with deeper water? Unfortunately, no detailed in situ observations of downstream swimming behavior of American eel have been reported (EPRI 1999). For European eels, Adams and Schwevers (1997) reported three forms of downstream behavior: passive drift, controlled drift and active movements. A possible explanation for the St. Lawrence Estuary is that eels use controlled drift and selective tidal transport to travel downstream and they use the shallow waters for osmoregulation purposes. Salinity gradient increases more gradually close to the shore because freshwater from the watershed flows along the south shore (Centre Saint-Laurent 1996). When eels leave the area, salinity is 30 ppt and osmoregulation adjustment has to be completed by this time.

Data collected by the index fishermen permits estimation of the total annual catch with a high degree of confidence. As total harvest could fluctuate with effort, annual CPUE was preferred to assess the fishery for years before 1995 (Axelsen 1997). Axelsen (1997) observed a major drop in the CPUE for the estuarine fishery between 1984 and 1995. Axelsen stated that a longtime decrease in CPUE reflected a long-term downward trend in eel abundance. This seems correct for those years when effort varied only slightly among years, as our results show that CPUE and effort are not independent. In recent years (1996–2000), effort decreased sharply but CPUE did not show the same trend. During our study, fishing effort declined by 39.4% while CPUE revolves around a mean of 3.88 kg/m. This reduction in effort was a consequence of export market failure and lower prices for eel. Surprisingly, yields were only reduced by 5.5%. Yield reduction affected the strength of the activity. Indeed, the fishermen from the downstream zone, those who had the lowest yield, abandoned eel fishing more rapidly than those upstream. Moreover, we did not observe significant correlation ($r^2 = 0.006$) between fishing effort and CPUE for the last five years, although a significant correlation ($r^2 = 0.714$) existed between those parameters before 1996. This situation raises some questions about the effectiveness of exploitation parameters to assess the fishery and the exploited stock trends. The assumed linear relation between CPUE and abundance has long been debated (Garrod 1964; Swain and Sinclair 1994). When a fish sub-population is concentrated, like migrating eels in the estuary, and effort is rapidly decreasing, misinterpretation of elevated CPUE contributes to overestimating sub-population size and does not reflect actual abundance. Direct measurement of abundance is preferred to indirect evaluation. Until a new period of stability for fishing effort occurs, CPUE must be interpreted with great care.

Total capture for a year ranged between 72.8 and 133.5 metric tons for an estimated number of eels ranging from 57,600–106,800 individuals. This amount seems important, especially when this stock is composed exclusively of females heading for spawning grounds. It must be regarded in a wider perspective; eels are present in many watersheds connected to the St. Lawrence Estuary and Gulf in Quebec, downstream of the fishery. Total area of these growth habitats is very large and productive. Eels from these watersheds are not exploited and anthropogenic pressure is light or nonexistent. We do not know the contribution of eels from these watersheds and more research must be conducted before targeting any escapement rate for this valuable resource.

Acknowledgments

Quebec's Parks and Wildlife Society through the commercial fishing research fund financially supported this study. We thank J.-F. Gaudrault for field sampling and index fishermen who allowed us to sample their catches and kindly gathered and provided to us their daily catch statistics. M. Castonguay and D. K. Cairns made constructive comments on a first draft of this paper.

References

Adams, C. C. and D. U. Schwevers. 1997. Behavioral surveys of eels (*Anguilla anguilla*) migrating downstream under laboratory conditions. Institute of Applied Ecology, Neustader Weg 25, 36320 Kirtorf-Wahlen, Germany.

Axelsen, F. 1997. The status of the American eel (*Anguilla rostrata*) stock in Quebec. *In* Peterson R. H. editors The American eel in Eastern Canada: stock status and management strategies. Proceedings of Eel Workshop, January 13–14, 1997, Quebec City, (QC). Canadian Technical Report of Fisheries and Aquatic Sciences No. 2196: 121–133.

Barbin, G. P., and J. D. McCleave. 1997. Fecundity of the American eel *Anguilla rostrata* at 45°N in Maine. Journal of Fish Biology 51:840–847.

Bouillon, D. R., and R. L. Haedrich. 1985. Growth of silver eels, *Anguilla rostrata*, in two areas of Newfoundland Journal of Northwest Atlantic Fisheries Science 6:95–100.

Brennan, M. 1976. The biology of *Anguilla rostrata*, with reference to the commercial fishery. Master's Thesis. Marine Sciences Centre, McGill University, Montreal, Quebec, Canada

Casselman, J. M., L. A. Marcogliese, and P. V. Hodson. 1997. Recruitment index for the upper St. Lawrence River and Lake Ontario eel stock: a re-examination of eel passage at the R. H. Saunders Hydroelectric Generating Station at Cornwall, Ontario, 1974–1995. *In* R. H. Peterson, editor. The American eel in eastern Canada: stock status and management strategies. Proceedings of Eel Workshop, January 13–14, 1997, Quebec City, (QC). Canadian Technical Report of Fisheries and Aquatic Sciences No. 2196:161–169.

Castonguay, M., J. D. Dutil and C. Desjardins. 1989. Distinction between American eels (*Anguilla rostrata*) of different geographic origins on the basis of their organochlorine contaminant levels. Canadian Journal of Fisheries and Aquatic Sciences 46(5):836–843.

Castonguay, M., P. V. Hodson, C. M. Couillard, M. J. Eckersley, J. D. Dutil and G. Verreault. 1994. Why is recruitment of the American eel, *Anguilla rostrata*, declining in the St. Lawrence River and Gulf? Canadian Journal of Fisheries and Aquatic Sciences 51(2):479–488.

Centre Saint-Laurent. 1996. Rapport-synthèse sur l'état du Saint-Laurent. Volume1: *L'écosystème du Saint-Laurent*. Environnement Canada—région du Québec, Conservation de l'environnement— et Éditions MultiMondes, Montréal Coll., Bilan Saint-Laurent.

Cochran, G. 1977. Sampling techniques. Wiley. New York.

Couillard, C. M., P. V. Hodson and M. Castonguay. 1997. Correlations between pathological changes and chemical contamination in American eels, *Anguilla rostrata*, from the St. Lawrence River. Canadian Journal of Fisheries and Aquatic Sciences 54(8):1916–1927.

d'Anglejean, B.-F. 1990. Recent sediments and sediment transport process in the St. Lawrence estuary. Pages 109–129 *in* M. I. El-Sabh and N. Silverberg editors, Oceanography of a large-scale estuarine system, the St. Lawrence. Coastal and Estuarine Studies. Springer-Verlag, Berlin, Germany.

Dutil, J. D., B. Legaré and C. Desjardins. 1985. Discrimination d'un stock de poisson, l'anguille (*Anguilla rostrata*), basée sur la présence d'un produit chimique de synthèse, le mirex. Canadian Journal of Fisheries and Aquatic Sciences 42(3):455–458.

EPRI (Electrical Power Research Institute). 1999. American eel (*Anguilla rostrata*) scoping report: a literature and data review of life history, stock status, population dynamics, and hydroelectric impacts. EPRI Report TR-111873. Palo Alto, California.

Garrod, D. J. 1964. Effective fishing effort and catchability coefficient. Rapport et Procès-Verbaux des Réunions de la Commission Internationale pour l'Exploration Scientifique de la Mer Méditerranée. Paris, France 155:66–70.

Ghaniné, L., J.-L. Desgranges and S. Loranger. 1990. Les régions biogéographiques du Saint-Laurent. Lavalin Environnement inc. pour Environnement Canada et Pêches et Océans, Région du Québec. Rapport final.

Gray, R., and C. W. Andrews. 1971. Age and growth of the American eel (*Anguilla rostrata* (LeSueur)) in Newfoundland waters. Canadian Journal of Zoology 49:121–128.

Jessop, B. M. 1987. Migrating American eels in Nova Scotia. Transactions of the American Fisheries Society 116:161–170.

Ouellet, Y and J. Cerceau. 1976. Mélange des eaux douces et salées du Saint-Laurent: circulation et salinité. Les cahiers de Centreau, Université Laval, Québec, Canada. Vol 1, n. 4.

Pankhurst, N. W. 1982. Relation of visual changes to the onset of sexual maturation in the European eel *Anguilla anguilla* (L). Journal of Fish Biology 21:127–140.

Sagard, G. 1636. Histoire du Canada et voyages que les Frères Mineurs Récollets y ont faits pour la conversion des Infidèles, Republished by Edwin Tross, Paris, France 1866.

Swain, D. P., and A. F. Sinclair. 1994. Fish distribution and catchability: what is the appropriate measure of distribution? Canadian Journal of Fisheries and Aquatic Sciences 51:1046–1054.

Vøllestad, L. A. 1992. Geographic variation in age and length at metamorphosis of maturing European eel: environmental effects and phenotypic plasticity. Journal of Animal Ecology 61:41–48.

Estimation of the Population Size, Exploitation Rate, and Escapement of Silver-Phase American Eels in the St. Lawrence Watershed

FRANÇOIS CARON*

*Société de la faune et des parcs du Québec, 675, boul. René-Lévesque Est, Québec,
Québec G1R 5V7, Canada
Corresponding author: francois.caron@fapaq.gouv.qc.ca

GUY VERREAULT

Société de la faune et des parcs du Québec, 506, rue Lafontaine, Rivière du Loup,
Québec G5R 3C4, Canada

ERIC ROCHARD

*Cemagref, Unité Ressources Aquatiques Continentales, 33612,
Cestas cedex, France*

Abstract.—The St. Lawrence River is one of the largest watersheds colonized by the American eel *Anguilla rostrata*. Upstream from Québec City, eels are present in the St. Lawrence main stem and tributaries, including the Lake Ontario watershed. Recently, much concern has been raised concerning declining recruitment of young eels in Lake Ontario and declining catch of silver eels in the St. Lawrence estuary fishery, but no previous study attempted to estimate current silver eel production. In 1996 and 1997, the number of silver eels migrating from the freshwater St. Lawrence watershed was estimated by a mark–recapture technique. Calculations conducted with pooled data (Petersen estimate) and with stratified data (Schaefer and Darroch-Plante estimates) estimated the migrating population at approximately 488,000 fish in 1996 and 397,000 in 1997. The exploitation rate by the commercial fishery in the estuary during these two years was 19% and 24%, leaving a spawner escapement of approximately 396,000 and 302,000 eels, respectively. This study marks the first silver eel population estimation for a large watershed. Considering that 99% of the St. Lawrence silver eels are female and their mean weight is from 1.2 to 1.3 kg, their contribution of the North American spawning stock must be very important for the entire species.

Introduction

The two species of catadromous eel present in the North Atlantic—the American eel *Anguilla rostrata* and the European eel *A. anguilla* have declined markedly in recent years (Castonguay et al. 1994a; Ritter et al. 1997; Moriarty and Dekker 1997; Richkus and Whalen 2000). The International Council for the Exploration of the Sea has recently concluded that the European eel stock is not within safe biological limits (International Council for the Exploration of the Sea 1998). In a similar way, a review of available data confirmed either declining or neutral abundance of American eel (International Council for the Exploration of the Sea 2000). Continuous exploitation and unknown oceanographic effects support the adoption of a precautionary approach in management (International Council for the Exploration of the Sea 2000).

The panmixia hypothesis, referring to the idea that each species migrates to the Sargasso Sea for reproduction and comprises a single randomly-mating population, was recently challenged with regard to European eel (Wirth and Bernatchez 2001), but is still largely accepted for American eel (Williams et al. 1973; Avise et al. 1986, 1990). Consequently, the entire species of American eel should be managed as a single stock; any loss of production or overexploitation in an important part of its range could affect spawner escapement and, possibly, recruitment throughout the whole species' range.

American eel distribution may extend from the northern portion of South America to Greenland (Jenkins and Burkhead 1993). In North America, the Gulf of St. Lawrence is in the northern part of its distribution. Eels colonize the main stem of the St. Lawrence River and its tributaries up to Lake Ontario. The freshwater runoff of the St. Lawrence River represents approximately 19% of the total freshwater runoff in the species' range (Castonguay et al. 1994b). Lake Ontario, located in headwater of the St. Lawrence, has suffered a severe decline in eel recruitment since 1985 (Casselman et al. 1997; Richkus and Whalen 2000). Moreover, captures of silver eels (silver-phase eels) in the St. Lawrence, from which more than 70% come from the principal St. Lawrence–Lake Ontario axis (Dutil et al. 1985; Couillard et al. 1997), have also decreased in the past twenty years due to the constant fishing effort (Robitaille and Tremblay 1994; Axelsen 1997).

Prior to this study, the downstream migration of eels had never been quantified in the St. Lawrence. In fact, until now, research has mainly been conducted in smaller watersheds (Rossi and Cannas 1984; Vøllestad and Jonsson 1988; Adam 1997; Therrien and Verreault 1998; Fournier and Caron 2001). This study was conducted in 1996 and 1997 with the goals of estimating the population of silver eels leaving the freshwater part of the St. Lawrence River, the exploitation rate in brackish waters and the escapement of silver eels from the fishery.

Methods

Study Area

The St. Lawrence River is one of the three largest rivers on the North American continent. It extends from the Laurentian Great Lakes to the Gulf of St. Lawrence (Figure 1) draining a catchment of 1.6×10^6 km^2 that holds over 25% of the world's freshwater (Vincent and Dodson 1999). The 400-km fluvial section, extending from Lake Ontario to Lake Saint-Pierre, is followed by a long estuary that extends from Lake Saint-Pierre to the Gulf of St. Lawrence (Figure 1). The estuary is composed of three distinct sections commonly referred to as: the fluvial estuary (freshwater) extending from river kilometer (rkm) 160 to rkm 0, the upper estuary (brackish water) extending from rkm 0 to rkm –150, and the lower estuary (salt water; St. Lawrence Center 1996).

The study area covers a part of the fluvial estuary and the upper estuary. The fluvial estuary is generally 3–5 km wide. At Québec City (rkm 40), the average annual discharge is 12,600 m^3/s and mean tides reach 4.1 m in amplitude (St. Lawrence Center 1996). The upper estuary zone is 15–25 km wide; tidal marshes on the south shore are mostly flat and muddy, and are 0.5–3 km wide. At the salinity front (rkm 0), maximum tides reach 6.9 m (St. Lawrence Center 1996). The salinity in the study area ranges from 0 to 25‰ (Gagnon et al. 1993). This zone is characterized by high turbidity levels associated with sediment re-suspension due to the mixing of fresh and brackish waters from rkm 0 to rkm –50, and strong tidal currents.

Estimates of the Silver Eel Population Leaving Freshwater

Mark–recapture methods were used to estimate the number of silver eels leaving the St. Lawrence. Marked eels came from two sources: eel traps in the intertidal zone, submerged at each tidal cycle, used by commercial fishers and river traps considered as nonselective (20 mm mesh on the side) used to capture all silver eels descending the Rimouski and Sud-Ouest rivers.

The marking zone is located in the freshwater estuary, near Québec City, from rkm 35 to rkm 25 (Figure 1). All eels were marked and released on the same shore on which they were captured. Marking was done by hot branding on the underside, between the pectoral fins and the head; this method allowed the use of numerous codes, in addition to generating markings that are easily verified and do not affect the catchability of the fish (MacKeown 1984; Moring 1990).

In 1996, two eel groups were subject to different forms of handling. Eels coming from commercial eel traps were marked and released the same day, within the hours following their capture; marks used were specific to a release location and a period of time from one to seven days. Eels coming from river traps were kept in captivity for several days before being anesthetized in ice water between 6° and 10°C, and marked and released at night on the south shore on two occasions. In 1997, only commercial eel traps were used. The eels were transferred to a tank before being marked and released within the

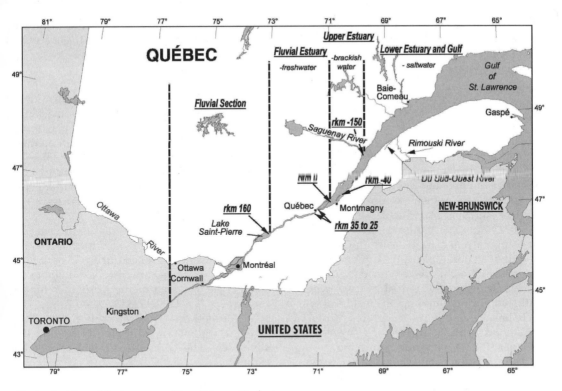

Figure 1. Map of St. Lawrence River Estuary Study Area.

hours that followed, or up to a maximum of three days.

The recapture zone is located on the south shore from rkm –40 to rkm –150 where 89% of silver eel landings in Québec take place (Verreault and Dumont, this volume). There is no important river likely to produce eels in the area between the marking and recapture zones. Searching for marked eels was done at a fish plant where eels coming from the recapture zone were processed. A sampling of eels (control group) was done on a regular basis; eels were counted, verified for markings and, if marked, identified by their code.

Certain postulates have to be respected to permit the estimation by mark–recapture (Ricker 1975; Seber 1982). The mark must remain visible on the animal during the length of the experiment and the marking should not affect survival during the study period. Six eels subject to the same form of handling as those coming from eel traps were marked and kept under observation for 50 days to verify these postulates.

Estimation of the population ($\hat{N}$) was done using three estimators: the regrouped Petersen method known as "Pooled Petersen" (Seber 1982; Arnason et al. 1996), the Schaefer (1951) method and the Darroch (1961) method completed by Plante (1990), which we refer to as the DP method. The Schaefer and DP methods require that markings and controls be regrouped into strata corresponding to different phases of the downstream migration. These estimates are based on the assumption that the marked animals mix randomly with the rest of the population or, at least, they resume the position they held within the population prior to being marked. In 1996, use of the same marking code for several days limited the stratification to nine one-week periods. In 1997, use of a daily code allowed for stratification of the run in a number of ways. We regrouped the data into ten periods corresponding to each fraction of 10% of landings done by commercial fishers in the recapture zone. Periods for which we had an insufficient number of recaptures were regrouped; the last marking periods were not always considered for this reason. Likewise, the number of eels taken into account in the calculations varies slightly according to the method and stratification. The calculations were performed using the SPAS software (Stratified Population Analysis System) from Arnason et al. (1996). All confidence intervals (C.I.) are for $\alpha = 0.05$.

Exploitation Rate and Escapement From the Fishery

The last fishery that the silver eel meets during the course of its migration to the sea is found in the brackish water zone. The number of fishery licenses is limited and the fishers must declare the total weight of their capture. The details of capture estimates are presented in Verreault et al. (this volume). Technical staff visited fishers on a regular basis to measure the individual weight of a subsample of the catch. Total catch recorded by weight was converted into numbers of eels using the individual average weights gathered during sampling, the results of which were balanced in terms of captures carried out weekly according to the method suggested by Frontier (1983) for stratified sampling.

The exploitation rate was calculated using the total catch estimate in number divided by the Pooled Petersen estimate. Lower limits were calculated as the lower value of the C.I. of the catch divided by the higher value of the C.I. of the estimate. Likewise, the higher limits were calculated as the higher value of the C.I. of the catch over the lower value of the C.I. of the estimate. Escapement from the fishery was obtained by subtracting the annual catch from the population estimate.

Results

Population Estimates of Migrating Silver Eel

All eels maintained in captivity survived and marks remained easily visible after 50 days. This suggests that the marking procedure did not affect survival and that all marked eels could be detected during the study period.

In 1996, 28,714 eels were sub-sampled at the fishing plant and controlled for markings; marked eels from the north shore ($n = 516$), from the south shore ($n = 102$) and those from river traps ($n = 429$) that underwent a different marking and release protocol had similar recapture rates of, respectively, 5.6%, 8.8% and 5.6%. In 1997, 51,704 eels were controlled; the recapture rate for eels from the north shore ($n = 641$) and from the south shore ($n = 726$) was similar at 13% and 12%, respectively. There is no significant difference between these groups (χ^2 test). Consequently, data were pooled to enable estimation.

In 1996, a total of 1,047 eels were marked during 34 occasions (minimum = 1, maximum = 144) and 28,714 were controlled for markings during 23 occasions (minimum = 128, maximum = 4,536; Table 1). The Pooled Petersen method estimates population at $\hat{N} = 492,845$ (383,693–633,091). The Schaefer method leads to an estimate of $\hat{N} = 470,585$. The DP method gives an estimate of $\hat{N} = 502,059$ (standard error = 277,254).

In 1997, 1,351 eels were marked during 35 occasions (minimum = 8, maximum = 58) and 51,704 eels were controlled during 25 occasions (minimum = 46, maximum = 6,603) (Table 1). The Pooled Petersen method provides for an estimate of $\hat{N} = 410,895$ (353,591–477,492). The Schaefer method with stratification according to deciles gives an estimate of $\hat{N} = 382,476$ silver eels. The DP method failed to give an estimate.

Silver Eel Harvesting in Brackish Water

Total fishing effort was similar for the two years. In 1996, significant reported catches went from mid-September to mid-November with a single peak on October 11. The mean weight of eels was 1.297 kg (1.253–1.340; $n = 1,081$). In 1997, catches were concentrated over a shorter period from the end of September to early November with two peaks, the first one on October 1 and the second one during the third week of October. The mean weight of eels was 1.182 kg (1.157–1.206; $n = 1,492$).

Total captures weighed 119.5 metric tons (107.5–133.3) in 1996, and 111.9 metric tons (102.1–121.6) in 1997. Reported in numbers, the captures represent 92,136 eels (80,224–106,385) in 1996, and 94,670 (84,660–105,099) in 1997.

The exploitation rate was 19% (13–28%) in 1996, and 24% (18–30%) in 1997. Consequently, the number of silver eels that escaped from the fishery was 396,333 (353,626–428,009) in 1996, and 302,010 (278,966–326,553) in 1997.

Discussion

Several methods which take into account stratified populations have served to estimate the numbers of adult salmon *Oncorhynchus nerka* in anadromous migration (Schaefer 1951) or of smolts *Salmo salar* and *O. nerka* in downstream migration (MacDonald and Smith 1980; Dempson and Stansbury 1991; Plante et al. 1998). However, such methods have not yet been used for eels. This study demonstrates that it is possible to quantify a population of silver eels in downstream migration in a large watershed using mark–recapture methods.

Postulates related to the estimation by mark–recapture were respected: marked eels kept in captivity had all survived during the time of the study and marks remained visible after 50 days. The similar recapture rate observed for eels from the north shore and for the two lots of eels on the south shore suggest a good mixing of all marked eels in the population. The fishery with its stationary, passive gear is independent of the fishers' behavior and, therefore, we consider that the captures reflect the general tendency of the downstream migration of silver eels in this sector. The use of a nonselective river trap excludes the possibility of bias related to different catchability of a fraction of the population as suggested by Cone et al. (1988). According to the criterion put forward by Robson and Regier (1964), the number of marked and controlled fish allows for an unbiased estimation by the Petersen method. Indeed, the product of the number of marked fish multiplied by the number of controlled fish is clearly higher than the estimate of the numbers.

In 1996, the three estimates are similar, approximately 488,000 silver eels. The Schaefer's and DP estimates are within the confidence interval of the Pooled Petersen estimate. In 1997, the two estimates are also similar, approximately 397,000 silver eels. Furthermore, Schaefer's estimate is again within the Pooled Petersen estimate. The DP estimate, regardless of the stratification used, could not provide acceptable results in 1997 because of the negative capture probability in some strata. The bi-modality observed in the commercial catches for that year could be an explanation. Although the estimate of the downstream migration seems to suggest a smaller downstream migration in 1997, the confidence intervals for the two years in question overlap and do not indicate a significant difference between the two years.

According to empirical observation by commercial fishers, eel captures in downstream migration are known to be partly dependent on meteorological conditions, particularly wind-related factors. The exploitation rate observed is quite close for those two years—19% in 1996, and 24% in 1997—which were years of average weather conditions for the fishery. Considering the only slight modification in gear used and the deployed fishery effort over the past thirty years, it is conceivable that this is an average exploitation rate for this population of silver eel for this period.

In the absence of a study indicating the production of eel expected for the area of a watershed comparable to the St. Lawrence, it is difficult to judge the value of our estimates. It is nevertheless apparent that the number of silver eels in downstream migration was greater in the past than observed during the period of this study. During the last two decades, the estimated number of silver eel captures in the St. Lawrence culminated in 1980 and 1990, with 335,000 and 239,600 captures, respectively, which is approximately three times greater than observed during the period of this study. It is, therefore, likely that the number of eels in migration over these years was definitely higher than what we had estimated during the years covered by our study.

This study represents the first quantification of a silver eel population in downstream migration in a large river, and it provides the first estimates of the exploitation rate of silver eel in the St. Lawrence estuary. These results demonstrate that it is possible, under particular conditions, to perform estimations in an open environment of numbers of eels in downstream migration. This information enables establishing targets for escapements which should be the basis for managing the species' stocks in future years.

The annual spawning stock biomass for the American eel population is unknown (Atlantic State Marine Fisheries Commision 2000). An essential recommendation which emerges for the two species of Atlantic eel is undoubtedly to let enough spawners leave in order to ensure the maintenance of the species. We do not yet have enough criteria to allow us to specify the necessary number of spawners, but it seems essential that we try to estimate the silver eel numbers leaving the principal watersheds. Silver eel escapement produced in the freshwater part of the St. Lawrence River was estimated at approximately 396,000 and 302,000 for the two-year study. Given that 99% of the silver eels leaving the St. Lawrence River are females of large size (Vladykov 1966) with a high fecundity (Verreault

2002) and considering that there is a single spawning stock for American eel, their contribution to the spawning stock could be crucial to this species. The need for sound management of the silver eel is therefore underlined, both with respect to its habitat and to its fishery.

Acknowledgments

We thank the technical team that participated in this project: Bruno Baillargeon, Simon Blais, Denise Deschamps, Denis Fournier, Jean François Gaudreault, Conrad Groleau, Denis Labonté, Louis Mathieu, Gontrand Pouliot, David Routhier and Rémi Tardif. We also wish to express our appreciation with all the commercial fishers who acted as index fishers and to Pêcheries Gingras for its collaboration. Louise Buisson (SFA Québec), Ginette Morel (SAEF Montréal), Chantal Gardes (Cemagref Bordeaux) and Pedro Nilo (UQAM) helped us with documentary research. We also appreciate the technical comments of Willem Dekker and Douglas Dixon on earlier drafts of this manuscript that helped to strengthen the analyses and presentation of results. This project was in part realized thanks to the collaboration set out in a Franco–Québec agreement between the Cemagref and the Conseil Supérieur de la Pêche and Faune et Parcs Québec.

References

Adam, G. 1997. L'anguille européenne (*Anguilla anguilla* L. 1758): dynamique de la sous-population du lac de Grand-Lieu en relation avec les facteurs environnementaux et anthropiques. Thèse de Doctorat à l'Université Paul Sabatier, Toulouse III, spécialité hydrobiologie, France.

Arnason, A. N., C. W. Kirby, C. J. Schwarz and J. R. Irvine. 1996. Computer analysis of data from stratified mark—recovery experiments for estimation of salmon escapements and other populations. Canadian Technical Report of Fisheries and Aquatic Sciences No. 2106.

Atlantic State Marine Fisheries Commission. 2000. Interstate fishery management plan for American eel (*Anguilla rostrata*). Fishery Management Report No. 36, Washington, D.C.

Avise, J. C., G. S. Helfman, N. C. Saunders, and L. S. Hales. 1986. Mitochondrial DNA differentiation in North Atlantic eels: population genetics consequences of an unusual life history pattern. Proceedings of the National Academy of Science of the USA 83:4350–4354.

Avise, J. C., W. S. Nelson, J. Arnold, R. K. Koehn, G. C. Williams, and V. Thornsteinsson. 1990. The evolutionary genetic status of Icelandic eels. Evolution 44:1254–1262.

Axelsen, F. 1997. The status of the American eel (*Anguilla rostrata*) stock in Québec. Pages 121–133 *in* R. H. Peterson, editor. The American Eel in Eastern Canada: Stock Status and Management Strategies. Proceedings of Eel Workshop, January 13–14, 1997, Quebec City, Quebec. Canadian Technical Report of Fisheries and Aquatic Sciences No. 2196.

Casselman, J. M., L. A. Marcogliese and P. V. Hodson. 1997. Recruitment index for the upper St. Lawrence River and Lake Ontario eel stock: a re-examination of eel passage at the R. H. Saunders hydroelectric generating station at Cornwall, Ontario, 1974–1995. Pages 161–169 *in* R. H. Peterson, editor. The American Eel in Eastern Canada: Stock Status and Management Strategies. Proceedings of Eel Workshop, January 13–14, 1997, Quebec City, Quebec. Canadian Technical Report of Fisheries and Aquatic Sciences No. 2196.

Castonguay, M., P. V. Hodson, C. M. Couillard, M. J. Eckersley, J.-D. Dutil, and G. Verreault. 1994a. Why is recruitment of the American eel, Anguilla rostrata, declining in the St Lawrence River and Gulf? Canadian Journal of Fisheries and Aquatic Sciences 51:479–488.

Castonguay, M., P. V. Hodson, C. Moriarty, K. F. Drinkwater, and B. M. Jessop. 1994b. Is there a role of ocean environment in American and European eel decline? Fisheries Oceanography 3(3):197–203.

Cone, R. S., D. S. Robson, and C. C. Krueger. 1988. Failure of statistical test to detect assumption violations in the mark-recapture population estimation of brook trout in Adirondack ponds. North American Journal of Fisheries Management 8:489–496.

Couillard, C. M., P. V. Hodson, and M. Castonguay. 1997. Correlations between pathological changes and chemical contamination in American eels, Anguilla rostrata, from the St Lawrence River Canadian Journal of Fisheries and Aquatic Sciences 54:1916–1927.

Darroch, J. N. 1961. The two-sample capture-recapture census when tagging and sampling are stratified. Biometrika 48:241–260.

Dempson, J. B., and D. E. Stansbury. 1991. Using partial counting fences and a two sample stratified design for mark-recapture estimation of an Atlantic salmon smolt population. North American Journal of Fisheries Management 11:27–37.

Dutil, J.-D., B. Légaré, and C. Desjardins. 1985. Discrimination d'un stock de poisson, l'anguille (*Anguilla rostrata*), basée sur la présence d'un produit chimique de synthèse, le mirex. Canadian Journal of Fisheries and Aquatic Sciences 42:455–458.

Fournier, D. and F. Caron. 2001. Travaux de recherche sur l'anguille d'Amérique (*Anguilla rostrata*) de la Petite rivière de la Trinité en 1999 et 2000. Société de la faune et des parcs du Québec. Direction de la recherche sur la faune, Québec City, Quebec, Canada

Frontier, S., editor. 1983. Stratégies d'échantillonnage en écologie. Masson, Paris and Presses de l'Université Laval, Québec, Canada

Gagnon, M., Y. Ménard, J.-F. La Rue. 1993. Caractérisation et évaluation des habitats du poisson dans la zone de transition saline du Saint-Laurent. Canadian Technical Report of Fisheries and Aquatic Sciences No. 1920.

International Council for the Exploration of the Sea. 1998. European eel. Extract of the Report of the Advisory Committee on Fishery Management to the European Commission, n°11. May 1998. Copenhagen, Denmark.

International Council for the Exploration of the Sea. 2000. Report of the EIFAC/ICES Working Group on eels. ICES CM 2001/ACFM:03. Copenhagen, Denmark.

Jenkins, R. E., and N. M. Burkhead. 1993. Freshwater fishes of Virginia. American Fisheries Society, Bethesda, Maryland.

MacDonald, P. D. M., and H. D. Smith. 1980. Mark-recapture estimation of salmon smolt runs. Biometrics 36:401–417.

MacKeown, B. A. 1984. Fish Migration. Croom Helm and Timber Press Publishers, Beaverton, Oregon.

Moriarty, C., and W. Dekker editors. 1997. Management of the European eel. Fisheries Bulletin (Dublin) 15.

Moring, J. R. 1990. Marking and tagging intertidal fishes: review of techniques. Pages 109–116 *in* N. C. Parker et al., editors. Fish-Marking Techniques. American Fisheries Society, Symposium 7, Bethesda, Maryland.

Plante, N. 1990. Estimation de la taille d'une population animale à l'aide d'un modèle de capture-recapture avec stratification. Masters thesis, Université Laval, Québec, Canada

Plante, N., L.-P. Rivest, and G. Tremblay. 1998. Stratify capture-recapture estimation of the size of a close population. Biometrics 54:47–60.

Ricker, W. E. 1975. Computation and interpretation of biological statistics of fish populations. Fisheries Research Board of Canada, Bulletin 191.

Richkus, W. A., and K. Whalen. 2000. Evidence for a decline in the abundance of the American eel, Anguilla rostrata (Le Sueur), in North America since the early 1980s. Dana 12:83–97.

Ritter, J. A., M. Stanfield and R. H. Peterson. 1997. Final discussion. Pages 170–174 *In* R. H. Peterson editors. The American Eel in Eastern Canada: Stock Status and Management Strategies. Proceedings of Eel Workshop, January 13–14, 1997, Quebec City, Qc. Canadian Technical Report of Fisheries and Aquatic Sciences No. 2196.

Robson, D. S., and H. A. Regier. 1964. Sample size in Petersen Mark-Recapture Experiments. Transactions of the American Fisheries Society 93: 215–226.

Robitaille, J. A., and S. Tremblay. 1994. Problématique de l'anguille d'Amérique ériquezz (*Anguilla rostrata*) dans le réseau du Saint-Laurent. Ministère de l'environnement et de la faune. Direction de la faune et des habitats. Rapport Technique IX. Québec City, Quebec, Canada.

Rossi, R., and A. Cannas. 1984. Eel fishing in a hypersaline lagoon of southern Sardinia. Fisheries Research 2:285–298.

Schaefer, M. B. 1951. Estimation of the size of animal populations by marking experiments. U. S. Fish and Wildlife Service Fishery Bulletin 52:189–203.

Seber, G. A. F. 1982. The estimation of animal abundance and related parameters. 2nd edition. Oxford University Press, New York.

St. Lawrence Center. 1996: State of Environment Report on the St. Lawrence River. volume 1 The St. Lawrence Ecosystem. Environment Canada. Québec City, Quebec, Canada.

Therrien, J. and G. Verreault. 1998. Évaluation d'un dispositif de dévalaison et des populations d'anguilles en migration dans la rivière Rimouski. Ministère de l'Environnement et de la Faune, direction régionale du Bas-Saint-Laurent, Service de l'aménagement et de l'exploitation de la faune et Groupe-conseil Génivar Inc., Québec City, Quebec, Canada.

Verreault, G. 2002. Dynamique de la sous-population d'anguilles d'Amérique ériquezz (Anguilla rostrata) du bassin versant de la rivière du Sud-Ouest. MS Thesis. Université du Québec à Rimouski, Rimouski, Québec, Canada.

Vincent, W. F., and J. J. Dodson. 1999. The St. Lawrence River, Canada-USA: the Need for an Ecosystem-Level Understanding of Large Rivers. The Japanese Journal of Limnology 60:29–50.

Vladykov, V. D. 1966. Remarks on the American eel (*Anguilla rostrata* LeSueur). Sizes of elvers entering streams; the relative abundance of adult males and females; and present economic importance of eels in North America. Verhandlungen Internationale Vereinigung fur Theoretische und Angewandte Limnologie 16:1007–1017.

Vøllestad, L. A., and B. Jonsson. 1988. A 13-year study of the population dynamics and growth of the European eel *Anguilla anguilla* in a Norwegian river: evidence for density-dependent mortality, and development of a model for predicting yield. Journal of Animal Ecology 57:983–997.

Williams, G. C., R. K. Koehn, and J. B. Mitton. 1973. Genetic differentiation without isolation in the American eel, *Anguilla rostrata*. Evolution 27:192–204.

Wirth, T., and L. Bernatchez. 2001. Genetic evidence against panmixia in the European eel. Nature (London) 409:1037–1040.

An Estimation of American Eel Escapement from the Upper St. Lawrence River and Lake Ontario in 1996 and 1997

Guy Verreault*

*Faune et Parcs Québec, Direction de l'aménagement de la faune du Bas-Saint-Laurent,
506 Lafontaine, Rivière-du-Loup, Québec G5R 3C4, Canada
Corresponding author: guy.verreault@fapaq.gouv.qc.ca

Pierre Dumont

*Faune et Parcs Québec, Direction de l'aménagement de la faune de la Montérégie,
201 Place Charles-Lemoyne, Longueuil, Québec J4K 2T5, Canada*

Abstract.—In North America, a high but decreasing proportion of American eel landings has historically come from the St. Lawrence River (SLR) system. Available information indicates that the SLR stock is composed almost exclusively of females and that Lake Ontario, the only Great Lake where the species is commonly found, is an important zone of settlement for juvenile eel. From this growing habitat, eels eventually migrate to the estuary and back to the Atlantic Ocean for spawning. This stock is successively exposed to three major sources of mortality: two large hydroelectric projects, Moses-Saunders (built between 1954 and 1958) and Beauharnois-Les Cèdres (gradually equipped between 1912 and 1961), and a commercial fishery from Lake Saint-Pierre to the upper St. Lawrence Estuary. Estimates of mortality rates caused by these three factors and the number of migrating eels in the estuary were combined with a new analysis of available data on eel organic chemical contamination, which can be used to identify the geographic origin of the catch. These data allowed the first evaluation of eel escapement from the upper St. Lawrence River and Lake Ontario. In 1996 and 1997, less than half a million eels were estimated to have left this sector. In the first 500-km between Lake Ontario and the lower Estuary, eels were subjected to an estimated cumulative anthropogenic mortality of 53%. Three quarters of this loss was caused by fish passage through turbines. Impacts of these sources of mortality on eel escapement in the SLR system are discussed.

Introduction

A large stock of American eel, *Anguilla rostrata*, inhabit the St. Lawrence River watershed (Figure 1) to complete the freshwater part of their life cycle. Castonguay et al. (1994) hypothesize that most juveniles entering the river eventually settle in the upper St. Lawrence River and Lake Ontario (USLRLO), the only Great Lake where eels are commonly found (Scott and Crossman 1973). Over a period of 1–4 years, young eels migrate to USLRLO, where they reside for 6–14 years before emigrating back to the sea as maturing yellow or silver-phase eel (Castonguay et al. 1994).

The USLRLO represents an important area for growth of juvenile eels and is considered to be especially important in maintaining the overall abundance of this panmictic species for at least four reasons: the freshwater runoff of the St. Lawrence River (10,100 m³/s) represents approximately 19% of the total freshwater runoff in the species' range (Castonguay et al. 1994); available data indicate that the upper St. Lawrence stock is composed almost exclusively of female eels (Vladykov 1966; Dutil et al. 1985; Couillard et al. 1997); size at migration increases with latitude and mature eels from USLRLO are some of the largest specimens known to occur (Oliveira 1999; Nilo and Fortin 2001); and fecundity increases with an increase in size (Barbin and McLeave 1997). Thus, this run of eels could represent a major contribution to the reproductive potential of this panmictic species.

Historically, this stock supported important commercial fisheries in Lake Ontario (Hurley 1973; Kolenosky and Hendry 1982; Stewart et al.

1997) and, during emigration, in the St. Lawrence Estuary (Robitaille and Tremblay 1994; Axelsen 1997). Although eel harvest and abundance have varied over the years, this stock is now experiencing a dramatic decline in abundance (Castonguay et al. 1994; Axelsen 1997; Casselman et al. 1997a). The decline has been observed since the beginning of the 1990s in Lake Ontario and the St. Lawrence Estuary, and has occurred following a reduction in recruitment at the Cornwall eel ladder after 1983 (Axelsen 1997; Casselman et al. 1997a). Casselman et al. (1997b) report that, in recent years, record-low numbers of juvenile eels have ascended this ladder. The lowest value of their recruitment index, observed in 1999 (Mathers 2000), represents a thousand-fold decrease from 1982.

The number of eels migrating from the USLRLO has never been estimated. In addition to mortality from the fall commercial fishery between Lake Saint-Pierre and the upper Estuary (Figure 1), migrants through the upper St. Lawrence River are exposed to at least two other major sources of mortality: the large hydroelectric projects, Moses-Saunders (built between 1954 and 1958) and Beauharnois-Les Cèdres (gradually equipped between 1912 and 1961).

Recent research to estimate mortality caused by these two hydropower dams and the number of migrating eels in the upper estuary allows the first evaluation of eel escapement from the USLRLO. The purpose of our study was to develop the first estimate of the escapement of eels from the USLRLO.

Methods

Model Construction and Source of Data

We utilized a simple opportunistic model to estimate escapement of eels from the USLRLO in 1996 and 1997 to the marine spawning grounds. This model considers eel passage, commercial landings, percent of migrating eels in the commercial catch, and turbine survival rate. Our approach is based on five assumptions: between Moses-Saunders hydropower project and Lake Saint-Pierre, the harvest of migrating eel is negligible; during the period of eel migration in the USLRLO (end of May to end of September), water flow is almost completely used for electricity production; the proportion of migrating and nonmigrating eels in the fall

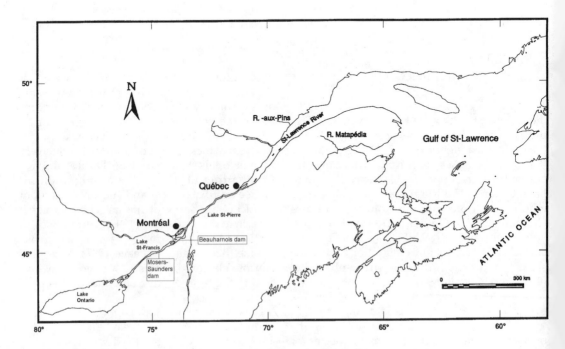

Figure 1. Map of the study area (adapted from Castonguay et al. 1994).

commercial catch did not change between 1982 and 1996–1997; water flow is directed to eel migration on a 1:1 basis; and the proportion of eels originating from the USLRLO in the fall commercial catch did not change during the same period.

Four of these assumptions are realistic. First, in Quebec, an eel is a rare, incidental, and undesirable capture for sportfishing (Fournier et al. 1987) and the only other commercial fishery in the River is limited to Lake St. Francis, a summer longline fishery with the catch composed of small yellow stage eels (Dumont et al. 1998). Second, the spring run-off period is generally before migration and, except for periods when turbines are shut down due to mechanical problems, most of the St. Lawrence River flow is successively diverted through the two power projects. For example, between 1 June and 30 September 1996 and 1997, only 0.6% of the total St. Lawrence River flow did not pass through the turbines of the Beauharnois-Les Cèdres Complex (Hydro-Québec, unpublished data). Based on the period of record (from 1900 through 1998), annual flow at the Moses-Saunders Project averaged 7,063 m³/s. The sum of all the diversions (Seaway locks and Cornwall Canal) typically ranges from approximately 11–21 m³/s and represents only about 0.1–0.3% of the total flow in the river (Kevin McGrath, New York Power Authority, personal communication). This information was used to estimate the survival rates of eels during their passage through the two power projects. Third, similar to the 1980s, the fall commercial catch is still mainly composed of large silver eels (Verreault et al. and Caron et al., both this volume). Fourth, no detailed in situ observations of downstream swimming behavior have been reported (EPRI 1999), but recent studies using acoustic and radio telemetry found that maturing eels occupied a variety of depths and locations when migrating (Haro et al. 1999). Preliminary observations at both projects suggest that migrating eels from USLRLO show the same behavior (Kevin McGrath, New York Power Authority, and Richard Verdon, Hydro-Québec, personal communication).

The fifth assumption is likely not valid. It is not correct to assume that the proportion of migrating eels coming from USLRLO has not changed from 1982 to 1996–1997. A comparison of the proportion of eels coming from the USLRLO in 1982 (Desjardins et al. 1983) and again in 1990 (Couillard et al. 1997) suggests that the contribution of the USLRLO to the upper Estuary fishery may have decreased. The percentage was estimated at 88% (91/104) in 1982 and 71% (10/14) in 1990. The 1982 data set of Desjardins et al. (1983) is the only one covering the different sections of the River that were included in the model. Using these data likely overestimates the proportion of eels originating from the USLRLO in the fall commercial catch of 1996 and 1997. Effects of this bias on the estimation of eel escapement are discussed.

The model used for our analysis is expressed as follows:

$$\text{USLRLO escapement} = \{(N_{QUE} + L_{QUE}) \times P_{QUSLRLO} + (L_{LSP} \times P_{SPUSLRLO})\} \times 1/S_{BL} \times 1/S_{MS}$$

N_{QUE} is the number of eels migrating past Québec City during fall in 1996 and 1997, based on the capture–recapture study of Caron et al. (this volume); these estimates, including 95% confidence intervals (CI), are, respectively, 466,617 (CI = 372,051–596,150) and 369,382 (CI = 322,331–426,055). According to the distribution of oocyte diameters reported by Dutil et al. (1985), 98.6% of these fish are migrating silver eels.

L_{QUE} is the number of migrating eels landed during fall (September to November) 1996 and 1997 in the fluvial estuary, between Québec City and Lake Saint-Pierre. These harvests were calculated by dividing the biomass of eels landed (24,255 and 26,313 kg in 1996 and 1997, respectively), as registered in the commercial landings database of the Ministère de l'Agriculture, des Pêcheries et de l'Alimentation du Québec, by the average weight of a migrating eel (1.3 and 1.18 kg in 1996 and 1997, respectively), as measured by Verreault et al. (this volume). These estimates were also adjusted for the percentage (98.6%) of migrating eels in the harvest (Dutil et al. 1985).

L_{LSP} is the number of migrating eels landed in Lake Saint-Pierre during fall of 1996 and 1997; the biomass landed (2,749 and 2,752 kg, respectively) was adjusted by the same average weights as used for L_{QUE}. According to Dutil et al. (1985), the percentage of migrants in this catch is estimated at 68.5%.

$P_{QUSLRLO}$ and $P_{SPUSLRLO}$ are the percentage of eels originating from the USLRLO and landed, respectively, in the upper estuary and in Lake Saint-Pierre. The origin of these eels was determined by referring to the level of Mirex contamination as described in the Desjardins et al. (1983)

dataset for 1982. Dutil et al. (1985) reported that Mirex presence could be considered as a reliable criterion to identify migrating eels having growth in the USLRLO. Castonguay et al. (1989) concluded that this criterion probably overestimates the contribution of this stock to the lower St. Lawrence fishery. However, the presence/absence criterion in the 1982 dataset was confirmed by a new interpretation of the Desjardins et al. (1983) dataset applying the graphical projection proposed by Couillard et al. (1997) on which the $\log_{10}$ concentration of Mirex is related to body mass of eels collected in the lower estuary in 1990 (Figure 2). $P_{QUSLRLO}$ and $P_{SPUSLRLO}$ were estimated at 66.2 and 62.2%, respectively.

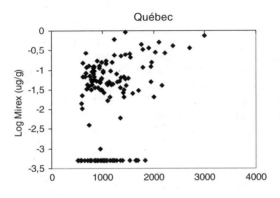

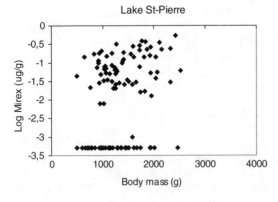

Figure 2. Concentration of Mirex relative to body mass in migrating eel captured in the St. Lawrence River, near Québec city, and in Lake Saint-Pierre in 1982. Data are from Desjardins et al. (1983). Mirex values below the limit of detection (0.001 ug/g) were replaced by a value equal to one half this limit. $\log_{10}$ Mirex concentrations higher than –3 differentiate eel migrating from USLRLO from eels from less contaminated areas.

S_{BL} is the migrating eel survival rate after passage through the turbines of the Beauharnois-Les Cèdres hydropower project. The 48-hour mean survival was estimated at 76.1% (CI = 68.3%–83.9%) for the fixed blade propeller turbines and 84.2% (CI = 76.7%–91.7%) for the Francis type turbines (Desrochers 1995). Maturing eels migrate through the upper part of USLRLO over an extended period of time. In 1999, dead or injured eels appeared in the tailwaters of the Moses-Saunders Power dam by the end of May to the beginning of October (McGrath et al.[1], this volume). A similar summer migration was also observed in the Richelieu River from 1974 to 1982 where June, July, August and September commercial landings averaged, respectively, 16.9%, 23.8%, 34.3% and 19.2% of total annual landings (Dumont et al. 1998). A weighted estimate of mortality rate was calculated after taking into account the flow distribution between the two dams from 1996 and 1997 (Beauharnois: 7,524 and 7,799 m^3/s; Les Cèdres: 532 and 956 m^3/s; locks and sluices: 47.8 and 52.7 m^3/s) (Hydro-Québec, unpublished data) and the type of turbines (Beauharnois: 10 propeller turbines and 26 Francis turbines; Les Cèdres: 17 Francis turbines) (Robitaille 1994). Extrapolating this information, in 1996, 73.6% of the flow had been used in the 43 Francis turbines, 25.8% in the 10 propeller turbines and 0.6% bypassed, for a weighted estimate of S_{BL} of 82.2%. For 1997, S_{BL} was estimated at 82.3%.

S_{MS} is the migrating eel survival rate after passage through the turbines of the Moses-Saunders Hydropower Project, equipped with 32 fixed-blade propeller turbines. In a 1997 study in the U.S. half of this project, survival has been estimated at 73.5% (90% CI = 67.9%–79.1%) for an 88-hour period (Normandeau Associates Inc. and Skalski 1998). Since these turbines are nearly identical in design in the Canadian side of the project (Kevin McGrath, New York Power Authority, personal communication), we applied this estimate to the entire project. Furthermore, the survival rate estimated for the U.S. half of the complex is within the CI of the rate estimated at Beauharnois for the same type of turbine. Since 0.3% of the flow is bypassed, S_{MS} is estimated at 73.6%.

---

[1]McGrath, J. Bernier, S. Ault, J.-D. Dutil, and K. Reid

Table 1. Calculated values of the parameters used to estimate the escapement of eels from the Upper St. Lawrence River and Lake Ontario.

Year	N_{QUE} (number of eels)	L_{QUE} (number of eels)	$P_{QUSLRLO}$ (%)	L_{LSP} (number of eels)	$P_{SPUSLRLO}$ (%)	S_{BL} (%)	S_{MS} (%)	USLRLO escapement (number of eels)
1996	460,084	18,396	66.2	1666	62.2	82.2	73.6	525,281
1997	364,211	21,987	66.2	1598	62.2	82.3	73.6	423,717

Results

The calculated values of the parameters included in the model are shown in Table 1. The number of maturing eels migrating past Québec city (N_{QUE}) during fall 1996 was estimated at 460,084. To this estimate, we added 18,396 migrating eels landed between Québec and Lake Saint-Pierre during the fall (L_{QUE}). Of this total of 478,480, we estimated that 316,754 originated from USLRLO to which we added 1,037 migrating eels landed in Lake Saint-Pierre. From this point, we added 68,816 eels that died during turbine passage in Beauharnois-Les Cèdres Project and 138,674 that died 80 km upstream at the Moses-Saunders Project. The 1996 escapement of migrating eels from USLRLO was then estimated at 525,281. The same calculations were performed for 1997 and escapement was estimated at 423,717.

Discussion

All the known anthropogenic causes of mortality were considered in our research. Sportfishing is negligible and migrating eels in USLRLO are harvested only by commercial fishermen in the lower part of the system, from Lake Saint-Pierre to the estuary. There is no fishing for these migrants upstream from this point to the Moses-Saunders complex.

According to our computations, about half a million migrating eels left the USLRO in 1996 and 1997. Is this estimate likely? Migrating eel annual landings in the St. Lawrence Estuary historically exceeded 250 metric tons (Robitaille 1994), equivalent to about 210,000 eel, and was estimated at 119.5 and 111.9 metric tons in 1996 and 1997, respectively (Verreault et al., this volume). USLRLO is a major zone of growth for eel in the St. Lawrence watershed, contributing more than 70% of the river's flow. Even after the construction of large dams by the end of the 1990s, it has also been one of the zones most accessible to eels that likely used locks and, more recently, a ladder to complete their upstream migration. Along the Québec part of the St. Lawrence River, in most of the tributaries, eel migration is blocked by natural (mainly falls in the northern shore of the River, on the foothills of the Canadian shield) or man made obstacles (dams). For example, open access to Lake Champlain had been almost completely obstructed by the end of the 1960s after the rebuilding of two old dams on the Richelieu River (Verdon et al.[2], this volume).

In 1982, Mirex concentration indicated that 88% of eels landed in the estuary originated from Lake Ontario (Desjardins et al. 1983). More recent but more fragmentary data suggest that this proportion was still high in 1990, about 71%, but decreasing (Couillard et al. 1997). Our model is quite sensitive to the values of $P_{QUSLRLO}$ and $P_{SPUSLRLO}$, the proportions of USLRO eels in the landings of the Québec City and Lake St-Pierre sectors. It is quite likely that these proportions, estimated at 66.2% and 62.2% in the 1982 samples, also decreased. Applying a reduction of 20% to these proportions, as observed in the estuary, would mean that $P_{QUSLRLO}$ and $P_{SPUSLRLO}$ were likely near 50% in 1990 and maybe lower in 1996 and 1997, and that escapement probably did not exceed 400,000 in 1996 and 320,000 in 1997.

Each year, 39.5% of the migrating eels died after their passage through turbines, or 207,500 in 1996 and 167,400 in 1997. This means a large number of dead eels, and even if a high proportion of them sink, a large number of carcasses downstream of the two complexes.

---

[2] Verdon, R., D. Desrochers and P. Dumont

Three different sources of information indicate that the number of eel carcasses downstream of the Moses-Saunders Project is high. In May 1978, in a letter on eel mortality at the Project addressed to J. R. Mongeau, Regional Biologist for the Montreal District of the Québec Fish and Wildlife Service, G. R. Whitney, himself Regional Biologist for the Ottawa and Cornwall District of the Ontario Ministry of Natural Resources, reported that 15,300 kg (33,600 lb) of dead fish were picked up by the Ontario Hydropower Project staff downstream of the Canadian half of the Project. In his letter, he assumes "that this poundage is an annual value which is probably an underestimate since fish with swimbladder damage would not be recoverable along the shore or on the surface". Kolenosky (1976) reported that, "during a two-day investigation, an estimated 225 kg (500 pounds) of eels, most of which were pieces rather than whole, were collected by Ontario Hydro's employee along the shoreline as far as 8 km below the dam. Only six of these specimens did not show obvious fatal injuries upon cursory examination". More recently, McGrath et al. (this volume) used the number of dead or injured eels collected by hand netting in the tailwaters of the two halves of the Project as an index of eel seasonal migration. They used a standard survey area and standard survey techniques and they sampled twice a week to make instantaneous counts of eel carcasses. At least 1,970 dead eels were collected between the beginning of June and the end of September 1999 during these partial counts. Similar surveys, however, have not been made at the Beauharnois-Les Cèdres Project (Richard Verdon, Hydro-Québec, personal communication). Dead or injured eels are observed but they rarely wash up upon the shore because the tailrace is deep (i.e., 12–29 m) and water velocity is quite high (> 2 m/s). These Figures suggest that our estimations of the number of eels killed or injured during their passage in the turbines of the two projects are likely and that our estimations of escapement are realistic.

Annual escapement estimates for eel are scarce in North America. For shallow lakes in the Canadian Atlantic provinces, Smith and Saunders (1955) evaluated annual escapement between 0.7 and 2.0 kg/ha. For a coastal lake in New Brunswick, Smith (1966) found a mean annual escapement of 2.3 kg/ha (1.5–5.1 kg/ha) for a 15-year period. In comparison with those evaluations, estimates for Lake Ontario seem very low. Based on our findings, annual escapement of spawning stock biomass for the two years studied is near 0.3 kg/ha for the total surface area of the Lake Ontario (19,604 km^2). Although we do not have information on tributary colonization for eel in this lake watershed, we assume that the bulk of growth habitat is found in the lake itself. Productivity could be higher without losses from fishing and incidental mortality. We do not have any information on anthropogenic impacts such as chemical spills or hazardous chemical exposures in Lake Ontario, but direct fishing mortality was evaluated. Commercial harvest is done almost exclusively on yellow stage eel and records for Canadian waters are available through the Ontario Ministry of Natural Resources database.

The population estimates for migrating eels in 1996 and 1997 do not represent the maximum escapement for Lake Ontario. Historically, the number of eels leaving the lake was probably greater. Casselman et al. (1997a) showed highly significant correlation between the Cornwall eel ladder recruitment index and the commercial harvest in the lake. The best correlation with the overall commercial harvest involved an eight year time lapse ($P = 0.009$). If two additional years are needed for eels to mature and escape from the lake, mean time lapse between ascending the eel ladder and escapement from Lake Ontario as silver eel can reasonably be estimated at 10 years. Recruitment index values for 1986 and 1987 were, respectively, 5,380 and 9,277, well below previous years when this index averaged 16,538 (SD = 6,195) for the 1974–1985 period. If the correlation between recruitment for the 1974–1985 period and subsequent escapement ten years later was the same as during our study, migrating eel average escapement from these earlier recruitment years was probably twice or more than our estimate for 1996 and 1997.

Escapement from Lake Ontario does not mean direct contribution to the spawning stock; there are important losses between the lake and the estuary. Migrating eels are exposed to three known sources of anthropogenic mortality as they migrate down the St. Lawrence River: two hydropower projects and a commercial fishery. A 26.4% mortality rate was observed at the first dam eels encounter (Normandeau Associates, Inc. and Skalski 1998) and 17.8% at the second dam (Desrochers 1995). The estuarine commercial fishery adds approximately 22% (Caron et al., this volume) for an overall cumulative mortality rate estimated at 53%. Turbine passage is responsible for three-quarters of this loss.

We believe the survival rates presented are conservative and that mortality during the spawning run could be even greater. Other causes, such as predation, may play an important role. Migrating eels likely compose a part of the annual diet of the beluga whale *Delphinapterus leucas* in the St. Lawrence Estuary (Hickie et al. 2000). Other causes could reduce the quantity or quality of spawning females. Couillard et al. (1997) reported that 1.9% of the migrating females sampled in the estuary for macroscopic external pathological examination had vertebral malformations and that 19% of a subsample of eels submitted to radiographic examination had a variety of vertebral lesions as fusions, compressions, dislocations and deformities. We still do not know the effects of malformations and vertebral lesions, some of them likely caused by turbines (Kolenosky 1976: Larinier and Dartiguelongue 1989), on the estimated 4,500 km spawning migration.

This research is the first attempt of which we are aware to estimate escapement from Lake Ontario. Weak recruitment of juveniles upstream into Lake Ontario could decrease future escapement even further. Providing better access to Lake Ontario through increasing fish passage facilities would be an important action. However, above all, decreasing mortality during downstream passage must be considered to increase eel escapement for this important stock of large and fecund females.

Acknowledgments

We thank M. Binet and R. Galego from the Ministère de l'Agriculture, des Pêcheries et de l'Alimentation du Québec for providing eel landing statistics in the St. Lawrence River, R. Verdon, from Hydro-Québec, and Kevin McGrath of the New York Power Authority for constructive comments and for providing flow data and technical information on the two power projects, and J. Leclerc for technical and graphical support. Scott Ault and an anonymous referee made constructive comments on a first draft of this paper.

References

Axelsen, F. 1997. The status of the American eel (*Anguilla rostrata*) stock in Quebec. Pages 121–131 *in* R. H. Peterson, editor. The American eel in Eastern Canada : stock status and management strategies. Proceeding of Eel Workshop, January 13–14, 1997, Québec City, Quebec, Canada. Canadian Technical Report of Fisheries and Aquatic Sciences No. 2196.

Barbin, G. P., and J. D. McLeave. 1997. Fecundity of the American eel *Anguilla rostrata* at 45°N in Maine. Journal of Fish Biology 51:840–847.

Casselman, J. M., L. A. Marcogliese, T. Stewart, and P. V. Hudson. 1997a. Status of the upper St. Lawrence River and Lake Ontario American eel stock—1996. Pages 106–120 *In* R. H. Peterson, editor. The American eel in Eastern Canada : stock status and management strategies. Proceeding of Eel Workshop, January 13–14, 1997, Québec City, Quebec, Canada. Canadian Technical Report of Fisheries and Aquatic Sciences No. 2196.

Casselman, J. M., L. A. Marcogliese, and P. V. Hudson. 1997b. Recruitment index for the upper St. Lawrence River and Lake Ontario eel stock : a re-examination of eel passage at the R. H. Saunders hydroelectric generating station at Cornwall, Ontario, 1974–1995. Pages 161–169 *in* R. H. Peterson, editor. The American eel in Eastern Canada : stock status and management strategies. Proceeding of Eel Workshop, January 13–14, 1997, Québec City, QC. Canadian Technical Report of Fisheries and Aquatic Sciences No. 2196.

Castonguay, M., J.-D. Dutil, and C. Desjardins. 1989. Distinction between American eels (*Anguilla rostrata*) of different geographic origins on the basis of their organochlorine contaminant levels. Canadian Journal of Fisheries and Aquatic Sciences 46:836–843.

Castonguay, M., P. V. Hodson, C. M. Couillard, M. J. Eckersley, J.-D., and G. Verreault. 1994. Why is recruitment of the American eel, *Anguilla rostrata*, declining in the St Lawrence River and Gulf? Canadian Journal of Fisheries and Aquatic Sciences 51:479–488.

Couillard, C. M., P. V. Hodson, and M. Castonguay. 1997. Correlations between pathological changes and chemical contamination in American eels, *Anguilla rostrata*, from the St Lawrence River. Canadian Journal of Fisheries and Aquatic Sciences 54:1916–1927.

Desjardins, C., J. D. Dutil, and R. Gélinas. 1983. Contamination de l'anguille (*Anguilla rostrata*) du bassin du fleuve Saint-Laurent par le mirex. Canadian Industry Report of Fisheries and Aquatic Sciences 141.

Desrochers, D. 1995. Suivi de la migration de l'anguille d'Amérique (*Anguilla rostrata*) au complexe Beauharnois, 1994. Rapport préparé par MILIEU & Associés inc. pour la vice-présidence Environnement, Hydro-Québec. Montréal, Québec, Canada.

Dumont, P., M. LaHaye, J. Leclerc, and N. Fournier. 1998. Caractérisation des captures d'anguilles d'Amérique dans des pêcheries commerciales de la rivière Richelieu et du lac Saint-François en 1997. P. 97–106. *In* Bernard, M. and C. Groleau, editors. Compte-rendu du troisième atelier sur les pêches commerciales, Duchesnay, 13–15 janvier 1998. Québec, Ministère de l'Environnement et de la faune, Direction de la faune et des habitats et Direction des affaires régionales.

Dutil, J. D., B. Légaré, and C. Desjardins. 1985. Discrimination d'un stock de poisson, l'anguille (*Anguilla rostrata*), basée sur la présence d'un produit chimique de synthèse, le mirex. Canadian Journal of Fisheries and Aquatic Sciences 42:455–458.

EPRI (Electric Power Research Institute). 1999. American Eel (*Anguilla rostrata*) scoping study: a literature and data review of life history, stock status, population dynamics, and hydroelectric impacts. Report TR-111873, Palo Alto, California.

Fournier, P., M. Beaudoin, and L. Cloutier. 1987. Suivi de la pêche sportive dans les eaux de la région de Montréal. Montréal, Québec, Ministère du Loisir, de la Chasse et de la Pêche. Rapport Technique 06–42.

Haro, A., T. Castro-Santos, and J. Boubée. 1999. Behavior and Passage of Silver-phase American Eels at a small Hydroelectric Facility. International Council for the Exploration of the Sea. EIFAC meeting, Silkeborg, September 1999.

Hickie, B. E., M. C. S. Kingsley, P. V. Hodson, D. C. G. Muir, P. Béland, and D. Mackay. 2000. A modeling-based perspective on the past, present, and future polychlorinated biphenyl contamination of the St. Lawrence beluga whale (*Delphinapterus leucas*) population. Canadian Journal of Fisheries and Aquatic Sciences 57 (Supplement 1):101–112.

Hurley, D. A. 1973. The commercial fishery for American eel *Anguilla rostrata* (LeSueur) in Lake Ontario. Transactions of the American Fisheries Society 102:369–377.

Kolenosky, D. P. 1976. Eel mortality resulting from Robert H. Saunders St. Lawrence Generating Station. Lake Ontario Fisheries Assessment Unit. Technical Report. Richmond Hill, Ontario, Canada.

Kolenosky, D. P., and M. J. Hendry. 1982. The Canadian Lake Ontario fishery for American eel (*Anguilla rostrata*). P. 8–16 *in* K. H. Loftus, editor. Proceedings of the 1980 North American eel conference. Ontario Fisheries Technical Report Series 4. Toronto, Ontario, Canada.

Larinier, M., and J. Dartiguelongue. 1989. La circulation des poissons migrateurs : le transit à travers les turbines des installations hydro-électriques. Bulletin Français de Pêche et Pisciculture 312–313:1–90.

Mathers, A. 2000. St. Lawrence River fish community. Pages 4.1–4.5 *in* Lake Ontario Fish Communities and Fisheries: 1999 Annual Report of the Lake Ontario Management Unit. Ontario Ministry of Natural Resources, Picton, Ontario, Canada.

Nilo, P., and R. Fortin. 2001. Synthèse des connaissances et établissement d'une programmation de recherche sur l'anguille d'Amérique (*Anguilla rostrata*). Université du Québec à Montréal, Département des Sciences biologiques pour la Société de la faune et des parcs du Québec, Direction de la rercherche sur la faune. Québec, Canada.

Normandeau Associates Inc., and J. R. Skalski. 1998. Draft final report. Estimation of survival of American eel after passage through a turbine at the St. Lawrence-FDR power project, New York. Prepared for New York Power Authority, White Plains, New York.

Oliveira, K. 1999. Life history characteristics and strategies of the American eel, *Anguilla rostrata*. Canadian Journal of Fisheries and Aquatic Sciences 56:795–802.

Robitaille, J. 1994. Problématique de la migration de l'anguille d'Amérique (*Anguilla rostrata*) aux ouvrages hydroélectriques. Rapport du Bureau d'écologie appliquée à la Vice-Présidence Environnement, Hydro-Québec, Montréal, Québec, Canada.

Robitaille, J., and S. Tremblay. 1994. Problématique de l'anguille d'Amérique (*Anguilla rostrata*) dans le réseau du Saint-Laurent. Québec, Ministère de l'Environnement et de la Faune, Direction de la faune et des habitats. Rapport technique. Québec, Québec, Canada.

Scott, W. B., and E. J. Crossman. 1973. Freshwater fishes of Canada. Fisheries Research Board of Canada Bulletin 184.

Smith, M. W. 1966. Amount of organic matter lost to a lake by migration of eels. Journal of Fisheries Research Board of Canada 23:1799–1801.

Smith, M. W., and J. W. Saunders. 1955. The American eel in certain freshwaters of the maritime provinces of Canada. Journal of Fisheries Research Board of Canada 12:238–269.

Stewart, T. J., J. M. Casselman, and L. A. Marcogliese. 1997. Management of the American eel, *Anguilla rostrata*, in Lake Ontario and the upper St. Lawrence River. Pages 54–61 *In* R. H. Peterson editors. The American eel in Eastern Canada: stock status and management strategies. Proceeding of Eel Workshop, January 13–14, 1997, Quebec City, QC. Canadian Technical Report of Fisheries and Aquatic Sciences No. 2196.

Vladykov, V. D. 1966. Remarks on the American eel (*Anguilla rostrata* LeSueur). Size of elvers entering streams; the relative abundance of adult males and females; and present economic importance of eels in North America. Verhandlungen Internationale Vereinigung for Theoretische und Angewandte Limnologie 16:1007–1017.

Eel Fishing in the Great Lakes/St. Lawrence River System During the 20th Century: Signs of Overfishing

JEAN A. ROBITAILLE*
Bureau d'écologie appliquée, Lévis, Canada
Corresponding author: jrobitai@globetrotter.net

PIERRE BÉRUBÉ, SERGE TREMBLAY, AND GUY VERREAULT
Société de la Faune et des Parcs, Québec, Canada

Abstract.—Trends in eel fishing in the Great Lakes and St. Lawrence River (GLSLR) provide signs of overfishing. Downstream migrating eels in this system are large, their average age is 16 years, and over 99% are female. The system seems to be a major contributor of female spawners at sea. Eel catches in GLSLR have reached two peaks since 1920, and each peak was followed by a period of low abundance in the watershed. Catch levels are negatively correlated with the landings reported 13 to 20 years before, and the highest r-value is for a lag of 16 years. The first peak in catch occurred in the 1930s, during the Great Depression, and was mostly due to an increase in fishing of silver eels along the estuary. This period of intense fishing did not last for a long time; it ended at the onset of World War II. When fishing resumed after the war, eel abundance remained low for about 10 years. The second period of intense fishing lasted for a longer time. It was mostly due to the development of yellow eel fishing on its feeding grounds in the upper basin, caused by an increase in dockside prices paid for the fish. Until the 1960s, eel fishing was important only in a few North American rivers, including the St. Lawrence, and filled mostly a regional demand. After 1960, the activity has become increasingly linked to the demand from export markets. Eel fishing and catch have grown steadily over time, and spread across most of the species range, including upstream feeding grounds, as landings data from the GLSLR system show. We suggest that the more widespread and intense eel fishing in most North American eel-producing rivers, and a larger proportion of immatures in the catch, have been detrimental to escapement of seaward migrants and subsequent recruitment.

Introduction

The ongoing decline of American eel *Anguilla rostrata* in most North American rivers has been a subject of concern for fisheries biologists and managers (Peterson 1997; Atlantic States Marine Fisheries Commission 2000). Various types of explanations for the decline have been suggested, including changes in sea habitat, loss of access to growth areas in the upper basin of eel-producing rivers, contaminants, parasites, and overfishing (reviewed in Electric Power Research Institute 1999). Lack of long time series data has undermined efforts to confirm that fishing mortality might have contributed to the decline. In most parts of the species range, data on eel fishing effort and catch are fragmented among various jurisdictions, and cover only the last thirty to forty years (Atlantic States Marine Fisheries Commission 2000), making it difficult to point out indices of overfishing for a long-lived, supposedly panmictic species.

This paper reviews fishing and landings data from the Great Lakes/St. Lawrence River (GLSLR) system (Figure 1) to gather possible evidence of a link between peak catches and subsequent periods of low abundance.

Methods

Most data analyzed in this paper have been taken from reports on commercial fisheries published since 1920 by government agencies in charge of natural resources and economics (Dominion Bureau of Statistics 1920–1944; Bureau de la Statistique du Québec 1962; Bérubé 1992; Bérubé and Yergeau 1992; G. Johnson and D. Hébert, Ministère de l'Agriculture, des Pêcheries et de l'Alimentation du Québec, personal communication; P. Smith and L. Marcogliese, Ontario

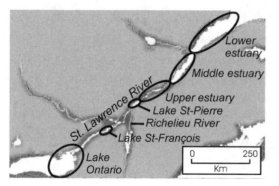

Figure 1. Main eel fishing areas in the GLSLR system.

Ministry of Natural Resources, personal communication). Fisheries data were collected essentially to describe economic activities, and thus have some inherent limitations to biological interpretation (see description in *Results* section).

No reliable, uniform data on catch per unit effort (CPUE) cover all years and all regions of the GLSLR system; thus no attempt was made to relate CPUE statistics between regions. However, autocorrelograms of the eel catch were generated (Legendre and Legendre 1984), either for the whole system or for some important eel fishing areas, in order to find possible cycles (e.g., to determine if there was a constant lag between high catch and low catch levels). For periods when eel catches were low in the whole system, CPUE series available for some regions were used to determine whether this was due to low eel densities or to a decrease in the fishing effort.

Results

Some characteristics of the eel fishery in the GLSLR system and in its most important eel fishing regions are relevant to the analysis presented here. Since no overview on this subject is available, this information is summarized below.

Historical Background

When the Europeans explored the St. Lawrence River and valley in the 16th and 17th centuries, eel fishing was already a seasonal activity of the aboriginal people (Bourget 1984). Native people came to the estuary near Quebec in the fall to catch eels with spears or with weirs set up on the tidal flats (Le Jeune 1634). The first French settlers learned to use this resource, and relied on eels as a subsistence food until they had cleared their land plot and grown their first crops. Under the seigniorial regime, land was allocated to new settlers in narrow strips from the river to deep inland (Figure 2), emphasizing the importance of an access to the river, both for transportation and for subsistence fishing. During the 18th and 19th centuries, farm production and increased trade with other regions allowed a more diverse diet (Provencher 1988); however, eel fishing has continued to the present time in the St. Lawrence River and estuary.

Descriptors of Eel Fishing and Their Limitations

Government agencies in charge of natural resources and economics have collected commercial fisheries statistics by counties since 1920, mostly to describe economic activities. Thus the data have some inherent limitations to biological interpretation. The only fishery data common to all sections of GLSLR system are the total catch by species or species group for each year. Some fish categories in the statistics have changed over the years or have been lumped together for some periods. However, eels have always been recorded separately.

The statistics cover only fish caught for sale. Subsistence catch is not included, but in the case of eels, it is thought to be unimportant during most of the 20th century, except for the economic crisis of the 1930s. Most people in Quebec do

Figure 2. Typical land subdivision along the south shore of the St. Lawrence estuary, inherited from the seigniorial system, under the French rule (Source: Aerial photograph Q80107–152, June 1990, Ministère de l'Énergie et des Ressources du Québec).

not enjoy eating eel, which is considered a cheap meal or subsistence food. Eating eel is still associated by the elders with the hard times of the Great Depression.

Only coarse indicators of effort or intensity of fishing are available throughout the 20th century: either the number of fishermen or the number of gear by types. The men counted as fishermen are the license holders and their regular helpers over thirteen years of age. In Lake Ontario, eels were mostly a by-catch of impoundment gear until an important hookline fishery developed to fill the demand for the export market in the 1960s and 1970s (Hurley 1973). Developing a standard measure of fishing effort with hooklines has not been possible (Kolenosky and Hendry 1982). The best available indicator of the fishing effort for eels seems to be the dockside price paid to fishermen (Casselman et al. 1997).

In the other parts of the GLSLR system, the number of active fishermen may provide a measure of fishing effort through time within a section, but comparisons between sections are not always possible, because gears used and the labor requested to operate them may differ from one section to another. There are also differences in the proportion of fishing effort assigned to eels versus other species between sections of the GLSLR.

Most fishing along the estuary aims primarily at eels, and is done with a specific impoundment gear set on the tidal flats, the eel weir. Most of the catch is made during September and October on maturing downstream migrants, known as silver or fall eels. Setting up the gear before eel migration, and dismantling it afterwards must be done at low tide and requires a few days of labor. The most important, full-time fishermen, set their weir year after year. But the effort spent by smaller, part-time fishermen seems to be more variable. Before they decide to put up their weir, they usually guess, from their catch of previous seasons and from the catch of other fishermen, if eels will be numerous enough to cover their expenses and make some profit. They may fish for only part of the season or decide not to put up their weir.

Since 1963, the commercial fishery along the St. Lawrence River and estuary has become more closely regulated, with fishing seasons, a smaller number of species allowed, and the number of fishing licenses kept at a steady level (Robitaille et al. 1988). These regulations included the control on the maximum number of eel weirs set in the estuary. This probably contributed to stabilizing landings for three decades, from 1960 to 1990, because saturation seems to set a maximum annual catch to each individual weir. Eels are typically caught in low numbers during most of the summer; the bulk of the catch for any year is made simultaneously in all weirs of the same region during a few fall nights of very high abundance. A typical pattern of daily catches in neighbouring weirs during the peak of the run is shown in Figure 3. For ten weirs operating at the end of September, the highest daily catch occurred on 29 or 30 September or on 1 October. The other two weirs (W11 and W12) started fishing at the beginning of October, and had their highest catch on 5 and 14 October.

Immature yellow eels are fished in non-tidal parts of the system, especially Lake St-Pierre, Lake St-François, and Lake Ontario. In the latter two fishing areas, eels were initially present among the catch of trapnets and hoop-nets. Hookline fisheries for eels have developed in Lake Ontario after 1960 (Hurley 1973), and in Lake St-François in the mid-1980s. In Lake St-Pierre, located above the head of tide, eels are caught mostly in hoop-nets used for a multispecific fishery. Eels are only a part of the total catch there, thus fishing goes on even in years of low eel abundance.

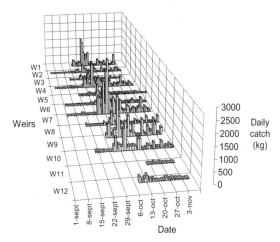

Figure 3. Daily eel catch in weirs near Rivière-Ouelle, St. Lawrence estuary, in 1986. All weirs are located along a 10 km stretch of shoreline (Source: Ministère de l'Agriculture, des Pêcheries et de l'Alimentation du Québec).

More than 99% of eels caught in the GLSLR are female (Vladykov 1966; Larouche et al. 1974; Dutil et al. 1985). The age of maturing eels caught in the estuary while migrating to sea ranges from 7 to 25 years, with the mode and mean at 16 years (Figure 4). Their average length and weight are 84 cm and 1.20 kg (Verreault et al., this volume). By comparison, yellow eels caught in Lake St-François have an average length of 70 cm; their average weight is 0.68 kg, about half the weight of silver eels (Dumont et al. 1998).

Trends in the GLSLR Eel Catch

Eel landings for the whole GLSLR system have gone through two main peaks since 1920 (Figure 5). The first one occurred during the Great Depression, in the middle of the 1930s, and was mostly due to an increase in fishing in the upper estuary. The second peak occurred between 1975 and 1980, and was more spread over time than the first one. During this second period of intense fishing, the landings remained quite stable in the estuary, because regulations did not allow an increase in the number of weirs after 1963. However, yellow eel fishing became more important in upstream parts of the GLSLR system, especially in Lake Ontario. Both peaks in landings have been followed by periods of low catches, the first one at the end of the 1940s and the second one during the 1990s.

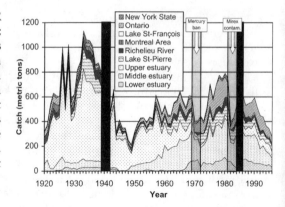

Figure 5. Area chart of the reported eel landings stacked by sections of the St. Lawrence River and Lake Ontario, 1920 to 1996. Black rectangles hide years where data are missing or incomplete; gray rectangles are for years of unusual fishing conditions: either the mercury ban in 1970–1971, or the closure of some export markets due to mirex contamination in 1982–1983 (Sources: Dominion Bureau of Statistics, Annual Reports 1920–1944; Bureau de la Statistique du Québec 1962; Bérubé 1992; Bérubé and Yergeau 1992; Lary and Busch 1997; Z. Bérubé, Bureau de la Statistique du Québec, personal communication; G. Johnson, D. Hébert, Ministère de l'Agriculture, des Pêcheries et de l'Alimentation du Québec, personal communication; P. Smith, L. Marcogliese, Ontario Ministry of Natural Resources, personal communication).

Correlation Analysis

The annual landings of eels in the GLSLR system show a positive autocorrelation up to lags of four or five years (Figure 6), a feature common to eel fisheries. However, r-values for lags over five years do not stay close to zero but rather reach significant negative values at lags of 13 to 20 years, with a maximum r-value of –0.60 at 16 years. When computed separately by sections of the GLSLR system, the landings autocorrelograms do not show the same pattern. The estuary and Richelieu River show positive autocorrelation for four or five years only, and then the r-values become non-significant.

Lake St-Pierre and Lake Ontario show positive, then negative autocorrelation with increasing lag, but they do not reach their maximum negative r-values at the same interval. Thus, the patterns in the landings and in the correlograms are not consistent with the hypothesis that catch

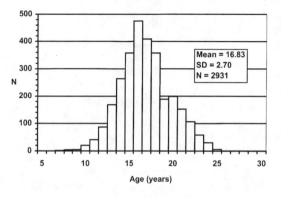

Figure 4. Age distribution of downstream migrating eels caught along the St. Lawrence upper and middle estuary in 1970 (Source: Larouche et al. 1974).

levels are driven mostly by cycles in eel abundance. However, they are consistent with an hypothesis that low catches in the whole GLSLR lag behind periods of very high catches in one section or another.

Historical evidence, summarized in the next sections, explain the periods of high catches by an increase in the fishing effort for eels, mostly in the St. Lawrence estuary for the first peak in the mid-1930s, and mostly in Lake Ontario, for the second peak in the 1970s. However, the low catch levels following both peaks were common to most sections of the GLSLR system. Data on catch and effort for some regions allow relating both periods of lower landings to lower eel densities, rather than a decreased effort.

Peak Catch In the Mid-1930s

When the Great Depression struck the province of Quebec in the 1930s, many people turned to fishing for their own subsistence or to earn an income. There was almost no control on the fishery. Although the price of fish was low, there was a demand for cheap food.

The commercial catch for all fish species in the province of Quebec reached its highest level in the middle of the 1930s (Robitaille et al. 1988). The species in the catch varied depending on the region. Along the St. Lawrence estuary, the most abundant, easily caught fish were eels. No boat or special equipment was needed to set up an eel weir and collect fish. This type of gear could be built on the tidal flats with some poles and twigs. Farmers who owned land along the river would set eel weirs in front of it or allow somebody else to do so. Thus, a fisherman could provide some food to his family, earn a modest income or get something to barter for other goods.

The total reported eel catch in the system rose over 1000 metric tons from 1933 to 1936, with the largest part of the increase from the upper, freshwater estuary (Figure 5). These numbers probably underestimate the real catch because the statistics covered only marketed fish, while eels were commonly sold for cash by itinerant merchants in the villages, bartered for other goods, or eaten by the fisherman's family. Moreover, fishing also increased markedly in the upper basin of major tributaries of the St. Lawrence, such as Ottawa, Richelieu, Yamaska, St-François, St-Maurice, Chaudière, and Saguenay Rivers, where eels were present (Dominion Bureau of Statistics 1920–1944). Some estuary fishermen tried to get better prices for their catch, and traveled in the fall to larger cities in the United States with live eels for sale (Bourget 1984). Others established connections with fish merchants in New England and Germany (Martin 1980).

The improvement of the economy just before World War II ended this period of intense fishing. The war made commercial shipping through the Atlantic risky, and cut the links with the European market. Above all, there were fewer people available for eel fishing because well-paying jobs in wartime industries and factories became available.

In the years following the end of the war, commercial fishing resumed gradually (Bourget 1984) but eels remained in very low abundance until the mid-1950s in most fishing areas (Figure 5). This lower abundance is apparent in the estuary catch from 1945 to 1950 (Figure 7).

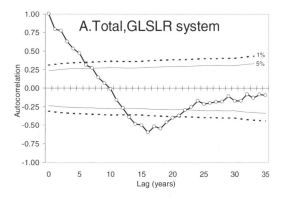

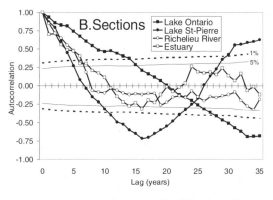

Figure 6. Autocorrelograms of eel landings for the whole GLSLR system (A) and for some important eel fishing sections (B). Dotted lines give the 1% and 5% critical r-values.

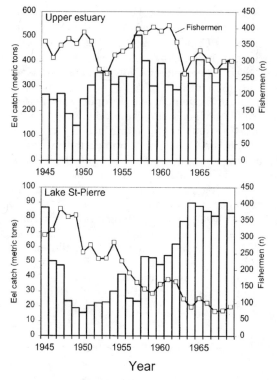

Figure 7. Eel catch and number of fishermen in the St. Lawrence estuary and Lake St-Pierre, 1945 to 1969 (Sources: Bureau de la Statistique du Québec 1962; Bérubé 1992; Bérubé and Yergeau 1992; Z. Bérubé, Bureau de la Statistique du Québec, personal communication).

Further increase in the number of downstream migrants after 1952 might have been concealed by gear saturation, as mentioned previously. However, the most important, steady fishermen in the estuary remember well that eels were unusually sparse after World War II (Lizotte 1993).

The scarcity of eels is more obvious in the commercial catch data from Lake St-Pierre, just upstream from the estuary (Figure 7). Lake St-Pierre has remained accessible from the sea and is still today the most important multi-species commercial fishing area along the river. Commercial landing data from Lake St-Pierre for post-war years show a marked increase in the number of fishermen, most of them using hoop-nets. However, very few eels were reported among their multi-species catch during those years.

Peak Catch In 1975–1980

In non-tidal waters of the GLSLR system, yellow eel has long been considered a low value by-catch in impoundment gear set for other, more popular species. However, rising price due to the development of the export market, beginning in the 1960s, was an incentive to fish for eels in Lake Ontario and Lake St-François (Hurley 1973; Kolenosky and Hendry 1982; Robitaille and Tremblay 1994; Stewart et al. 1997).

In Lake Ontario, the development of eel fishing was accelerated by the collapse of commercial fisheries for lake trout *Salvelinus namaycush*, lake whitefish *Coregonus clupeaformis*, lake herring *Coregonus artedii*, burbot *Lota lota*, walleye *Stizostedion vitreum*, and white perch *Morone americana* in the 1960s and 1970s (Hurley 1973; Stewart et al. 1997). The increase in fishing effort for eels came mostly from additional fishing licenses for hooklines. Toward the end of the 1970s, about 65% of the eel catch was made with this gear (Kolenosky and Hendry 1982). Hooklines are cheaper than other fishing gear, they are easy to set and operate by a single person, and can be moved to follow the fish, if necessary. The fishing effort with this type of device is difficult to standardize because its efficiency decreases when fish get caught on the hooks (Hurley 1973). During the best fishing years of the 1970s, some Lake Ontario fishermen were frequently visiting their lines to gather the eels and bait the hooks during the peak of the season. In the best fishing spots, a 30-hook line could catch about 45 kg of eels per day (Kolenosky and Hendry 1982).

Dockside price for eels and catch level in Lake Ontario seem to have been closely related over the past 50 years. They show very similar trends over time (Figure 8), and have been shown to be correlated (Casselman et al. 1997). With a larger fishing effort aimed at eels, the Lake Ontario fishery rapidly developed to reach its maximum catch of 230 metric tons in 1978 (Stewart et al. 1997), while the average size of the eels decreased (Kolenosky and Hendry 1982). The yearly catch started to go down in 1979, when prices were still high. Mirex contamination brought the prices and the catch down in 1982 (Marcogliese et al. 1997). From 1982 to 1992, prices remained at a steady level, without any definite trend up or down. A breaking point was reached in 1992, when landings started to decline, although the price paid for eels was increasing.

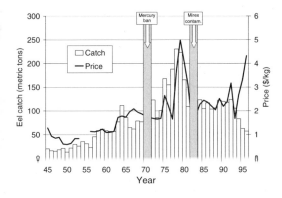
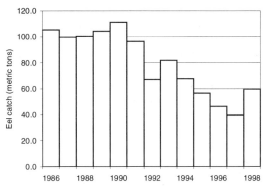

Figure 8. Dockside price (constant Canadian dollars, base 1986) for eel and total eel catch (metric tons) in Canadian waters of Lake Ontario, 1945 to 1996 (Sources: P. Smith, L. Marcogliese, Ontario Ministry of Natural Resources, personal communication).

Figure 9. Total yearly catch of 23 eel weirs set along the middle estuary from 1986 to 1998. The catch of fishermen who quit fishing during the period is not included (Source: Ministère de l'Agriculture, des Pêcheries et de l'Alimentation du Québec).

The increase in catch of immature eels in growth areas of GLSLR system during the 1970s and 1980s is less obvious than the peak catch of large silver eels in the estuary in the mid-1930s, because the landings are computed in metric tons, and yellow eels are about half the weight of silver eels. Available data suggest a marked increase in number of immature eels removed from the system during those years. Sampling of the yellow eel catch in Fishing District 6 of Lake Ontario have shown that the estimated number of eels caught by hooklines increased from an average 17,000 in the mid-1960s to more than 100,000 eels in 1978 (Kolenosky and Hendry 1982). The impact on the numbers of maturing fish migrating downstream each year has not been assessed.

In the St. Lawrence estuary, there was no decrease in the yearly catch, suggesting no reduction in number of seaward migrating eels, until 1990. The reason for this might be that the number of migrants had remained high enough to saturate the eel weirs during the few nights when most of the yearly catch is made. This appears to have occurred until 1990; thereafter, the eel catch began to decrease in the St. Lawrence estuary (Figure 9).

Discussion

Interpretation of commercial fishing statistics has some limits because the data were not collected for biological analysis, however, historical information on the eel fishery and other data provide some keys to interpret past changes in the catch. Commercial catch data from the GLSLR system from 1920 to 1996 suggest that two intense eel fishing periods have occurred, and that both were followed by a lower abundance of eels 13 to 20 years later.

Landings data show positive autocorrelation up to lags of four or five years. This is typical of the eel fishery and seems to be due to the biological characteristics of the fish, especially the long process of upstream migration to the feeding grounds by numerous successive year-classes of eels, the length of freshwater residence, and the high age at maturity. The catch is made on many successive year-classes. The full-time eel fishermen know that changes in eel abundance take place slowly over the years, and thus the fishing effort does not change drastically from one year to the next. However, r-values for lags over 5 years do not stay close to zero but rather reach significant negative values at lags of 13 to 20 years, with a maximum r-value of –0.60 at 16 years. Since this lag is similar to the average age of maturing eels caught in the estuary, the autocorrelation seems to agree with the hypothesis that overfishing might have limited reproduction.

The drop in eel abundance might have been short in time in the 1950s because it had been caused by a very intense but short-lived peak catch, mostly on maturing migrants. The homogeneous structure of the eel group in the St.

Lawrence—only large female fish—might be a reason why a marked reduction in spawner escapement in the mid-1930s was more clearly perceivable in this system than anywhere else as a lower abundance in the next generation of eels.

No consistent data from other North American regions are available to confirm that the decline in 1950 was widespread, however, the Great Depression struck the whole North American continent. It seems likely that the hard times caused heavy subsistence fishing in other parts of the eel's range, especially in places where the fish was easy to catch. Since eel was considered a low-value species in most areas before 1960, those caught by commercial fishermen were usually not recorded separately, but lumped together with other fish. The importance of the commercial and subsistence eel catch in most North American rivers during the Depression and the following years have not been assessed.

Available data from most eastern states and provinces in North America suggest that the current decline of American eel abundance is continent-wide (Peterson 1997; Electric Power Research Institute 1999; Atlantic States Marine Fisheries Commission 2000). Based on a similar downward trend in European eels, suggestions have been made for factors acting at sea (Castonguay et al. 1994). An alternate explanation would be that the international demand for eels has simultaneously increased fishing effort on all life stages of both species in most of their range, and has caused similar effects on recruitment.

Other changes within the freshwater habitat of eels have presumably lowered production of mature eels in the GLSLR system (Robitaille 1994; Verreault and Dumont, this volume; Verdon et al., this volume), as well as in other large systems (Electric Power Research Institute 1999). Structures restraining migration to and from the feeding grounds have been built during the 20th century. Most studies on this aspect of eel management have focused on major structures, such as hydroelectric projects, and dams controlling flow and water level in the St. Lawrence main stem and its largest tributaries. However, numerous flow-control structures built along many smaller rivers and streams have also hindered access to many eel-producing freshwater habitats. The gradual build-up of these inconspicuous habitat losses might have intensified the detrimental effects of overfishing on spawner escapement. Then the maximum number of eels that could be collected in the system without impairing recruitment would have been lower in the 1970s than it was in the 1930s.

Other changes that remain unexplained occurred in GLSLR eel during the 20th century. For instance, the best fishing sites for eels in the St. Lawrence estuary have gradually shifted downstream, from fresh water to sea water, during the 1950s and 1960s (Robitaille and Tremblay 1994). Mass mortality during transition to sea has also been observed, mostly during the 1970s. It was found to be associated with premature onset of salt excretion by chlorine cells in the gills, but the triggering factor has not been identified (Dutil et al. 1987).

The idea that overfishing within the GLSLR system might have impacted recruitment in the system itself seems puzzling at first sight, because it does not seem to fit with the concept of complete random mating of a whole, panmictic American eel population. Genetic evidence against panmixia in European eels has been reported (Wirth and Bernatchez 2001), and this may support the idea that some form of spatial or temporal segregation might exist among eel spawners, in American eels as well as in European eels.

Alternate hypotheses could also explain how the status of the GLSLR eel stock relates to the whole American eel population. The first one would be that GLSLR system is a major producer of large female spawners, and has a major weight in the production of offspring. A second explanation would be that the trends in GLSLR data reflect similar changes in fishing intensity and in eel abundance that occurred in the whole range, but were not conspicuous in other rivers for reasons given previously; i.e., either a lack of landings data for eels (because it was then considered a secondary, uninteresting resource), or a heterogeneous structure of local eel groups concealing a temporary drop in recruitment. The current, longer-lasting decline in recruitment, however, is now becoming evident across the entire American eel's range (Electric Power Research Institute 1999; Atlantic States Marine Fisheries Commission 2000).

Up to now, the unusual lifecycle and biological characteristics of eels may have contributed to blur the relation between overfishing and the resulting impairment of recruitment. However, the review of catch data from the GLSLR systems provides reasons to keep recruitment overfishing among the possible explanations of the current decline. We suggest that overfishing of silver and yellow eels in large producing systems like the GLSLR, together with more widespread fishing in most of its range, have been detrimental to eel recruitment.

Acknowledgments

The authors wish to acknowledge the useful suggestions on data treatment by Yves Grégoire and Gaétan Daigle. René Lesage, director of research at Société de la Faune et des Parcs du Québec, provided funds for this work. Dr. Pierre Dumont, from Société de la Faune et des Parcs du Québec, and an anonymous reviewer gave valuable comments on a first version of this paper. We appreciate the help received from Dr. Doug Dixon during the final revision.

References

Atlantic States Marine Fisheries Commission. 2000. Interstate Fishery Management Plan for American Eel *(Anguilla rostrata)*. Prepared by the American Eel Development Team, Washington, D.C.

Bérubé, Z. 1992. La pêche maritime au Québec, 1956–1985. Bureau de la statistique du Québec. Ministère de l'Industrie et du Commerce, Québec, Canada.

Bérubé, Z., and R. Yergeau. 1992. La pêche maritime au Québec, 1917–1955. Bureau de la statistique du Québec. Ministère de l'Industrie et du Commerce, Québec, Canada.

Bourget, C. 1984. Le fleuve et sa rive droite. 2. La pêche à l'anguille: une tradition. GIRAM (Groupe d'initiatives et de recherches appliquées au milieu). CEGEP Lévis-Lauzon, Lévis, Québec, Canada.

Bureau de la Statistique du Québec. 1962. Pêcheries intérieures 1945–1960. Ministère de l'Industrie et du Commerce, Québec, Canada.

Casselman, J. M., L. A. Marcogliese, T. Stewart, and P. V. Hodson. 1997. Status of the upper St. Lawrence River and Lake Ontario American eel stock. Pages 106–120 *In* R. H. Peterson (Editor). 1997. The American eel in Eastern Canada: Stock status and management strategies. Proceedings of the Eel Workshop, January 13–14, 1997, Quebec City, Quebec. Canadian Technical Report of Fisheries and Aquatic Sciences 2196

Castonguay, M., P. V. Hodson, C. Moriarty, K. F. Drinkwater, and B. M. Jessop. 1994. Is there a role of ocean environment in American and European eel decline? Fisheries Oceanography 3:197–203

Dominion Bureau of Statistics. 1920–1944. Fisheries Statistics, Ottawa, Ontario, Canada.

Dumont, P., M. LaHaye, and J. Leclerc. 1998. Caractérisation des captures d'anguille d'Amérique dans des pêcheries commerciales de la rivière Richelieu et du lac Saint-François en 1997. Pages 97–107 *in* M. Bernard and C. Groleau, editors. 1998. Compte-rendu du troisième atelier sur les pêches commerciales, Duchesnay (Québec), 13–15 janvier 1998. Société de la faune et des Parcs du Québec. Québec, Canada.

Dutil, J. D., B. Légaré, and C. Desjardins. 1985. Discrimination d'un stock de poisson, l'anguille *(Anguilla rostrata)*, basée sur la présence d'un produit chimique de synthèse, le mirex. Canadian Journal of Fisheries and Aquatic Sciences 42:455–458

Dutil, J. D., M. Besner, and S. D. McCormick. 1987. Osmoregulatory and ionoregulatory changes and associated mortalities during the transition of maturing eels to a marine environment. Pages 175–190 *in* Common Strategies of Anadromous and Catadromous Fishes. American Fisheries Society, Symposium 1, Bethesda, Maryland.

Electric Power Research Institute. 1999. American Eel *(Anguilla rostrata)* Scoping Study: a literature review of life history, stock status, population dynamics, and hydroelectric impacts. EPRI Report TR–111873, Palo Alto, California.

Hurley, D. A. 1973. The commercial fishery for American eel *(Anguilla rostrata* LeSueur) in Lake Ontario. Transactions of the American Fisheries Society 102:369–377

Kolenosky, D., and J. Hendry. 1982. The Canadian Lake Ontario fishery for American eel *(Anguilla rostrata)*. Pages 8–16 *In* K. H. Loftus (Editor). 1982. Proceedings of the 1980 North American Eel Conference. Ontario Ministry of Natural Resources, Ontario Fisheries Technical Report Series, 4. Toronto, Ontario, Canada.

Larouche, M., G. Beaulieu, and J. Bergeron. 1974. Quelques données sur la croissance de l'anguille d'Amérique *(Anguilla rostrata)* de l'estuaire du Saint-Laurent. Ministère de l'Industrie et du Commerce, Direction générale des Pêches maritimes, Direction de la Recherche, Rapport annuel 1973: 109–116. Québec, Canada.

Lary, S. J., and W. D. N. Busch. 1997. American eel *(Anguilla rostrata)* in Lake Ontario and its tributaries: Distribution, Abundance, Essential Habitat and Restoration Requirements. United States Department of the Interior, Fish and Wildlife Service. Administrative Report 97–01. Amherst, New York.

Le Jeune, s. j. 1634. Relation of 1634. 6: 309–310. In: The Jesuit Relations, and Allied Documents. In Reuben Gold Twaites, editor. Travel and Explorations of the Jesuit Missionaries in New France (1610–1791), Cleveland, Ohio.

Legendre, L., and P. Legendre. 1984. Écologie numérique. Tome 2. La structure des données écologiques. Presses de l'Université du Québec. Québec

Lizotte, G. H. 1993. Problématique de l'anguille: perception de l'Association des pêcheurs d'anguilles et de poissons d'eau douce du Québec. Pages 59–64 *in* S. Tremblay, editor. 1993. Compte-rendu de l'atelier sur l'anguille d'Amérique *(Anguilla rostrata)*. Sainte-Foy, mars 1993. Ministère du Loisir, de la Chasse et de la Pêche, Direction de la faune et des habitats. Québec, Canada.

Marcogliese, L. A., J. M. Casselman, and P. V. Hodson. 1997. Dramatic declines in recruitment of American eel *(Anguilla rostrata)* entering Lake Ontario: Long-term trends, causes, and effects. EMAN (Ecological Monitoring and Assessment Network), 3rd National Meeting, Saskatoon, Saskatchewan, 22 January 1997.

Martin, R. 1980. L'anguille. Leméac, Traditions du geste et de la parole, V. Montréal, Québec, Canada.

Peterson, R. H., editor. 1997. The American eel in Eastern Canada: Stock status and management strategies. Proceedings of the Eel Workshop, January 13–14, 1997, Quebec City, Quebec, Canada. Canadian Technical Report of Fisheries and Aquatic Sciences 2196

Provencher, J. 1988. Les quatre saisons dans la vallée du Saint-Laurent. Les Editions du Boréal. Montréal, Québec, Canada.

Robitaille, J. A., Y. Vigneault, G. Shooner, C. Pomerleau and Y. Mailhot. 1988. Modifications physiques de l'habitat du poisson dans le Saint-Laurent de 1945 à 1984 et effets sur les pêches commerciales. Rapport technique canadien des sciences halieutiques et aquatiques 1608.

Robitaille, J. A. 1994. Problématique de la migration de l'Anguille d'Amérique *(Anguilla rostrata)* aux ouvrages hydroélectriques. Rapport du Bureau d'écologie appliquée à Hydro-Québec, Vice-Présidence Environnement. Montréal, Québec, Canada.

Robitaille, J. A. and S. Tremblay. 1994. Problématique de l'Anguille d'Amérique *(Anguilla rostrata)* dans le réseau du Saint-Laurent. Ministère du Loisir, de la Chasse et de la Pêche, Direction de la gestion des espèces et des habitats. Rapport technique. Québec, Canada.

Stewart, T. J., J. M. Casselman, and L. A. Marcogliese. 1997. Management of the American Eel, *Anguilla rostrata,* in Lake Ontario and the upper St. Lawrence River. Pages 54–61 *in* R. H. Peterson, editor. The American eel in Eastern Canada: Stock status and management strategies. Proceedings of the Eel Workshop, January 13–14, 1997, Quebec City, Quebec. Canadian Technical Report of Fisheries and Aquatic Sciences 2196.

Vladykov, V. D. 1966. Remarks on the American eel *(Anguilla rostrata* Le Sueur): sizes of elvers entering streams; the relative abundance of adult males and females; and present economic importance of eel in North America. Internationale Vereinigung für Theoretische und Angewandte Limnologie Verhandlungen 16: 1007–1017.

Wirth, T., and L. Bernatchez. 2001. Genetic evidence against panmixia in the European eel. Nature 409:1037–1040.

PART III

Spawning Migration and Protection

Lunar Cycles of American Eels in Tidal Waters of the Southern Gulf of St. Lawrence, Canada

David K. Cairns*

*Department of Fisheries and Oceans, Box 1236,
Charlottetown, Prince Edward Island C1A 7M8, Canada
Email: cairnsd@dfo-mpo.gc.ca*

Peter J.D. Hooley

*Faculty of Medicine, Dalhousie University,
Halifax, Nova Scotia B3H 4J1, Canada
Email: peterhooley@hotmail.com*

Abstract.—Lunar cycles in catch per unit effort (CPUE) have been widely reported in *Anguilla* eels. CPUE of sub-legal (nearly all yellow) and legal (mix of yellow and silver) American eels *A. rostrata* in commercial fisheries in tidal bays and estuaries of Prince Edward Island, Canada, was markedly lower during full moon than at other times of the lunar cycle. CPUE was negatively correlated with an index of nighttime illuminance based on moon fullness, proportion of the night that the moon was above the horizon, and cloud cover. However, CPUE was also negatively correlated with proportion of moon fullness when only dark nights were considered, and CPUE was uncorrelated with the index of illuminance on nights when the moon was greater than 90% full. This suggests that illuminance did not control eel CPUE. It also suggests that lunar CPUE patterns were not an artifact of visually-based gear avoidance. Eel CPUE was negatively correlated with tidal amplitude in an area of relatively weak tides, but eel CPUE was not correlated with tidal amplitude in an area of stronger tides. Light and tide variables did not appear to trigger lunar cycles in eel CPUE. Endogenous rhythms may be the proximate cause of depressed eel CPUE during full moon in the study area.

Introduction

Eel *Anguilla* spp. activity is highly cyclical. In addition to daily (LaBar et al. 1987; Dutil et al. 1988) and seasonal (Walsh et al. 1983; Jessop 1987) rhythms, eels commonly exhibit lunar periodicity in activity. Activity peaks corresponding with lunar phases have been reported for American *A. rostrata*, European *A. anguilla*, shortfinned *A. australis*, and longfinned *A. dieffenbachii* eels, and in all continental life-stages, including elver, yellow, and silver eels (Winn et al. 1975; Tesch 1977; Deelder 1984; Todd 1981; Sorensen and Bianchini 1986).

The observed correspondence between eel activity and lunar cycles requires a proximate triggering mechanism. Moon phase is not consistently detectable by an animal on earth because the moon is often obscured by clouds. However, moon phase influences nighttime light regimes and tidal amplitudes. Thus animals could potentially maintain lunar activity cycles by using light or tide variation as approximate indicators of moon phase.

Eels show little activity during daytime hours (van Veen et al. 1976; Vøllestad et al. 1986), and the downstream migrations of silver eels may be delayed or interrupted by artificial lights (Lowe 1952; Hadderingh et al. 1999). Moreover, maturing eels swimming in the ocean tend to descend to lower levels during sunlit or moonlit periods (Tesch 1978, 1989). This light-avoidance behavior suggests that lunar cycles could control eel activity through changes in light levels. This is supported by Lowe's (1952) observation that European silver eels migrated during full moon only during heavy fog or when the water was unusually turbid. Light as a trigger of lunar cycles is also supported by reports that American and European eels are most active between sunset and moonrise in the period following full moon, when the moon rises progressively later in the evening (Hain 1975; Winn et al. 1975; Tesch 1977). However, European eels

---

*Present Address: Faculty of Medicine, Dalhousie University, Halifax, Nova Scotia B3H 4H7, Canada.

confined to darkened tanks also showed lunar cycles (Boëtius 1967), and eels in a nontidal river maintained their lunar cycles even in overcast conditions (Jens 1953). Many of the reported associations between activity and lunar cycles are based on capture rates in various fishing gears. If eels are better able to see and avoid fishing equipment on moonlit nights, some reported lunar activity cycles could simply be artifacts of methodology (Lindroth 1979).

Eels respond to tide-induced water currents during both resident (yellow) (Helfman et al. 1983; Dutil et al. 1988) and migratory (silver) (Ciccotti et al. 1995; Parker 1995; Parker and McCleave 1997) life-stages. Since tidal amplitude, and therefore tide-induced current, increases at full moon, lunar activity cycles could be triggered by changes in the tidal regime in tidal waters. However, lunar activity cycles are also common in natural waters (Jens 1953; Pursiainen and Tulonen 1986; Adam and Elie 1994) and experimental tanks (Boëtius 1967; Westin and Nyman 1979) where tide is absent.

Aquatic animals are profoundly influenced by cycles of light and (where present) of tide, and many have developed endogenous clocks that correspond to circadian and tidal cycles (Korringa 1947; Morgan 1996; Palmer 1997). Experiments in tanks have shown that eels exhibit lunar cycles in the absence of both tidal and natural light signals (Boëtius 1967; Tesch et al. 1992). This suggests that endogenous rhythms could trigger lunar activity cycles in eels.

This paper uses commercial fishing logbook data from the southern Gulf of St. Lawrence to examine lunar cycles in American eels and to determine whether moon-induced variation in light and tide levels are the proximate cause of lunar cycles in catch rates.

Methods

Eel catch per unit effort (CPUE) was calculated from the logbook reports of seven commercial fyke net fishermen, who fished in bays and estuaries along the North Shore and the eastern portion of the South Shore of Prince Edward Island, Canada, in 1996–1998. Log-keepers recorded the summed weights of eels of legal size ($\geq$ 46 cm in 1996–1997; $\geq$ 50.8 cm in 1998) and tallied numbers of sub-legal eels. The fyke net fishery was open from 16 August to 31 October, although most fishing ended by mid-October.

Logkeepers did not record eel color. Samples from concurrent commercial fisheries indicated that nearly all eels less than 50.8 cm long were yellow (99.5% yellow, 0.5% silver, N = 376). Larger eels included a substantial fraction of silver eels (79.2% yellow, 20.8% silver, N = 154).

Log-keepers fished from one to six different sites, with two to 51 nets per site, and reported a total of 28,219 net-days of fishing. CPUE was calculated for each site as weight (or numbers) of eels per net-day. To eliminate the effects of inter-site differences in catch rate, CPUE was standardized by dividing raw CPUE for each day by the mean seasonal CPUE for that site.

Atmospheric data were recorded at the Environment Canada weather station in Charlottetown, which is located 15–90 km from the fishing sites. Values for wind velocity, air pressure, and sky clearness (proportion of sky that is free of clouds) were from hourly measurements taken during nighttime hours. Moon phase was expressed as proportion of moon fullness (proportion of moon's face that is visible on a cloudless night); data were supplied by Environment Canada. Illuminance is the amount of light received per unit area on the ground. Because the moon does not act as a perfect mirror, lunar illuminance is not a linear function of the proportion of moon fullness. The proportion of moon fullness was converted to illuminance as a proportion of illuminance during full moon using tables supplied by Allen (1973). Stars and other astronomical phenomena also provide nighttime light. These sources yield about 0.5% of the illuminance provided by a full moon on a clear night (R. Garstang, University of Colorado, personal communication).

Total nighttime illuminance depends on lunar illuminance, the amount of time the moon is above the horizon, cloud cover, and light from nonlunar sources. We calculated a nighttime illuminance index that indicates illuminance relative to cloudless nights with a full moon:

$$I = (0.995 \times MI \times MT \times CM) + (0.005 \times CE),$$

where I is the illuminance index, MI is the illuminance of the moon relative to that of a full moon, MT is the proportion of nighttime hours that the moon is above the horizon, CM is the mean proportion of the sky that is free of clouds during the time the moon is above the horizon, and CE is the mean proportion of the sky covered by clouds during the entire night.

Nighttime tidal amplitudes were derived for the North and South Shores from predictions for Rustico and Pictou, where mean tidal amplitudes are 0.7 and 1.2 m, respectively (Anon. 1995, 1996, 1997).

Because nets were typically fished every two to three days, net-hauls often reflected catches over several nights. Analyses relating CPUE to illuminance indices were based only on net-hauls for which illuminance index varied by less than 0.1 during the nights whose catches contributed to the net-haul.

Proportion of moon fullness influenced illuminance index ($r = 0.559$, $P < 0.001$, $N = 183$), tidal amplitude on the North Shore ($r = 0.512$, $P < 0.003$, $N = 486$), and tidal amplitude on the South Shore ($r = 0.197$, $P < 0.025$, $N = 178$). Effects of moon fullness and illuminance index on CPUE were examined by calculating correlations between these variables and CPUE in data subsets in which the other variable was constant or nearly so. Full moon nights were those in which proportion of moon fullness was greater than 0.9. Full moon nights were used for analysis because during the waxing and waning portions of the lunar cycle the proportion of moon fullness changes rapidly from day to day. Dark nights were those in which illuminance index was less than 0.001. Effect of tidal amplitude on CPUE was tested during quarters of the lunar cycle.

Because CPUE and environmental variables were not normally distributed, and resisted correction by transformation, bootstrap analysis was employed. In each bootstrap trial cases were randomly selected, with replacement, from the data set. The number of cases selected was equivalent to the sample size of the data set. Correlations were calculated as the median correlation from 10,000 bootstrap trials. Probability values for positive correlations were the number of trials giving correlations less than or equal to 0, divided by 10,000. Probability values for negative correlations were the number of trials giving correlations greater than or equal to 0, divided by 10,000. Probability values were Bonferroni corrected. Statistical significance was accepted as $P = 0.05$.

Results

Standardized CPUE decreased at full moon for both legal (by weight) and sub-legal (by number) eels in all three study years (Figure 1). Stan-

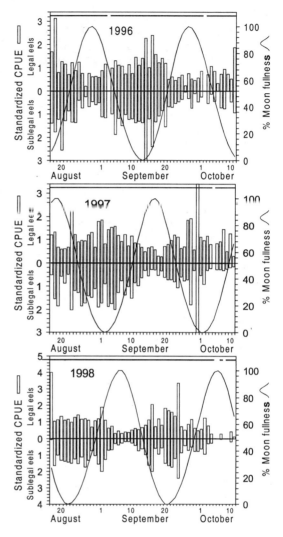

Figure 1. Standardized catch per unit effort (raw CPUE/mean) of legal, and sub-legal American eels, and percent moon fullness, fall. 1996–1998. Gaps in the horizontal lines at the top of each panel indicate absence of fishing effort.

dardized CPUE was negatively correlated with the proportion of moon fullness for legal eels ($r = -0.301$, $P < 0.003$, $N = 866$) and for sub-legal eels ($r = -0.279$, $P < 0.003$, $N = 866$) (Table 1; Figure 2). Standardized CPUE was also negatively correlated with the illuminance index for legal eels ($r = -0.182$, $P < 0.003$, $N = 656$) and sub-legal eels ($r = -0.186$, $P = 0.003$, $N = 656$).

Standardized CPUE showed significant negative correlations with tidal amplitude at North

Shore sites (legal eels: $r = -0.198$, $P < 0.003$, $N = 649$; sub-legal eels: $r = -0.253$, $P < 0.003$, $N = 649$), but not at South Shore sites (legal eels: $r = -0.167$, $P = 0.073$, $N = 217$; sub-legal eels: $r = -0.129$, $P = 0.479$, $N = 217$) (Table 1; Figure 2). Standardized CPUE for sub-legal eels was positively correlated with minimum air temperature ($r = 0.212$, $P < 0.003$, $N = 866$), and standardized CPUE for legal eels was positively correlated with precipitation ($r = 0.154$, $P = 0.039$, $N = 866$). Other correlations between CPUE and weather variables were non-significant.

Standardized CPUE was negatively correlated with proportion of moon fullness for both legal ($r = -0.302$, $P < 0.003$, $N = 142$) and sub-legal ($r = -0.328$, $P < 0.003$, $N = 142$) eels on dark nights (illuminance index < 0.001) (Figure 3). Standardized CPUE was uncorrelated with illuminance index for both legal ($r = 0.001$, $P = 1.0$, $N = 60$) and sub-legal ($r = 0.001$, $P = 1.0$, $N = 60$) eels during periods when moon fullness was greater than 0.9 (Figure 4).

Standardized CPUE for both legal and sub-legal eels was uncorrelated with tidal amplitude on the North Shore within each of the four quarters of the lunar cycle, categorized by moon fullness (proportion moon fullness less than 0.25, between 0.25 and 0.5, between 0.5 and 0.75, and greater than 0.75; range of correlations: -0.198 to 0.097, $P = 1.0$ for all tests) (Table 1).

Discussion

CPUE of American eels in Prince Edward Island bays and estuaries showed a marked lunar cycle, with sharply lower catches during full moon. This finding applied both to legal sizes, which are a mixture of yellow and silver eels, and to sub-legal sizes, which are almost all yellow eels.

Eel CPUE showed significant negative correlations with nighttime illuminance for both legal and sub-legal eels (Figure 2). However, moonlight could not have been the proximate trigger of the lunar CPUE cycle for two reasons. First, the negative correlations between standardized CPUE and proportion moon fullness were maintained with undiminished strength when only dark nights were included in the analysis (Figure 3). Second, standardized CPUE was uncorrelated with the illuminance index during full moon nights (Figure 4), indicating that eels showed similar activity on overcast nights and on bright moonlit nights during this period.

Table 1. Correlations between eel catch per unit effort and environmental variables.

| | | Standardized catch per unit effort | | | | |
| | | Legal eels | | Sublegal eels | | |
Variable	Data	R	P^a	r	P^a	N
Moon fullness (proportion)	All	−0.301	<0.003	−0.279	<0.003	866
Illuminance index	All	−0.182	<0.003	−0.186	<0.003	656
Minimum air temperature (°C)	All	0.082	0.918	0.212	<0.003	866
Precipitation (mm)	All	0.154	0.039	0.123	0.112	866
Air pressure (kP)	All	−0.029	1.0	0.037	1.0	866
Wind speed (km h^{-1})	All	0.019	1.0	−0.068	1.0	866
Moon fullness (proportion)	Illuminance index < 0.001	−0.302	<0.003	−0.328	<0.003	142
Illuminance index	Prop. moon fullness > 0.9	0.001	1.0	0.001	1.0	60
Tidal amplitude (m)	North Shore	−0.198	<0.003	−0.253	<0.003	649
Tidal amplitude (m)	South Shore	−0.167	0.073	−0.129	0.479	217
Tidal amplitude (m)	North Shore, prop. moon fullness < 0.25	−0.113	1.0	−0.133	0.599	246
Tidal amplitude (m)	North Shore, prop. moon fullness > 0.25 and < 0.5	0.097	1.0	−0.099	1.0	118
Tidal amplitude (m)	North Shore, prop. moon fullness > 0.5 and < 0.75	−0.111	1.0	−0.201	0.610	94
Tidal amplitude (m)	North Shore, prop. moon fullness > 0.75	0.016	1.0	0.084	1.0	191

[a] P values are Bonferroni adjusted

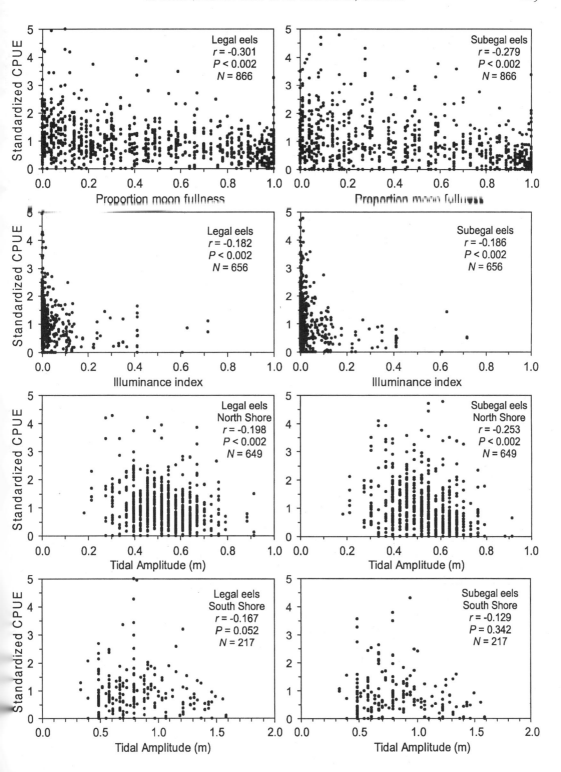

Figure 2. Standardized catch per unit effort (raw CPUE/mean) of legal and sub-legal eels versus moon fullness, illuminance index, and tidal amplitude.

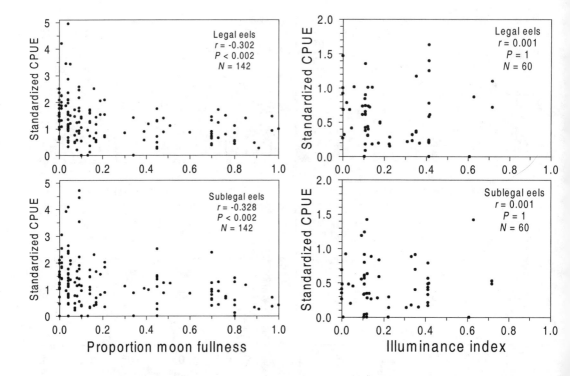

Figure 3. Standardized catch per unit effort (raw CPUE/mean) of legal and sub-legal eels versus proportion moon fullness on dark nights (illuminance index < 0.001).

Figure 4. Standardized catch per unit effort (raw CPUE/mean) of legal and sub-legal eels versus illuminance index, during full moon (moon fullness ≥ 0.9).

Because fyke nets have fixed locations and capture only moving eels, their CPUEs are commonly used as proxies for activity level (e.g., Jellyman 1991). However, if eels are better able to see and avoid fishing gear on moonlit nights, then reported lunar activity cycles would be an artifact of the sampling method. If this is so, then eel CPUE should vary directly with illuminance. Because CPUE is not linked with nighttime illuminance when moon fullness is held nearly constant, the hypothesis can be rejected.

Eel CPUE was negatively correlated with tidal amplitude on PEI's North Shore (Figure 2). However, when the lunar period was divided into four phases based on the proportion of moon fullness, correlations between CPUE and tidal amplitude within each phase were weak and insignificant (Table 1). Tidal amplitudes on the North Shore are strongly correlated with proportion moon fullness ($r = 0.512$, $P < 0.003$). Thus the correlation between eel CPUE and tidal amplitude on the North Shore may be due to the link between tidal amplitude and moon fullness. If eels cue on tidal amplitude (or strength of tidally induced currents, which depend on tidal amplitude) to determine activity levels, then linkages between tidal amplitude and CPUE should be strongest on the South Shore, where mean tidal amplitude is greater than on the North Shore (1.2 m versus 0.7 m). However, eel CPUE was not significantly correlated with tidal amplitude on the South Shore.

Our data allow rejection of moonlight, gear avoidance, and tidal amplitude as triggers of lunar cycles in CPUE of Prince Edward Island eels. Endogenous rhythms are an alternate potential trigger of lunar cycles in Prince Edward Island eels. Endogenous clocks are well developed in *Anguilla* eels, and can maintain both circadian (Edel 1976; van Veen et al. 1976) and lunar (Boëtius 1967; Tesch et al. 1992) rhythms in the absence of environmental cues. Lunar rhythms in captive silver eels are very durable, persisting up to nine months after removal from the natural environment (Boëtius 1967).

Despite the widespread documentation of lunar cycles in yellow and silver eels (Table 2), few workers have attempted to isolate the proximate trigger of such cycles. Lowe (1952), Hain (1975), and Vøllestad et al. (1986) argued that lunar cycles are a direct effect of moon-light. In contrast, Deelder (1984), citing Jens' (1953) observation that lunar cycles were maintained in cloudy periods and Boëtius' (1967) tank studies, suggested that endogenous rhythms trigger lunar activity cycles. Our data suggest that environmental cues do not trigger lunar cycles in tidal waters of Prince Edward Island, but controlled experiments would be required to determine if these cycles are endogenously regulated.

Table 2. Lunar activity cycles in yellow and silver *Anguilla* eels.

Species	Stage	Location	Habitat	Period of peak activity[a]	Source
rostrata	Yellow	Prince Edward I	Bays and estuaries	Last quarter to first quarter	This study
rostrata	Yellow and silver	Prince Edward I	Bays and estuaries	Last quarter to first quarter	This study
rostrata	Yellow and silver	Nova Scotia	Rivers	Last quarter to first quarter	Taylor 1979
rostrata	Yellow and silver	New Brunswick	Lake outlet	No lunar periodicity	Smith and Saunders 1955
rostrata	Yellow and silver	Rhode Island	Streams and estuaries	Between full and new moon	Winn et al. 1975
rostrata	Adult	Cuba	River	First quarter to full moon	Fernandez and Vasquez 1978
anguilla	Silver	Finland	Rivers	New moon	Pursiainen and Tulonen 1986
anguilla	Silver	Swedish east coast	Non-tidal sea	Between last and first quarters	Lindroth 1979
anguilla	Silver	Swedish east coast	Outdoor tanks	Full moon and last quarter	Westin and Nyman 1979
anguilla	Yellow	Swedish east coast	Outdoor tanks	Last quarter and new moon	Westin and Nyman 1979
anguilla	Silver	Norway	River	Times other than full moon	Vollestad et al. 1986
anguilla	Silver	Norway	River	First quarter	Haraldstad et al. 1985
anguilla	Silver	Denmark	Darkened experimental tanks	Last quarter	Boetius 1967
anguilla	Silver	Germany	River	Last quarter	Jens 1953
anguilla	Yellow and silver	Germany	Indoor tanks	New moon	Tesch et al. 1992
anguilla	Silver	England	Rivers	New moon	Lowe 1952
anguilla	Silver	Ireland	River	Between last and first quarters	Piggins 1985
anguilla	Yellow	France	Lake	New moon	Adam and Elie 1994
australis	Maturing adults	Tasmania	River	No lunar periodicity	Sloane 1984
australis, dieffenbachii	Adults	New Zealand	Outlets of lakes and rivers	Between full and new moon	Todd 1981
australis, dieffenbachia	Adults	New Zealand	Lake	Little or no lunar periodicity	Jellyman 1991

[a] The first quarter is the half moon following the new moon. The last quarter is the half moon following the full moon.

Eel activity is influenced by river discharge and temperature, which may exert a more powerful effect than lunar cycles during downstream migrations of silver eels (Smith and Saunders 1955; Vøllestad et al. 1986; 1994; Pursiainen and Tulonen 1986). In the present study, CPUE of sub-legal eels declined with temperature and CPUE of legal eels increased with precipitation.

This study showed a depression of eel activity centered at full moon (Figure 1). Investigations elsewhere indicate that activity peaks may occur anywhere in the lunar cycle, including new moon, first quarter, full moon, and last quarter (Table 2). The most common pattern appears to be activity peaks in the period following full moon (Table 2), when the moon rises at an increasing interval of time after sunset. Eels are commonly most active in the early part of the night (Vøllestad et al. 1986), which corresponds to moonless darkness in the postfull moon period. Smith and Saunders (1955), Sloane (1984), and Jellyman (1991) reported little or no relation with lunar cycles, but absence of lunar periodicity appears to be the exception among *Anguilla* eels.

Because *Anguilla* eels are panmictic across broad geographical areas (Avise et al. 1986, Wirth and Bernatchez 2001), local genetic variation should be limited or absent. Thus the variable patterns of lunar activity peaks cannot be attributed to local genetic traits. Instead such patterns are presumably locally appropriate responses to ambient conditions. Nocturnal activity in eels may reduce predation risk (Helfman et al. 1983). Lunar cycles which concentrate activity during moonless periods might serve to reduce risk from nocturnal predators, but if eels adopt lunar cycles to avoid predation, we would expect a consistent pattern of depressed activity near full moon. The ultimate cause of lunar cycles in eels remains obscure.

Acknowledgments

We are grateful to the logbook fishermen who provided the data which made this paper possible. We are greatly indebted to Roy Garstang of the University of Colorado (retired) for showing us how to calculate moon illuminance and for computing illuminance from nonlunar sources. We thank Brian Jessop, Manon Mallet, Gérald Chaput, and referees Bill Richkus and Gayle Zydlewski for comments that helped improve the manuscript.

References

Adam, G., and P. Elie. 1994. Mise en évidence des déplacements d'anguilles sédentaires (*Anguilla anguilla* L.) en relation avec le cycle lunaire dans le lac de Grand-lieu (Loire-Atlantique). Bulletin français de la pêche et de la pisciculture 335:123–132.

Allen, C. W. 1973. Astrophysical quantities, 3rd edition. Athlone Press, London, England.

Anon. 1995. Canadian tide and current tables 1996: Gulf of St. Lawrence. Department of Fisheries and Oceans, Ottawa, Canada.

Anon. 1996. Canadian tide and current tables 1997: Gulf of St. Lawrence. Department of Fisheries and Oceans, Ottawa, Canada.

Anon. 1997. Canadian tide and current tables 1998: Gulf of St. Lawrence. Department of Fisheries and Oceans, Ottawa, Canada.

Avise, J. C., G. S. Helfman, N. C. Saunders, and L. S. Hales. 1986. Mitochondrial DNA differentiation in North Atlantic eels: population genetic consequences of an unusual life history pattern. Proceedings of the National Academy of Science 83:4350–4354.

Boëtius, J. 1967. Experimental indication of lunar activity in European silver eels, *Anguilla anguilla* (L.). Meddelelser fra Danmarks Fiskeri-og Havundersogelser 6:1–6.

Ciccotti, E., T. Ricci, M. Scardi, E. Fresi, and S. Cataudella. 1995. Intraseasonal characterization of glass eel migration in the River Tiber: space and time dynamics. Journal of Fish Biology 47:248–255.

Deelder, C. L. 1984. Synopsis of biological data on the eel *Anguilla anguilla* (Linnaeus, 1758). Food and Agriculture Organization Fisheries Synopsis No. 80, Revision 1.

Dutil, J.-D., A. Giroux, A. Kemp, G. Lavoie, and J.-P. Dallaire. 1988. Tidal influence on movements and on daily cycle of activity of American eels. Transactions of the American Fisheries Society 117:488–494.

Edel, R. K. 1976. Activity rhythms of maturing American eels (*Anguilla rostrata*). Marine Biology 36:283–289.

Fernández, J., and J. Vásquez. 1978. Las pesquerías de angulas en la provincia de Holguín. Revista Cubana de Investigaciones Pesqueras 3:48–61.

Hadderingh, R. H., G. H. F. M. van Aerssen, R. F. L. J. de Beijer, and G. van der Velde. 1999. Reaction of silver eels to artificial light sources and water currents: an experimental deflection study. Regulated Rivers Research and Management 15:365–371.

Hain, J. H. W. 1975. The behaviour of migratory eels, *Anguilla rostrata*, in response to current, salinity, and lunar period. Helgolander Meeresuntersuchungen 27:211–233.

Haraldstad, O., L. A. Vøllestad, and B. Jonsson. 1985. Descent of European silver eels, *Anguilla anguilla* L., in a Norwegian watercourse. Journal of Fish Biology 26:37–41.

Helfman, G. S., D. L. Stoneburner, E. L. Bozeman, P. A. Christian, and R. Whalen. 1983. Ultrasonic telemetry of American eel movements in a tidal creek. Transactions of the American Fisheries Society 112:105–110.

Jellyman, D. J. 1991. Factors affecting the activity of two species of eel (*Anguilla* spp.) in a small New Zealand lake. Journal of Fish Biology 39:7–14.

Jens, G. 1953. Über den lunaren Rhythmus der Blankaalwanderung. Archiv für Fishchereiwissenschaft 4:94–110.

Jessop, B. M. 1987. Migrating American eels in Nova Scotia. Transactions of the American Fisheries Society 116:161–170.

Korringa, P. 1947. Relations between the moon and periodicity in the breeding of marine animals. Ecological Monographs 17:347–381.

LaBar, G. W., J. A. Hernando Casal, and C. F. Delgado. 1987. Local movements and population size of European eels, *Anguilla anguilla*, in a small lake in southwestern Spain. Environmental Biology of Fishes 19:111–117.

Lindroth, A. 1979. Eel catch and lunar cycle on the Swedish east coast. Rapports et Procès-verbaux des Réunions du Conseil international pour l'Exploration de la Mer 174:124–126.

Lowe, R. H. 1952. The influence of light and other factors on the seaward migration of the silver eel (*Anguilla anguilla* L.). Journal of Animal Ecology 21:275–309.

Morgan, S. G. 1996. Influence of tidal variation on reproductive timing. Journal of Experimental Marine Biology and Ecology 206:237–251.

Palmer, J. D. 1997. Dueling hypotheses: circatidal versus circalunidian battle basics. Chronobiology International 14:337–346.

Parker, S. J. 1995. Homing ability and home range of yellow-phase American eels in a tidally dominated estuary. Journal of the Marine Biology Association of the United Kingdom 75:127–140.

Parker, S. J., and J. D. McCleave. 1997. Selective tidal stream transport by American eels during homing movements and estuarine migration. Journal of the Marine Biology Association of the United Kingdom 77:871–889.

Piggins, D. J. 1985. The silver eel runs of the Burrishoole River system: 1959–84. Numbers, weights, timing and sex ratios. International Council for the Exploration of the Sea C.M. 1985:M:5.

Pursiainen, M., and J. Tulonen. 1986. Eel escapement from small forest lakes. Vie et Milieu 36:287–290.

Sloane, R. D. 1984. Preliminary observations of migrating adult freshwater eels (*Anguilla australis australis* Richardson) in Tasmania. Australian Journal of Marine and Freshwater Research 35:471–476.

Smith, M. W., and J. W. Saunders. 1955. The American eel in certain fresh waters of the Maritime Provinces of Canada. Journal of the Fisheries Research Board of Canada 12:238–269.

Sorensen, P. W., and M. L. Bianchini. 1986. Environmental correlates of the freshwater migration of elvers of the American eel in a Rhode Island brook. Transactions of the American Fisheries Society 115:258–268.

Taylor, J. E. S. 1979. Preliminary study of eel fishing (*Anguilla rostrata*) in Nova Scotia. Nova Scotia Department of Fisheries Manuscript and Technical Report Series No. 79-03.

Tesch, F.-W. 1977. The eel: biology and management of anguillid eels. Chapman and Hall, London, England.

Tesch, F.-W. 1978. Telemetric observations on the spawning migration of the eel (*Anguilla anguilla*) west of the European continental shelf. Environmental Biology of Fishes 3:203–209.

Tesch, F.-W. 1989. Changes in swimming depth and direction of silver eels (*Anguilla anguilla* L.) from the continental shelf to the deep sea. Aquatic Living Resources 2:9–20.

Tesch, F.-W., T. Wendt, and L. Karlsson. 1992. Influence of geomagnetism on the activity and orientation of the eel, *Anguilla anguilla* (L.), as evident from laboratory experiments. Ecology of Freshwater Fish 1:52–60.

Todd, P. R. 1981. Timing and periodicity of migrating New Zealand freshwater eels (*Anguilla* spp.). New Zealand Journal of Marine and Freshwater Resources 15:225–235.

van Veen, T., H. G. Hartwig, and K. Muller. 1976. Light-dependent motor activity and photonegative behavior in the eel (*Anguilla anguilla* L.). Journal of Comparative Physiology 111:209–219.

Vøllestad, L. A., B. Jonsson, N. A. Hvidsten, and T. F. Naesje. 1994. Experimental test of environmental factors influencing the seaward migration of European silver eels. Journal of Fish Biology 45:641–651.

Vøllestad, L. A., B. Jonsson, N. A. Hvidsten, T. F. Naesje, O. Haraldstad, and J. Ruud-Hansen. 1986. Environmental factors regulating the seaward migration of European silver eels (*Anguilla anguilla*). Canadian Journal of Fisheries and Aquatic Sciences 43:1909–1916.

Walsh, P. J., G. D. Foster, and T. W. Moon. 1983. The effects of temperature on metabolism of the American eel *Anguilla rostrata* (LeSueur): compensation in the summer and torpor in the winter. Physiological Zoology 56:532–540.

Westin, L., and L. Nyman. 1979. Activity, orientation, and migration of Baltic eel (*Anguilla anguilla* L.). Rapports et Procès-verbaux des Réunions du Conseil international pour l'Exploration de la Mer 174:115–123.

Winn, H. E., W. A. Richkus, and L. K. Winn. 1975. Sexual dimorphism and natural movements of the American eel (*Anguilla rostrata*) in Rhode Island streams and estuaries. Helgolander Meeresuntersuchungen 27:156–166.

Wirth, T., and L. Bernatchez. 2001. Genetic evidence against panmixia in the European eel. Nature (London) 409:1037–1040.

Do Stocked Freshwater Eels Migrate? Evidence from the Baltic Suggests "Yes"

Karin E. Limburg
SUNY College of Environmental Science & Forestry,
Syracuse, New York, USA

Håkan Wickström
Swedish Board of Fisheries, Institute of Freshwater Research,
Drottningholm, Sweden

Henrik Svedäng
Swedish Board of Fisheries, Institute of Marine Research,
Lysekil, Sweden

Mikael Elfman and Per Kristiansson
Department of Nuclear Physics, Lund Technical University,
Lund, Sweden

Abstract.—In response to declining catches of eels in the brackish Baltic Sea, the Swedish government stocks eels *Anguilla anguilla* (L.), both in lakes (mainly glass eels/elvers) and in the sea itself (mainly yellow eels). However, the degree to which these fish contribute to the spawning stock, if at all, was unknown. We collected silver eels at the exit of the Baltic Sea and analyzed indices of their maturity status. In addition, we used electron (WDS) and nuclear (microPIXE) microprobes to map out the strontium and calcium contents of their otoliths, as Sr:Ca correlates with salinity. As a calibration, we analyzed otoliths from eels collected around the Swedish coast and fresh water (0–25 psu) and derived a relationship between salinity and Sr:Ca. Our results show that, of 86 silver eels analyzed, 17 eels had Sr:Ca profiles consistent with having been stocked into fresh water, six showed patterns consistent with stocking directly into the Baltic from marine waters, and 10 showed patterns indicative of natural catadromy. In all, 31.4% of silver eels showed histories of freshwater experience, including 24% of those found outside the Baltic. Silver eels caught exiting the Baltic had higher fat contents (21.1% of body weight) than those collected in the southern Baltic near Denmark (18.6%), but differences were not significant between wild and presumed stocked fish within geographic areas. The conclusions of Tsukamoto et al. (1998), i.e. that freshwater eels are not supported by catadromous individuals, do not appear to hold for the Baltic, although it is clear that noncatadromous fish composed the majority of our silver eel samples.

Introduction

Baltic eel *Anguilla anguilla* (L.) catches have declined precipitously since the 1960s and a supplemental stocking program has been in effect in Sweden since the late 1970s (Svedäng 1996). The stocking program, at an annual cost of nearly 10 million SEK (~ US$1 million; H. Wickström, personal communication), has consisted either of transplanting small eels (ca 0.10–20 cm) collected as glass eels or elvers in the Severn estuary in England, or transplanting yellow eels (35–48 cm) caught off the Swedish west coast (Wickström 1993). Today, stocking provides an estimated 8–9% of all young eels in Sweden. Whereas much has been learned about the biology of stocked eels from experiments (e.g., Wickström et al. 1996), there has been no direct evaluation of the contribution of stocked eels to Baltic catches, nor to their ultimate contribution to the spawning stock. However, previous work raised two testable hypotheses.

In one study, experimentally stocked eels on Gotland, that were tagged and allowed to migrate as silver eels, did not migrate through the outlet of the Baltic at Öresund (Figure 1) as did natural migrants, but rather ended up around the Danish islands in the Belt Sea (Westin 1990). In a second study, Svedäng and Wickström (1997) found no correlation between maturity stage and muscle fat concentration, and suggested that silver eels (a non-feeding stage) with low fat concentrations may temporarily halt migration, revert to a feeding stage, and "bulk up" until fat reserves are sufficient to carry out successful migration to the spawning area. Thus, by sampling silver eels at the exit point from the Baltic, measuring their fat reserves, and identifying stocked versus natural fish, one should be able to test whether: stocked eels can find their way out of the Baltic (Westin 1990) and whether there is a correlation between stocked/natural status and low fat concentrations (Svedäng and Wickström 1997).

In a third study, Tsukamoto et al. (1998) used the method of strontium:calcium ratios in otoliths (for a review of the method and other otolith microchemistry, see Campana 1999) to determine whether or not catadromous European and Asian anguillids contribute to spawning stocks. Briefly, Sr entrained in aragonitic otoliths typically reflects environmental concentrations, and when normalized to otolith Ca it serves as a proxy for salinity because Sr is generally higher in marine waters than in fresh. Because otoliths accrete over time, a temporal record of Sr:Ca is maintained and can be assessed with various microprobe techniques

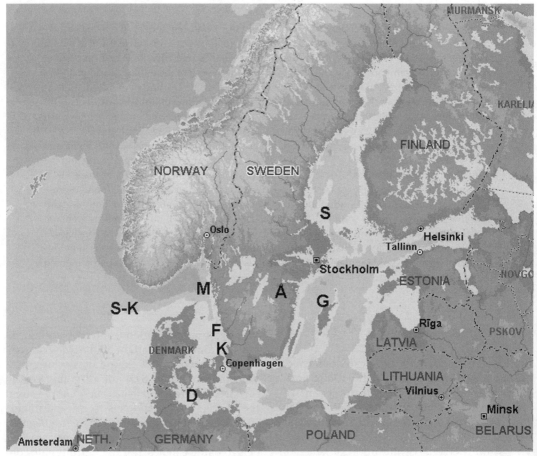

Figure 1. Map of the study region. Key: S = Salnö, G = Gotland, Å = Lake Ången, D = area sampled in Denmark for silver eels, K = Kullen, F = Fladen, M = Marstrand, and S-K = Skagerrak.

(Campana 1999). Otoliths from eels collected in the North Sea and the East China Sea revealed no evidence of freshwater experience, leading the investigators to question whether catadromous eels really do contribute their genes to future generations. Thus a third hypothesis—that only noncatadromous eels recruit to the spawning stock—should also be testable.

In the present study, we used the Sr:Ca method to trace the environmental histories of migrating Baltic eels. We first calibrated the method by measuring the Sr:Ca ratios in eel otoliths collected either in fresh water or along a stable salinity gradient, and then used these to examine and analyze otoliths of silver eels, collected either exiting the Baltic or down in its southern reaches. We found a substantial fraction of migrating silver eels whose Sr:Ca patterns are consistent with having been stocked, and further found fish showing catadromous patterns.

Methods

Study Area

Eels were collected at different places in the Baltic Sea, the Skagerrak-Kattegat area, and the Swedish west coast (Figure 1) either by trapping, trawling, or by purchase from fishermen. Eels that were experimentally stocked and monitored in two lakes (Lake Ången on the Swedish mainland, and Lake Fardume on the island of Gotland) served as known freshwater endmembers. Yellow eels collected from three areas of different salinity (Marstrand, approx. 25 psu, Gotland, approx. 7 psu, and Salnö, approximately 5 psu) were assayed to provide Sr:Ca ratios representative of those salinities. Finally, silver eels were collected at sites exiting the Baltic Sea (Kullen, Fladen, and in the Skagerrak) as well as in the southern Baltic in the vicinity of the Danish islands of Lolland and Falster. Eels were stored frozen until ready for analysis.

Laboratory Procedures

Morphometric measures (total length, weight, eye diameter, jaw length, and pectoral fin lengths) were made on defrosted fish. A sample of tissue was taken just anterior of the anal vent and analyzed for lipid content as described in Svedäng and Wickström (1997). Eye diameters and total lengths were used to calculate an index (I) of maturation status (Pankhurst 1982):

$$I = (25\pi/8TL) \times [(A + B)_L^2 + (A + B)_R^2],$$

where TL is total length, and A and B are the height and width of the left (L) and right (R) orbitals. Silver eels with values of I greater than 6.5 are classified as sexually maturing (Pankhurst 1982; Svedäng and Wickström 1997).

Sagittal otoliths were removed from the skull, cleaned, embedded in Spurr's epoxy, and sectioned in the sagittal plane by grinding to the core with a graded series of grinding papers, and finally polished to 0.5 μm surface fineness. Following carbon coating, otoliths were analyzed for Sr and Ca by using either a wavelength-dispersive (WDS) electron microprobe at the Department of Geosciences, Uppsala University, or nuclear microscopy combined with proton induced X-ray emission analysis (μPIXE) at the Department of Nuclear Physics, Lund Technical University. The latter method has the advantages of higher resolution and more rapid data collection and was therefore the method of choice, given availability of machine time. Thirty-eight otoliths were analyzed using WDS and 100 with μPIXE.

Analyses using WDS involved visually locating the core of the otolith with transmitted light, then laying out a transect of 30–50 points spaced 20–40 μm apart traversing the otolith from the core to the outer posterior edge (a so-called "life-history transect"). The parameters for operation were: accelerating voltage, 20 kV; current, 20nA; electron beam diameter, 15 μm. Strontium was counted for 40 s on the peak and 40 s on the background (only on one side, to avoid the strong interference from a second order Ca K-α peak). Calcium was counted until a precision of at least 0.1% was reached (usually < 20 s), and background was counted for 10 s on each side of the peak. Strontianite ($SrCO_3$) and calcite ($CaCO_3$) were used as calibration standards. The detection limits were 0.03 ± 0.004 weight percent for both elements.

μPIXE analyses were made at the Lund Nuclear Microprobe facility, using a standard 2.55 MeV proton beam. X-rays were detected with a Kevex Si(Li) detector of 5 mm² active area and a measured energy resolution of approximately 155 eV at the 5.9 keV Mn K_α peak. A thick absorber (mylar + aluminum) was used during the analysis to suppress the Ca X-ray peaks; this permitted increasing the current to enhance the signals of Sr and other trace elements. The total charge was approximately 1 micro-Coulomb.

The normal procedure for a scan was to raster as much of the otolith as possible in a grid of 128 × 128 pixels. Thus, we mapped out large areas of the otolith rather than being confined to a line transect, which facilitated interpretation of the data. Following data collection, the data sets were normalized to counts per charge.

Data Analysis

Strontium:calcium ratios were calculated, graphed, and examined for patterns. Yellow eel otoliths that had been analyzed with μPIXE were used to calibrate Sr:Ca to salinity. Subsamples (two replicate 6 × 6 groups of pixels) were chosen from near the outer growing edge on different parts of each otolith and mean Sr:Ca ratios were calculated. A nonlinear regression of Sr:Ca on salinity yielded the following relationship:

$$Sr:Ca_{PIXE} = a \times (1 - b\, e^{-K \times Salinity}), \text{ where}$$

$a = 3.467 \pm 0.191$ (standard error)
$b = 0.905 \pm 0.0254$
$K = 0.132 \pm 0.0251$
$N = 29, R = 0.97.$

Because of the mylar absorber used to suppress Ca peaks in the μPIXE analyses, the Sr:Ca ratios do not reflect true mass ratios. Therefore, otoliths from six fish were analyzed with both methods, and regression analysis was used to relate electron microprobe Sr:Ca measurements to μPIXE:

$$Sr:Ca_{WDS} = 0.477 + 1.531 \times Sr:Ca_{PIXE}$$

($R^2 = 0.84, p < 0.05$).

Based on otolith Sr:Ca and the associated estimated salinity histories, silver eels were classified into eight distinct groups. Five of these groups describe wild fish: entirely marine (M); marine moving into brackish water (MB); entirely brackish after the glass eel stage (B); "complex migration" (CM) when moving back and forth between waters of different salinities, but never into fresh water; and catadromous (CAT) when the movements following the glass eel stage included residence in fresh water. The last three groups are called "stocked," and show patterns that are consistent with either having been stocked as glass eels and remained virtually until capture in fresh water (S1), captured along the Swedish west coast as yellow eels and transferred into the Baltic (S2), or having been stocked as glass eels but migrating out to brackish or marine waters to feed and grow (S3). Examples of these types are given in Figure 2.

Analysis of variance (ANOVA) and goodness-of-fit tests were conducted on silver eels to test the hypotheses that differences exist between fish caught exiting the Baltic and those caught in the southern Baltic (Denmark), and that differences exist between wild and stocked eels with respect to lipid content. All statistical analyses were conducted with Statistica (Statsoft 1999).

Results

Collectively, the silver eels we analyzed displayed a wide repertoire of habitat use patterns. Many eels appeared to move around among different areas over the course of their lifetimes, among different salinity zones, up into fresh water, or both (Figure 2). Distributions of eel habitat use patterns differed between the eels caught exiting or outside the Baltic, and those collected in the Baltic around southeastern Denmark (Figure 3; goodness of fit $\chi^2 = 32.06$, df = 7, $p \ll 0.001$). Categories that differed the most included MB (fish moving from marine to brackish water), CM (movement back and forth among zones of different salinity), catadromous, S1 (entirely freshwater, presumed to be stocked), and S3 (stocked but moved into saline waters), with all being more numerous in the Danish samples except for CM and S3.

Overall, eels presumed to have been stocked composed 26.7% of the silver eels collected for the study, not counting several fish that were used in the calibration. Aggregating the three stocked categories, there was no statistical difference between the ratios of wild and stocked eels between the geographic areas (Figure 4; $\chi^2 = 0.723$, df = 1, $p < 0.39$). However, among the stocked categories, S1 fish were almost exclusively found in the Danish sample, and S3 fish were three-fold higher exiting the Baltic than in the Danish collection (Figure 3). Counts of eels classed as S2 were the same in both areas. In contrast, the proportion of silver eels that had spent time in fresh water, either stocked or wild, was marginally higher in the Danish sample (38% in Danish waters versus 24% outside; $\chi^2 = 3.20$, df = 1, $p < 0.07$). Eels with freshwater experience composed 31.4% of all silver eels (excluding individuals used in calibration).

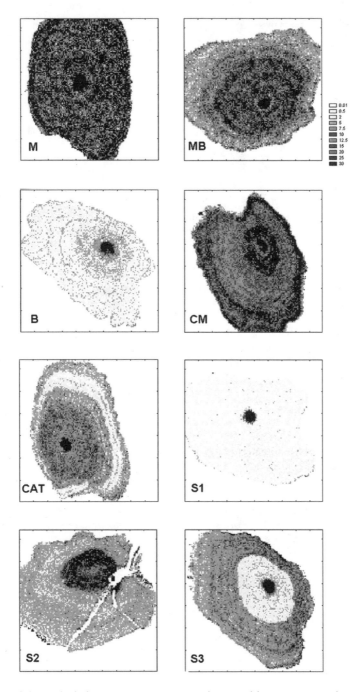

Figure 2. Examples of the eight habitat use categories as discerned by strontium:calcium ratios. Images are of μPIXE scans of otolith chemistry converted to estimated salinity values. Note that the high values in the otoliths' centers are due to high levels of Sr associated with the leptocephalus stage. Key to categories: M = entire life spent in marine waters; MB = moved from marine to brackish waters; B = spent entire postglass-eel life in brackish waters; CM = complex migration, moving back and forth between marine and brackish waters; CAT = catadromous; S1 = stocked into fresh waters and remaining there until maturation; S2 = stocked from marine waters into the Baltic Sea; S3 = stocked into fresh water and leaving to spend time feeding and growing in brackish and/or marine waters.

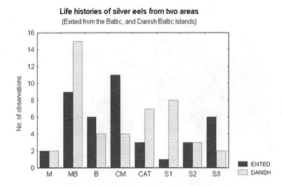

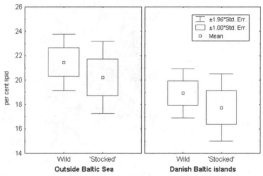

Figure 3. Distributions of habitat use patterns by geographic area. Key to categories: M = entire life spent in marine waters; MB = moved from marine to brackish waters; B = spent entire postglass-eel life in brackish waters; CM = complex migration, moving back and forth between marine and brackish waters; CAT = catadromous; S1 = stocked into fresh waters and remaining there until maturation; S2 = stocked from marine waters into the Baltic Sea; S3 = stocked into fresh water and leaving to spend time feeding and growing in brackish and/or marine waters.

Figure 5. Box plots of lipid contents of silver eels, by geographic area and stocking status.

Trends were observed in the lipid contents of silver eels. Lipids tended to be higher among wild versus presumably stocked eels, and greater in eels exiting the Baltic than those collected in Denmark (Figure 5). Although there was no statistical difference between wild and stocked eels, eels exiting the Baltic were significantly fatter than Danish-caught eels (ANOVA, $F_{1,77} = 4.14$, $p < 0.05$). Differences also existed among eels presumed to have different stocking histories ($F_{2,19} = 3.39$, $p < 0.055$). Of the nine eels that spent their entire lives in fresh water (S1), only one exceeded 20% lipid content, whereas ten out of thirteen S2 and S3 eels had lipid levels of at least 20% (Figure 6).

Silver eels collected in the Danish Baltic islands varied differently from exiting eels with respect to Pankhurst's maturation index (Table 1). In the Danish collection, catadromous eels had the highest mean I of 7.86, while two eels classified as marine had the lowest (mean I = 5.92). Exited catadromous eels, conversely, had the lowest I-values (mean 6.26), but were also the most fatty (mean per cent lipid = 27.5). I-values less than or equal to 6.5 are generally interpreted as sexually immature. I-values were

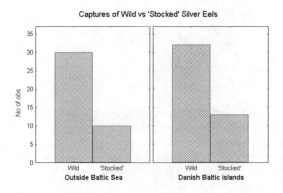

Figure 4. Wild versus stocked silver eels by geographic area.

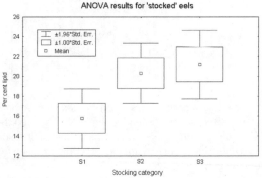

Figure 6. Box plots of lipid contents of silver eels classified as stocked.

Table 1. Mean (± s.e.) values of Pankhurst's index (I) of maturation status and mean muscle tissue lipid contents of silver eels collected either exiting the Baltic or in Danish Baltic waters. Habitat use categories are described in the Methods section.

Habitat use category	Exited eels			Danish eels		
	Maturation index (I)	Lipid content (%)	N	Maturation index (I)	Lipid content (%)	N
M	9.54 (0.05)	14.0 (4.0)	2	5.92 (0.58)	12.92 (4.23)	2
MB	8.28 (0.53)	21.1 (1.6)	9	6.60 (0.40)	19.63 (0.78)	14
B	8.74 (0.39)	22.4 (2.1)	6	6.71 (0.30)	18.38 (0.77)	4
CM	8.01 (0.77)	22.0 (2.8)	8	6.80 (0.49)	15.89 (4.00)	4
CAT	6.26 (1.21)	27.5 (1.5)	2	7.86 (0.55)	21.61 (2.59)	7
S1	8.56 (—)	16.0 (—)	1	6.67 (0.32)	15.81 (1.73)	8
S2	8.31 (0.69)	20.3 (1.4)	3	7.11 (0.92)	20.20 (3.17)	3
S3	8.16 (0.16)	21.0 (2.6)	5	6.57 (1.06)	21.68 (0.68)	2

significantly higher (mean 8.28 ± 0.24 s.e.) in the exited eels when compared to those collected in the Danish islands (mean 6.91 ± 0.18 s.e.; ANOVA $F_{1,81} = 21.7$, $p < 0.0001$). Four of the nine S1 stocked eels had I-values greater than 6.5, but all had low fat values (11–15%). There was no correlation between I-values and lipid contents for Danish silver eels (Pearson $r = 0.009$), although there was a very slight positive trend in the data. In contrast, there was a significant negative correlation (Pearson $r = -0.47$, $p < 0.001$) between these parameters in eels exiting the Baltic.

van Ginneken and van den Thillart (2000) report that the energy cost of swimming for eels is 0.137 cal per gram wet weight and kilometer. Based on data in their paper, the caloric value of eel fat is 10.68 kcal g^{-1}. Knowing an eel's wet weight and fat content, one can therefore compute a "migration potential," assuming that 60% of the lipids are reserved for gonadal development (van Ginneken and van den Thillart 2000). Results of such calculations indicate that catadromous eels in this study had the highest migration potential (mean = 7149 km) and stocked eels that spent their entire lives in fresh water (S1) had the lowest (mean = 4938 km; Table 2).

Discussion

The strontium-calcium "tag" provides a powerful means to examine salinity histories, given caveats. Although otolith Sr:Ca ratios have not been related to salinity under controlled experimental conditions for European eels, the calibration using geographic locations along a stable salinity gradient yielded a good fit to an asymptotic curve, similar in shape to one derived for striped bass by Secor and Piccoli (1996) and also similar to results presented for A. japonica (Tzeng 1996). Because of the asymptotic shape of the curve, it is difficult to distinguish between strongly brackish and marine salinities, but resolution improves at lower salinities. Other confounding influences on otolith Sr:Ca ratios include the effects of growth rate (Sadovy and Severin 1992) and temperature (Campana 1999). Aside from the slow-growing leptocephalus stage, during which time otolith Sr accumulates to very high levels (Figure 2), we could not distinguish the influence of growth rate per

Table 2. Mean (± s.e.) migration potentials (km) for silver eels collected exiting the Baltic and around the Danish islands in the southern Baltic, by habitat use type. Assumes caloric value of 10.68 kcal g^{-1} eel fat and that 40% of the lipid reserves may be used in migration (i.e., that 60% of reserves is required for spawning: van Ginneken and van den Thillart 2000). Habitat use categories are described in the Methods section.

Habitat use category	Migration potential (km)	N
M	5126 (1046)	3
MB	6292 (234)	22
B	6430 (533)	9
CM	6100 (745)	10
CAT	7149 (1091)	9
S1	4938 (476)	9
S2	6334 (486)	6
S3	6611 (552)	7

se in our specimens. On the other hand, Sr:Ca ratios could be seen to increase during winter, when temperatures drop and growth rate slows. These showed up as bands of higher Sr:Ca and were only visible outside of freshwater environments. Thus, classification of eels depended on examination of the entire "gestalt" pattern, and comparisons to known-origin eels. As with other studies (e.g., Tzeng 1996; Tzeng et al. 1997; Tsukamoto et al. 1998), however, the discrimination of freshwater and nonfreshwater residency was not a problem.

Recently it has been suggested (Tsukamoto et al. 1998) that catadromous forms of freshwater eels do not migrate to the spawning grounds, and therefore do not contribute to maintenance of populations. The conclusion was based upon a survey of 18 *A. anguilla* individuals from the North Sea, of which only seven were silver, and 12 *A. japonica* individuals (all silver stage). It is well known that eels of all ages and developmental stages are found in marine waters (Tesch 1977), and this was confirmed by our study as well. However, we found that over 30% of all the silver eels (total N = 86) in our collections had spent at least some time in fresh water, including 24% of eels exiting the Baltic.

Aside from otolith Sr:Ca patterns, we have no other independent means of telling whether an eel was stocked or wild; therefore we can only classify our eels as having habitat use patterns that are consistent with having been stocked. For eels displaying the S2 pattern, an alternative explanation is that the eels swam very quickly from marine to brackish water. An alternative for the S3 pattern could be that glass eels entered a river somewhere outside of the Baltic, spent several years in that system, and then migrated into the Baltic or marine waters. However, there is no good alternative explanation for the S1 pattern, because historically eels have reached the Baltic as bootlace eels (small and pigmented) rather than as glass eels (Svärdson 1976). Due to time and labor constraints, eel ages were not determined, but in other studies of Baltic eels, age did not correlate with maturation (Svedäng et al. 1996; Svedäng and Wickström 1997) or growth (Holmgren 1996).

Although all silver eels classified as stocked did not vary in their proportions whether inside the Baltic or out, most of the S1 eels were captured in the Danish Baltic islands. Whereas half of these fish are classified as mature by Pankhurst's index of maturity, all but one of them had lipid contents less than 20% of body weight (one of them had a lipid content of 27%, however). This means, if the assumption that 40% of muscle lipid is available for migration is valid (van Ginneken and van den Thillart 2000), that the S1 eels have shorter migration potentials and on the whole would not have the energy reserves to make the greater than 6000 km journey to the spawning grounds in the Sargasso Sea. Interestingly, whereas no relationship existed between Pankhurst's index and lipid content in the Danish eels, lipid content declined with increasing eye size in the exiting eels. This may be consistent with eels burning off energy reserves as they migrate into deeper waters, this last necessitating enlargement of their eyes.

One of the ongoing debates about stocked eels in the Baltic is whether or not they can orient their way out to spawn (Westin 1990; 1998; Tesch et al. 1991). In successive tagging studies, Westin found that stocked eels took months to years longer to reach the southern Baltic than eels presumed to be wild in origin, and also found that a high proportion of stocked individuals did not orient properly to locate the closest exit from the Baltic at Öresund. Rather, many stocked eels were recaptured in the southern Baltic along the coasts of Poland, Germany, and Denmark (Westin 1998). It was also shown that wild eels rendered anosmic by plugging up either both nostrils or only the left nostril had poorer navigation abilities. From this Westin (1998) suggested that stocked eels, lacking olfactory imprints for their way across the Baltic to freshwater sites, cannot use smell as a cue and instead must use the less reliable cue of declining water temperature.

Although we cannot tell how long the eels in this study had migrated in the silver stage, the fact that most of the S1 eels were collected at the Danish site appears to corroborate Westin (1990, 1998). However, the eels with highest fat contents (catadromous eels), and therefore highest migration potentials, were also mostly found at this site. Further, as eels are stocked in southern Baltic countries such as Poland (Moriarty et al. 1990; Bartel and Kosior 1991), it is also possible that the eels we identified as S1 did not originate in Sweden and were undertaking a coastal migration from Poland, Germany, or Denmark.

The results of our study are at odds with those reported by Tsukamoto et al. (1998) in that we did find eels with freshwater histories, and many of these showed evidence of maturity and of being able to undertake the long migration to the spawning grounds. Nevertheless, it is striking that nearly 69% of the silver eels showed no evidence of freshwater residency. One possible reason is suggested by Tsukamoto et al. (1998), i.e., that catadromy is but one of many habitat use strategies employed by eels, which have high plasticity in many of their life history characters (e.g., Panfili and Ximenes 1994; Holmgren 1996; Glova et al. 1998; Secor and Rooker 2000). Another possibility is that dams and other artificial obstructions within the Baltic Sea drainage may be preventing young eels from ascending many rivers. Such barriers have been implicated in reducing the numbers of eels migrating upstream in British rivers (White and Knights 1997). Although glass eels and elvers can climb vertical surfaces such as dam walls (Tesch 1977), bootlace eels, such as those entering the Baltic, cannot. The hypothesis that dams serve as barriers to eel recruitment might be tested by assessing the frequencies of catadromous silver eels in drainages differing with respect to barriers, distance from the spawning grounds (affecting the life stage at which eels arrive at rivers), or both.

Conclusions

A most striking result of this study was the confirmation of the remarkable plasticity of habitat use patterns among European eels. Eels that migrate in the silver stage show a life history (based on otolith microchemistry) of exploiting all forms of aquatic habitat ranging from marine to fresh waters. Our results indicate that Baltic drainage eels that spend all their lives in fresh water are less likely to be able to make the spawning migration, unless they remain in the system for a number of years to feed, as suggested by Svedäng and Wickström (1997). However, contrary to Tsukamoto et al. (1998), our study strongly suggests that catadromous eels indeed begin the spawning migration, and show strong potential for reproductive success.

Acknowledgments

We thank Hans Harryson, Department of Geosciences, Uppsala University, for assistance with the WDS microprobe analyses, and Klas Malmqvist, Department of Nuclear Physics, Lund University, for support. We are grateful to our colleague Lars Westin for his discussions and ideas; Lars also provided the eels from Denmark and Gotland. Leif Johansson prepared the otoliths, and Liselott Wilhelmsson conducted the lipid analyses. We also thank C. Moriarity, R. Kraus, and one anonymous reviewer for thoughtful and constructive comments. Financial support for this study came from the Swedish Council for Forestry and Agricultural Research.

References

Bartel, R., and K. Kosior. 1991. Migrations of tagged eel released into the lower Vistula and to the Gulf of Gdansk. Polish Archives of Hydrobiology 38:105–113.

Campana, S. E. 1999. Chemistry and composition of fish otoliths: Pathways, mechanisms and applications. Marine Ecology Progress Series 188:263–297.

Glova, G. J., D. J. Jellyman, and M. L Bonnett. 1998. Factors associated with the distribution and habitat of eels (Anguilla spp.) in three New Zealand lowland streams. New Zealand Journal of Marine and Freshwater Research 32:255–269.

Holmgren, K. 1996. On the sex differentiation and growth patterns of the European eel, Anguilla anguilla (L.). Doctoral dissertation. Uppsala University, Sweden.

Moriarty, C., M. Bninska, and M. Leopold. 1990. Eel, Anguilla anguilla L., stock and yield in Polish lakes. Aquatic and Fisheries Management 21:347–355.

Panfili, J., and M. C. Ximenes. 1994. Age and growth estimation of the European eel (Anguilla anguilla L.) in continental waters: methodology, validation, application in Mediterranean area and comparisons in Europe. Bulletin Francais de la Peche et de la Pisciculture 0 (335): 44–66.

Pankhurst, N. W. 1982. Relation of visual changes to the onset of sexual maturation in the European eel Anguilla anguilla (L.). Journal of Fish Biology 21:127–140.

Sadovy, Y., and K. P. Severin. 1992. Trace elements in biogenic aragonite: correlation of body growth rate and strontium levels in the otoliths of the white grunt, Haemulon plumieri (Pisces: Haemulidae). Bulletin of Marine Science 50:237–257.

Secor, D. H., and P. M. Piccoli. 1996. Age- and sex-dependent migrations of striped bass in the Hudson River as determined by chemical microanalysis of otoliths. Estuaries 19:778–793.

Secor, D. H., and J. R. Rooker. 2000. Is otolith strontium a useful scalar of life cycles in estuarine fishes? Fisheries Research 46:359–371.

Statsoft. 1999. Statistica for Windows (computer manual). Statsoft, Inc., Tulsa, Oklahoma.

Svärdson, G. 1976. The decline of the Baltic eel population. Report of the Institute of Freshwater Research 55:136–143.

Svedäng, H. 1996. The development of the eel (*Anguilla anguilla* L.) stock in the Baltic Sea: an analysis of catch and recruitment statistics. Polish-Swedish Symposium on Baltic Coastal Fisheries Resources and Management 255–267.

Svedäng, H., E. Neuman, and H. Wickström. 1996. Maturation patterns in female European eel: age and size at the silver eel stage. Journal of Fish Biology 48:342–351.

Svedäng, H., and H. Wickström. 1997. Low fat contents in female silver eels: indications of insufficient energetic stores for migration and gonadal development. Journal of Fish Biology 50:475–486.

Tesch, F.-W. 1977. Eel: biology and management of anguillid eels. Chapman and Hall, London, England.

Tesch, F.-W., H. Westerberg, and L. Karlsson. 1991. Tracking studies on migrating silver eels in the central Baltic. Meeresforsch 33:183–196.

Tzeng, W. N. 1996. Effects of salinity and ontogenetic movements on strontium:calcium ratios in the otoliths of the Japanese eel, *Anguilla japonica* Temminck and Schlegel. Journal of Experimental Marine Biology and Ecology 199:111–122.

Tzeng, W. N., K. P. Severin, and H. Wickström. 1997. Use of otolith microchemistry to investigate the environmental history of European eel *Anguilla anguilla*. Marine Ecology Progress Series 73–81.

Tsukamoto, K., I. Nakai, and W.-V. Tesch. 1998. Do all freshwater eels migrate? Nature 396: 635–636.

van Ginneken, V. J. T, and G. E. E. J. M. van den Thillart. 2000. Eel fat stores are enough to reach the Sargasso. Nature (London) 403:156–157.

Westin, L. 1990. Orientation mechanisms in migrating European silver eel (*Anguilla anguilla*): temperature and olfaction. Marine Biology 106:175–179.

Westin, L. 1998. The spawning migration of European silver eel (*Anguilla anguilla* L.) with particular reference to stocked eel in the Baltic. Fisheries Research 38:257–270.

White, E. M., and B. Knights. 1997. Dynamics of upstream migration of the European eel, *Anguilla anguilla* (L.), in the Rivers Severn and Avon, England, with special reference to the effects of man-made barriers. Fisheries Management and Ecology 4:311–324.

Wickström, H. 1993. Inför 1993-års ålutsättningar. PM Nr. 2. Swedish National Board of Fisheries, Institute of Freshwater Research. (in Swedish).

Wickström, H., L. Westin, and P. Clevestam. 1996. The biological and economic yield from a long-term eel-stocking experiment. Ecology of Freshwater Fish 5:140–147.

Life History Patterns of Japanese Eel *Anguilla japonica* in Mikawa Bay, Japan

WANN-NIAN TZENG* AND JEN-CHIEH SHIAO

*Department of Zoology, College of Science, National Taiwan University,
Taipei 106, ROC, Taiwan
*Corresponding author: 886-2-23639570, Fax: 886-2-23636837,
E-mail: wnt@ccms.ntu.edu.tw*

YOSHIAKI YAMADA AND HIDEO P. OKA

*IRAGO Institute, 377 Ehima Ghirah a, Atsumi-cho, Atsumi-gun,
Aichi-ken 441-3605, Japan*

Abstract.—The otolith microchemistry of female Japanese eels *Anguilla japonica* collected in Mikawa Bay, Japan, in June 1998 to June 1999 was examined with an electron probe microanalyzer. Total length ranged between 62.0 and 96.0 cm, and body weight between 291 and 2057 g. Gonadosomatic index (GSI) increased from less than 0.2 in August to greater than 4.0 in January. Temperature of the sampling site ranged between 6.6°C and 28.8°C, and salinity between 24.6‰ and 32.7‰. Otolith Sr:Ca indicated that the life history patterns of eels could be classified into three types: Type 1, Sr:Ca from approximately 150 µm from primordium to edge of the otolith maintained at the level of approximately 0–4×10^{-3}, indicating that eels after the elver stage stayed in fresh water until the silver eel stage; Type 2, the ratios were 4–10×10^{-3}, indicating that eels stayed in seawater from elver stage to the silver eel stage; Type 3, the ratios fluctuated between those of Types 1 and 2, indicating that eel migrated between freshwater and seawater before the silver stage. Results suggest that the growth phase of yellow stage Japanese eel does not have definite habitat selection and that the population is predominately estuarine.

Introduction

The Japanese eel *Anguilla japonica* is widely distributed in the northwestern Pacific, south from Taiwan, through China, Korea, and north to Japan (Tesch 1977). The eel's life cycle has five principal phases, i.e., leptocephalus, glass eel, elver, yellow eel, and silver eel (Bertin 1956). Japanese eel is presumed to be a panmictic population (Sang et al. 1994), spawning in the waters west of the Marian Islands, 15°N 140°E (Tsukamoto 1992). Leptocephali drift on the North Equatorial Current for approximately five to six months before arriving to the continental shelves (Tzeng 1990; Tzeng and Tsai 1992; Cheng and Tzeng 1996). Upon arriving to coastal waters they metamorphose to glass eels and become pigmented elvers in the estuary. Elvers are over-exploited for restocking and thus the eel population has decreased (Tzeng 1985, 1986, 1996a). The yellow eels spend approximately five to eight years growing in rivers before they metamorphose into silver eels (Tzeng et al. 2000a), which migrate downstream in the autumn to return to their spawning grounds.

In recent years, trace elemental strontium (Sr) in otoliths has been used to study the salinity history and, therefore, migratory movements of eels. The absolute concentration of Sr ranges from 9×10^{-7}M in freshwater to 8.7×10^{-5}M in seawater, an almost 100-fold difference (Campana 1999). The molar ratio of Sr to calcium (Sr:Ca) in seawater (8.6×10^{-3}) is about 4.8 times greater than that of freshwater (1.8×10^{-3}). Otolith Sr:Ca is positively correlated to ambient salinity (Secor et al. 1995; Tzeng 1996b; Kawakami et al. 1998). Because the eel is diadromous, otolith Sr:Ca has been used to investigate their migratory environmental history (Casselman 1982; Otake et al. 1994; Tzeng and Tsai 1994; Arai et al. 1997; Tzeng et al. 1997, 1999, 2000b; Tsukamoto et al. 1998).

In general, freshwater eels are presumed to grow in fresh water and spawn at sea. However, recent otolith Sr:Ca studies indicate that some of the eels skip freshwater growth entirely to complete their life cycle in the ocean or in brackish water (Tsukamoto et al. 1998; Tzeng et al. 2000b; Limburg et al., this volume). These recent observations indicate the life history pattern of the yellow stage eel is diversified. To understand the migratory movements of yellow stage Japanese eel between fresh and saline waters, otolith Sr:Ca of silver eels collected in the Mikawa Bay, middle Japan, were examined.

Methods

Japanese eels were collected in Mikawa Bay on the east coast of mid Japan, 137°E 34°45'N (Figure 1). The three rivers connected with the bay allow fish to migrate freely upstream into fresh water. Thus, the bay is an ideal area to test if they migrate opportunistically. Specimens for otolith microchemistry analysis were randomly selected from silver eels collected from the bay for artificial propagation by the Irago Institute in Japan. Eels were collected monthly using both set net and eel tubes during the period from June 1998 through June 1999. The fishing gear was set at 2–3 m depth, approximately 50 m from shoreline. The gear was set at dawn and captured eels were recovered the following morning. Water temperature and salinity at the sampling site were recorded.

Total length (TL) and body weight (BW) were measured after anesthetizing in a solution of 800 ppm 2-phenoxythanol for about five minutes.

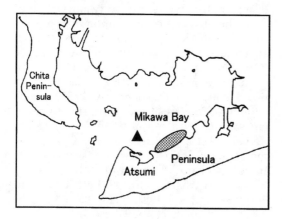

Figure 1. Sampling site of the eel (⟨⟩) and location of temperature, and salinity measurements (▲) in Mikawa Bay, Japan.

Maturity status was analyzed based on gonadosomatic index (GSI, ovary weight/body weight × 100%) and ovum diameter (OD). Otoliths were removed and preserved in a dry chamber for Sr:Ca analysis. A total of 45 otoliths of female silver eels were examined.

EPMA (Electron Probe Micro-analyzer, JEOL, JXA-8800M) was used to analyze Sr:Ca. $SrTiO_3$ and $CaMnO_4$ were used as standards for Sr and Ca measurements. Sr and Ca concentrations were measured along a transect from the primordium to the edge of otolith with a beam current of 3nA and 15 KeV accelerating voltage. At intervals of approximately 10 μm, the electron beam was focused on an area approximately 1 μm in diameter. The energy dispersive strength of Ca and Sr was evaluated using 10 seconds scanning periods. After ZAF (Z = atomic number effect; A = absorption factor; F = fluorescence effects) correction, Sr:Ca was calculated. In addition, a subsample of otoliths was selected for mapping Sr and Ca concentration to examine the temporal change of Sr:Ca. Otolith growth annuli were identified as described in Tzeng et al. (1994, 1999).

Results

Water Temperature and Salinity

Surface water temperature in the Mikawa Bay was highest in July (about 28.8°C) and lowest in January (about 6.6°C). Salinity was approximately 32°/oo in the winter but fluctuated considerably during the summer (range 24.6°/oo – 32.7°/oo). The lower salinity in the summer was probably due to high river discharge (Figure 2).

Seasonal Variation of Maturity

Gonadosomatic index (GSI) was lowest in August (approximately 0.2), but increased thereafter and reached a maximum in January (approximately 4.0) (Figure 3). This indicates that maturity was inversely related to water temperature. In addition, the maximum ovum diameter (OD) at the highest GSI was approximately 200 μm (Figure 4), indicating that silver eels collected from the bay were immature.

Otolith Sr:Ca

Sr:Ca increased from approximately 8×10^{-3} at the primordium to a peak of $16 \sim 18 \times 10^{-3}$ at

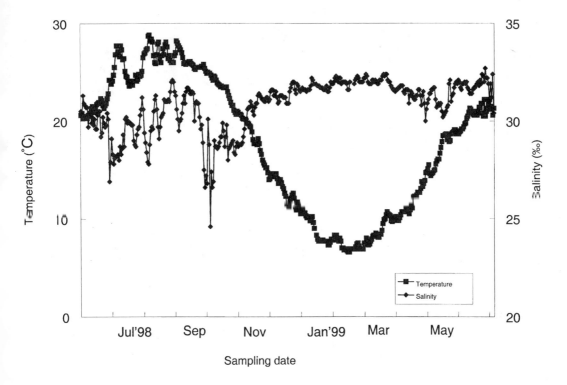

Figure 2. Seasonal variations of water temperature, and salinity in Mikawa Bay, Japan.

approximately 150 μm from the primordium. This pattern of Sr:Ca was similar among individuals, because the migratory environment at the marine leptocephalus stage was similar for all. However, from the elver stage, the Sr:Ca profiles divided the eels into three types (Figure 5), each of which are subsequently described.

Type 1 (freshwater resident): Sr:Ca in the layer approximately 150 μm from the primordium to the edge of the otolith fluctuated between 0 and 4 × 10^{-3}, indicating that the eels after the elver stage had migrated into fresh water to grow and later metamorphose to silver eel (Figure 5a). Otolith microchemistry mapping indicated that Ca was

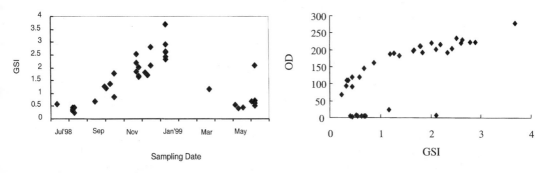

Figure 3. Seasonal variation of gonadosomatic index (GSI).

Figure 4. The relationship between ovum diameter (OD) and gonadosomatic index (GSI).

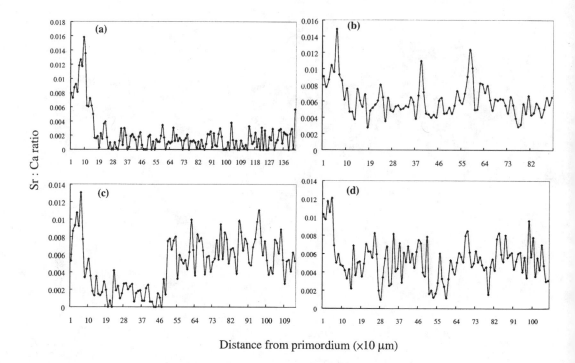

Figure 5. Classification of migratory environments of the eel based on the otolith Sr:Ca at yellow eel stage: (a) 66.8 cm TL, Type 1 (freshwater); (b) 76.6 cm TL, Type 2 (seawater); (c) 80.0 cm TL, Type 3a (freshwater then seawater); and (d) 76.0 cm TL, Type 3b (between fresh water, and seawater).

homogenous throughout the otolith (Figure 6a, b, and c). Sr concentration, however, markedly changed across the life history transect. Sr was highest in the core, which corresponds to the marine leptocephalus stage. It then decreased slightly which indicates a brackish water elver stage, with further decrease that indicates freshwater residency during the yellow eel stage. Sr concentration increased again in the edge of the otolith when the eel migrated from fresh to brackish water of the bay. The level of Sr was similar in the layer approximately 150 μm from primordium and in the edge of the otolith, indicating that the eel at elver stage and silver stage stayed in a similar saline environment (Figure 6a).

Type 2 (seawater resident): Sr:Ca 150 μm from the primordium to the edge of the otolith ranged from 4 to 10×10^{-3} and averaged approximately 6×10^{-3}, indicating that the elvers had remained in brackish water to grow and later metamorphose to the silver stage (Figure 5b). The transect concentration of Ca was also similar to that of Type 1 otoliths (Figure 6b). The concentration of Sr, however, was different from that of Type 1 except at the core. Sr concentration from the elver stage to silver stage were higher than those of Type 1 and four growth annuli of higher Sr concentration were identified during the yellow stage growth phase (Figure 6b). Accordingly, eels grew in the seawater for approximately four to five years with the Sr band being formed in the winter when water temperature was lowest and salinity was highest (Figure 2).

Type 3 (estuarine resident): an intermediate type between Types 1 and 2. Postelver Sr:Ca were lower than 4×10^{-3} in the early yellow stage, but thereafter increased to higher than 6×10^{-3} until the edge of otolith. This pattern indicates that the elvers had been migrating to freshwater and later returned to seawater at the yellow stage where they remained until metamorphosing to silver stage (Figure 5c). An alternative type, Sr:Ca changed between 2×10^{-3} and 8×10^{-3}, approximately a fourfold difference (Figure 5d). This indicates that the eel irregularly

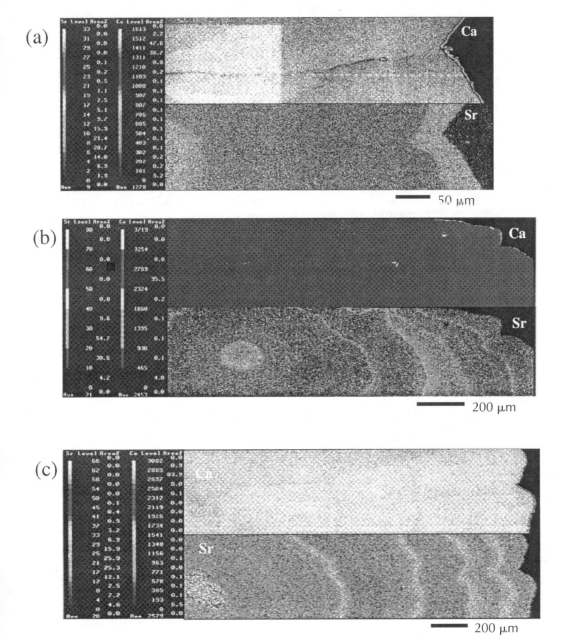

Figure 6. Ca, and Sr concentration maps of otoliths: (a) Type 1 or Type 3a; (b) Type 2: and (c) Type 3b. The eel (a) measured 73.0 cm TL, while the eels (b) and (c) are the same as (b) and (d) in Figure 5.

migrated between fresh and seawater throughout the yellow stage growth phase. There were also four growth annuli with higher Sr concentration, indicating that growth phase of the eel lasted for approximately four to five years (Figure 6c).

Composition of Life History Types

Based on Sr:Ca, the migratory environments during the growth phase of yellow stage eels was divided into three types, i.e., Type 1 (freshwater), Type 2 (seawater), Type 3a (freshwater then seawater) and Type 3b (between freshwater and seawater). Types 3a and 3b constituted nearly 58% of the total eels examined (Table 1), indicating that approximately half of the eels in the Mikawa Bay migrated between fresh water and seawater during the yellow stage. The remainder of the population was almost equally distributed among freshwater eel (20%) and seawater eel (22%).

Comparison of Length, Weight, and GSI Among Different Life History Types

Mean length, weight and GSI of female eels did not vary significantly among life history types (Kruskal-Wallis one way analysis of variance (ANOVA) on ranks, $P = 0.162 - 0.843$). This indicated that size distribution and maturation were independent of migratory environments (Figure 7).

Discussion

Our measure of otolith Sr:Ca beyond the elver stage was approximately $0-4 \times 10^{-3}$ in freshwater eel (Type 1) and approximately $4-10 \times 10^{-3}$ with a mean of 6×10^{-3} in seawater eel (Type 2). Campana (1999) reported that Sr:Ca is approximately 1.8×10^{-3} in fresh water and 8.6×10^{-3} in seawater. The difference in Sr:Ca between the eels of Types 1 and 2 was similar to those of fresh water and seawater. Accordingly, our classification of eels into freshwater and seawater types in this study is believed to be reliable. In addition, regular spaced bands of higher Sr concentration were found in otolith of the seawater eels (e.g., Figure 6b and c). This indicated that the seasonal change in salinity of Mikawa Bay likely influences the variable levels of otolith Sr:Ca of

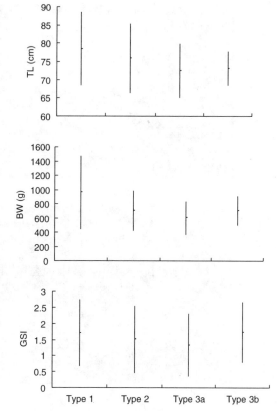

Figure 7. Comparison of mean (± SD) total length, body weight, and gonadosomatic index (GSI) of the eels among the four different life history types.

Table 1. Composition of life history types of female silver stage eels in Mikawa Bay, Japan.

Life history type	Number of eels	Percent of total
Type 1: freshwater	9	20.0
Type 2: seawater	10	22.2
Type 3a: freshwater then seawater	20	44.5
Type 3b: between freshwater and seawater	6	13.3
Total	45	100.0

the seawater eel. Similar observations have also been reported for the European eel, *A. anguilla* (Tzeng et al. 1999).

The relationship between otolith Sr:Ca of Japanese eel and ambient salinity was established using rearing experiments (Tzeng 1996b; Kawakami et al. 1998). The otolith Sr:Ca in relation to salinity was used to reconstruct the past salinity history of the striped bass *Morone saxatillis* in Chesapeake Bay (Secor et al. 1995; Secor and Rooker 2000). We did not, however, attempt to do a similar back-calculation of absolute salinity history as performed by these researchers for young-of-year striped bass, because we found that the mean otolith Sr:Ca of the eel reared in fresh water was 4.2×10^{-3} at the temperature of 22–28°C (Tzeng 1996b), and between $4.5 (\pm 0.91)$ and $5.0 (\pm 0.68) \times 10^{-3}$ at the temperature of 12–27°C (Kawakami et al. 1998), which was significantly higher than those of the freshwater eel of Type 1 in this study (0–4 × 10^{-3}) and fresh water (1.8×10^{-3}) (Campana 1999). This Sr:Ca discrepancy between freshwater-reared eels and the natural freshwater eel of Type 1 was probably due to the effect of Sr uptake from food. The Sr content of the food organism, tubifex worm *tubificid oligochaetes* was close to seawater and higher than fresh water (Kawakami et al. 1998). Enriched otolith Sr:Ca resulting from food uptake was also found in American shad *Alosa sapidissima* (Limburg 1995). On the contrary, the otolith Sr:Ca of the eel reared in the 35‰ seawater with the tubifex worm averaged 9.27×10^{-3} (Tzeng 1996b) and 8.5 (± 0.64) to 9.1 (± 1.27) × 10^{-3} (Kawakami et al. 1998), which was close to that of seawater (8.6×10^{-3}) (Campana 1999). This indicated that the effect of Sr from food uptake was not simply additive. Thus, the use of a standard curve of otolith Sr:Ca versus salinity established by rearing experiment to back-calculate the absolute salinity history of the eel should be made cautiously.

In addition, otolith Sr:Ca was not only influenced by ambient salinity, but also influenced by ontogenetic developmental stage (Tzeng and Tsai 1994; Otake et al. 1994; Tzeng 1996b). Sr:Ca was significantly higher in leptocephali than in elvers, particularly around metamorphosis from leptocephalus to glass eel (Figure 6). This indicated that the relationship between otolith Sr:Ca and salinity was different among development stage of the fish. Incorporation of Sr into the otolith is a complicated biochemical process influenced by physical factors such as temperature, salinity and water chemistry, as well as by biological factors such as genetics, developmental stage, growth rate, food and physiological condition of the fish (Campana 1999; Dodd 1967; Gallahar and Kingsford 1992; Kalish 1989; Radtke and Shafer 1992; Sadovy and Severin 1992).

The Japanese eel *A. japonica*, similar to other *Anguilla* species, is considered to be a catadromous fish; i.e., it spawns in the ocean and grows in freshwater environments. Otolith Sr:Ca measured in this study indicates that individual eels were widely distributed from freshwater to seawater during growth-phase yellow stage. Migration to freshwater for the yellow eel is not an obligate pathway, but rather is a facultative diadromy with seawater residents as an ecophenotype (Tsukamoto et al. 1998). The migration of diadromous fishes may be a direct response to the environment, especially temperature, water level, and food availability (Baker 1978; Northcote 1978; McDowall 1987; Tzeng et al. 2000b), and influenced by the actions of conspecific competitors (e.g., Gross 1985). Diadromy may, therefore, be a complex collection of life history traits, some of which are strategies and others simple in random distribution. The Mikawa Bay estuary, like other estuaries around the world, is among the most productive areas on earth with productivity comparable to that of coastal upwelling areas (Haedrich and Hall 1976). The food availability in the bay may lead to the eel of Type 3 to be dominant. On the contrary, if the freshwater migration of the eel is an obligate pathway, the freshwater eel in the population should be dominant; however, the freshwater eel constituted only 20% of the population. Most of the eel in the Mikawa Bay were brackish-water residents. Thus, the distribution of the eel population probably depends on the carrying capacity of the environment. The brackish-water eel likely dominated the study area because the Bay's carrying capacity far exceeded the capacity of habitat found in the Bay's adjacent rivers.

Acknowledgments

This study was financially supported by the National Science Council of the Republic of China (Project No. NSC 89–2611-B002–004). The authors are grateful to Ms S. Y. Tsai for helping electron probe microanalysis and Dr T. F. Tsai and two anonymous reviewers for helpful comments on previous drafts of the manuscript.

References

Arai T., T. Otake, and K. Tsukamoto. 1997. Drastic changes in otolith microstructure and microchemistry accompanying the onset of metamorphosis in the Japanese eel, *Anguilla japonica*. Marine Ecology Progress Series 161:17–22.

Baker, R. R. 1978. The evolutionary ecology of animal migration. Hodder and Stoughton, London, England.

Bertin, L. 1956. Eels—a biological study. Cleaver-Hume Press, London, England.

Campana, S. E. 1999. Chemistry and composition of fish otoliths: pathways, mechanisms and applications. Marine Ecology Progress Series 188:263–297.

Casselman, J. M. 1982. Chemical analyses of the optically different zones in eel otoliths. Proc. 1980 N. American Eel Conference, pages 74–82.

Cheng, P. W., and W-N. Tzeng. 1996. Timing of metamorphosis and estuarine arrival across the dispersal range of the Japanese eel *Anguilla japonica*. Marine Ecology Progress Series 131:89–96.

Dodd, R. J. 1967. Magnesium and strontium in calcareous skeletons: a review. Journal of Palaeontology 41:1313–1329.

Gallahar, N. K., and M. J. Kingsford. 1992. Patterns of increment width and strontium: calcium ratios in otoliths of juvenile rock blackfish, *Girella elevate* (M.). Journal of Fish Biology 41:749–763.

Gross, M. R. 1985. Disruptive selection for alternative life histories in salmon. Nature (London) 313:47–48.

Haedrich, R. L., and C. A. S. Hall. 1976. Fishes and estuaries. Oceanus 19(5):55–63.

Kalish, J. M. 1989. Otolith microchemistry: validation of the effects of physiology, age and environment on otolith composition. Journal of Experimental Marine Biology and Ecology 132:151–178.

Kawakami, Y., N. Mochioka, K. Morishita, T. Tajima, H. Nakagawa, H. Toh, and A. Nakazono. 1998. Factors influencing otolith strontium/calcium ratios in *Anguilla japonica* elvers. Environmental Biology of Fishes 52:299–303.

Limburg, K. 1995. Otolith strontium traces environmental history of subyearling American shad *Alosa sapidissima*. Marine Ecology Progress Series 119:25–35.

McDowall, R. M. 1987. The occurrence and distribution of diadromy among fishes. Pages 1–13 *in* M. J. Dadswell, R. J. Klauda, C. M. Moffitt, R. L. Saunders, R. A. Rulifson, and J. E. Cooper, editors. Common strategies of andromous and catadromous fishes. American Fisheries Society, Symposium 1, Bethesda, Maryland.

Northcote, T. G. 1978. Migratory strategies and production in fresh water. Pages 326–359 *in* S. D. Gerking, editor. Ecology of freshwater fish production. Blackwell Scientific Publications, Oxford, England.

Otake, T., T. Ishii, M. Nakahara, and R. Nakamura. 1994. Drastic changes in otolith strontium/calcium ratios in leptocephali and glass eels of Japanese eel *Anguilla japonica*. Marine Ecology Progress Series 112:189–193.

Radtke, R. L., and D. J. Shafer. 1992. Environment sensitivity of fish otolith microchemistry. Australian Journal of Marine and Freshwater Research 43:935–951.

Sadovy, Y., and K. P. Severin. 1992. Trace elements in biogenic aragonite: correlation of body growth rate and strontium levels in the otoliths of the white grunt, *Haemulon plumieri* (Pisces: Haemulidae). Bulletin of Marine Science 50:237–257.

Sang, T. K., H. Y. Chang, C. T. Chen, and C. F. Hui. 1994. Population structure of the Japanese eel, *Anguilla japonica*. Molecular Biology and Evolution 11(2):250–260.

Secor, D. H., A. Henderson-Arzapalo, and P. M. Piccoli. 1995. Can otolith microchemistry chart patterns of migration and habitat utilization in anadromous fishes? Journal of Experimental Marine Biology and Ecology 192:15–33.

Secor, D. H., and J. R. Rooker. 2000. Is otolith strontium a useful scalar of life cycles in estuarine fishes? Fisheries Research 46:359–371.

Tesch, F. W. 1977. The eel: biology and management of anguillid eels. Translated from German by J. Greenwood. Chapman and Hall/Wiley, New York.

Tsukamoto, K. 1992. Discovery of the spawning area for Japanese eel. Nature (London) 356:789–791.

Tsukamoto K., I. Nak, W. V. Tesch. 1998. Do all fresh water eels migrate? Nature 396:635–636.

Tzeng, W-N. 1985. Immigration timing and activity rhythms of the eel, *Anguilla japonica*, elvers in the estuary of northern Taiwan, with emphasis on environmental influences. Bulletin of Japanese Society of Fisheries Oceanography 47/48:11–28.

Tzeng, W-N. 1986. Resources and ecology of the Japanese eel *Anguilla japonica* elvers in the coastal waters of Taiwan. China Fisheries Monthly 404:19–24 (in Chinese).

Tzeng, W-N. 1990. Relationship between growth rate and age at recruitment of *Anguilla japonica* elvers in a Taiwan estuary as inferred from otolith growth increments. Marine Biology 107:75–81.

Tzeng, W-N. 1996a. Short- and long-term fluctuations in catches of elvers of the Japanese eel *Anguilla japonica* in Taiwan. Pages 85-89 *in* D.A. Hancok, D.C. Smith, A. Grand and J. P. Beumer, editors. Developing and sustaining world fisheries resources: the state of science and management. 2nd World Fisheries Congress Proceedings, CSIRO publishing, Collingwood, VIC 3006, Australia.

Tzeng, W-N. 1996b. Effects of salinity and ontogenetic movements on strontium: calcium ratios in the otoliths of the Japanese eel, *Anguilla japonica* Temminck and Schlegel. Journal of Experimental Marine Biology and Ecology 199:111–122.

Tzeng, W-N., and Y. C. Tsai. 1992. Otolith microstructure and daily age of *Anguilla japonica* Temminck & Schlegel elvers from the estuaries of Taiwan with reference to unit stock and larval migration. Journal of Fish Biology 40:845–857.

Tzeng, W-N., and Y. C. Tsai. 1994. Changes in otolith microchemistry of the Japanese eel, *Anguilla japonica*, during its migration from the ocean to the rivers of Taiwan. Journal of Fish Biology 45:671–683.

Tzeng, W-N., H. F. Wu, and H. Wickström. 1994. Scanning electron microscopic analysis of annulus microstructure in otolith of European eel *Anguilla anguilla*. Journal of Fish Biology 45: 479–492.

Tzeng, W-N., K. P. Severin, and H. Wickström. 1997. Use of otolith microchemistry to investigate the environmental history of European eel *Anguilla anguilla*. Marine Ecology Progress Series 149: 73–81.

Tzeng, W-N., K. P. Severin, H. Wickström, and C. H. Wang. 1999. Strontium bands in relation to age marks in otolith of European eel *Anguilla anguilla*. Zoological Studies 38(4):452–457.

Tzeng, W-N., H. R. Lin, C. H. Wang, and S. N. Xu. 2000a. Differences in size, and growth rates of male, and female migrating Japanese eels in Pearl River, China. Journal of Fish Biology 57(5):1245–53.

Tzeng, W-N., C. H. Wang, H. Wickström, and M. Reizenstein. 2000b. Occurrence of the semi-catadromous European eel *Anguilla anguilla* (L.) in Baltic Sea. Marine Biology 137:93–98.

Downstream Movement of Mature Eels in a Hydroelectric Reservoir in New Zealand

Erina M. Watene

The University of Auckland, School of Biological Sciences, Private Bag 92019, Auckland, New Zealand

Jacques A. T. Boubée

National Institute of Water and Atmospheric Research Ltd, Post Office Box 11-115, Hamilton, New Zealand

Alex Haro

S.O. Conte Anadromous Fish Research Center, Biological Resources Division, U. S. Geological Survey, Post Office Box 796, Turners Falls, Massachusetts, USA

Abstract.—This study investigates the behavior of migrant eels as they approached the Patea hydroelectric dam on the West Coast of the North Island, New Zealand. Seventeen mature migrant eels (870–1,240 mm; 2,000–6,380 g) were implanted with coded acoustic transmitters and released. Their movements in the reservoir were monitored for 14 months with stationary data logging and manual tracking receivers. The downstream migration of sexually maturing eels was found to occur mainly at night, usually during, or immediately after, rainfall events. Eels tended to travel at the surface, within the upper 4 m of the water column, at speeds ranging from 16 to 89 cm/s. Upon reaching the headrace, eels typically spent time searching, presumably for an unobstructed downstream route. In order to aid downstream passage of eels at the Patea Dam, power station operators began spillway opening trials during peak migration periods. Although this allowed some migrant eels to safely pass over the dam, information on the relative effectiveness and cost of this method over other possible mitigation methods is still required.

Introduction

Three species of freshwater eel occur in New Zealand, the endemic longfinned eel *Anguilla dieffenbachii*, the shortfinned eel *Anguilla australis*, which also occurs along the eastern coast of Australia, and the Australian longfinned eel *Anguilla reinhardtii*, which recently extended its range to the North Island of New Zealand. The endemic longfinned eel (longfin) is the top predator in New Zealand freshwaters and penetrates further inland than the shortfinned eel (shortfin), which is found predominantly in lowland regions of coastal catchments. The Australian longfinned eel has recently been discovered in New Zealand and seems to prefer estuarine habitats (Jellyman et al. 1996; Chisnall 2000; McDowall 2000).

Eels have always been an important and highly regarded traditional resource to the Maori people of New Zealand. Traditional customary fishing practices ensured conservation of the eel resource, allowing prolonged, sustainable harvesting of eels throughout New Zealand (Maniapoto 1998). All of the eel species support important commercial, recreational and traditional fisheries (McDowall 1990).

Anguillid eels are diadromous and must migrate to and from the sea to complete their life cycles. In spring, juvenile eels migrate into rivers from the sea. The climbing behavior of eels has been successfully exploited using a number of effective fish-pass designs to provide access to some highly productive hydroelectric reservoirs and extensive upstream habitats (Mitchell and Chisnall 1992; Chisnall and Hicks 1993; Boubée and Mitchell 1994; Beentjes et al. 1997). The downstream migration of mature eels (often referred to as silver eels) generally occurs in late summer and autumn (Cairns 1941; Burnet 1969; Todd 1981; Boubée et al. 2001), when the eels leave freshwater and migrate to their spawning grounds in the Pacific, which are thought to be in warm tropical seas between Fiji and Tahiti (McDowall 1990).

Protection of adult eel stocks and downstream passage of silver eels have become significant issues in New Zealand recently, as studies have shown a considerable decrease in the recruitment of glass eels and elvers into New Zealand waterways (Jellyman et al. 2000). Many factors may have contributed to this decline: Over-exploitation, loss of habitat and the presence of barriers on many rivers and streams individually, or collectively, may have reduced the number of mature eels reaching the sea to spawn. Protection of downstream migrating adults has, therefore, been advocated as a potential means of increasing recruitment (Jellyman et al. 2000; Boubée et al. 2001).

Hydroelectric dams (hydro dams) can be a major obstruction to downstream eel migration. Where intake screens are small-meshed, eels can impinge upon the mesh and be injured or killed. Coarser screens allow the eels to pass through to the turbines, where many are killed. Dead, mutilated eels are commonly found below hydroelectric dams during the autumn migration. Small turbines cause the most damage and long fish, like eels, are most affected (Larinier and Dartiguelongue 1989; Travade and Larinier 1992; Desrochers 1995). A review of the problem in New Zealand concluded that the survival of the large New Zealand migrant eels (over 80 cm) through turbines was likely to be nil (Mitchell and Boubée 1992).

Consequently, methods of passing adult eels over dams need to be developed. However, information is required on eel behavior at dams in order to develop and optimize these methods. Telemetry is a useful method to study the movement and behavior of migrant eels, because it allows detailed tracking of individual eels. Although several studies have used telemetry to study movements of migrant eels in estuaries, coastal waters (Helfman et al. 1983; Parker 1995; Parker and McCleave 1997), and in the open sea (Tesch 1974, 1978; McCleave and Arnold 1999), very few have used telemetry to examine the behavior of migrant eels in freshwater. Haro et al. (1999) used acoustic and radio telemetry to study the behavior of migrant eels *Anguilla rostrata* at a small hydroelectric facility in North America.

The first objective of this study was to use acoustic telemetry to determine the timing of migration, vertical swimming depth, distribution, and migration speed of mature eels in a New Zealand hydroelectric reservoir. The second objective was to examine the effect of rainfall and moon phase upon the timing of migrations. The next objective was to examine how the behavior of migrant eels changed when passage was obstructed. Finally, the feasibility of spillway openings to assist downstream passage was assessed. These findings were used to formulate downstream passage mitigation techniques for eels in New Zealand.

Methods

Study Area

The study was carried out in Lake Rotorangi, located on the Patea River in the Taranaki Region, New Zealand (Figure 1). The Patea River originates on the slopes of Mount Taranaki (2,500 m) and is the second largest river in Taranaki in terms of flow (33 m^3/s) and catchment area (871 km^2).

Lake Rotorangi (Figure 2) was formed in 1984 by the construction of an earth fill dam 41 km upstream from the mouth of the Patea River. The lake has an average depth of 28 m, an average width of 130 m and is 46 km long. The lake is riverine in some aspects, in that it can be substantially affected by flood events. It is also mildly eutrophic and exhibits thermal stratification during the summer (Burns 1995).

The hydroelectric plant has a capacity of 30 megawatts (MW) and consists of three 10 MW Vertical Francis Turbines. Water from the lake is delivered to the turbines via an intake structure incorporating two radial guard gates, and three 2.8 m diameter steel penstocks. In addition to the main turbines, a small 700 kW turbine generator provides a residual flow (1.4 m^3/s) down the river, when the main turbines or spillway are not operating (Anon 1988).

Experimental Design

The study was conducted on wild migrant eels captured from Lake Rotorangi by commercial eel fishers using standard fyke nets (12 mm stretched mesh) employed in the fishery. Eels were classified as migrants on the basis of coloration, timing of capture and morphological features. Pankhurst's (1982) eye index (1), which relates the external surface area of the eye to

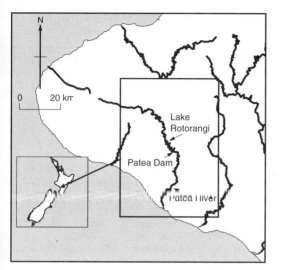

Figure 1. Location of the Patea River, Taranaki, New Zealand.

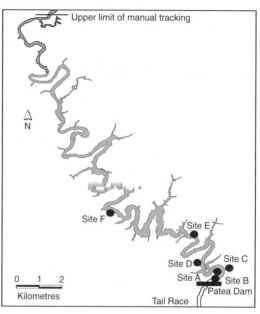

Figure 2. Location of the acoustic receiver stations, Lake Rotorangi, Patea River.

total body length, was also used to confirm that the eels were migrants. Eels with an eye index of greater than 6.5 were classed as sexually maturing adults.

$$(1) \quad I = \frac{\left[\frac{(v+h)^2}{4}\right] \pi \times 100}{L}$$

I = eye index
v = vertical eye diameter
h = horizontal eye diameter
L = Total body length

The eye indicies for the eels used in this study ranged between 6.5 and 17.1.

On 20 May 1999, 12 large migrant eels were surgically implanted with V16 coded acoustic transmitters (Vemco Limited, Nova Scotia, Canada). Three additional migrant eels from Lake Rotorangi were similarly tagged on 8 February 2000, and two more on 10 April 2000. All tagged eels weighed between 2,000 and 6,380 g.

Each transmitter had a life expectancy of 540 days, weighed 39 g in air (18 g in water) with a diameter of 16 mm and a length of 108 mm, and operated on a frequency of 69 kHz. Pulse rates were factory-set, at a random rate of 25–45 seconds. The transmitters weighed less than 2% of the body weight of the eels used in this study, and were relatively short in comparison to the body length of the migrant eels (870–1240 mm). Characteristics of the eels used are shown in Table 1.

Surgery

Eels were placed into a 50 L bin containing a slurry of lake water and ice, and anesthetized with tricane methanesulfonate (60 mg/L), buffered with 200–250 mg/L sodium bicarbonate. Once sedated, the eels were measured for total length, total weight, and vertical and horizontal eye diameter. The eels were then placed ventral side up in a tray containing anesthetic. A soft foam pad was used to stabilize the eel's body during surgery. A 20 mm incision was made in the body wall on the mid ventral line in the posterior quadrant of the body cavity, and 1 mL of oxytetracycline (200 mg/mL) was injected into the incision. The transmitters were inserted into the body cavity and positioned 30–40 mm forward of the incision. To provide further identification, a 32 mm PIT tag (Texas Instruments model RI-TRP-RC2B) was also placed in the body cavity. The incision was closed with three sterile synthetic absorbable sutures (Ethicon model BH4667) and the incised area was

Table 1. Length, weight, capture/release dates and migratory route of the eels implanted with acoustic transmitters in Lake Rotorangi, May 1999–June 2000. LF = longfin eel; SF = shortfin eel.

Eel ID	Species	Total length (mm)	Weight (g)	Eye index	Tag weight (% of eel weight)	Tagging/ release Date	Fate of eel and date
LF1	LF	1000	2340	9.5	1.6	20 May 1999	Turbines (21.5.99)
LF2	LF	1050	3780	17.1	1.0	20 May 1999	Spillway (10.4.00)
LF3	LF	1240	6380	13.8	0.6	20 May 1999	Spillway (15.5.00)
LF4	LF	1030	2060	12.3	1.8	20 May 1999	Turbines (23.3.00)
LF5	LF	870	2020	12.3	1.9	20 May 1999	Turbines (4.11.99)
LF6	LF	1040	2080	12.1	1.8	8 Feb 2000	Upstream (30.6.00)
SF1	SF	960	2420	6.5	1.6	20 May 1999	Turbines (20.5.00)
SF2	SF	970	2380	6.5	1.6	20 May 1999	Turbines (10.4.00)
SF3	SF	890	2000	13.5	1.9	20 May 1999	Upstream (30.6.00)
SF4	SF	965	2060	6.7	1.6	20 May 1999	Turbines (13.11.99)
SF5	SF	1080	2140	6.7	1.8	20 May 1999	Upstream* (30.6.00)
SF6	SF	920	2010	9.4	1.9	20 May 1999	Upstream* (30.6.00)
SF7	SF	1030	2120	10.5	1.8	20 May 1999	Turbines (20.5.99)
SF8	SF	930	2100	6.6	1.8	8 Feb 2000	Upstream (30.6.00)
SF9	SF	1090	2700	6.7	1.4	8 Feb 2000	Upstream (30.6.00)
SF10	SF	1140	2500	6.5	1.5	10 Apr 2000	Turbines (10.4.00)
SF11	SF	1115	2600	6.7	1.5	10 Apr 2000	Turbines (11.4.00)

*no movement detected for over a year

swabbed with an anti-inflammatory cream containing a local anesthetic (Topigol antibiotic). Each implantation took about five minutes. After surgery, each eel was placed in a soft mesh holding bag in the lake for recovery, and then released approximately 1 km upstream from the dam within half an hour of surgery.

Data Loggers and Monitoring

Initially three stationary datalogging receivers (Vemco VR1) were installed near the headrace of the Power Station (Site A, 0 km from the dam; Site B, 0.1 km from the dam; Site C, 0.5 km from the dam; Figure 2) and a fourth receiver was positioned in the tailrace. The receivers continually recorded the time and depth of eels that were within acoustic range. In addition to the stationary receivers, the eels were tracked manually from a boat. Manual tracking was performed on a bi-weekly basis, using a Vemco VR60 receiver and an omnidirectional hydrophone to initially detect the transmitter signal, and then a directional hydrophone to determine its location.

Within three months of implanting the first 12 transmitters, it was noticed that the majority of eels had moved upstream beyond the reception area of the uppermost receiver (Site C). Three additional receivers were therefore deployed further up the lake (Site D, 2.5 km from the dam; Site E, 5.5 km from the dam; Site F, 12.5 km from the dam; Figure 2). Another receiver was deployed 6.2 km downstream of the dam in the Patea River in February 2000, prior to the autumn 2000 migration season.

Spillway Opening

To evaluate the potential use of spillway openings as a means of assisting downstream passage of mature eels, one of the two bottom opening spillway gates was intentionally opened 150 mm for two hours (1930–2130 hours) on three consecutive nights during a rain event in autumn 2000 (8–11 April 2000). On the first night the spillway was opened, a large fyke net (12 mm mesh) was set across the entire width of the river (50 m), 2 km downstream of the tailrace. The aim was to capture any eels that may have passed over the spillway and moved down river.

Rainfall Data

Rainfall data (nearest mm) was collected daily by automated loggers maintained by the

Taranaki Regional Council at the Mangaehu Site, in the upper headwaters of the Patea River. This site is approximately 15 km upstream from Patea Dam.

Results

Tagging Effects

Surgical implantation was successful for fifteen eels, whose movement within the lake could be followed. One of the eels (LF2) was recaptured 11 months after implantation and showed no external sign of the surgery. However, the eel had lost condition (25% of its initial weight). Two eels (SF5, SF6) probably died or expelled their transmitters, as the signal location remained stationary for over a year.

Seasonal Migration Patterns

Over 14 months, 10 downstream migration pulses were observed, when at least one tagged eel attempted to pass the station (Figure 3). During four of these pulses, 12 out of the 17 tagged eels passed over the dam. Three of the four successful passes over the dam occurred during the new moon. When the eels did not pass over the dam, they swam back upstream 2–12 kilometres, often to localities they had resided in prior to the migration downstream.

Nearly all of the migration pulses coincided with rainfall events (> 35 mm rain in 24 hours). The first pulse on 20–21 May 1999, occurred two days after heavy rainfall (147 mm rain over four days). LF1 and SF7 most likely passed through the turbines on that occasion as their transmitters remained in the tailrace. LF5 and SF4 attempted to pass during that event, but were unsuccessful and returned upstream.

During November 1999, two eels passed through the turbines. LF5 passed through at the beginning of the month, which coincided with the onset of a nine day rainfall event, after which SF4 passed through the turbines. SF2 came into the headrace on two separate occasions during November 1999, in both cases during rain events. This eel did not pass over the dam, opting to head back upstream instead.

The main migration pulse coincided with the heaviest rainfall of autumn (125 mm of rain over five days), when six tagged eels came into the headrace (8–11 April 2000). Of the six eels, four passed through the turbines and were killed, and

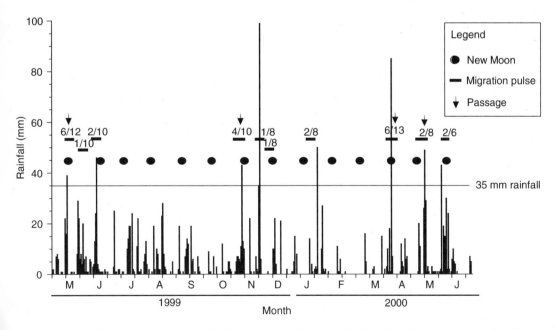

Figure 3. Relationship between tagged eel detection at the Patea hydroelectric power station headrace, rainfall, and moon phase. Proportion of eels detected to total number of tagged eels in the lake is also shown.

the fifth (LF2) passed via the spillway, which was opened 150 mm for the spillway trial. The sixth eel (SF3) returned upstream, after making 23 attempts to pass the dam.

LF2 was recaptured in the net set downstream of the dam and returned to the upper lake where it was subsequently tracked. It passed over the dam for the second time two days later, and was recorded going past the lower river receiver station three hours after last being recorded in the lake. It may have passed over the spillway again (as it was open during its second passage) or traveled through the turbines.

The final migration pulse occurred in May 2000, when LF3 passed over on the 15th (following 104 mm rain over three days). LF3 was recorded moving past the lower river receiver station two hours after being recorded in the upper lake for the last time. The spillway was open from 12–16 May, so it is possible that this eel also used the spillway. SF1 went through the turbines five days later on the 20th.

Diel Migration Patterns

During migration events, the majority of eels traveled during the night (between 30 minutes after sunset, to 30 minutes before sunrise, Figure 4). LF2, LF3 and SF1 were exceptions. LF2 and LF3 started their downstream movement during darkness, but as morning approached, these eels continued to migrate. On these days, conditions were stormy with low light conditions. LF2 entered the headrace (Site A) in the morning, searched for seven minutes then swam 500 m back upstream. It returned to the headrace at nightfall for a short period, and eventually went over the spillway. LF3 moved through the headrace rapidly during the day (1300 hours) and was not recorded by the headrace receiver. However, it was recorded moving past the lower river receiver station two hours after last being recorded in the lake. SF1 began its downstream migration during the afternoon and continued through the night until it reached the dam.

Migration Speeds

Migration speeds for five eels were calculated using the distance between two receiver stations and the time it took the eel to travel between the two points (the time of the last recording at an upstream station to the time of the first recording at the downstream station). These speeds

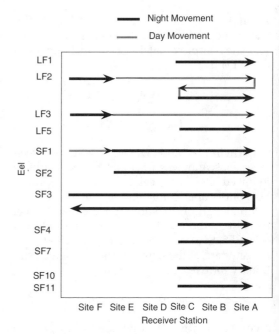

Figure 4. Diurnal movement patterns of eels between receiver sites Lake Rotorangi, Patea River. Black shading indicating night movement, double lines day movement. Location of receiver stations shown in Figure 2.

may therefore be minimum rates, as eels may not move directly between points. The migration speeds in the lake ranged from 16 to 62 cm/s (mean = 44 cm/s, SD = 15 cm/s; Table 2). The slowest speeds were obtained between the two most widely spaced receivers (7 km between Site F and Site E).

Movement Behavior

During their downstream migration in the lake, eels swam mainly at the surface. LF3, SF2 and SF3 primarily traveled in the upper 2 m. All of the eels tracked, migrated mostly within the top 5 m (Figure 5).

As the eels approached the headrace and found their pathway blocked by the dam, they responded by diving repeatedly. Most eels utilized the entire depth range (9.0–12.0 m maximum depth, depending on lake level) of the headrace during these dives (Figure 6). The mean depth of the eels in the headrace was deeper than the mean depth of the same eels in the lake, probably indicating searching behavior (Table 3).

Table 2. Migration speeds between receiver stations, Lake Rotorangi, Patea River.

Eel ID	Direction of Movement	Migration Speeds (cm/s)			
		Site F-E (7 km)	Site E-D (3 km)	Site D-C (2 km)	Tailrace-lower river (6.2 km)
LF2	downstream		51	62	89
LF3	downstream	16	49	55	50
SF1	downstream	25	45	52	
SF3	downstream		20	62	
SF3	upstream		49	45	

Five out of ten eels made more than one approach to the headrace before finally going through the turbines or over the spillway. The amount of time spent in the headrace prior to passing through the turbines varied between individuals. Seven eels spent 16 minutes or less in the headrace per visit, whereas LF1 spent an average of 219 minutes in the headrace per visit.

Spillway Opening Trial

During the trial spillway opening period (8–11 April, 2000) six eels attempted to pass the dam, of which four passed through the turbines, one returned upstream, and one passed over the spillway. The single tagged eel that passed over the spillway was recaptured in the net along with 118 other migrant eels (51 longfins and 67 shortfins).

Mortality

Of the twelve eels that went through the turbines or over the spillway, ten were probably killed, as the transmitters were subsequently located in the tailrace and failed to move thereafter. Two other eels (LF2, LF3) possibly survived passage by going over the spillway, as they were recorded moving past the lower river receiver station. In addition to mortality effects, two sublethal effects were observed during this

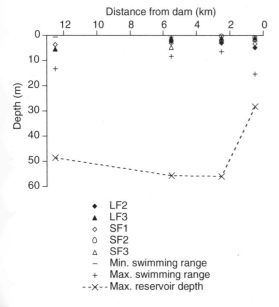

Figure 5. Vertical movement patterns of tagged migrant eels in Lake Rotorangi. (note: most eels did not reach the uppermost recording stations)

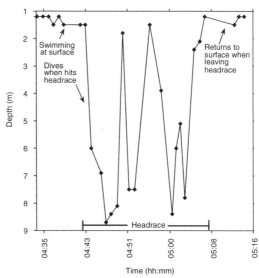

Figure 6. Vertical movement of SF3 in the headrace, showing diving behaviour in front of the Patea Dam.

Table 3. Time spent in headrace and swimming depth of migrant eel in front of the Patea Dam intake.

Eel ID	Number of approaches	Mean time in headrace (minutes)	Depth range (m)	Mean depth (m)	Depth at final detection (m)
LF1	2	219	2.0–9.0	7.9	9.0
LF2	4	16	3.8–9.0	7.3	7.9
LF2*	3	58	3.0–9.4	7.2	**
LF3	1	**			
SF1	1	10	3.7–7.4	5.2	5.6
SF2	1	1	2.5–3.7	3.1	2.5
SF3	23	7	1.2–9.0	6.3	***
SF4	1	3	2.0–9.0	5.0	9.0
SF10	5	8	0.4–8.6	5.8	5.6
SF11	1	8	3.7–10	6.8	7.8

* LF2 was recaptured in the river below the dam on 8 April 2000 and returned to the lake.
** Not recorded in the headrace during final passage
*** Returned upstream after making several attempts to pass the dam.

study. These were delayed migration (SF3) and loss of condition (LF3).

Discussion

There was not enough data to compare differences in the behavior or movement of the longfin and shortfin eels, so for the purpose of this study the two have been treated as one. The main migration period during this telemetry study occurred between April and May 2000, when 10 out of the remaining 13 tagged eels passed over or through the dam. This period concurs with other studies which, in general, set the main migration periods between February and April for shortfins, and April to May for longfins (Cairns 1941; Hobbs 1947; Todd 1981; Boubée et al. 2001). The two eels that migrated in November 1999 did not fit this general pattern. However, the downstream migration of eels in a tributary of the Waimakariri River (Canterbury region of the South Island, New Zealand), occurred between October and February (Burnet 1969). The downstream migration period of mature eels in New Zealand may therefore extend from October to May, but it is mainly restricted to the autumn period (February to May).

The migration of eels in Lake Rotorangi was strongly related to rainfall. The downstream migration of the tagged eels occurred in 10 pulses, during four of these pulses 12 out of 17 tagged eels passed over the dam. Each migration pulse occurred either during or within two days of a rainfall event. This finding concurs with those of Boubée et al. (2001) who found that increases in water level and flow associated with rainfall were key triggers in the downstream migration of shortfin and longfin eels in the Rangitaiki River. The importance of water flow and rainfall during the downstream migration of mature eels is also supported by other New Zealand studies (Cairns 1941; Burnet 1969; Todd 1981). As with the European eel *A. anguilla*, floods may be a guiding factor which mechanically assists the eels in their downstream migration (Lowe 1952; Deelder 1954; Tesch 1977).

The amount of rainfall appears to be a significant factor in determining whether eels will migrate or not. In Lake Rotorangi, migration pulses were observed on, or soon after, every occasion that rainfall exceeded 35 mm in a day. Boubée et al. (2001) suggested that a 40 mm rainfall trigger could be used to predict the downstream migration of eels on the Rangitaiki River.

In contrast to some other studies (Burnet 1969; Todd 1981; Sloane 1984; Vøllestad et al. 1986), here it was found that moon phase was important in the downstream migration of mature eels in Lake Rotorangi. Ten migration pulses in total were observed, seven of which coincided with the new moon. Passage over the dam only occurred on four occasions, and three out of these four migration pulses also coincided with a new moon. This relationship between the new moon and peak downstream eel migration is well known among Maori traditional fishers and commercial eel fishers (Best 1929; Boubée et al. 2001). Lowe (1952) suggests that the phase of the moon probably influences eel migrations more through the effect of light rather than a periodic effect, since she found that eels will run on the full moon when the water is turbid.

The majority of eel movement in Lake Rotorangi occurred at night, which is consistent with telemetry studies of other species of *Anguilla* (Helfman et al. 1983; LaBar and Facey; 1983; Parker and McCleave 1997). However, two eels continued to migrate during the day, albeit both under dark, cloudy skies. Daylight movement has also been reported by McGovern and McCarthy (1992), who found swimming speeds for the European eel during overcast conditions were similar to those recorded at night.

The migration speeds calculated for the eels in Lake Rotorangi ranged from 16 to 62 cm/s and averaged 44 cm/s. These speeds are similar to those reported by McCleave and Arnold (1999), who found in their movement study of European eels in the North Sea that eels swam at modest speeds through the water, between 35 and 58 cm/s. They did note, however, that their mean calculated speeds were minimum estimates, because the eel was assumed to have swum on a straight path. Our swimming speed calculations are also minimum estimates, as they were determined using the time it took eels to move the distance between two receiver stations. Tesch (1989) found that female silver European eels *Anguilla anguilla* L. had an average swimming speed of 60 cm/s when they moved through deep water (> 350 m) in the ocean. This is slightly faster than the swimming speeds recorded in Lake Rotorangi. Bell (1986) reported cruising swimming speeds for the American eel *Anguilla rostrata* as between 90 and 210 cm/s, which is also faster than speeds recorded in Lake Rotorangi.

The swimming speeds were calculated for two eels that passed over the spillway and continued to migrate in the Patea River. The swimming speed of one of the eels below the dam was similar to speeds calculated in the lake (50 cm/s). The swimming speed of another eel in the lower river was considerably faster (80 cm/s). The second eel was moving in the river during a flood event, and this may account for the faster downstream speed. Our field observations, therefore, concur with the laboratory study by Adam and Schwevers (1997) who found that eels moving downstream usually swam at approximately the same speed as the water flow.

In Lake Rotorangi, eels swam mostly at the surface within the top 2 m of the water column during their migration through the lake. This conforms with Adam and Schwevers (1997) laboratory study, who noted that many eels preferred to travel in the upper water column (upper third) of the artificial hydraulic channel they constructed. Similarly McCleave and Arnold (1999) tracked three silver eels out at sea, and found that two of these eels swam mostly near the surface or in the upper portion of the water column.

Management Implications

Under the current operating regime of most hydro dams in New Zealand, no provision for the downstream passage of mature migrant eels has been made, and consequently, the mortality rate is extremely high. In addition to direct mortality, two sub-lethal effects (loss of condition and delayed migration) were observed and provide additional cause for concern. These sub-lethal effects are almost inevitably linked (i.e., delayed migration may result in loss of condition) and are likely to further compromise the chance for mature eels to successfully reach spawning grounds.

To minimize the impacts of the dam, operators need to be able to predict migrations accurately and to implement effective mitigation measures. Based on the information collected in both this study and Boubée et al. (2001), rainfall could be used by dam operators to implement mitigation measures. These mitigation measures may take the form of a bypass, which would exploit the searching and diving behavior exhibited by eels when faced with a dam. Alternatively, since the eels migrate mainly in the surface waters, an intensive netting effort above the dam could be concentrated in the upper 2–3 m to intercept and collect surface migrants and manually transfer them below the dam. Opening the spillway gate at selected times during peak migration may also be a simple, yet effective means of allowing migrant eels to pass over dams. Our preliminary trial to assess the effectiveness of partial spillway opening had some success, but the effectiveness and cost of this procedure over other potential mitigation activities needs to be carefully assessed before it can be recommended as a viable means of mitigation.

Acknowledgments

We thank Trust Power Ltd, in particular Ron Coleman and Ian Paul, for their assistance throughout this study. We are also grateful to all those who assisted with field work. Grant Williams supplied the eels and provided much valued support. Thanks also to Brendan Hicks and Bruce Pease who provided valuable comments on an earlier draft of this paper. The Treaty of Waitangi Fisheries Commission, Tainui Development Limited and Mighty River Power provided travel funds. This study was funded by Foundation for Research, Science and Technology (New Zealand).

References

Adam, B., and D. Schwevers. 1997. Summary of the report on the behavioural survey of eels (*Anguilla anguilla*) migrating downstream under laboratory conditions. Institute of Applied Ecology, Neustädter Weg 25, 36320 Kirtorf-Whalen, Germany.

Anon. 1988. Lake Rotorangi Monitoring a new hydrolake. Taranaki Catchment Board, PO Box 159, Stratford, New Zealand.

Beentjes, M. P., Chisnall, B. L., Boubée J. A. T., and D. J. Jellyman. 1997. Enhancement of the New Zealand eel fishery by elver transfers. New Zealand Fisheries Technical Report 45.

Bell, M. C. 1986. Fisheries Handbook of Engineering Requirements & Biological Criteria. Fish Passage Development and Evaluation Program of Engineers, North Pacific Division Portland, Oregon.

Best, E. 1929. Fishing methods and devices of the Maori. Dominion Museum Bulletin 12:1–230.

Boubée, J. A. T., and C. P. Mitchell. 1994. The eels that climb over a dam. Water and Waste in New Zealand Sept: 23–26.

Boubée, J. A., Mitchell, C. P., Chisnall, B. L., West, D. W., Bowman., E. J., and A. Haro. (2001). Factors regulating the downstream migration of mature eels (*Anguilla* spp.) at Aniwhenua Dam, Bay of Plenty, New Zealand. New Zealand Journal of Marine, and Freshwater Research 35:121–134.

Burnet, A. M. R. 1969. Migrating eels in a Canterbury river, New Zealand. New Zealand Journal of Marine and Freshwater Research 3:230–244.

Burns, N. M. 1995. Results of Monitoring the Water Quality of Lake Rotorangi. NIWA Research Report for Taranaki Regional Council. TRC302. National Institute of Water and Atmospheric Research, Hamilton, New Zealand.

Cairns, D. 1941. Life-history of the two species of New Zealand fresh-water eel. Part I. Taxonomy, age and growth, migration, and distribution. New Zealand Journal of Science and Technology, series B, 23(2B):53b-72b.

Chisnall, B. L., and B. J. Hicks. 1993. Age and growth of longfinned eels (*Anguilla dieffenbachii*) in pastoral and forested streams in the Waikato River basin, and in two hydro-electric lakes in the North Island, New Zealand. New Zealand Journal of Marine and Freshwater Research 27:317–332.

Chisnall, B. L. 2000. The Australian longfinned eel, *Anguilla reinhardtii*, in New Zealand. Conservation advisory science notes, 302. Department of Conservation.

Deelder, C. L. 1954. Factors affecting the migration of the silver eel in Dutch inland waters. Journal du Conseil Permanent International pour l'Exploitation de la Mer 20: 177–185.

Desrochers, D. 1995. Suivi de la migration de l'anguille d'Amérique (*Anguilla rostrata*) au complexe Beauharnois, (1994). Milieu et Associes Inc., Montreal, Quebec, Canada.

Haro, A., Castro-Santos, T., and J. Boubée. 1999. Proceedings of the ICES/EIFAC meeting, 20-24 September, 1999, Silkeborg, Denmark.

Helfman, G. S., Stoneburner, E. L., Boseman, P. A., Whalen, C., and R. Whalen. 1983. Ultrasonic telemetry of American eel movements in a tidal creek. Transactions of the American Fisheries Society 112:105–110.

Hobbs, D. F. 1947. Migrating eels in Lake Ellesmere. Transactions of the Royal Society of New Zealand 77(5):228–232.

Jellyman, D. J., Chisnall, B. L., Dijkstra, L. H., and J. A. T Boubee. 1996. First record of the Australian longfinned eel, Anguilla reinhardtii, in New Zealand. Marine and Freshwater Research 47:1037–1040.

Jellyman, D. J., Graynoth, E., Francis, R. I. C. C., Chisnall, B. L., and M. P. Beentjes. 2000. A review of evidence for a decline in the abundance of longfinned eels (*Anguilla dieffenbachii*) in New Zealand. Final Research Report for Ministry of Fisheries Research Project EEL9802. National Institute of Water, and Atmospheric Research.

LaBar, G. W., and D. E. Facey. 1983. Local movements and inshore population sizes of American eels in Lake Champlain, Vermont. Transactions of the American Fisheries Society 112:111–116.

Larinier, M., and J. Dartiguelongue. 1989. La circulation des poissons migrateurs: le transit a travers les turbines des installations hydroelectrique. Bulletin Francais de la pêche et de la pisciculture 312.

Lowe, R. H. 1952. The influence of light and other factors on the seaward migration of the silver eel (*Anguilla anguilla* L.). Journal of Animal Ecology 21:275–309.

Maniapoto, H. 1998. Nature and extent of the Maori customary access and use of Tuna (*Anguilla* species) in the Ngati Maniapoto region (King Country). A Ministry of Fisheries Research Project. On behalf of Maniapoto Maori Trust Board, October 1998). PO Box 301 Te Awamutu, New Zealand.

McCleave J. D., and G. P Arnold. 1999. Movements of yellow- and silver-phase European eels (*Anguilla anguilla* L.) tracked in the western North Sea. ICES Journal of Marine Science 56:510–536.

McDowall, R. M. 1990. New Zealand Freshwater Fishes A Natural History and Guide. Heinemann Reed MAF Publishing Group, Wellington, New Zealand.

McDowall, R. M. 2000. The Reed Field Guide to New Zealand freshwater fishes. Reed publishing (NZ) Ltd, Auckland, New Zealand.

McGovern, P., and T. K. McCarthy. 1992. Local movements of freshwater eels (*Anguilla anguilla* L.) in western Ireland. In I.G. Priede and S.M. Swift editors, Wildlife Telemetry: Remote Sensing and Monitoring of Animals. Ellis Horwood, Chichester, United Kingdom:319–327.

Mitchell, C. P., and J. A. T. Boubée. 1992. Impacts of turbine passage on downstream migrating eels. New Zealand Freshwater Fisheries Miscellaneous Report 112:47.

Mitchell, C. P., and B. L. Chisnall. 1992. Problems facing migratory native fish populations in the upper Rangitaiki River system. Report to Bay of Plenty Electric Power Board. New Zealand Freshwater Fisheries Miscellaneous Report 119:21.

Pankhurst, N. W. 1982. Relation of visual changes to the onset of sexual maturation in the European eel *Anguilla anguilla* (L.). Journal of fish biology 21:279–296.

Parker, S. J. 1995. Homing ability and home range of yellow-phase American eels in a tidally dominated estuary. Journal of the Marine Biological Association of the United Kingdom 75:127–140.

Parker, S. J., and J. D. McCleave. 1997. Selective tidal stream transport by American eels during homing movements and estuarine migration. Journal of the Marine Biological Association of the United Kingdom 77:871–889.

Sloane, R. D. 1984. Preliminary observations of migrating adult freshwater eels (*Anguilla australis australis* Richardson) in Tasmania. Australian Journal of Marine and Freshwater Research 35:471–476.

Tesch, F. W. 1974. Speed and direction of silver and yellow eels, *Anguilla anguilla*, released and tracked in the open North Sea. Berichte der deutschen wisenschaftlichen Kommission für Meeresforchung 23:181–197.

Tesch, F. W. 1977. The eel: biology and management of anguillid eels. Chapman and Hall, London, England.

Tesch, F. W. 1978. Telemetric observations on the spawning migration of the eel (*Anguilla anguilla*) west of the European continental shelf. Environment Biology of Fishes 3:203–209.

Tesch, F. W. 1989. Changes in swimming depth and direction of silver eels (*Anguilla anguilla* L.) from the continental shelf to the deep sea. Aquatic Living Resources 2:9–20.

Todd, P. R. 1981. Timing and periodicity of migrating New Zealand freshwater eels (*Anguilla* spp.). New Zealand Journal of Marine and Freshwater Research 15:225–235.

Travade, F., and M. Larinier. 1992. La migration de devalaison: problèmes et dispositifs. Bulletin Français de la pêche et de la piciculture 326:165–176.

Vøllestad, L. A., Jonsson, B., Hvidsten, N. A., Naesje, T. F., Haraldstad, Ø., Ruud-Hansen J. 1986. Environmental factors regulating seaward migration of European silver eels (*Anguilla anguilla*). Canadian Journal of Fisheries and Aquatic Sciences 43:1909–1916.

Surface and Midwater Trawling for American Eels in the St. Lawrence River

KEVIN J. MCGRATH*

New York Power Authority, 123 Main Street, White Plains, New York 10601, USA
*Corresponding author: mcgrath.k@nypa.gov

JOSEPH W. DEMBECK IV

Kleinschmidt Associates, 75 Main Street, Pittsfield, Maine 04967, USA

JAMES B. MCLAREN, ALAN A. FAIRBANKS, AND KEVIN REID

Stantec, 140 Rotech Drive, Lancaster, New York 14086, USA

STEPHEN J. CLUETT

State University of New York at Stony Brook, 139 Discovery Hall, Stony Brook, New York 11794, USA

Abstract.—The feasibility of trawling was investigated as a technique to capture downstream migrating adult American eels in the St. Lawrence River near Massena, New York. The surface/midwater trawl, with a 9.1 m horizontal × 7 m vertical mouth opening, was deployed from June through August of 1999 from both a single tow vessel and paired tow vessels. The gear was successful in catching a total of 34 mature, migrating eels. All eels were alive and in good condition when caught. More eels were caught by single-boat trawling (23 eels in 121 tows) than by paired-boat trawling (11 eels in 133 tows). More eels were caught with midwater tows then with surface tows. Effective tow speeds ranged from 1.6 to 2.3 m/s. The highest catch rates were associated with the dark phases of the moon and periods of greater cloud cover.

Introduction

The American eel *Anguilla rostrata* is a catadromous panmictic species that spawns in the Sargasso Sea (Schmidt 1923; Avise et al. 1986). Leptocephali larvae utilize oceanic currents to distribute themselves along the Atlantic and Gulf coasts (McCleave and Kleckner 1987). As the leptocephali enter continental waters they transform into glass eels (Wang and Tzeng 1998). The glass eels transform into elvers, which migrate into rivers, lakes and tributaries where they develop and mature (Dutil et al. 1989; Haro and Krueger 1991).

In the St. Lawrence River many eels migrate into Lake Ontario and its tributaries, a distance of greater than 800 km. Yellow, resident eels live in the River or Lake Ontario for 8–15 years, then migrate back to the Atlantic Ocean as maturing yellow or silver eels, to complete the cycle (Casselman 2001). Eels utilize an eel ladder to pass upstream of the Moses-Saunders Power Dam at the St. Lawrence-FDR Power Project at Massena, New York (Casselman et al. 1997). It is thought that almost all downstream migrating eels must pass through the Power Dam turbines in their return migration to the Sargasso Sea.

The eel ladder at the Moses-Saunders Power Dam was constructed in 1974. Average annual counts at the ladder were about 890,000 eels through 1985 but have declined markedly since that time (Castonguay et al. 1994; Casselman et al. 1997) and have been less than 4,000 over the past three years (Casselman 2001).

The New York Power Authority has been conducting studies on American eel as part of the relicensing of the St. Lawrence-FDR Power Project. These studies have focused on learning the relative abundance and distribution of juvenile eels as they approach the Power Dam (McGrath et al.[1], this volume) and on learning

---

[1]McGrath, K. J., D. Desrochers, C. Fleury, and J. W. Dembeck IV

the when, where, and how of downstream migrants in the vicinity of the Power Dam. The ultimate goal is to try to pass eels safely around the Dam. The primary means of gathering information on the downstream migrants has focused on telemetry studies (McGrath et al.[2], this volume). As part of these efforts there was a need to collect downstream migrants.

The objective of this study was to determine if a midwater/surface trawl could be used as a method for capturing downstream migrating American eels in the upper freshwater portion of the St. Lawrence River near Massena, New York. Trawling has not been used commercially to capture silver eels in the St. Lawrence River; most are collected in the estuarine portions of the River using large weirs (Eales 1968). Work completed in previous years in the same area indicated that electrofishing, hoop netting, and eel pots were effective for collecting resident eels but were ineffective for collecting downstream migrating eels. We believed that migrating eels were probably in the main part of water channel in the mid to upper portions of the water column (Tesch 1983 and 1994 in Thon 1999) and, therefore, were inaccessible to these sampling gears. The trawl was selected as a gear capable of sampling this habitat.

The specific objectives for the 1999 investigation of trawling techniques were to: design trawling gear that could be effectively used in Lake St. Lawrence; develop techniques for deploying, retrieving, and general handling of the trawls; capture migrating eels; and determine the biological characteristics of all eels.

Methods

The study area for the investigation was the section of Lake St. Lawrence from immediately above the Moses-Saunders Power Dam (near Massena, New York) upstream to the Long Sault Islands, approximately 8 km (Figure 1).

The net used for sampling was a 9.1 m × 7.0 m × 33.5 m modified French midwater trawl (Figure 2). The vessel used for single-boat trawling was a 24.4 m, 500 hp vessel. A 21.3 m, 500 hp vessel was used in conjunction with the primary vessel for paired boat trawling. During single boat trawling, 1.5 m Superkrub midwater trawl doors were attached to the sweeplines of the net. Development of trawl deployment and fishing techniques involved the adjustment of the trawl gear (net floats, net weights, trawl doors, etc.) to achieve optimal performance in the relatively shallow waters (< 35 m) of the study area. The objectives for gear testing were to reduce gear avoidance and increase gear efficiency by maintaining a large net mouth opening (6.1–6.7 m) and a fast tow speed (> 1.5 m/s: 0.8 m/s to 1.8 m/s over land speed into a 0.5–1.0 m/s current). The performance of the net was monitored electronically with a trawl monitoring system (Scanmar of Norway depth sensor [model HC4], trawl sounder [model HC4-TS150], and processor [model RX 400]).

Trawling was conducted over a five week period from early July to early August. Migration through this portion of the St. Lawrence River occurs from mid-June through mid-September with a broad peak during July and August (McGrath et al.[3], this volume). All sampling was conducted at night, between 0.5 hours after sunset and 0.5 hours before sunrise (Vøllestad et al. 1986; Eales 1968). Sampling was conducted on four nights each week (Monday-Thursday). The sampling design consisted of two tows per night at each of six sampling stations, one tow at the surface and one at middepth. Trawl direction was upstream and sample duration was 10 minutes. Paired boat sampling occurred on the first two nights and single boat sampling occurred on the last two nights. The sampling station serving as the starting point during a particular night was chosen randomly. The order of subsequent sampling followed the sequence of stations either clockwise or counterclockwise (determined randomly) around the study area

External morphological characteristic data recorded from all collected eels included: total length; total weight; girth; color of caudal fin and skin on dorsal, lateral and ventral surfaces; pectoral fin size and coloration; pectoral fin aspect ratio (Tesch 1977); gape length (Wenner 1973; Wenner and Musick 1974); horizontal and vertical diameter of the eye; head width; presence/absence of food in the stomach; and anal condition (tightness of the anus indicating atrophy of the gut).

---

[2]McGrath, K. J., S. Ault, K. Reid, D. Stanley, and F. Voegeli

[3]McGrath, K. J., J. Bernier, S. Ault, J.-D. Dutil, and K. Reid

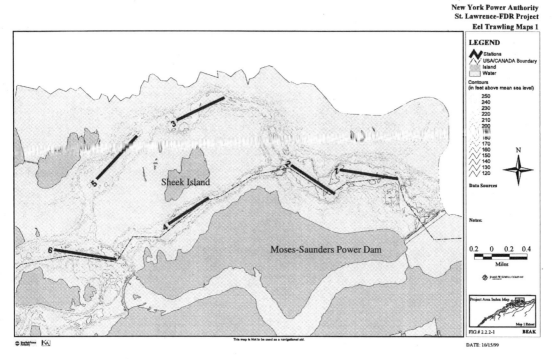

Figure 1. Study area and trawling stations in the vicinity of the Moses-Saunders Power Dam, St. Lawrence River, Massena, New York.

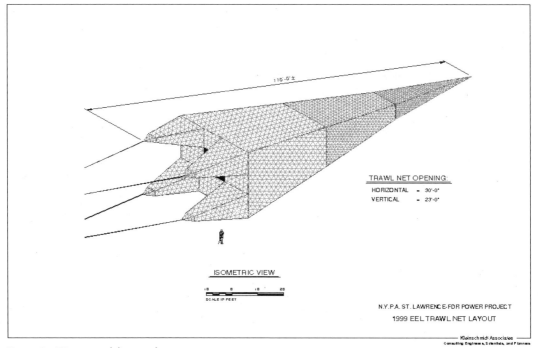

Figure 2. Diagram of the trawl net.

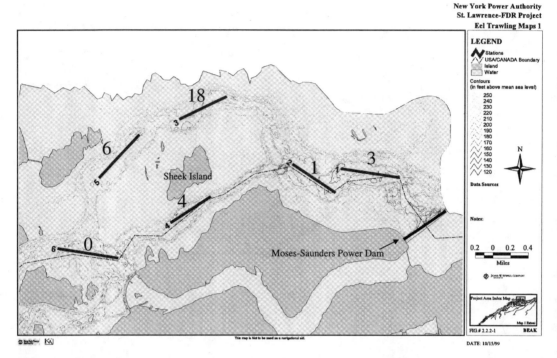

Figure 3. Number of eels collected at each trawling station. (Note: two additional eels were collected at a trawling station 45 km upstream.)

Results

A total of 34 eels were captured in 254 tows (Figure 3; catch rate of 0.082 eels/10–minute tow). Single-boat trawling was more effective than paired-boat trawling with 23 eels in 121 tows versus 11 eels in 133 tows (catch rates of 0.173 eels/10-minute tow for single boat trawling and 0.080 eels/10-minute tow for paired-boat trawling). The majority (71%) of all eels were captured at two stations (#3 and #5), both located along the north side of Sheek Island. These two stations had catch rates of 0.353 and 0.120 eels/10-minute tow, respectively, while the remainder of the stations ranged from 0.095 to 0.000 eels/10-minute tow. The highest catch rates were observed at some of the deepest depths sampled, which was approximately middepth in the water column for most of the sampling stations (Figure 4). The majority (92%) of eels collected was larger than 900 mm (Figure 5). The highest catch rates were observed during the darker phases of the moon and when cloud cover was greatest (Figure 6 and Figure 7, respectively).

Discussion

Trawling proved to be an effective means of capturing eels from the mid to upper portion of the water column, an area not accessible by other gear types. Overall, the catch rate was relatively low, however one sampling location did have a high enough catch rate (0.353 eels/10-minute tow) to consider it feasible to collect larger number of eels if a considerable amount of trawling effort was expended. It is not known if this location is more productive due to a greater abundance of eels or if some unique bathymetric feature or current/flow condition concentrated eels and made them more susceptible to capture.

Based upon the overall large size (> 800 mm) of the eels collected (Hurley 1972; Facey and LaBar 1981) and their silver-like appearance (whitish ventral surface, some bronzing along the lateral line and gill covers, longer pectoral fins and somewhat enlarged eyes) they were presumed to be mature migrants (i.e., silver eels; Pankhurst 1982; Winn et al. 1975).

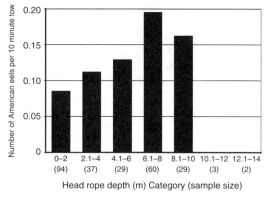

Figure 4. Catch per unit effort at specified trawl depths.

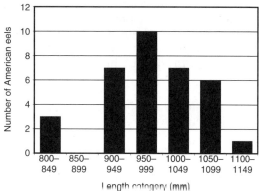

Figure 5. Length frequency of eels collected.

The highest catch rates were observed at middepth (6–10 m), however we are uncertain as to the significance of this finding since bottom obstructions precluded sampling close to and along the bottom (depth approximately 18–24 m), making comparison throughout the entire water column impossible. Tesch (1983) reported that silver European eel *A. anguilla* migrate primarily at intermediate depths while others have indicted that they are bottom oriented (Lowe 1952). Haro et al. (2000) reported that the American eel can move at a variety of depths and can quickly alter their depth of travel. We anticpated that eels migrating near the surface would avoid the trawling vessel during single boat trawling and, therefore, paired trawling would have been a more effective deployment technique. However, we observed just the opposite. The fact that the highest catch rates occurred at middepth may indicate that either more eels were migrating at these depths or that migrating eels traveling near the surface sounded at the approach of the trawling vessel, regardless of trawl deployment technique, and were captured deeper in the water column as a result of the avoidance behavior.

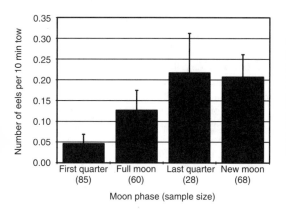

Figure 6. The relationship between moon phase and the mean catch of American eels per 10-minutes trawl tow in the St. Lawrence River at stations 1–6 during 1999, with the standard error bars.

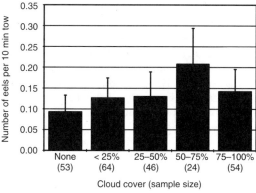

Figure 7. The relationship between cloud cover and the mean catch of American eels per 10-minutes trawl tow in the St. Lawrence River at stations 1–6 during 1999, with the standard error bars.

We noted that the catch rates were somewhat higher during the dark phases of the moon and when cloud cover was greatest. Many other researchers have noted similar patterns (Winn et al. 1975; Tesch 1977), but this correlation needs to be further investigated with additional sampling that spans two or more lunar cycles during the out-migration season and periods with variable cloud cover.

Acknowledgments

We would like to thank the New York Power Authority for support and funding; Tom Tatham of the New York Power Authority for support and sound scientific guidance; Scott Ault of Kleinschmidt Associates for all his assistance in study planning, coordination of study logistics, and review of this paper; Jim Minor, Davey Fidler, and Henry Chusing, the crew members of Dan Minor and Sons for all their efforts and long hours in trawling; and Steve LaPan of New York State Department of Environmental Conservation and Pat Geer of the Virginia Institute of Marine Science for their technical comments on an earlier draft of this paper. Lastly we want to thank Doug Dixon of EPRI for his never ending patience in allowing us to finalize this paper.

References

Avise, J. C., G. S. Helfman, N. C. Saunders, and L. S. Hales. 1986. Mitochondrial DNA differentiation in North Atlantic eels: population genetic consequences of an unusual life history pattern. Proceedings of the National Academy of Science. USA 83:4250–4354.

Casselman, J. M. 2001. Dynamics of American Eel, *Anguilla rostrata*, Resources: Declining Abundance in the 1990s. International Symposium, Advances in Eel Biology, held at the University of Tokyo, Tokyo, Japan, September 2001.

Casselman, J. M., L. A. Marcogliese, and P. V. Hodson. 1997. Recruitment index for the Upper St. Lawrence River and Lake Ontario Eel Stock: a Re-examination of Eel Passage at the R.H. Saunders Hydroelectric Generating Station at Cornwall, Ontario, 1974–1995. *In* R. H. Peterson, editor. The American Eel in Eastern Canada: Stock Status and Management Strategies, Proceedings of Eel Management Workshop, January 13–14, 1997, Quebec City, Quebec, Canada.

Canadian Technical Report on Fishery Aquatic Science 2196: v + 174 pp.

Castonguay, M., P. V. Hodson, C. M. Couillard, M. J. Eckersley, J. -D. Dutil, and G. Verreault. 1994. Why is recruitment of the American eel, *Anguilla rostrata*, declining in the St. Lawrence River and Gulf? Canadian Journal of Fisheries and Aquatic Sciences 51:479–488.

Dutil, J.-D., M. Michaud, and A. Giroux. 1989. Seasonal and diel patterns of stream invasion by American eels, *Anguilla rostrata*, in the northern Gulf of St. Lawrence. Canadian Journal of Zoology 67:182–188.

Eales, J. G. 1968. The Eel Fisheries of Eastern Canada. Fisheries Research Board of Canada, Bulletin 166, Ottawa, Ontario, Canada.

Facey, D. E., and G. W. LaBar. 1981. Biology of American eels in Lake Champlain, Vermont. Transactions of the American Fisheries Society 110:396–402.

Haro, A. J., and W. H. Krueger. 1991. Pigmentation, size and migration of elvers, *Anguilla rostrata*, in a coastal Rhode Island stream. Canadian Journal of Zoology 69:812–814.

Haro, A., T. Castro-Santos, and J. Boubee. 2000. Behavior, and passage of silver-phase American eels, *Anguilla rostrata* (LeSueur), at a small hydroelectric facility. Dana 12:33–42.

Hurley, D. A. 1972. The American eel (*Anguilla rostrata*) in eastern Lake Ontario. Journal of the Fisheries Research Board of Canada 29:535–543.

Lowe, R. H. 1952. The influence of light and other factors on the seaward migration of the silver eel, *Anguilla anguilla* L. Journal of Animal Ecology 21:275–309.

McCleave, J. D., and R. C. Kleckner. 1987. Distribution of leptocephali of the catadromous anguilla species in the Western Sargasso Sea in relation to water circulation and migration. Bulletin Marine Science 41(3):789–806.

Pankhurst, N. W. 1982. Relation of visual changes to the onset of sexual maturation in the European eel *Anguilla anguilla* (L.). Journal of Fish Biology 21:127–140.

Schmidt, J. 1923. The breeding places of the eel. Philosophical Transactions of the Royal Society of London Series B - Biological Sciences 211:179–208.

Tesch, F. W. 1977. The eel. Chapman and Hall, London, England.

Tesch, F. W. 1983. Der Aal. –Verlag Paul Parety, Hamburg, Germany.

Tesch, F. W. 1994. Verfolgung von Blankaalen in Weser und Elbe. –Fischokologie 7:47–59.

Thon, M. 1999. Protection of Migrating Silver Eels (Anguilla anguilla L.) in Regulated Rivers—literature study –Hamburg – translated by M.N.L. Seaman –Eel Consevation Initiative Rheinland-Pfalz (Germany)/RWE Energie AG.

Vøllestad, L. A., B. Jonsson, N. A. Hvidsten, T. G. Naesje, O. Haraldstad, and J. Rudd-Hansen. 1986. Environmental factors regulating the seaward migration of European silver eels (*Anguilla anguilla*). Canadian Journal of Fisheries and Aquatic Science 43:1909–1916.

Wang, C-H., and W-N. Tzeng. 1998. Interpretation of geographic variation in size of American eel *Anguilla rostrata* elvers on the Atlantic coast of North America using their life history and otolith ageing. Marine Ecology Progress Series 168:35–43.

Wenner, C. A. 1973. Occurrence of American eel (*Anguilla rostrata*) in waters overlying the eastern North American continental shelf. Journal of the Fisheries Research Board, Canada 30:1752–1755.

Wenner, C. A., and J. A. Musick. 1974. Fecundity and gonad observation of the American eel, *Anguilla rostrata*, migrating from Chesapeake Bay, Virginia. Journal of the Fisheries Research Board, Canada 31:1391.

Winn, H. E., W. A. Richkus, and L. K. Winn. 1975. Sexual dimorphism and natural movements of the American eel (*Anguilla rostrata*) in Rhode Island streams and estuaries. Helgolander Wissenschaftliche Meeresuntersuchungen 27:156–166.

Differentiating Downstream Migrating American Eels *Anguilla rostrata* from Resident Eels in the St. Lawrence River

KEVIN J. MCGRATH*

New York Power Authority, 123 Main Street, White Plains, New York 10601, USA
*Corresponding author: mcgrath.k@nypa.gov

JULIE BERNIER**

Environnement Illimite inc., 1453 Saint-Timothee, Montreal, Quebec H2L 3N7, Canada

SCOTT AULT

Kleinschmidt Associates, 33 Main Street West, Strasburg, Pennsylvania 17579, USA

JEAN-DENIS DUTIL

Institut Maurice-Lamontagne, 850 route de la Mer, C.P. 1000, Mont Joli,
Quebec G5H 3Z4, Canada

KEVIN REID***

Stantec, 14 Abacus Road, Brampton, Ontario L6T 5B7, Canada

Abstract.—Studies were conducted in 1998 and 1999 to understand the maturation process of American eels *Anguilla rostrata* in the St. Lawrence River. Known and presumed migrants and immature eels were sampled from May through October over a 550 km range extending from Ogdensburg, New York to Kamouraska, Quebec, Canada. Morphological and physiological characteristics were measured from 720 eels. A cluster analysis was performed on the physiological parameters (gonadosomatic index, oocyte developmental stages and oocyte diameter) and eels were separated into four clusters representative of advancing maturation stages. Three of the clusters were combined, representing immature resident eels while the remaining cluster formed the mature migrating group. A discriminant analysis was then used to differentiate these two groups on the basis of five external morphological characteristics: length, weight, girth, ocular index, and pectoral fin index. The discriminant functions resulted in the correct classification of 87% of the mature migrants and 83% of the immature resident eels. Detailed morphological and physiological characteristics are described for each of the maturation stages of American eel in the St. Lawrence River.

Introduction

American *Anguilla rostrata* and European eels *A. anguilla* undergo a complex series of morphological and physiological changes as they transform from their freshwater yellow form to their seagoing silver form in preparation for migration and spawning in the Sargasso Sea. While

---

**Current address: Fisheries and Oceans Canada, Maurice Lamontagne Institute, 850 route de la Mer, P.O. Box 1000, Mont-Joli, Québec G5H 3Z4, Canada.

***Current address: Ontario Commercial Fisheries Association, Box 2129, 45 James Street, Blenheim, Ontario N0P 1A0, Canada.

many of the morphological and physiological changes have been documented (Winn et al. 1975; Tesch 1977; Han et al. 2001), the complete maturation process in natural populations has not been described for American eel. Some of the morphological changes associated with the maturation process for *A. rostrata* or *A. anguilla* include a change in the skin color from a yellow to silver or bronze (Tesch 1977; Eales 1968; Bouillon and Haedrich 1985), thickening of the integument (Tesch 1977; Pankhurst and Lythgoe 1982), increased eye diameter (Vladykov 1973; Wenner and Musick 1974; Winn et al. 1975; Pankhurst 1982), increased length and weight (Hurley 1972), and relative lengthening of the pectoral fins

(Tesch 1977; Durif et al. 2000). Physiological changes include: degeneration of the alimentary canal (Pankhurst and Sorensen 1984; Durif et al. 2000); an increase in gonadal weight relative to the body weight (Jessop 1987; Durif et al. 2000); an increase in oocyte diameter (Dutil et al. 1985); change in the gill structure and cell composition (Dutil et al. 1987; Fontaine et al. 1995); changes in the muscle properties (Ellerby et al. 2001); an increase in the later developmental stages of oocytes (Columbo et al. 1984; Couillard et al. 1997); and biochemical changes (Lewander et al. 1974; Johansson et al. 1974; Dave et al. 1974; Durif et al. 2000; Cottrill et al. 2001).

The New York Power Authority has been conducting studies on American eel as part of the relicensing of the St. Lawrence-FDR Power Project near Massena, New York (Power Project). These studies have focused on learning the relative abundance and distribution of juvenile eels as they approach the Power Project (McGrath et al.[1], this volume) and on learning the basic migratory biology of downstream migrants in the vicinity of the Power Project. The purpose of this study was to develop criteria to differentiate maturing outmigrant American eels from resident yellow eels and to understand the morphological, physiological and gonadal changes that occur in maturing American eels in the St. Lawrence River. The need to understand the maturation process and the need to differentiate resident yellow eels from maturing outmigrating eels evolved out of a need to utilize only downstream migrants for a telemetry study at the Power Project (McGrath et al.[2] and McGrath et al.[3], both this volume).

In the upper freshwater portion of the St. Lawrence River, all eels are female (Hurley 1972; Dutil et al. 1985; Couillard et al. 1997) and occur in different stages of maturity ranging from immature resident yellow eels to maturing migrant silver eels. Dutil et al. (1985) and Couillard et al. (1997) utilized oocyte diameter and oocyte stage development to determine the maturity status of eels. This required sacrificing the animals. However, because of the need to minimize handling effects and to use only downstream migrants as opposed to resident eels in our planned telemetry study, it was necessary to develop methods to identify maturity status based upon external morphological characteristics.

Methods

Eel Collection

For this differentiation study we collected a large number of eels in different stages of maturity and measured a broad array of external morphological and internal physiological parameters, as subsequently described. These eels were captured over a broad geographical area and over a time period spanning their presumed migration in this portion of the St. Lawrence River. Eels ($n = 720$) were collected from August through October in 1998 and May through October in 1999 (Table 1), bracketing the downstream migration period of American eels in the St. Lawrence River (see Results section, Figure 2). Eels were collected at eight sites distributed over a 550 km long reach of the St. Lawrence River comprising the freshwater fluvial portion and the brackish middle estuary (Figure 1). Eels were collected in Lake St. Lawrence using electrofishing and hoop nets in shallow areas and trawling in deeper portions (McGrath et al.[3], this volume). The Lake St. Lawrence eels were expected to show a broad range of maturity given the habitats sampled, the extended sampling period and because maturing eels from Lake Ontario and the upper St. Lawrence River migrate through this region as they progress towards the Sargasso Sea. Presumed immature eels were purchased from commercial eel fishers using hoop nets in shallow portions of Lake St. Francis and presumed mature migrating silver eels were purchased from commercial eel fishers using weirs in the estuarine portion of the lower St. Lawrence River near Quebec City and Kamouraska, Quebec, Canada. Additionally, dead and injured eels that had passed through the hydroelectric turbines of the Moses-Saunders Power Dam were systematically (once per week along a prescribed route) collected by hand netting in the tailwaters of the Dam (NYPA 2000).

Eels from Lake St. Lawrence, Quebec City and Kamouraska were collected using a stratified random design. To the extent possible, twenty eels were collected in the following length categories 700–750, 800–850 and 900–950 mm total length from each of the collection regions. This was done to provide comparable size eels from each region (eel length has been associated

---

[1]McGrath, K. J., D. Desrochers, C. Fleury, and J. W. Dembeck IV.

[2]McGrath, K. J., S. Ault, K. Reid, D. Stanley, and F. Voegeli.

[3]McGrath, K. J., J. W. Dembeck IV, J. B. McLaren, A. A. Fairbanks, K. Reid, and S. J. Cluett.

Table 1. Collection location, method of capture, date, and number of eels used for analysis of morphological and physiological parameters.

Location of capture	Method of capture	Study site	Year	Number of specimens collected						
				May	June	July	Aug.	Sept.	Oct.	Total
Chimney and Galop Island	Electrofishing	Lake St. Lawrence (shallow)	1998				155	59	69	283
Chimney and Galop Island	Electrofishing	Lake St. Lawrence (shallow)	1999	35	30	21	15			101
Sheek Island (shallow areas)	Electrofishing Hoop nets, Eel pots	Lake St. Lawrence (shallow)	1999	25	31	20	10	7		93
Sheek Island (channels)	Trawling	Lake St. Lawrence (Trawling)	1999				16	11		27
Downstream	Hand netted	Moses Saunders Tailwater	1999		6	11	25	8		50
Lake St. Francis	Hoop nets	Lake St. Francis	1999					19		19
Quebec City	Weir	Quebec City	1998					67		67
Quebec City	Weir	Quebec City	1999					20		20
Kamouraska	Weir	Kamouraska	1998						60	60
Total				60	67	68	216	180	129	720

Note: The number of eels used in the cluster analysis totaled 689. For 31 eels, one or more of the necessary physiological parameters was missing.

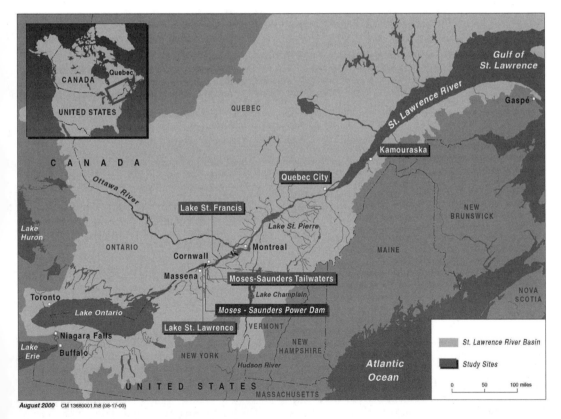

Figure 1. Stations location and general study area.

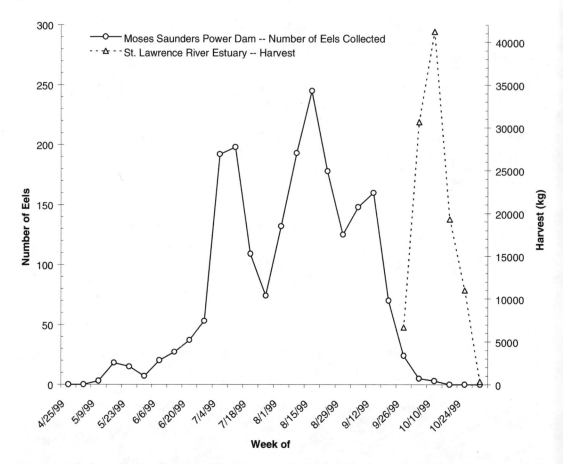

Figure 2. Number of eels collected in the tailwater at the Moses-Saunders Power Dam and weight of eels harvested in the commercial fishery in the St. Lawrence Estuary (Verreault et al. 2000), by week. The tailwater eels were obtained by systematically collecting dead or injured eels in the tailwater (NYPA 2000).

with maturity/migration [Hurley and Christie 1982; Columbo and Grandi 1996]) and to ensure a large enough sample size in each length group for statistical analyses. The eels from Lake St. Francis were collected specifically to provide some eels known to be immature residents. They were all of a smaller size. The eels collected by trawling and in the tailwaters were based upon what could be collected; most were thought to be maturing migrants.

Study Parameters

Eels were sacrificed by immersion in a 10 ppm solution of clove oil/ethanol in water up to one hour (Woody et al. 2002). This method resulted in minimal stress and is not known to affect any of the physiological or morphological characteristics used in this study.

Morphological parameters measured on each eel included: total length (TL), weight (Tw), girth (GL), pectoral fin length (Pf), and eye diameter (Ev: vertical and Eh: horizontal). The pectoral fin length was used to calculate the pectoral fin index (Pi) according to the formula (Durif et al. 2000):

$$Pi = \left(\frac{Pf}{TL}\right) \times 100$$

The eye diameter measurements were used to calculate an ocular index (Ei) using Pankhurst's (1982) formula:

$$Ei = \left(\frac{Ev + Eh}{4}\right)^2 \times \frac{\pi}{TL} \times 100$$

The following physiological parameters were measured: ovary weight (Ow), ovary texture, oocyte diameter, four stages of oocyte development, and stomach content weight (Sc). A sample from the center of the left ovary was fixed in 10% phosphate-buffered formalin; embedded in paraffin; and 5 μm thick sections were prepared and examined. The maturity stage of oocytes was determined following Couillard et al. (1997), examining 200 randomly selected oocytes. Oocyte diameter was determined by measuring and averaging the 10 most mature oocytes. The texture of the ovary tissue was described using three categories, smooth, filamentous and granular (Buellens et al. 1997; Colombo and Grandi 1996).

The eels from the furthest downstream sampling locations (Quebec City and Kamouraska) had no food in their stomachs and their stomachs had undergone considerable atrophy; therefore, the gonadosomatic index (GSI) for all eels was calculated on a stomach free body mass as well as a gonad free body mass. This avoided bias due to stomach size and stomach contents and differences in degrees of maturity. The GSI was calculated using the following formula:

$$GSI = [Ow/(Tw - Ow - Sc)] \times 100.$$

Minor differences between the 1998 and the 1999 histological preparations of the gonadal tissue resulted in slight differences in oocyte shrinkage. Thus, an adjustment factor was applied to the 1998 oocyte diameter data (36.893 added to each of the values) to make these data consistent with the 1999 data.

Data Analyses

The analytical approach involved two major steps. First cluster analysis, based upon physiological parameters, was used to separate the eels in specific groupings showing varying degrees of maturity. Physiological parameters, particularly oocyte diameter and oocyte stage development, are considered to provide the most reliable measures of maturity (Dutil et al. 1985; Colombo and Grandi 1996; Couillard et al. 1997). The cluster analysis allowed us to quantitatively establish a cutting point to differentiate maturing migrants from immature resident eels. Finally, discriminant analysis was used to develop functions to differentiate maturing migrants from immature residents based upon external morphological characteristics.

Preliminary statistical analyses were conducted using SYSTAT software (SPSS 1999) while the cluster and discriminant analyses were performed using SAS software (SAS 2002).

Cluster analysis (SAS Procedure FASTCLUS; SAS 2002) was performed on 689 eels that had a complete set of all six physiological parameters (oocyte diameter, GSI and the percentage of the four stages of oocyte development). The data were standardized prior to analysis (Legendre and Legendre 1998). The procedure was repeated 100 times from random starting configurations for partition into $k = 2$ to 10 groups. The minimum pseudo F-statistic was used as the measure of the best partitioning in a least squares sense (Legendre and Legendre 1998). Discriminant analysis (SAS Procedure DISCRIM) was then conducted to derive discriminant functions based on external morphological characteristics for the two maturity stages derived from the cluster analysis. Length, weight and girth were log10 transformed while pectoral ratio and the ocular index were not transformed since preliminary analyses showed they met the appropriate statistical criteria.

A detailed examination of the data showed that the eels collected from Quebec City and Kamouraska differed substantially from the eels collected from the upstream sites (Figure 3). The eels from Quebec City and Kamouraska were significantly ($p < 0.05$) shorter, had significantly ($p < 0.05$) smaller girths and weighed significantly ($p < 0.05$) less than those from upstream. At the same time, the eels from Quebec City and Kamouraska exhibited higher indicators of maturity, including significantly ($p < 0.05$) greater average oocyte diameter and significantly ($p < 0.05$) higher GSI. It is thought that a portion of the eels collected in Quebec City and Kamouraska come from multiple drainages (Ottawa River, Richelieu River and other downstream tributaries of the St. Lawrence River) and that these eels may be smaller in size than those emanating from Lake Ontario and the upper St. Lawrence River. Differences in average size of female silver American eels and female silver European eels emigrating from different drainages have been documented by Bouillon and Haedrich (1985) and Svedang et al. (1996). Although the Quebec City and Kamouraska eels were the most mature based upon physiological parameters, these data would have confounded the subsequent discriminant analysis because of their smaller size (i.e., length, width, weight). Because of this difference, eels from Quebec City and Kamouraska collection areas were excluded from the discriminant analysis.

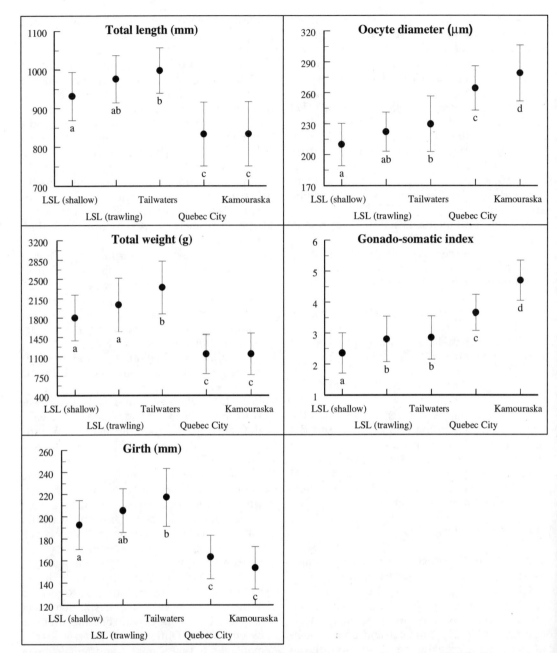

Figure 3. Mean values for morphological and physiological parameters of eels collected from five locations on the St. Lawrence River. Standard deviation (+/–) depicted by vertical error bars. The letter under each plot indicates significant different groups (Tukey, $p < 0.05$).

Results

In 1999, eels migrated through the upstream freshwater portion of the St. Lawrence River from the end of June through the beginning of October while the primary period of migration in the lower estuarine portion of the River (approximately 550 km downstream) occurred in October (Figure 2). No comparable data existed for 1998; however, annual collections in the Power Project tailwaters in 2000–2002 (NYPA 2001, 2002a, 2002b) each show a very similar trend, indicating that sampling for this study was conducted at the appropriate time to ensure collection of maturing migrants.

Cluster Analysis

The first step in the cluster analysis was to determine the optimum number of clusters. Based on 100 random starts for each number of clusters (2–10), it was determined that the optimum number of clusters in the least squares sense was 4 (Table 2).

The groupings of the cluster analysis are shown in Figure 4, the top portion of the figure shows the results of the groupings for all sampling sites (4a), while the smaller figures show the results for the individual collection locations (4b–g). The top left portion of Figure 4a contains the least mature eels (clusters 1 and 2), primarily those from Lake St. Francis and a portion of those from Lake St. Lawrence. These eels had lower GSI's, lower oocyte diameters and substantially fewer stage 3 and stage 4 oocytes (Table 3). The third cluster, in the middle of the Figure 4, is comprised largely of the resident eels from Lake St. Lawrence and a portion of the eels collected in trawling and the tailwaters. These eels had higher GSI's, higher oocyte diameters and many of the eels had a substantial portion of oocytes in stage 3 development. The eels from clusters 1–3 were considered to be immature residents and were combined into an immature resident category for the subsequent discriminant analysis. The fourth cluster was comprised largely of eels from Quebec City and Kamouraska with some eels from trawling, the tailwater collections and Lake St. Lawrence. They were all considered to be mature migrants for the discriminant analysis. These eels showed a substantial portion of stage 3 and 4 oocytes; had the largest oocyte diameters and had the highest GSI values (Table 3).

Discriminant Analysis

Discriminant analysis was then performed, based on morphological parameters, on the remaining eels (539), exclusive of the eels from Quebec City and Kamouraska (see above). The derived discriminant functions are presented in Table 4. In application, the appropriate morphological parameters of each eel are inputted into both functions. The function that results in the highest value designates the appropriate maturation category (immature or mature).

The discriminant functions were then applied to the 539 eels and the results were compared to findings from the cluster analysis. The discriminant functions accurately classified 87% of the mature migrants, while 83% of the immature resident eels were correctly classified (Table 5).

Discussion

Seasonal Migration Pattern

Maturing eels migrate through the upper part of the St. Lawrence River over a broad time period, primarily from July through September. A similar broad summer migration has also been observed in the Richelieu River, a tributary of the St. Lawrence River approximately 150 km downstream from Lake St. Lawrence (Facey and Helfman 1985). When these eels migrate through the estuarine portions of the River, approximately 550 km downstream of Lake St. Lawrence, they move through in a much shorter period of time, approximately a month, primarily in October (Verreault et al., this volume). This would suggest that as eels migrate down the St. Lawrence River and its tributaries that some staging may be occurring, perhaps as the eels undergo some of the physiological changes prior to movement from freshwater to saltwater. However, no concentration of eels that one would associate with a staging area has ever been found.

Comparison to Other Studies

Hurley (1972), in his tagging studies on eels migrating from Lake Ontario, determined that most migrating eels were greater than 830 mm.

Table 2. Maximum pseudo-F statistic from 100 iterations of the cluster analysis, each iteration based upon a different random starting configuration.

Number of Clusters	Maximum Pseudo-F
2	615.56
3	850.63
4	1042.93*
5	965.93
6	946.70
7	899.79
8	862.23
9	824.50
10	792.59

*maximum value at 4 clusters.

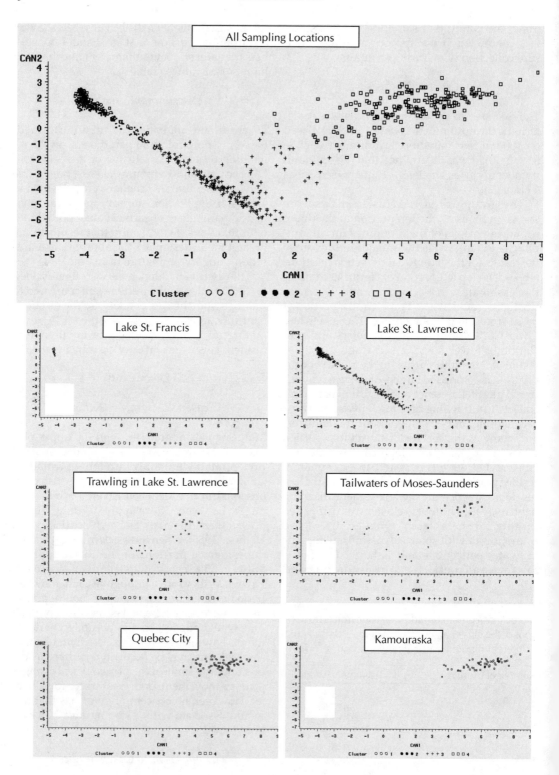

Figure 4. Plots of cluster analysis for all sampling locations (graph a) and individual sampling locations (graphs b–g). Clusters 1–3 are immature resident eels, cluster 4 are mature migrant eels.

Table 3. Mean, standard deviation and range of morphological and physiological characteristics of the four clusters of eels collected in the St. Lawrence River, excluding those from Quebec City and Kamouraska.

	Immature			Mature
	Cluster 1 ($n = 119$)	Cluster 2 ($n = 177$)	Cluster 3 ($n = 190$)	Cluster 4 ($n = 53$)
Total length (mm)	750.3 ± 83 (556 – 934)	848 ± 102 (537 – 1098)	912 ± 72 (740 – 1081)	976 ± 70 (824 – 1154)
Girth (mm)	131 ± 19 (92 – 176)	158 ± 26 (89 – 224)	181 ± 24 (130 – 251)	208 ± 21 (164 – 245)
Total weight (g)	780 ± 261 (248 – 1361)	1245 ± 470 (237 – 3011)	1623 ± 464 (1036 – 3330)	2110 ± 530 (1013 – 3350)
Ocular index	5.5 ± 1.0 (2.8 – 7.8)	6.2 ± 1.3 (3.0 – 11.2)	7.1 ± 1.4 (4.3 – 12.2)	8.7 ± 1.4 (5.4 – 12.4)
Pectoral fin index	4.2 ± 0.4 (3.3 – 5.3)	4.4 ± 0.5 (3.4 – 6.0)	4.6 ± 0.5 (2.3 – 5.8)	5.1 ± 0.5 (4.2 – 6.4)
Gonado-somatic index	0.7 ± 0.3 (0.2 – 2.0)	1.3 ± 0.5 (0.2 – 2.7)	1.9 ± 0.6 (0.8 – 3.7)	3.0 ± 0.7 (1.8 – 5.1)
Oocyte stage 1 (%)	65.2 ± 16.6 (36.0 – 100)	24.0 ± 10.5 (0 – 46.0)	12.4 ± 7.6 (0 – 39.0)	7.2 ± 5.0 (0 – 23.0)
Oocyte stage 2 (%)	32.2 ± 14.4 (0 – 57.0)	70.1 ± 12.8 (39.0 – 95.0)	23.7 ± 9.3 (5.0 – 56.0)	18.1 ± 6.9 (0 – 35.0)
Oocyte stage 3 (%)	3.1 ± 6.7 (0 – 35.0)	5.7 ± 8.8 (0 – 33.0)	60.3 ± 14.4 (28.0 – 90.0)	23.6 ± 11.7 (3.0 – 49.0)
Oocyte stage 4 (%)	0.0 ± 0.2 (0 – 2.0)	0.0 ± 0.1 (0 – 1.0)	4.2 ± 9.5 (0 – 56.0)	51.7 ± 12.7 (31.0 – 75.0)
Oocyte diameter (μm)	113 ± 17 (70 – 164)	148 ± 27 (85 – 214)	175 ± 28 (124 – 243)	233 ± 24 (177 – 293)

This measure corresponds very closely with our results for mature migrants, where 52 of the 53 (98%) eels we classified as migrants were greater than 830 mm. However, 61.8% of the eels that we categorized as being immature had lengths greater than 830 mm, so length alone does not appear to be a good determinant of migrants originating from Lake Ontario and the upper St. Lawrence River.

In another study, Cottrill et al. (2002) used three physiological parameters, based on known migrants from Quebec City and Kamouraska, to differentiate resident from migrant eels in the St. Lawrence River. They stated that if an eel met two out of three of the following criteria: GSI greater than 2.0, mean oocyte diameter greater than 0.13 mm, proportion of stage 3 and 4 oocytes greater than 75%; then an eel was a suspected migrant. Comparison of these parameters with the 53 eels designated in this study as migrants shows that 95.2% would have met two out of the three criteria and been classified as suspected migrants. However, 22.4% ($n = 486$) of the eels we categorized as immature would have been classified as suspected migrants utilizing the criteria of Cottrill et al. (2002).

Lastly, Dutil et al. (1985) developed a criterion for differentiating female migrants from female resident eels in the St. Lawrence River based upon oocyte diameter. Their criterion was an average oocyte diameter greater than or equal to 85 μm. The average oocyte diameter of the 53 eels we grouped as mature migrants was substantially higher, 233 μm (minimum average size 177 μm). The substantial difference between Dutil et al. (1985) criterion and those grouped as migrants in our study stems from a methodology difference. In our study, oocyte diameter

Table 4. Discriminant functions based on morphological characteristics for the two developmental categories, excluding the specimens from Quebec and Kamouraska, Canada.

Maturity Category	Constant	Length mm	Weight mm	Girth mm	Ocular Index	Pectoral Fin Index
Mature Migrating =	−11123	+ 8866 log10(L)	− 3507 log10(W)	+ 3009 log10(G)	− 5.52297 O	+ 85.62090 P
Immature Resident =	−11026	+ 8845 log10(L)	− 3490 log10(W)	+ 2976 log10(G)	− 6.28432 O	+ 83.98988 P

was estimated based upon the average of the 10 most mature oocytes, while Dutil et al. (1985) averaged 10 randomly selected oocytes.

Differentiating Resident from Migrating Eels

These analyses demonstrate that external morphological characteristics can be used to differentiate resident eels from migrating eels with a relatively high degree of reliability. In this study, we were able to correctly classify 87% of the mature migrants and 83% of the immature residents. However, we believe that these analyses are conservative in that they err towards maturity. That is, those eels that are classified as mature migrants are very likely to be mature migrants but some eels are classified as immature when they are borderline mature. For example, a careful examination of the cluster plots shows that a substantial portion of the eels collected in trawling (14 out of 34) and in the tailwaters (16 out of 27) are classified as immature. Based on our knowledge of eel biology and experience on the upper St. Lawrence River, we believe that almost all eels collected by trawling or in the tailwaters of the Power Project are mature. Eels captured by trawling were caught in the upper portion of the water column in the deeper parts of the River. Immature resident eels are more bottom/substrate oriented, primarily feed on benthic organisms and show limited movement, thus, few would be expected in the upper parts of the water column (Eales 1968; Ogden 1970; LaBar and Facey 1983). All eels collected in the tailwaters of the Moses-Saunders Power Dam were collected only during the outmigration season, not prior to or following the outmigration season, thus leading us to believe they also are mature migrants.

Another factor that could contribute to some of our inability to differentiate immature resident eels from maturing migrants is the increase in maturation that occurs as the season progresses. Dutil et al. (1987) noted in a study on female American eel in the lower portion of St. Lawrence River (Quebec City to Kamouraska) that the average oocyte diameter increased from 151 to 214 µm from mid-August to mid-October. Durif et al. (2000) documented similar seasonal progression in maturation in the European eel for both males and females in France. Eels in the current study were purposely collected over a broad period of time, from May through October, over two years, to provide a broad range of maturity. Thus, an eel of comparable size earlier in the season would not be expected to have the same degree of maturation characteristics as one collected later in the season. In our analyses, we combined all samples across time. This approach likely contributes to overall variance and to some blurring between various maturity groupings.

Development

The eels from this study were purposely collected to show a broad range of maturity. The

Table 5. Classification accuracy of the two developmental categories based upon the discriminant analysis of external morphological characteristics.

Based Upon Cluster Analysis	n	Based Upon Discriminant Analysis		Correct
		Immature resident	Mature migrating	
Immature Resident	486	403	83	83%
Mature Migrating	53	7	46	87%

data clearly show a gradual increase in several key morphological (such as pectoral fin index and eye index) and physiological characteristics (such as GSI, oocyte diameter and oocyte stage development) going from the least mature resident yellow eels to the most mature migrating silver eels (Table 3). Couillard et al. (1997) noted comparable increases in physiological parameters. In this study, each of the parameters shows considerable overlap among the various maturity stages suggesting that no one parameter, such as length, can be used for differentiation purposes, but rather a suite of characteristics is required. Similar trends were observed by Cottrill et al. (2002) for American eels in the St. Lawrence River and by Durif (2000) in a comparable study on European eels *Anguilla anguilla* in France.

During examination of ovaries, a distinct difference in the texture was observed between the ovaries of immature resident and mature migrating eels. Immature resident eels had ovaries with a smooth or filamentous appearance, while mature migrating eels exhibited ovaries with a distinctly granular appearance. We presume the granular appearance is due to the increasing number of larger oocytes (Pankhurst 1982; Dutil et al. 1985). These observations are qualitative in nature but appear to hold promise for further assisting in the differentiation of immature from maturing/migrating eels in the field assuming that a small incision can be tolerated, such as when one is implanting a transmitter into the coelomic cavity.

Conclusions

These results demonstrate that external morphological characteristics of American eels can be used to distinguish between resident and migratory eels captured from the Lake St. Lawrence region (i.e., migrating from Lake Ontario and the upper reaches of the St. Lawrence River) of the St. Lawrence River. However, these methods should be applied with some caution since there seems to be a tendency to categorize some maturing eels as immature. Other researchers can adapt the methodologies and techniques, using their site-specific data, to develop comparable methodologies and strategies for differentiating downstream migrant maturing eels from resident yellow eels.

Acknowledgments

We would like to thank A. Cottrill, J. Keene, G. Burchill, G. Guay, S. Chevarie and C. Champagne for substantial contributions in the field and lab; and C. Couillard for the excellent work she performed in the analyses of the gonadal tissue. C. Durif and G. Tremblay provided valuable comments on an earlier draft of this paper. We are most appreciative of P. Lengendre who provided helpful comments and significant guidance in the conduct of the statistical analyses. Lastly, we want to thank D. Dixon for his great patience in allowing us the time to revise and complete this paper.

References

Bouillon, D. R., and R. L. Haedrich. 1985. Growth of silver eels (*Anguilla rostrata*) in two areas of Newfoundland. Journal of Northwest Atlantic Fisheries Science 6:95–100.

Beullens, K., E. H. Eding, F. Ollevier, J. Komen and C. J. J. Richter. 1997. Sex differentiation, changes in length, weight and eye size before and after metamorphosis of European eel (*Anguilla anguilla* L.) maintained in captivity. Aquaculture 153:151–162.

Columbo, G., G. Grandi, and R. Rossi. 1984. Gonad differentiation and body growth in *Anguilla anguilla* L. Journal of Fish Biology 24:215–228.

Colombo, G., and G. Grandi. 1996. Histological study of the development and sex differentiation of the gonad in the European eel. Journal of Fish Biology 48:493–512.

Cottrill, R. A., R. S. McKinley, G. Van Der Kraak, J. D. Dutil and K. J. McGrath. 2001. Plasma non-esterified fatty acid profiles and 17 beta-oestradiol levels of juvenile immature and maturing adult American eels in the St Lawrence River. Journal of Fish Biology 59:364–379.

Cottrill, R. A., R. S. McKinley, and G. Van Der Kraak. 2002. An examination of utilizing external measures to identify sexually maturing female American eels, *Anguilla rostrata*, in the St. Lawrence River. Environmental Biology of Fishes 65:271–2287.

Couillard, C. M., P. V. Hodson, and M. Castonguay. 1997. Correlations between pathological changes and chemical contamination in American eels, *Anguilla rostrata*, from the St. Lawrence River. Canadian Journal of Fisheries Aquatic Science 54:1916–1927.

Dave, G., M. Johansson, A. Larsson, K. Lewander and U. Lidman. 1974. Metabolic and haemotological studies on the yellow and silver phases of the European eel, *Anguilla anguilla* (l.),- II fatty acid composition. Comparative Biochemistry and Physiology 47B:583–591

Dutil, J. D., B. Legare; and C. Desjardins. 1985. Discrimination d'un stock de poisson, l'anguille, *Anguilla rostrata*, basee sur la presence d'un produit chimique de synthese, le mirex. Canadian Journal of Fisheries Aquatic Science 42: 455–458.

Dutil, J. D., M. Besner, and S. D. McCormick. 1987. Osmoregulatory and ionoregulatory changes and associated mortalities during the transition of maturing American eels to a marine environment. Pages 175–190 *in* M. J. Dadswell, R. J. Klauda, C. M. Moffitt, R. L. Saunders, E. F. Rulifson, and J. E. Cooper, editors. Common strategies of anadromous and catadromous fishes. American Fisheries Society, Symposium 1, Bethesda, Maryland.

Durif, C., P. Elie, S. Dufour, J. Marchelidon and B. Vidal. 2000. Analysis of morphological and physiological parameters during the silvering process of the European eel (*Anguilla anguilla*) in the Lake of Grand-Lieu (France). Cybium 24(3):63–74.

Eales, G. J. 1968. The Eel Fisheries of Eastern Canada. Fisheries Research Board of Canada, Bulletin 166, Ottawa, Ontario, Canada.

Ellerby, D. J., I. L.Y. Spierts and J. D. Altringham. 2001. Slow muscle power output of yellow- and silver-phase European eels (*Anguilla anguilla* L.): changes in muscle performance prior to migration. Journal of Experimental Biology 204: 1369–1379.

Facey, D. E., and G. S. Helfman. 1985. Reproductive migrations of American eels in Georgia. 1985 Proceedings of the Annual Conference of the Southeast Association of Fish and Wildlife Agencies 39:132–138.

Fontaine, Y. A., M. Pisam, C. Le Moal, and A. Rambourg. 1995. Silvering and gill "mitochondria-rich" cells in the Eel, *Anguilla anguilla*. Cell and Tissue Research 281: 465–471.

Han, Y. S., W. N. Tzeng, Y. S. Huang, and I. C. Liao. 2001. Silvering in the eel: changes in morphology, body fat content, and gonadal development. Journal of Taiwan Fisheries Research 9(1&2):119–127.

Hurley, D. A. 1972. The American eel *Anguilla rostrata* in Eastern Lake Ontario. Journal of the Fisheries Research Board of Canada 535–543.

Hurley, D. A. 1973. The commercial fishery for American eel, *Anguilla rostrata* in Lake Ontario. Transactions of the American Fisheries Society 2:369–377.

Hurley, D. A., and W. J. Christie. 1982. A re-examination of statistics pertaining to growth, yield and escapement in the American eel *Anguilla rostrata* stocks of Lake Ontario. Pages 83–85 in K. H. Loftus (Editor), Proceedings of the 1980 North American Eel Conference. Ontario Fisheries Technical Report Series No. 4. Ontario Ministry of Natural Resources: Toronto, Ontario, Canada.

Jessop, B. M. 1987. Migrating American eels in Nova Scotia. Transactions of the American Fisheries Society 116:161–170.

Johansson, M. L., G. Dave, A. Larsson, K. Lewander and U. Lidman. 1974. Metabolic and haematological studies on the yellow and silver phases of the European eel, *Anguilla anguilla* (L.) — III Haematology. Comparative Biochemistry and Physiology 47B:593–599.

LeBar, G. W., and D. E. Facey. 1983. Local movements and inshore population sizes of American eels in Lake Champlain, Vermont. Transactions of the American Fisheries Society 112:111–116.

Legendre, P., and L. Legendre. 1998. Numerical Ecology. Elsevier Science B. V. Amsterdam, Netherlands.

Lewander, K., G. Dave, M. L. Johansson, A. Larsson, and U. Lidman. 1974. Metabolic and hematological studies on the yellow and silver phases of the European eel, *Anguilla anguilla* (L.) – I. Carbohydrate, lipid, protein and inorganic ion metabolism. Comparative Biochemistry and Physiology 47B:571–581

New York Power Authority (NYPA). 2000. Survey of fish in the tailwater and areas downstream of the St. Lawrence FDR Power Project. Prepared by Riveredge Associates. Massena, New York. New York Power Authority, White Plains, New York.

New York Power Authority (NYPA). 2001. 2000 Survey of American eel in the tailwater of the International St. Lawrence Power Project. Prepared by Kleinschmidt Inc., Strasburg, Pennsylvania. New York Power Authority, White Plains, New York.

New York Power Authority (NYPA). 2002a. 2001 Survey of American eel in the tailwater of the St. Lawrence Power Project. Prepared by Milieu, Inc., Saint-Bernard-de-Lacolle, Quebec, Canada. New York Power Authority, White Plains, New York.

New York Power Authority (NYPA). 2002b. 2002 Survey of American eel in the tailwater of the International St. Lawrence Power Project. Prepared by Kleinschmidt Inc., Strasburg, Pennsylvania. New York Power Authority, White Plains, New York.

Ogden, J. C. 1970. Relative abundance, food habits, and age of the American eel, *Anguilla rostrata* (LeSueur), in certain New Jersey streams. Transactions of the American Fisheries Society 1:54–59.

Pankhurst, N. W. 1982. Relation of visual changes to the onset of sexual maturation in the European eel, *Anguilla anguilla*. Journal of Fish Biology 21:127–140.

Pankhurst, N. W., and J. N. Lythgoe. 1982. Structure and color of the integument of the European eel *Anguilla anguilla* (L). Journal of Fish Biology 21:279–296.

Pankhurst, N. W., and P. W. Sorensen. 1984. Degeneration of the alimentary canal in maturing European *Anguilla anguilla* (L) and American eels *Anguilla rostrata* (LeSeur). Canadian Journal of Zoology 62:1143–1149

SAS. 2002. Version 6.12 for Windows. SAS Institute, Inc. SAS Campus Drive, Cary, North Carolina 27513

SPSS, Inc. 1999. SYSTAT 9.0. Chicago, Illinois 60606-6307.

Svendang, H., E. Eeuman, and H. Wickstrom. 1996. Maturation patterns in female European eel: age and size at the silver eel stage. Journal of Fish Biology 48:342–351.

Tesch, F. W. 1977. The eel: biology and management of anguillid eels. Chapman and Hall, London.

Verreault, G., G. Pouliot, and P. Pettigrew. 2000. L'exploitation de l'anguille d'Amerique dans le Bas-Saint-Laurent en 1999. Pages 193–196 *In* M. *Bernard and C. Groleau (Editors)*, Compte Rendu Du Cinquieme Atelier Sur Les Peches Commerciales, tenu a Quebec du 18 au 20 Janvier 2000, Societe de la faune et des parcs du Quebec, Quebec City, Quebec, Canada.

Vladykov, V. D. 1973. Macrophthalmia in the American eel (*Anguilla rostrata*). Journal of the Fisheries Research Board of Canada 30.689–693.

Winn, H. E., W. A. Richkus, and L. K. Winn. 1975. Sexual dimorphism and natural movements of the American eel (*Anguilla rostrata*) in Rhode Island streams and estuaries. Helgolander wiss. Meeresunters 27:156–166.

Wenner, C. A., and J. A. Musick. 1974. Fecundity and gonad observations of the American eel, *Anguilla rostrata*, migrating from Chesapeake Bay, Virginia. Journal of the Fisheries Research Board of Canada 31:1387–1391.

Woody, C. A., J. Nelson, and K. Ramstad. 2002. Clove oil as an anesthetic for adult sockeye salmon: field trials. Journal of Fish Biology 60:340–347.

Development of Hydrosonic Telemetry Technologies Suitable for Tracking American Eel Movements in the Vicinity of a Large Hydroelectric Project

KEVIN J. MCGRATH*

New York Power Authority, 123 Main Street, White Plains, New York 10601, USA
*Corresponding author: mcgrath.k@nypa.gov

SCOTT AULT

Kleinschmidt Associates, 2 East Main Street West, Strasburg, Pennsylvania 17579, USA

KEVIN REID AND DAVID STANLEY

Stantec, 14 Abacus Road, Brampton, Ontario L6T 5B7, Canada

FRED VOEGELI

Vemco, Inc., 100 Osprey Drive, Shad Bay, Nova Scotia B3T 2C1, Canada

Abstract.—The New York Power Authority investigated the development of hydrosonic telemetry equipment to monitor the behavior of downstream migrating American eels *Anguilla rostrata* in an area of high ambient noise near the Moses-Saunders Power Dam (Power Dam). The study consisted of: conducting a sound spectrum analysis to characterize background noise in the forebay of the Power Dam; evaluating existing and customized hydrosonic equipment that would function under the high noise conditions encountered; evaluating prototype receivers, depth sensitive transmitters, and hydrophones in the field; and evaluating a full-scale prototype telemetry system in the forebay of the Power Dam utilizing 10 eels surgically implanted with transmitters. The sound spectrum analysis indicated that background noise levels in the frequency range used by the standard hydrosonic equipment (60 to 80 kHz) were very high, approximately 75 dB, while sound levels were approximately an order of magnitude lower in the 150 to 200 kHz range. It was determined that hydrosonic equipment operating at 200 kHz would provide the optimum balance between transmission power and interference from background noise. Extensive field evaluations conducted with prototype equipment designed at 200 kHz indicated that an area 800 m long by 1 km wide in front of the Power Dam could be monitored with 19 receivers deployed in two transects. A full-scale prototype telemetry system was evaluated by releasing 10 downstream migrating eels with surgically implanted depth sensitive transmitters. The movements of nine of the test specimens were monitored as they passed through the system. Data indicated that movement could be monitored continuously while an eel was in the reception area and that the position of the eel in the reception area, including depth, could be determined. With continued fine tuning of the equipment and development of data analysis techniques, it should be possible in future studies to determine the three dimensional route of travel for a telemetered eel through the forebay of the Power Dam.

Introduction

The Moses-Saunders Power Dam (Power Dam) on the St. Lawrence River is one of the largest hydroelectric facilities in North America. The Power Dam consists of two power houses that constitute one continuous structure spanning the St. Lawrence River and the international border between Massena, New York and Cornwall, Ontario. The New York Power Authority (NYPA) operates and maintains the Robert Moses Power Dam on the U.S. side of the St. Lawrence River, while Ontario Power Generation (OPG) operates and maintains the Robert H. Saunders Generating Station on the Canadian side. Currently, NYPA is in the process of relicensing the U.S. portion of the Power Dam with the Federal Energy Regulatory Commission. One of the environmental issues identified during the relicensing effort was the effect of project operations on the downstream passage of adult American eels migrating from Lake Ontario, its tributaries, and the St. Lawrence River upstream of the Power Dam.

As a first step in determining the impact of project operations on downstream migrating eels, NYPA conducted a turbine passage survival study for adult eels at the Power Dam in 1997. Results from this study indicated a survival rate of approximately 74% (Normandeau Associates and Skalski 2000). Subsequently, NYPA began investigations to determine the behavior of downstream migrating eels approaching the Power Dam as the next logical step in evaluating potential measures to improve downstream passage survival. The major objectives were to determine how to collect, identify, and monitor the behavior of downstream migrating eels upstream of the Power Dam.

This paper primarily discusses our efforts to develop techniques for monitoring the movements of downstream migrating eels. The techniques used were based on ultrasonic transmitter and tag technology. The first documented use of ultrasonic tags to monitor fish movements was in 1956 for a study on the movement of adult chinook and coho salmon (Trefethen 1956). Westerberg (1975) as reported in Tesch (1977) used ultrasonic tags to document water temperature, depth, and location of European silver eels during migration. Tesch (1976) reported tracking three European silver eels and recording depth for one of the eels utilizing ultrasonic tags. Dutil et al. (1988) utilized externally (dorsally) attached ultrasonic transmitters to track 14 American eels in the tidal portion of the St. Lawrence River. Haro and Castro-Santos (2000) were able to track and record depth for eight American eels in the vicinity of the small hydroelectric facility on the Connecticut River utilizing external (dorsal) ultrasonic transmitters operating at 50–69 kHz. Steig (1999) reported difficulties in using the more standard lower acoustic frequencies (50–100 kHz) near a hydroelectric facility because of high background noise levels and developed higher frequency (300 kHz) acoustic tags and receivers to monitor juvenile salmonid movements.

Our study was iterative in nature with the results of one set of experiments used in the experimental design of subsequent work. In reporting on this effort, we believe that the best means to communicate this step-wise process is to present the methods and results of each task together in a fashion that clearly relates how the study progressed.

Methods and Results

Study Area

The study area included the St. Lawrence River from the Power Dam to approximately 1 km upstream (i.e., the forebay; Figure 1). The Power Dam is approximately 975 m long and has an installed capacity of about 1900 MW generated from 32 turbines at an operating head of 24 m. The average annual flow of the St. Lawrence River at this location is approximately 7,000 m^3/s. The St. Lawrence River in this region is up to 2 km wide and 35 m deep. The northern shoreline of the River is in Canada and the southern shoreline is in the U.S.

1998 Studies

Studies conducted in 1998 examined techniques that could be used to monitor the behavior of downstream migrating eels. They provided the following findings:

1. High conductivity (> 280 microhos/cm^3) in the St. Lawrence River precluded the effective use of radio telemetry techniques at depths greater than 7 m;
2. Standard "off the shelf" hydrosonic telemetry equipment operating in the range of 60 to 80 kHz provided a reception range of approximately 1 km in areas away from the Power Dam; however, high levels of background noise in the forebay of the Power Dam precluded the use of this equipment near the Dam;
3. The effectiveness of manual tracking was minimized by the behavior of eels to commonly inhabit submerged aquatic vegetation and substrate which blocked reception of hydrosonic signals (Nielsen and Johnson 1983);
4. Specialized depth sensitive hydrosonic transmitters provided a feasible means to collect data on the depth eels occupied with sub-meter accuracy, and;

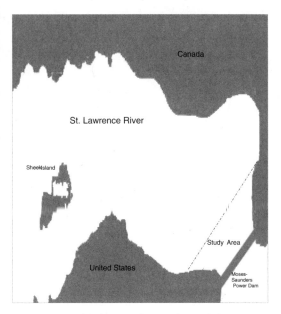

Figure 1. Study Area in the Forebay of the Moses-Saunders Power Dam on the St. Lawrence River near Massena, New York and Cornwall Ontario, Canada.

5. Surgical implantation of a transmitter into the coelomic cavity, compared to external attachment and gastric insertion, proved the best means of fitting specimens with transmitters.

Based on these results, studies in 1999 focused on determining the feasibility of developing a hydrosonic telemetry system that could function under conditions of extreme background noise that precluded the use of standard equipment.

1999 Studies

The 1999 study included four major components: evaluate and characterize the sound spectrum in the forebay of the Power Dam; develop prototype hydrosonic equipment based on the results of the sound spectrum analysis; evaluate prototype hydrosonic equipment in the forebay of the Power Dam; and deploy and evaluate a full-scale hydrosonic telemetry system.

All hydrosonic telemetry equipment used in this evaluation was supplied by VEMCO Limited of Shad Bay, Nova Scotia. References in this manuscript to receiver model number follow VEMCO's nomenclature.

Study component 1—evaluate and characterize the sound spectrum in the forebay of the power dam: The objective was to characterize background noise in the forebay of the Power Dam at frequencies ranging from 0 to 300 kHz. The goal was to determine if a specific band or range of frequencies was present where background noise was significantly lower (i.e., a quiet spot in the background noise). If a band of frequencies with low background noise was observed we planned on investigating the possibility of modifying existing technology or developing new equipment to operate at these frequencies.

A sound spectrum analysis was conducted in April 1999 using an Advantest/Tektronix U4941/N Field Spectrum Analyzer with a DC power supply and a broad-band spherical hydrophone. Data were collected in the forebay from the intake deck of the Power Dam and from a vessel at 65 locations along seven transects up to 1,000 m upstream of the Dam. Depth of measurement was either 1 m or 5 m when obtained from the intake deck or the vessel, respectively. The frequency of measurements along transects decreased with distance upstream from the Power Dam since preliminary measurements indicated that the intensity of background noise decreased as distance upstream from the Power Dam increased. Distance from the Power Dam was measured with a laser range finder and all sampling locations were geo-referenced with a differential global positioning system (DGPS). The intensity (β) of sound at each frequency is presented in this manuscript in decibels (dB) with respect to (re) 1μPascal at 1 Hz bandwidth (hereafter referred as dB only) with dB defined as $\beta = 10 \log A/B$ where A is the intensity corresponding to the level β and B is a reference level.

No distinct band of quiet frequencies was found to be present although background noise substantially decreased as frequency increased (Figure 2 top graph). Regardless of location, the intensity of background noise at frequencies between 60 to 80 kHz was approximately 10 dB greater (i.e., 10 times the power) than the noise measured at 150 kHz and 15 dB greater (31 times the power) than the intensity of noise measured at 200 kHz. The intensity of background noise at all frequencies decreased as distance upstream from the Power Dam increased, with the exception that noise levels peaked at a distance of 50 m. This is thought to be an area where noise emanating from operating equipment in the Power Dam converges, whereas a location

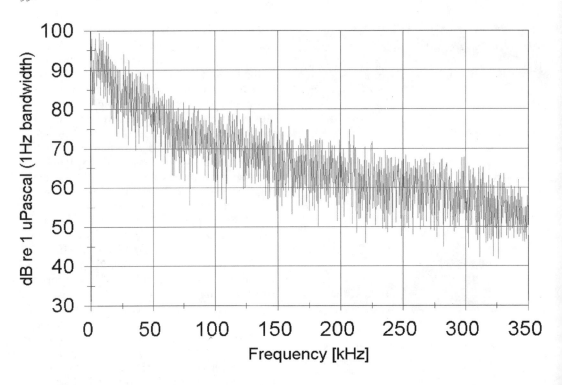

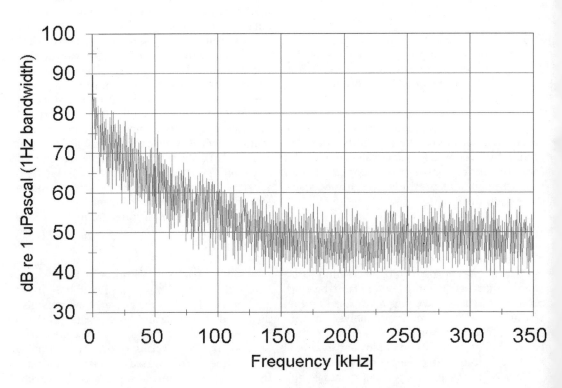

Figure 2. Background noise level by frequency (kHz) measured in the forebay 50 m upstream (top graph) and 1,000 m upstream (bottom graph) of Moses-Saunders Power Dam in April 1999.

immediately in front of the Power Dam is somewhat shielded from noise by the structure itself. The decrease in the intensity of noise with an increase in distance from the Power Dam was substantial, dropping from approximately 65 dB at 50 m to approximately 45 dB at 1,000 m, a decrease of about 0.5 dB per 100 m. (Figure 2 both graphs). While we believe the results of this survey are unique to the study area, the occurrence of high levels of background noise in the forebay of hydroelectric stations has also been reported by other investigators (Steig 1999; C. Whitmus, Pentec, personal communication and R. Verdon, Hydro Quebec, personal communication).

Some preliminary reception range tests were also conducted in April 1999 to confirm the results of studies in 1998 and to determine if transmissions at higher frequencies could provide adequate reception distances. Reception range tests were conducted with transmitting devices at 78, 150, and 266 kHz and a standard VR60 manual receiver without special low frequency noise rejection circuitry and an omnidirectional hydrophone. Range tests were conducted in the forebay in a high background noise location close to the Power Dam and in a location with lower background noise further upstream. Results confirmed that background noise at 78 kHz would preclude the use of relatively low frequencies commonly used for hydrosonic telemetry equipment. Reception range at 150 and 266 kHz was found to be acceptable (300 m to 1 km), although range at 266 kHz was approximately one-third less than at 150 kHz due to the rapid rate of signal attenuation for high frequencies in water.

Based on the results of the noise spectrum analysis and preliminary signal propagation tests, it was concluded that a frequency of 150 or 200 kHz would provide the best opportunity for success in development of equipment that could function in the high background noise environment. Selection between these two frequencies would depend on determining which frequency had the best overall reception range when all conditions in the study area were considered. The next task was to build and test prototype equipment at frequencies of 150 and 200 kHz.

Study component 2—development of prototype hydrosonic equipment based on the results of the sound spectrum analysis: Two types of prototype hydrosonic transmitters were developed that operated at both 150 and 200 kHz, simple pingers and coded transmitters. Simple pingers were developed for signal propagation tests and to measure detection distances throughout the study area. Coded transmitters were developed to determine the ability of current receiver technology to decode a string of signal pulses, encoded with data such as tag number and depth, in the high background noise environment. Depth was one of the primary data interests in studying outmigration behavior of eels at the Power Dam; therefore, the decoding of encoded data (termed "electronic detection") was a critical element to developing a system that could study the behavior of outmigrating eels. A VR60 manual receiver modified to operate at 150 and 200 kHz and to reject low frequency noise was used as the primary receiving unit. The VR60 was coupled to an omnidirectional hydrophone.

A combined total of 16 trials at 150 and 200 kHz were conducted in the forebay to compare performance of the two frequencies under similar conditions. Two types of trials were generally conducted to determine how receiver deployment location, transmitter location, and background noise affected signal reception. One type of trial was where the transmitter was deployed at a fixed location and the receiver moved toward the transmitter and the other was where the receiver was deployed at a fixed location and the transmitter moved toward the receiver. Trials included various scenarios at each frequency with the receiver and transmitters deployed in low and high background noise locations and at various depths. Some trials also included the use of a conical 45-degree reflector mounted on the hydrophone to determine if shielding a portion of the hydrophone would help reduce interference from background noise. Data recorded included distance to the transmitter/receiver, signal strength, percent signal detection, and background noise.

Four major findings were documented from these trials:

1. Reception range at 200 kHz was greater in the high background noise area close to the Power Dam than at 150 kHz because lower noise levels at 200 kHz permitted the use of higher gain settings on the receiver. Reception range at 200 kHz with omni-directional hydrophones was approximately 230 m in locations close to the Power Dam and 400 m in upstream areas;

2. The best reception for receivers located on the Power Dam was obtained when receivers were attached to the face of the large concrete piers (nose piers) that separate the intake bays of each turbine. These structures help block noise emanating from the Power Dam thereby providing a "quiet area" that optimizes receiver reception;
3. The optimal depth for receiver deployment, regardless of location in the study area was 4 m based on percent signal detection and signal strength;
4. A conical 45-degree reflector increased reception range at 200 kHz by about one-third in the high background noise area.

Study component 3—design and evaluation of prototype equipment at 200 kHz: Results of the comparison tests between 150 and 200 kHz indicted that hydrosonic equipment designed to function at 200 kHz would perform better than equipment functioning at 150 kHz. These data provided specific information for continued development of prototype equipment at 200 kHz and were used to redesign existing receiver configuration and housings, develop a 200 kHz depth sensitive coded transmitter suitable for surgical implantation in large adult eels, and design directional hydrophones to increase reception range over the range expected by an omni-directional hydrophone. Directional hydrophones receive signals from only one direction and when pointed away from the noise source can minimize reception interference from background noise.

A self-contained hydrosonic receiver designed for remote underwater deployment was developed using the basic internal electronic architecture of the VR20 remote heterodyne receiver. The new receiver, termed the VR25, was designed to operate at 200 kHz and featured front end noise reduction filters, a stainless steel case to shield the receiver from electromagnetic interference common at electric generating facilities, data download through an induction probe, and an experimental fiberglass nose cone to streamline flow and shed macrophytes. Expected battery life of the receiver was 30 days on a single replaceable battery pack. Additionally, two types of mounting devices were designed to mount the VR25 receiver in either a horizontal or vertical position.

Transmitter design and specification included operation at 200 kHz, depth sensitive capabilities with sub-meter accuracy, a cylindrical shape with a maximum size of 20 mm diameter by 75 mm in length, weight of less than 77 g in air [based on a goal for transmitter weight of 1.25% to 2% of the fish's weight out of water (Nielsen and Johnson 1983], and a battery life of 30 to 60 days. The final prototype transmitters met or exceeded each of these criteria, measuring 19 mm in diameter by 74 mm in length, weighing 33 g in air (15 g in water) and having a battery life of 30 days. A battery life of longer than 30 days would have resulted in a transmitter length of over 85 mm which we considered too long for surgical implantation in the size of eels we were working with.

Two types of directional hydrophones were designed to operate at 200 kHz, a radial line array directional hydrophone for use on receivers mounted in a vertical position and an end mounted line array directional hydrophone for use with receivers mounted in a horizontal position. The theoretical beam width of each hydrophone type was 30 degrees.

Reception range tests with receivers deployed at a variety of locations in the forebay of the Power Dam were conducted to evaluate the prototype equipment. Three prototype VR25 receivers and three transmitters were constructed specifically for this task based on specifications detailed above. Both directional and omni-directional hydrophones were utilized during trials.

Receivers were deployed from moorings in the forebay or from mounting brackets attached to nose piers on the face of the Power Dam. Nine moorings were deployed in three transects at distances of 300 m, 450 m, and 600 m upstream from the Power Dam. Moorings were separated by a distance of approximately 100 m in each of the three transects. Moorings consisted of a 1.2 m spar buoy tethered to a 320 kg concrete anchor. Receivers were deployed from the buoy in either a horizontal or vertical position depending on hydrophone configuration. Horizontal mounting brackets included fins to provide stability in the current and to orient the receiver in an upstream direction. The mounting brackets for deploying a VR25 receiver on the nose piers consisted of a steel tube 2.5 cm square by 4.8 m long nested in a U-channel attached to the face of the nose pier. Mounting brackets were attached to six adjacent nose piers across the Power Dam, each one approximately 100 m apart.

A total of nine trials were conducted to evaluate performance of the receivers at different locations, with different deployment techniques, with different types of hydrophones, and with transmitters at a variety of depths. During each trial, three VR25 receivers were deployed at one of the three mooring transects or at three adjacent nose piers. A transmitter was lowered from a vessel to a predetermined depth and the vessel was allowed to drift toward the receivers. Drifts started approximately 500 m to 300 m upstream from the receiver line. Distance from the vessel to the receivers was measured with a laser range finder and the drift path of the vessel was documented with DGPS. For each electronic detection, the receivers recorded date and time, transmitter identification number, transmitter depth, signal strength, receiver gain setting, and a relative background noise value in millivolts. Similar trials were conducted with both directional and omni-directional hydrophones.

In all trials, directional hydrophones outperformed omni-directional hydrophones with respect to distance, rate, and consistency of electronic detections. Generally, directional hydrophones increased reception range by approximately 200 m for moored receivers (horizontal mount) and 100 m for receivers mounted on nose piers. Reception range was 400 m to 600 m for moored receivers and 200 m to 300 m for receivers mounted on nose piers depending on depth of the transmitters and proximity of the receiver to the Power Dam. Receivers close to the Power Dam exhibited the shortest reception ranges due to interference from background noise. Regardless of location, at least a 100% overlap in reception area was observed between adjacent receivers spaced 100 m apart. Additional gains in performance were achieved with the experimental fiberglass nose cone originally designed only to protect the hydrophone from damage due to debris. We believe that the nose cone streamlined flow past the hydrophone and reduced interference from turbulence.

Deploying three receivers adjacent to one another provided an opportunity to evaluate the relationship between relative signal strength as recorded by the VR25 receiver and distance between the transmitter and receivers. It was hoped that an understanding of this relationship would eventually enable us to determine the position of a tagged eel in relation to each receiver where the signal was logged. By knowing the position of the receiver and coupling the signal strength information with depth data from the transmitter we could then determine a three dimensional track of a telemetered animal. Figure 3 shows relative signal strength at each receiver during three trials when a transmitter was drifted past a group of three receivers moored at a transect 300 m upstream of the Power Dam. The transmitter was deployed at 1 m below the surface, mid-depth, and 1 m above the bottom. On each drift, the transmitter passed within 10 m of the middle receiver (Buoy 2). These trials produced three important findings.

1. Relative signal strength as recorded by each receiver increased and diminished in relation to the distance between the transmitter and the receivers with the greatest signal strength recorded at the receiver closest to the transmitter (Buoy 2). This result indicated that continued development and refinement of reception capabilities could increase the accuracy of estimating transmitter position based on relative signal strength;
2. For each trial, electronic detection rate was similar at each receiver indicating significant overlap in reception area. Understanding the extent of this overlap was a critical element in developing and deploying an array to cover the large study area, and;
3. The rate of electronic detections and relative signal strength recorded at individual receivers for each of the three trials indicated that for practical purposes, reception range was similar regardless of depth.

Study component 4—deploy and evaluate a full-scale hydrosonic telemetry system: Based upon the results of the prototype equipment tests, it was determined that a full-scale prototype system should be constructed and tested. With this system we could determine if tracking the movements of a downstream migrating eel in the forebay of the Power Dam was feasible. The full scale prototype system consisted of an array of 10 nose pier mounted VR25 receivers with directional hydrophones and nine horizontally oriented receivers with directional hydrophones deployed on moorings 300 m upstream of the Power Dam. Receivers were approximately 100 m apart in each location and deployed at 4 m

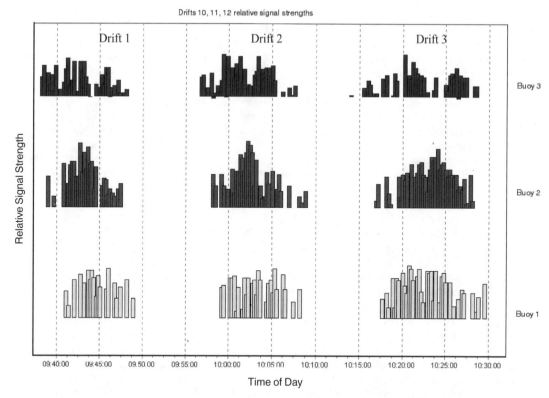

Figure 3. Relative signal strength (height of bars) for electronic detections by three VR25 receivers with omni-directional hydrophones at a mooring transect 300 m upstream of the Moses-Saunders Power Dam. Each receiver was moored at 4 m depth. Transmitter depths were 1 m for drift 1, 4 m for drift 2, and 1 m from the bottom for drift 3.

depth. Based on previous test results it was expected that the double line of receivers would provide continuous coverage of a telemetered eel at distances up to approximately 800 m upstream of the Power Dam.

The full scale prototype array was tested by releasing 10 eels with surgically implanted depth sensitive transmitters. Adult migrating eels ranging from 915 mm to 1,095 mm total length and 1,650 g to 3,100 g were obtained from the commercial weir fishery in the St. Lawrence River near Quebec City, Quebec. It is important to note that the fish used for these trials were known migratory animals collected over 300 km downstream from the study area. These fish are more sexually mature than eels in the study area and have entered a euryhaline environment. We used these fish because they were readily available and because we expected them to move through the telemetry system and thereby test the ability of the system to monitor the movements of live specimens. The type of behavior elicited by these eels may or may not be similar to eels naturally migrating through the study area. Further, it should be noted that this study made no attempt to determine any potential effects of surgical implantation on eel behavior. We recognize that stress associated with tagging and the presence of the tag in the coelomic cavity may affect an eels' behavior, but determining these effects was not an objective of this work.

Surgeries were performed under red light on the evening of release. Eels were anesthetized in a solution of clove oil based on Andersen et al. (1997). The transmitter was implanted in the coelomic cavity slightly anterior to the anal opening and the incision was closed with three to four sutures (Baras and Jeandrain 1998; Haro and Castro-Santos 2000). Eels were released at several locations up to 2 km upstream of the Power Dam. Two eels per night were released on five nights between 22 and 29 September 1999. Manual tracking was conducted from a vessel to monitor the behavior of eels immediately after release and then periodically until the end of the study to inventory telemetered animals. Data were downloaded from the VR25 receivers daily, from 23 September until 18 October 1999, weather permitting. Raw data files for individual eels were combined and post processed with a software program developed to produce graphic representations of relative signal strength and depth.

The prototype remote array successfully monitored the movement of nine test specimens. Manual tracking indicated that the tenth specimen was alive and moving but did not migrate through the array before the end of the study period. The extent of information recorded on each eel depended largely on the amount of time an eel took to pass through the receiver array. As expected, eels which spent more time in the receiver array produced more electronic detections on more receivers than eels that moved through quickly. In all cases, the eels' movement was monitored continually by at least two receivers (in most cases four or more receivers from each array logged data) from the time it entered the array until it passed the Power Dam. The extent of information was sufficient to determine the travel route and depth of each eel when passing through the array. Data for eel number 8 and 11 provide good examples of data collected by the prototype array and are used here to illustrate how the system functioned. Data from the additional seven eels are not discussed:

Eel number 8 was first detected on the north side of the River by the receiver located at mooring number 10 (Figure 4). As the eel approached, receivers on either side of mooring 10 began to decode signals until the receiver at mooring 6 was recording the highest relative signal strength. Greatest signal strength was recorded at approximately 20:20 hours. Thereafter, signal strength decreased, indicating the eel had likely crossed the line of moored receivers heading toward the Power Dam. Data on depth indicate that the eel was primarily surface oriented, making short forays to depths of 20 m. Relative signal

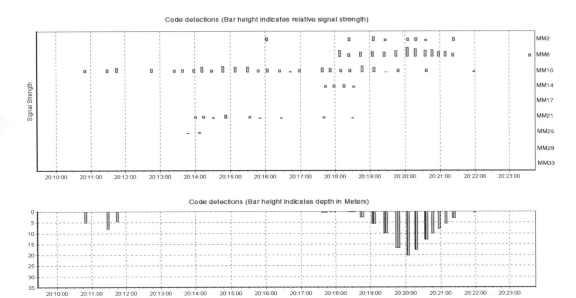

Figure 4. Relative signal strength, depth, and time of electronic detections for Eel 8 from VR25 receivers along a mooring line 300 m upstream of Moses-Saunders Power Dam. The top graph shows relative signal strength recorded for each receiver by time and the bottom graph shows a corresponding depth for each electronic detection. Depth was 0 m where signal detections are indicated in the upper graph but depth data is not visible on the lower graph.

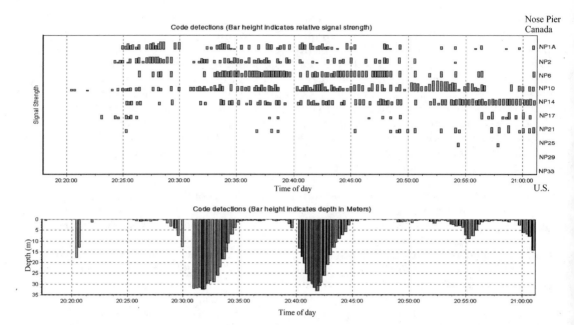

Figure 5. Relative signal strength, depth, and time of electronic detections for eel 8 from VR25 receivers mounted on nose piers at the Moses-Saunders Power Dam. The top graph shows relative signal strength recorded for each receiver by time and the bottom graph shows a corresponding depth for each electronic detection. Depth was 0 m where signal decodes are indicated in the upper graph but depth data is not visible on the lower graph.

strength indicates that the eel approached the Power Dam on a straight route toward nose pier 6 from approximately 20:33 to 20:50 hours (Figure 5). During this time the eel made two forays from the surface to depths of 33 m. After 20:50 hours the diminishing number of detections at nose pier 6 and the increase in detections at nose pier 10 and 14 indicate that the eel moved laterally across the forebay and likely passed the Power Dam in the vicinity of Unit 14 on the north side of the Power Dam. Depth during the 10 minutes prior to passing the Power Dam was generally less than 10 m with the eel passing the Power Dam at 15 m.

Eel number 11 approached the mooring transect on the south side of the River at about mooring 29. Movement was lateral across the River, crossing the line of moored receivers near mooring 21 or 17 and continuing to mooring 2 near the northern shoreline (Figure 6). The eel was primarily surface oriented, descending to depths of 10 m to 25 m on only a few occasions. The eel was first detected by the nose pier mounted receivers at about 23:52 hours upstream of nose pier 25 and 29 (Figure 7). It is important to note the overlap in the time of detection between receivers at the mooring transect and receivers on the nose piers. The overlap was approximately 15 minutes and data from each set of receivers show the same movement pattern (i.e., lateral movement from south to north, Figure 6 and Figure 7). After traversing the forebay from the south to north, the eel then reversed direction, and headed back across the forebay, passing through the Power Dam in the vicinity of Unit 17 (Figure 7). Depth of the eel varied continually between the surface and the bottom, approximately 35 m.

Summary and Conclusions

Utilizing a step-wise approach we were able to successfully develop hydrosonic telemetry equipment capable of functioning in an environment with background noise that precluded the

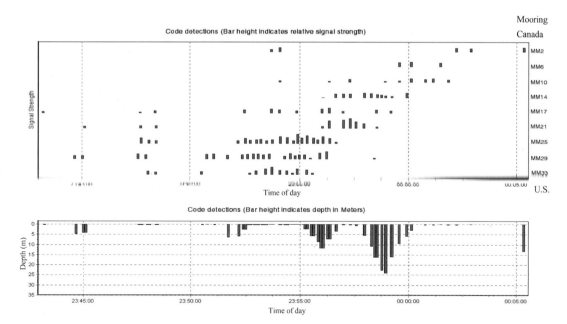

Figure 6. Relative signal strength, depth, and time of electronic detections for Eel 11 from VR25 receivers along a mooring line 300 m upstream of Moses-Saunders Power Dam. The top graph shows relative signal strength recorded for each receiver by time and the bottom graph shows a corresponding depth for each electronic detection. Depth was 0 m where signal detections are indicated in the upper graph but depth data is not visible on the lower graph.

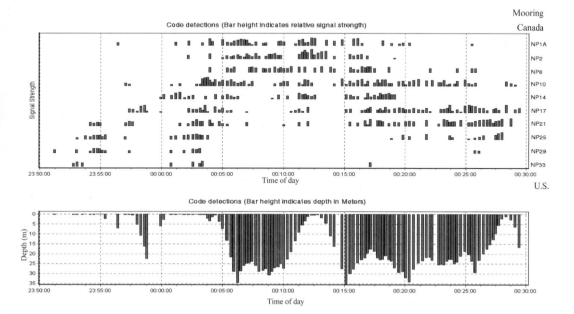

Figure 7. Relative signal strength, depth, and time of electronic detections for Eel 11 from VR25 receivers mounted on nose piers at the Moses-Saunders Power Dam. The top graph shows relative signal strength recorded for each receiver by time and the bottom graph shows a corresponding depth for each electronic detection. Depth was 0 m where signal decodes are indicated in the upper graph but depth data is not visible on the lower graph.

use of standard equipment. Development of this equipment was ultimately a balancing act between the longer signal transmission distances experienced at lower frequencies, reduced background noise at higher frequencies, transmitter size, and time available for equipment development. Major findings of this study were:

1. Background noise levels in the forebay of the Moses-Saunders Power Dam decrease substantially as frequency increased. The intensity of background noise was 15 dB lower at 200 kHz than at 60 to 80 kHz. The intensity of background noise also decreased substantially as distance increased upstream from the Power Dam, dropping approximately 20 dB over a distance of 1000 m;
2. Reception range of hydrosonic telemetry equipment operating at 200 kHz varied from 200 m to 600 m depending on the level of background noise. This range was considered sufficient for monitoring the behavior of downstream migrating eels in the large study area. Reception range was similar regardless of transmitter depth;
3. Reception range was substantially increased by using directional hydrophones and orienting receivers in an upstream facing position to reduce exposure to background noise emanating from the Power Dam;
4. A full-scale array of 19 receivers had a reception area of approximately 800 m long by 1 km wide. The movements of all telemetered eels that entered the reception area were continually monitored by two or more receivers simultaneously, and;
5. The travel route, in three dimensions, of a telemetered eels through the reception area could be determined based on relative signal strength and depth data recorded by the receivers.

The findings of this study largely parallel the experiences reported by Steig (1999) in developing a similar system in the vicinity of a hydroelectric facility on the Columbia River. They too found background noise levels to be high and adapted their system to operate at higher, less noisy frequencies to accurately document the movement of juvenile salmonids in three dimensions.

Based on the findings of the current study it appeared that the hydrosonic equipment developed could be scaled-up to cover a larger reception area and could be fine tuned to provide better resolution on transmitter position. A large scale study would provide information on the movement patterns of migrating eels in the vicinity of the Power Dam and will be helpful in determining ways to minimize the impacts of passage through the Power Dam. A large scale study utilizing this equipment was conducted in 2000. The 2000 study was able to determine the three dimensional position of 62 eels as they migrated through the forebay of the Power Dam. A report and paper are being prepared on this study.

Acknowledgments

We would like to thank the New York Power Authority for support and funding; Tom Tatham of the New York Power Authority for support and sound scientific guidance; and Bill Richkus of Versar Inc., Alastair Mathers of Ontario Ministry of Natural Resources and François Travade of Electricité de France for their technical comments on an earlier draft of this paper.

References

Anderson, G. W., S. R. McKinley, and M. Colavecchia. 1997. The use of clove oil as an anesthetic for rainbow trout and its effects on swimming performance. North American Journal of Fisheries Management 17:301–307.

Baras, E. and D. Jeandrain. 1998. Evaluation of surgery procedures for tagging eel *Anguilla anguilla* (L.) with biotelemetry transmitters. Hydrobiologia 371/372:107–111.

Dutil, J. D., A. Giroux, A. Kemp, G. Lavoie, and J. P. Dallaire. 1988. Tidal influence on movements and on daily cycle of activity of American eels. Transactions of the American Fisheries Society 117: 488–494.

Haro, A. and T. Costro-Santos. 2000. Behavior and passage of silver-phase American eels, *Anguilla rostrata* (LeSueur), at a small hydroelectric facility. Dana. 12:33–42.

Nielsen, L. A., and D. L. Johnson. 1983. Fisheries Techniques. American Fisheries Society, Bethesda, Maryland.

Normandeau Associates and J. Skalski. 2000. Estimation of survival of American eel after passage through a turbine at the St. Lawrence—FDR Power Project. March 1998. Final Report of Normandeau Associates to the New York Power Authority, White Plains, New York.

Steig, T. W. 1999. The use of acoustics tags to monitor the movements of juvenile salmonids approaching a dam on the Columbia River. Presented at the 15th International Symposium on Biotelemetry, Juneau Alaska, USA. 9–14 May, 1999.

Tesch F. W. 1977. The eel. Chapman and Hall. London, England.

Tesch, F. W. 1976. Tracking of silver eels (*Anguilla anguilla*) in different shelf areas of the North West Atlantic. International Council for the Exploration of the Sea, Rapp. Proc. Verb.

Trefethen, P. S. 1956. Sonic equipment for tracking individual fish. Special Scientific Report—Fisheries Number 179. United States Fish and Wildlife Service, Washington, D.C.

Westerberg, J. 1975. Counter current orientation in the migration of the European eel (*Anguilla anguilla* L.). Goteborgs University Oceanography Institution Report No. 9:1–18.

Behavioral Study of Downstream Migrating Eels by Radio-Telemetry at a Small Hydroelectric Power Plant

CAROLINE DURIF AND PIERRE ELIE

Cemagref, Unité Ressources Aquatiques Continentales, 50 avenue de Verdun, 33612 Cestas cedex, France

CLAUDE GOSSET AND JACQUES RIVES

INRA, Laboratoire d'Ecologie des Poissons, BP3, 64310, Saint-Pee-sur-Nivelle, France

FRANÇOIS TRAVADE

EDF, Etudes et Recherches, 6 Quai Watier, 78401 Chatou cedex, France

Abstract.—Eels, because of their size and life cycle, are among the most vulnerable species regarding the presence of obstacles on waterways. During a study of the efficiency of two types of fish passes, the behavior of migrating European eels *Anguilla anguilla* was investigated using telemetry and trapping at a small hydroelectric power plant (southwest of France). Radio-tracking was conducted manually and by stationary receivers in the turbine area and downstream and upstream from the power plant. Sixteen eels were tagged by surgical implantation of transmitters and released upstream of the power station. Results provide insight on eel behavior during the downstream run (swimming rates and delayed migration) as well as behavior in front of both exits to the trap. Almost all tagged individuals moved upstream after the release. Most of these eels migrated downstream after a heavy rainfall, avoiding the power station by crossing the overflowing dam. They were tracked down to the estuary (16 km) over several days during which time several periods of nonmovement occurred. Descending nontagged eels transiting through either of the two tested forebay bypasses were trapped. Daily catches corresponded to movements of radio-tagged individuals. Environmental parameters were recorded and compared to the downstream run. Results clearly showed that silver eel migration was closely linked to certain environmental parameters (flow rate, turbidity, and luminosity) and that downstream migration is inhibited if favorable environmental conditions are not met, such as during daytime when turbidity is low. Direct comparison of daily catches through the bottom and surface bypasses as well as observations of radio-tagged eels in the forebay both suggest that a bottom bypass may be appropriate for safely transiting downstream migrating eels.

Introduction

Numerous fish-passes for upstream migration have recently been installed in France. Protection of downstream migrating juvenile and adult diadromous fish of different species worldwide has also become a major concern in the last decade. Recent studies mainly concern salmonids. Eels, because of their body form and life cycle, are among the most vulnerable species regarding the presence of obstacles on waterways (Berg 1986; Larinier and Dartiguelongue 1989; Richkus and Whalen 1999). Mortality of eels passing through turbines of hydroelectric facilities can be quite significant (Travade and Larinier 1992; Hadderingh and Baker 1998). Very few studies have dealt with behaviors of eels around hydroelectric facilities. It is not known whether eels are simply entrained through the turbines or if they actively search alternate passages. It is generally assumed that eels are bottom dwellers; therefore, their behavior during catadromous migration may differ from more pelagic fishes (Haro and Castro-Santos 1997) and consequently specific bypasses for eels must be developed.

The objective of our study was to examine the behavior of European eels *Anguilla anguilla* during downstream migration in order to find solutions for eels to avoid passage through turbines. Moreover, the efficiency of two types of bypasses (bottom and surface) was tested. The study was conducted at the hydroelectric power plant of Halsou in the southwest of France, from October to December 1999. A trap was installed at the outlet of the bypasses in order to capture eels during their downstream migration. A telemetry survey was conducted on transmitter equipped individuals to test the attractiveness of the bypasses as well as to obtain information on behavior of eels during their downstream run.

Being able to precisely predict downstream runs of eels in response to environmental changes at different time scales may be another way of reducing the impact of hydroelectric facilities by lowering or ceasing turbine generation during migration peaks. Information on the duration of runs and environmental cues is necessary to consider this type of solution. Thus, behavior of radio-tagged eels was analyzed in the vicinity of the power plant downstream to the estuary zone.

Methods

Project Description

The EDF (Electricité de France) hydroelectric power station of Halsou is located 23 km from the sea on the Nive river in the Southwest of France. Its watershed is about 1000 km². The Nive results from inputs of small rivers that originate in the Pyrenean Mountains and flows approximately 80 km to the Adour estuary. The mean daily flow of the river is very unsteady depending on environmental conditions and varies between 6 and 300 m³ s⁻¹. A dam, 172 m long and 2.5 m high, located 2 km upstream of the power station diverts the water into a power canal 925 m long and 11 m wide (Figure 1). Both the power canal and forebay area (Figure 2) are 3–4 m deep. The projectors which usually lighten the forebay area were turned off during the study period. There were no other sources of light in the vicinity. The power plant is equipped with 3 double, horizontal Francis turbines which pass a maximum flow of 30 m³ s⁻¹ over a vertical drop of 4.25 m. The maximum power generated by the plant is 900 kW. A trashrack is located in front of the intakes, with openings between bars measuring 3 cm. Two bypasses are located at the end of the canal in

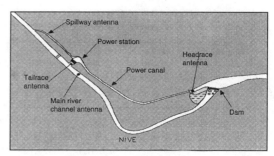

Figure 1. Study site showing the Halsou power station and the locations of radio-telemetry antennas.

the forebay area. The surface bypass measures 1.38 m in width and its opening (maximum depth of 17 cm) can be adjusted by a motorized lever. The bottom bypass (4 m deep) is 1.30 m wide and 1.20 m high and is located 2.70 m from the surface bypass (Figures 3 and 4).

Trapping Conditions

Migrating eels were trapped between 7 October and 6 December 1999. A reception pool fitted with railings (0.5 cm mesh size) was built at the outlet of the spillway in order to collect fish entrained by both bypasses (Figure 3). A discharge tower (Figure 4) was built in the reception pool against the weir (3 m high), in order to maintain an opening wide enough for eels to transit through the bottom bypass without inducing unfavorable hydrodynamic conditions for eels (i.e., small opening and high velocity gradient).

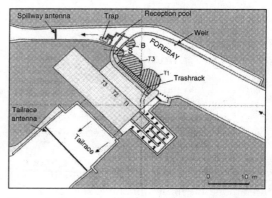

Figure 2. Top view of the power plant. Monitored areas by fixed antennas are represented by gray crossed zones: T1: turbine 1; T3: turbines 2 and 3; S: surface bypass; B: Bottom bypass. Receiver antennas in the spillway and at the tailrace are indicated.

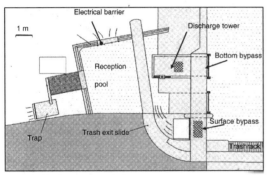

Figure 3. Top view of the trap area.

Bottom and surface bypasses were opened alternatively for 24-hour periods. Thus, fish entered either bypass, transited through the reception pool (via the discharge tower for the bottom bypass), then were entrained into the trap at the outlet. Both bypasses were adjusted so that the flow rate was set at 0.6 m³ s⁻¹, although due to changes in water discharge, it sometimes varied between 0.4 and 1 m³ s⁻¹. Leaves often clogged the trap entrance and the railings of the reception pool causing water to overflow, thus a possible escape of fish; therefore, an opening fitted with an electrical barrier was made in the pool so that leaves were evacuated without losing the eels. The trap was visited twice a day, morning and evening, to collect eels.

Environmental Parameters

Five environmental parameters were recorded continuously (every four minutes). Probes were placed near the water intake and temperature, conductivity, turbidity, water level and flow rate were measured in the canal. Flow rate in the Nive was obtained from the DDE (Direction Departementale de l'Equipement). Rainfall and light level (radiance) were recorded. Barometric air pressure was obtained from Meteo-France. All data are expressed in universal time (GMT).

Eel Measurements

Measurements were taken on all eels caught in the trap: total length (Lt), weight (W), horizontal and vertical diameters of the eye (Dh, Dv), and head width. Eye index (EI) was calculated according to Pankhurst (1982):

$$EI = \left(\frac{Dv + Dh}{4}\right)^2 \times \frac{\pi}{Lt} \times 100$$

Eels were also identified as yellow or silver eels according to their skin color and eye index (Durif et al. 2000). Clove oil was used as an anesthetic for all measurements and tagging (Peake 1998).

Telemetry, Tagging, and Release of Eels

Radio-telemetry was used on fifteen of the eels caught in the trap. Two different types of datalogging radio receivers were used: ATS 2000B and Lotek SRX-400. Two models of transmitters (ATS) were used depending on the size of eels: "10/28" (i.e., length = 45 mm, diameter = 11 mm, and weight = 8 g) and "10/35" (i.e., length = 56 mm, diameter = 12 mm, and weight = 11 g). Both were equipped with motion sensors, which emitted a different signal when eels had not moved for more than eight hours transmitters were implanted by surgery in the abdominal cavity (Baras and Jeandrain 1998). An incision was made in the posterior part of the abdomen and stitched with nylon thread. A hole was made through the body wall 2 cm behind the incision in order to leave an exit for the antenna. The eels were released within 12 hours after the surgery. One eel recaptured eight days after surgery exhibited no infection due to the transmitter. Nine eels were released in the power canal, six other eels were released in the forebay area. Certain eels were also marked with PIT-tags on their dorsal surface. These individuals were used to test the efficiency of the electrical barrier.

Radio-Tracking

Fixed stations were established at eight locations (Figures 1 and 2) around the power plant. Stations T1 and T3 were fixed along the water intake in front of the trashrack and corresponded to turbines 1, 2, and 3. Stations S and B were

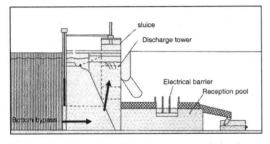

Figure 4. Side view of the trap area and discharge tower. Fish going through the bottom bypass swim up in the discharge tower and fall into the reception pool when the sluice is open.

located respectively in front of the surface and bottom bypasses. The four other receivers covered a distance between 20 and 40 m long; one was placed at the head of the canal. A second antenna was stretched across the main channel of the river Nive. A third receiver detected eels in the tailrace, thus eels that had transited through the turbines and the fourth antenna was set downstream from the trap in the spillway. Tracking was also conducted by foot with a portable receiver on a towpath, 2 km upstream of the dam down to the estuary (23 km). The telemetry monitoring was continued after trapping had ceased, until 15 December.

Results

Timing of Downstream Runs and Characteristics of Silver Eels

A total of 66 eels was collected in the trap between 6 October and 6 December. Downstream runs were irregular and 77% of the eels arrived between 14 and 21 November. 90% of the eels migrated at night. Trapping was inefficient on the night of 18 November as floods caused damage to the trap installation.

Lengths of eels ranged from 30 to 93 cm. Silver phase eels were all over 42 cm so we assumed all were females (Bertin 1951; Rossi and Villani 1980; Vøllestad and Jonsson 1988; Lecomte-Finiger 1990). Eels either had a typical silver eel appearance: white belly, dark back, well separated by the lateral line; in other cases the color was more bronze-like on the side and on the ventral surface rendering it difficult to identify these individuals as silver by color alone. Eye index is related to the degree of maturation of eels (Pankhurst 1982). Durif et al. (2000) showed for *A. anguilla*, that this is also true at an early stage of silvering, before changes of color actually occur. Four eels were obviously yellow as their eye indices were lower than four. Eye indices of silver eels ranged between 6 and 13.5 (mean of 9.5). Accordingly, eels migrated at different phases of the metamorphosis process.

Electrical Barrier Efficiency

Small samples of eels marked with PIT-tags were released in the reception pool in order to test the efficiency of the electrical barrier. Four tests were conducted with both bypasses as hydraulic circulation in the reception pool changed according to the bypass in use, the flow from the surface bypass being directed towards the electrical barrier. Number of recaptured eels (efficiency) varied between 60 and 100% (Figure 5). Changes in conductivity were not important enough to affect the efficiency of the barrier. This is supported by the fact that, unexpectedly, the number of recaptured eels was higher when conductivity was lower. Efficiency (with the bottom bypass) was lower when small eels were tested.

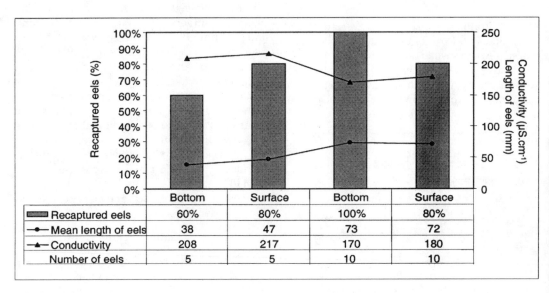

Figure 5. Efficiency (percentage of recaptured eels) of the electrical barrier according to the bypass in use, conductivity and length of eels.

Behavioral Study of Downstream Migration

Migration routes: Radio-tagged eels were released either in the power canal or the forebay. Ten eels out of 15 swam upstream out of the canal. Once upstream of the study site, eels had two possibilities to migrate: they either entered the canal towards the power station or passed over the dam to reach the main channel of the river; the latter was only possible when the dam was overflowing. The majority swam over the dam as only 5 eels out of 15 passed through either the turbines or the bypass.

Downstream runs in relation to environmental parameters: Radio-tagged eels did not migrate during the new moon nor did the number of trapped eels increase. In contrast, strong relationships between downstream runs and changes in other environmental conditions were observed through captures in the trap and through individual behaviors of radio-tagged eels. Downstream runs increased when water temperature dropped from 15 to 10°C over a period of four days (Figure 6). Major run periods corresponded to heavy rainfall, an increase in flow rate, turbidity and conductivity, especially

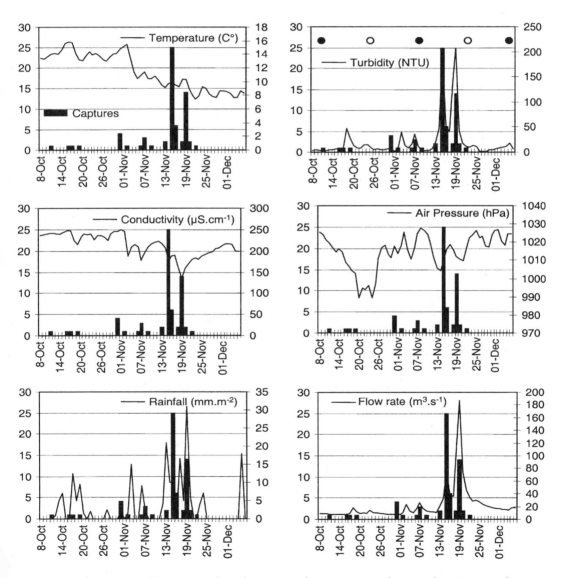

Figure 6. Daily captures of migrating eels in the trap in relation to moon phases and environmental parameters (x-axis: sample date; y-axis (left): daily number of captures; y-axis (right): value of measured parameter).

between 14 and 21 November. Captures of eels were significantly correlated ($P < 0.05$) with rainfall, turbidity, flow rate, and conductivity (Table 1). On 1 November, four eels were captured two days before the actual changes in rainfall, flow rate, turbidity, and conductivity. The only noticeable change coinciding with these captures was a slight rainfall and a drop in air pressure. No migration occurred between 20–30 October even though air pressure had notably decreased (Figure 6).

Observations of radio-tagged eels characterized individual behaviors before and during their migration down to the estuary. After release, eels all swam to two specific locations either in the forebay area or 2–3 km upstream of the facility where they remained for various periods of time. They moved very little and as transmitters were equipped with a motion sensor, it was noted that they often remained still for four to five days at a time. Timing between release of eels and directed downstream movements ranged from 1 to 28 days. Thus, departure did not seem to be related to the day of release but corresponded to specific environmental conditions (Figure 7). Downstream runs of tagged fish took place on three occasions. Not all eels made a continuous run to the estuary. Seven eels (out of 15) stopped less than 24 hours after they had left. Their second run took place 1–26 days later (Figure 7; Table 2) when environmental conditions were favorable. One eel was recaptured in the trap. Migration episodes occurred when air pressure dropped (1005 hPa minimum) but not all decreases in air pressure resulted in downstream movements. Conductivity also decreased to at least 175 mS cm^{-1} during these episodes. High turbidity and flow rate peaks appear to be the parameters most related to peaks in eel runs, but on 16 and 17 November, eels stopped moving when turbidity and flow rate decreased suddenly. They resumed their migration when again these two parameters increased. Two individuals (tags 410 and 871) did not follow the same pattern in their behavior as their migration occurred outside of the environmental window, no eels were caught in the trap on those days, and environmental parameters were not favorable. Tag signals of eel 871 were lost on the 24th, but because this eel was not looked for on the following days we cannot be sure it had really left on that day. Eel 410 was the only recaptured tagged eel, and we cannot make conclusions regarding its migrating behavior.

Because time of passage had been recorded at fixed receivers, we analyzed onsets of migration on a 24 hour scale. Before their run tagged eels were less than 3 km away from the receivers; thus the recorded time corresponded approximately to their time of migration. Departures of eels are related to environmental parameters that showed significant variations over a 24 hour period on days when migrations occurred (Figure 8). Since continuous data on flow rate in the Nive was not available, we used turbidity as a surrogate. Almost all eels departed when radiance was lower than 100 μmol s^{-1} m^{-2}. One eel left at midday when radiance was around 300 μmol s^{-1} m^{-2}; however, turbidity was maximum at that moment. Migration occurred when turbidity was at least 60 NTU except for eel 410 on 30 November (Figure 8).

Behaviors of Migrants In the Forebay

Radio-tagged eels: The first nine radio-tagged eels were released in the power canal. Six of them swam upstream and did not enter the forebay area at all. It was further decided to release the six other eels closer to the power station. In total,

Table 1. Pearson correlation matrix of daily catches of eels in the trap and values of environmental parameter. Correlation coefficients in bold are statistically significant ($P < 0.05$).

	Eels	Temperature	Turbidity	Conductivity	Air Pressure	Rainfall	Flow rate
Eels	1						
Temperature	–0.091	1					
Turbidity	**0.824**	–0.147	1				
Conductivity	**–0.395**	**0.734**	**–0.602**	1			
Air Pressure	–0.078	**–0.405**	–0.121	–0.276	1		
Rainfall	**0.412**	–0.092	**0.587**	**–0.403**	–0.084	1	
Flow rate	**0.637**	–0.325	**0.890**	**–0.757**	0.004	**0.591**	1

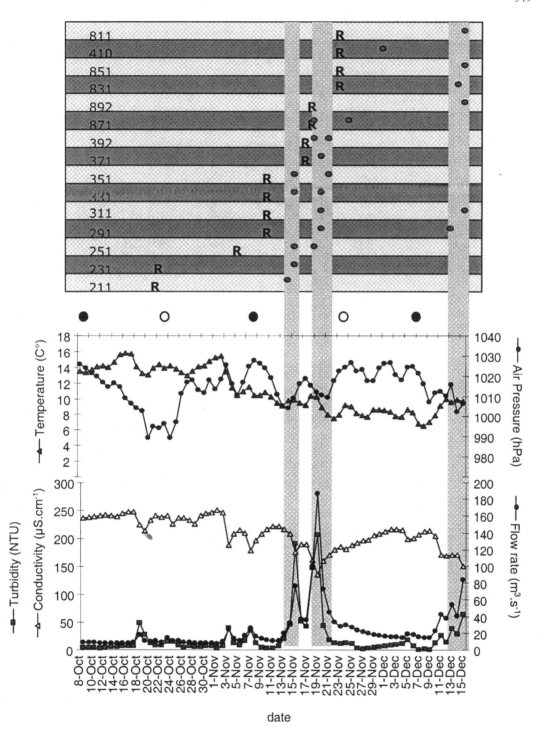

Figure 7. The graphic on the top shows for each tagged individual their date of release (R) and the day(s) on which they made a downstream run. All migration episodes lasted less than 24 hours and were thus represented by a dot. Tag numbers are listed on the left. Departures of eels are related to moon phases and environmental parameters (bottom graphics). Time intervals represented by shaded rectangles indicate the time periods (favorable meteorological windows) when almost all eels migrated.

Table 2. Movements of the 15 radio-tagged eels. Nine eels were released in the power canal and 6 in the forebay. For the first downstream run, date and time (hours) were obtained from receivers. Their second run usually occurred further downstream, out of the surveillance zone and date was obtained from manual tracking.

Release location	Tag number	Length of eels (cm)	Date of release	Upstream movement	Downstream passage	Rest period before 1st run	First run		Rest period before 2nd run	Second run
Power Canal	211	57	20 Oct.	Yes	Dam	23 d	14 Nov.	23:28		
	231	63	21 Oct.	Yes	Dam	23 d	15 Nov.	11:57		
	251	59	4 Nov.	Yes	Dam	11 d	15 Nov.	18:27	3 d	18 Nov.
	291	68	9 Nov.	Yes	Dam	10 d	19 Nov.	1:28	24 d	13 Dec.
	311	65	9 Nov.	Yes	Dam	10 d	19 Nov.	3:20	26 d	15 Dec.
	331	68	9 Nov.	Yes	Dam	6 d	15 Nov.	4:32	3 d	18 Nov.
	351	73	9 Nov.	Yes	Dam	6 d	15 Nov.	5:32	6 d	21 Nov.
	371	65	16 Nov.	Yes	Dam	3 d	19 Nov.	18:16		
	392	61	16 Nov.	No	Turbines	2 d	18 Nov.	19:34	3 d	21 Nov.
Forebay	871	81	17 Nov.	No	Bottom bypass	1 d	18 Nov.	14:30	5 d	24 Nov.
	892	91	17 Nov.	Yes	Dam	28 d	15 Dec.	18:34	28 d	
	831	93	22 Nov.	No	Bottom bypass	22 d	14 Dec.	16:42		
	851	82	22 Nov.	No	Bottom bypass	23 d	15 Dec.	18:38		
	410	71	22 Nov.	No	Bottom bypass	8 d	30 Nov.	18:02		
	811	90	22 Nov.	Yes	Dam	23 d	15 Dec.	15:32		

the behavior of nine eels was observed in the forebay area (Table 2). Only one individual (Tag 392; head width of 3.5 cm) transited through the turbines (T1). It was not detected by any other receiver and we therefore conclude that it headed straight for the turbines and did not search for any other means of passage. At that moment the generated flow rate was at its maximum (approximately 28 m^3 s^{-1}). Four eels went through the bottom bypass at moments when the generated flow varied between 15 and 28 m^3 s^{-1}. They all first made incursions in front of the trashrack. Their head width ranged between 4.5 and 6 cm so we assumed that the narrow trashrack blocked their passage. Two eels (Tags 811 and 892; head widths of 5 and 5.5 cm) spent only five minutes in the forebay area before swimming upstream. The generated flow was approximately 10 and 20 m^3 s^{-1}, respectively. Both were detected by the trashrack and bottom bypass receivers. Eel 211 (head width of 3.5 cm) stayed 34 hours in the forebay area before the upstream movement. It moved on rare occasions to the bottom bypass and trashrack but did not go through either even though its size permitted passage through the bars. The generated flow during that period was approximately 5 m^3 s^{-1}. Eel 311 (head width of 4 cm) returned twice to the power plant (Figure 9). It made several incursions in front of the trashrack and the bottom bypass but did not transit through either structure. The generated flow at that time was approximately 3.9 m^3 s^{-1}. It left the canal on 1 December and migrated over the dam on 15 December. The flow generated by the plant was also low on that day: 5 m^3 s^{-1}. For all observed eels, incursions were most frequent around the trashrack then at the bottom bypass. Very few movements were detected around the surface bypass.

Bottom versus surface bypass: Monitoring of radio-tagged individuals showed that eels were more attracted to the bottom bypass than the surface bypass. Attractiveness of both bypasses was also evaluated by daily captures according to the bypass in use. During the first part of the study, we were able to maintain a 24 hour cycle until 15 November for each bypass. The number of eels trapped with the bottom bypass was considerably and significantly greater (94%) than with the surface bypass (6%) (chi-square test, $P < 0.1$). However, as trapping conditions were considerably ameliorated with the bottom bypass because fewer leaves and debris reached the reception pool, the direct comparison of

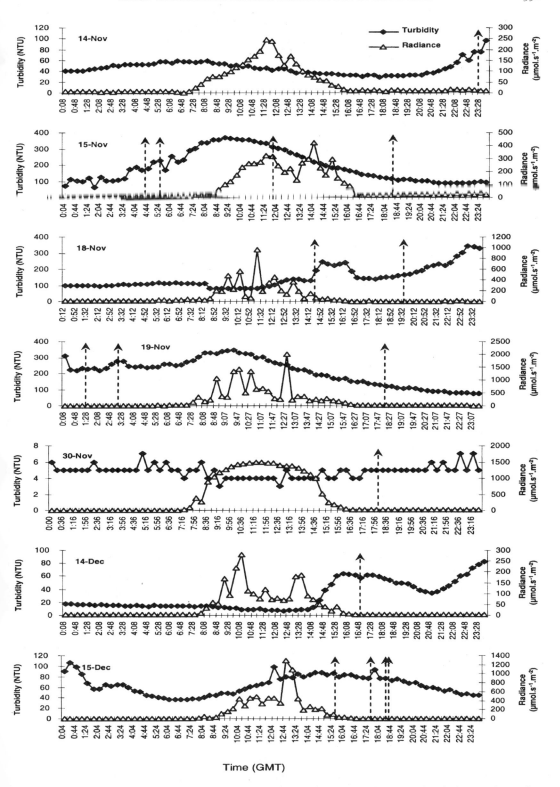

Figure 8. Migrating behavior of eels at a 24 hour scale in relation to radiance and turbidity. Each graphic represents one day. Arrows correspond to their timing of departure (time at which they were detected by receivers).

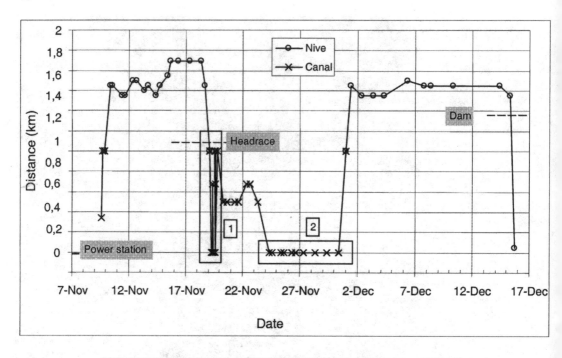
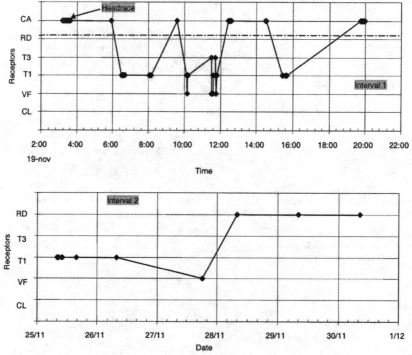

Figure 9. Behavior of eel 311 through radio-tracking data. The two lower graphics represent enlargements of the time the eel spent in the forebay area, materialized in the top graphic by rectangles 1 and 2.

Discussion

Characteristics of Silver Eels

No male eels were captured in the trap during the study period. Observations of males migrating earlier than females have often been made (Tesch 1979; Haraldstad et al. 1985; Jessop 1987). This would be due to a difference in geographical location of the two sexes: as size of eels increases with distance from the sea, males would be located further downstream (Deelder 1954; Tesch 1979; Helfman et al. 1984; Vøllestad et al. 1986; Moriarty 1986; Helfman et al. 1987; Krueger and Oliveira 1997). The fact that the larger eels arrived later at the study location supports this. Either, our study site was located too far upstream to catch any migrating males, or we missed a peak of migration as trapping started in early October, whereas migrating season may start sooner. Another possible explanation concerning the absence of males in our sample, is that smaller eels (i.e., males) were entrained in the turbines. However, 26% of the total catch had a head width less than 3 cm (clear distance between rails of the trashrack) so the lack of males cannot be entirely attributed to passage through turbines.

Behavior of Downstream Migrants In Relation to Environmental Parameters

Downstream migration was analyzed by two means: telemetry and trapping of eels. Effect of handling and radio-tagging of eels on their behavior is difficult to assess (Richkus and Whalen 1999). We detected no observable influence of tagging on migrating behavior over the course of the study. Upstream movement of eels after their release may have been a short term response to capture, handling, and tagging. Although, this behavior could also be attributed to the sudden change of environment when eels approached the power plant and the lack of a suitable area for them to remain until they resumed their migration. Moreover, in the present study the delay between release of eels and onset of migration was extremely variable and strongly related to environmental parameters. All departures of radio-tagged eels corresponded to peaks of migrants in the trap. This supports the assumption that the potential bias induced by surgical implantation of transmitters on a certain number of individuals is limited.

Temperature has been largely identified as a probable trigger for downstream migration. Although no definite threshold value can be given for downstream runs of eels, it is well known that migration coincides with a drop in temperature (Lowe 1952; Westin and Nyman 1979; Westin and Nyman 1979). In our study, an increase in the number of migrating individuals actually took place when temperatures dropped significantly from 14 to 10°C. A sudden decrease in temperature, rather than a threshold may be a trigger for downstream migration. This would explain the wide variability in thermal preferenda at the onset and during migration as well as the unexpected spring/summer silver eel runs observed by fishermen and other researchers (Frost 1950; Boëtius 1967; Haro 1991). Migration episodes may also correspond to drops in barometric pressure (Lowe 1952; Deelder 1954; Hvidsten 1985). Here again, it seems that there is no threshold value as migration did not increase when pressure was at its lowest, but that the effect may have been induced by the sudden change.

Four environmental parameters were significantly correlated to eel catches: turbidity, conductivity, rainfall and flow rate. These variables were all correlated. Observations of radio-tagged eels gave us clues as to which parameter had an influence on migration. The 24 hour scale observations suggest that eels wait for the darkest conditions to migrate when radiance is low and turbidity is high. This parameter is not a trigger but a requirement as luminosity will inhibit migration since eels rarely migrated during daytime. It is probably through the effect of light that moon phases influence migration events as there was no link between lunar cycle and downstream runs. Conductivity is proportional to salinity. The maximum change in conductivity during the study period was 100 mS cm^{-1}. This corresponds approximately to a salinity lower than one. Whether eels can detect such small salinity differences is unknown and needs further testing.

Both approaches of the migration phenomenon—telemetry and trapping—indicate that water discharge seems to be one of the most influential parameters such as stated by fishermen and other researchers (Frost 1950; Lowe 1952; Deelder 1954; Hain 1975; Tesch 1979; Haraldstad et al. 1985; Hvidsten 1985; Vøllestad et al. 1986; Vøllestad et al. 1994; Hadderingh et al. 1999; Boubée et al. 2001). We believe it acts as a trigger as well as a migratory vector such as for glass-eels (Elie and Rochard 1994) that will allow migration if the above environmental conditions are met.

Variations in barometric pressure and consequently water temperature appear necessary for the onset of the silver eel migration and may have a "wakening" effect on the eel. These changes may stimulate the first movements of eels after their sedentary period (yellow phase) during which metamorphosis occurs. If conditions in light (luminosity/turbidity) are favorable, then adequate water discharge will trigger eels to migrate. Thus, intensity of runs will depend on the synergy of these parameters which will determine the onset and the persistence of the phenomenon.

As we have seen in our study it may take several favorable "meteorological windows" for eels to complete their descent of the river even over a short distance as in the case of the Halsou system (23 km). Eels alternated between periods of migration and rest. Observations were made on a limited number of individuals as radio-monitoring can only involve small samples at a time. Moreover, during a mark–recapture experiment, Therrien and Verreault, (this volume) observed that migration speed was highly variable as eels took 2–44 days to travel a distance of 7800 m. In our study, all eels did manage to reach the estuary before the end of the migratory season, however, in a larger catchment with greater distances to swim, and if unfavorable migrating conditions persist, one can hypothesize that silver eels may not complete their descent in time, particularly when obstacles (dams, hydroelectric facilities) are present. The silvering process would continue as eels descend the river. This would explain the variability in stages of metamorphosis observed in silver eels.

Behavior of Eels At Obstacles and Bypasses

Our results show that bottom types of bypasses may be more appropriate for this species according to their behavior during downstream migration. However, as we have already mentioned, the comparison between daily catches with either bypasses was biased: first of all, because trapping efficiency was considerably improved with the bottom bypass and second, because the number of eels varied from day to day depending on environmental conditions. Thus, monitoring of radio-tagged eels appears to be the most appropriate method. In this way, we did observe that eels were more attracted to the bottom bypass, but these observations were made on a limited number of individuals and this point needs further investigation. Turbine intake flow is still very attractive for downstream migrating eels compared to the low bypass flow. Moreover, it seems that only body size will prevent eels from passing through trashracks when approach velocities are high. Male and female silver stage European eels shorter than 65 cm have a head width equal to, or smaller than 3 cm, and are thus likely to pass through turbines. Under these circumstances, the only demonstrated solution to turbine entrainment of eels is to decrease the bar spacing in front of water intakes. Further studies should lead towards defining a threshold value for bar-spacing according to size of migrants. Still, high approach velocities can result in impingement and death/injury of eels, regardless of bar spacing (Haro, personal communication).

In our study, the majority of eels arriving in the forebay chose to pass over the dam upstream of the canal. During a similar study on the behavior of silver American eels at a small hydroelectric facility, (Haro et al. 2000) observed that telemetered eels moved upstream several times and seemed reluctant to pass the power plant. Watene et al. (this volume) indicate the same type of behavior on telemetered individuals, as one eel made up to 23 attempts to pass the facility. This particular behavior of eels could be an advantage as bypasses could be installed further upstream; in the case of Halsou, at the dam level. Finally, another potential solution is to cease power generation during migration episodes as the runs seem to be restricted to short intervals corresponding to specific "meteorological windows." This would be feasible in small hydroelectric plants such as Halsou. Boubée et al. (2001) indicate that rainfall can be a good predictor as it anticipates water flow. Regardless of the mitigation measure (alternate generation schedule, installation of bypasses and behavioral or physical barriers) site specific models including several practical environmental parameters (temperature, rainfall, turbidity and luminosity) must be developed to predict migrations of silver eels and optimize their safe passage of hydroelectric facilities.

Acknowledgments

We thank Claude Garaïcoechea for all the technical support as well as for telemetry monitoring. We are grateful to Jacques Pinte for taking part in the lay out of the project and for his support throughout the study and to the EDF staff at the hydroelectric station of Halsou for their collaboration. We also appreciate the valuable comments of Alex Haro and Terry Euston on an earlier draft of this paper.

References

Baras, E., and D. Jeandrain. 1998. Evaluation of surgery procedures for tagging eels *Anguilla anguilla* (L.) with biotelemetry transmitters. Hydrobiologia 371/372:107–111.

Berg, R. 1986. Fish passage through Kaplan turbines at a power plant on the River Neckar and subsequent eel injuries. Vie Milieu 36:307–310.

Bertin, L. 1951. Les anguilles. variation, croissance, euryhalinité, toxicité, hermaphrodisme juvénile et sexualité, migrations, métamorphoses. Payot, Paris, France.

Boëtius, J. 1967. Experimental indication of lunar activity in European silver eels, *Anguilla anguilla* (L.). Meddelelser fra Danmarks Fiskeri Havunders 6:1–6.

Boubée, J. A., C. P. Mitchell, B. L. Chisnall, D. W. West, E. J. Bowman, and A. Haro. 2001. Factors regulating the downstream migration of mature eels (*Anguilla* spp.) at Aniwhenua Dam, Bay of Plenty, New Zealand. New Zealand Journal of Marine, and Freshwater Research 35:121–134.

Deelder, C. L. 1954. Factors affecting the migration of the silver eel in Dutch inland waters. Journal du Conseil International pour l'Exploration de la Mer 20:177–185.

Durif, C., P. Elie, S. Dufour, J. Marchelidon, and B. Vidal. 2000. Analysis of morphological, and physiological parameters during the silvering process of the European eel (*Anguilla anguilla* L.) in the lake of Grand-Lieu (France). Cybium 24 (3):63–74.

Elie, P., and E. Rochard. 1994. Civelle (*Anguilla anguilla*) migration in estuaries, process and specimens characteristics. Bulletin Français de la Pêche et de la Pisciculture 335:81–98.

Frost, W. E. 1950. The eel fisheries of the river Bann, Northern Ireland and observations on the age of silver eel. Journal du Conseil Permanent International pour l'Exploration de la Mer 16: 358–393.

Hadderingh, R. H., and H. D. Baker. 1998. Fish mortality due to passage through hydroelectric power stations on the Meuse and Vecht rivers. Fish Migration and Fish Bypasses. M. Jungwirth, S. Schmutz and S. Weis. Oxford, Fishing News Books: 315–328.

Hadderingh, R. H., G. H. Van Aerssen, R. F., De Beijer, and G. Van der velde. 1999. Reaction of silver eels to artificial light sources and water currents: an experimental deflection study. Regulated Rivers: Research and Management 15:365–371.

Hain, J. H. W. 1975. The behaviour of migratory eels, *Anguilla rostrata*, in response to current, salinity and lunar period. Helgoländer wiss Meeresunters 27:211–233.

Haraldstad, O., L. A. Vøllestad, and B. Jonsson. 1985. Descent of European silver eel, *Anguilla anguilla* L., in a Norwegian watercourse. Journal of Fish Biology 26:37–41.

Haro, A. J. 1991. Thermal preferenda and behavior of Atlantic eels (genus *Anguilla*) in relation to their spawning migration. Environmental Biology of Fishes 31:171–184.

Haro, A., and T. Castro-Santos. 1997. Downstream migrant eel telemetry studies, Cabot Station, Connecticut River, 1996. Conte Anadromous Fish Research Center, Turners Falls, Massachusetts.

Haro, A., T. Castro-Santos, and J. Boubée. 2000. Behavior, and passage of silver-phase American eels, *Anguilla rostrata* (LeSueur) at a small hydroelectric facility. Dana 12:33–42.

Helfman, G. S., E. L. Bozeman, and E. B. Brothers. 1984. Size, age, and sex of American eels in a Georgia River. Transactions of the American Fisheries Society 113:132–141.

Helfman, G. S., D. E. Facey, and L. S. Hales. 1987. Reproductive ecology of the American eel. Pages 42–56 *in* M. J. Dadswell, R. J. Klauda, C. M. Moffitt, R. L. Saunders, R. A. Rulifson, and J. E. Cooper, editors. Common strategies of anadromous and catadromous fishes. American Fisheries Society, Symposium 1, Bethesda Maryland.

Hvidsten, N. A. 1985. Yield of silver eel and factors effecting downstream migration in the stream Imsa, Norway. Institute of Freshwater Research 62:75–85.

Jessop, B. M. 1987. Migrating American eels in Nova Scotia. Transactions of the American Fisheries Society 116:161–170.

Krueger, H., and K. Oliveira. 1997. Sex, size, and Gonad Morphology of Silver American Eels *Anguilla rostrata*. Copeia 2:415–420.

Larinier, M., and J. Dartiguelongue. 1989. The movement of migratory fish: transit through turbines of hydroelectric installations. Bulletin Français de la Pêche et de la Pisciculture 312: 1–94.

Lecomte-Finiger, R. 1990. Métamorphose de l'anguille jaune en anguille argentée *Anguilla anguilla* L. et sa migration catadrome. Année Biologique 29:183–194.

Lowe, R. H. 1952. The influence of light and other factors on the seaward migration of the silver eel (*Anguilla anguilla* L.). Journal of Animal Ecology 21:275–309.

Moriarty, C. 1986. Riverine migration of young eels *Anguilla anguilla* (L.). Fisheries Research 4:43–58.

Pankhurst, N. W. 1982. Relation of visual changes to the onset of sexual maturation in the European eel *Anguilla anguilla* L. Journal of Fish Biology 21:127–140.

Peake, S. 1998. Sodium bicarbonate and clove oil as potential anesthetics for nonsalmonid fishes. North American Journal of Fisheries Management 18:919–924.

Richkus, W., and K. Whalen. 1999. American eel (*Anguilla rostrata*) scoping study: A literature and data review of life history, stock status, population dynamics, and hydroelectric impacts. EPRI, Palo Alto, California: 1999, TR-111873.

Rossi, R., and P. Villani. 1980. A biological analysis of eel catches, *Anguilla anguilla* L., from the lagoons of Lesina and Varano, Italy. Journal of Fish Biology 16:413–423.

Tesch, F. W. 1979. The Eel: Biology and management of anguillid eels. Chapman and Hall, London, England.

Travade, F., and M. Larinier. 1992. Downstream migration problems and facilities. Bulletin Français de la Pêche et de la Pisciculture 326-327:165–176.

Vøllestad, L. A., B. Jonsson, N. A. Hvidsten, T. F. Naesje, O. Haralstad, and J. Ruud-Hansen. 1986. Environmental factors regulating the seaward migration of European silver eels (*Anguilla anguilla*). Canadian Journal of Fisheries and Aquatic Sciences 43:1909–1916.

Vøllestad, L. A., and B. Jonsson. 1988. A 13-year study of the population dynamics and growth of the European eel *Anguilla anguilla* in a Norwegian river: evidence for density-dependent mortality, and development of a model for predicting yield. Journal of Animal Ecology 57: 983–997.

Vøllestad, L. A., B. Jonsson, N. A. Hvidsten, and T. F. Naesje. 1994. Experimental test of environmental factors influencing the seaward migration of European silver eels. Journal of Fish Biology 5:641–651.

Westin, L., and L. Nyman. 1977. Temperature as orientation cue in migrating silver eels, *Anguilla anguilla* (L.). Contributions from the Askö Laboratory 17:1–16.

Westin, L., and L. Nyman. 1979. Activity, orientation, and migration of baltic eel (*Anguilla anguilla* L.). Rapports et procés-verbaux des réunions, Conseil International pour l'Exploration de la Mer 174:115–123.

Simulated Effects of Hydroelectric Project Regulation on Mortality of American Eels

ALEX HARO AND THEODORE CASTRO-SANTOS

S. O. Conte Anadromous Fish Research Center, Biological Resources Division, U. S. Geological Survey, Post Office Box 796, Turners Falls, Massachusetts, USA

KEVIN WHALEN

Bureau of Land Management, 1620 L Street, N.W., Washington, D.C. 20426, USA

GAIL WIPPELHAUSER

Maine Department of Marine Resources, 21 State House Station, Augusta, Maine 04333, USA

LIA MCLAUGHLIN

U. S. Fish and Wildlife Service,10950 Tyler Road, Red Bluff, California 96080, USA

Abstract.—We used six years of silver-phase American eel *Anguilla rostrata* catch data from a weir on a Maine river to determine if river flow and rainfall were predictive of eel migration timing and simulate operational modifications at a hypothetical hydroelectric project to mitigate eel turbine and spill-induced mortality, using eel run timing characteristics and environmental data. We found a significant positive correlation between daily river flow and daily proportion of the run, and significantly more eels were captured on days when rain events occurred than on days when no rain occurred. There was no correlation between mean flow during the run and: timing of run initiation (first 5% of the run); number of days between 5 and 25% cumulative migration; or total number of eels captured over the entire run. Simulations showed that mortality of the entire run decreased with increasing spill flow, and also decreased significantly when generation was suspended on days with significant rainfall. Suspending hydro project operations on dates encompassing 25–75% of the cumulative eel catch caused a reduction in eel mortality of two-thirds to one-half relative to normal operation. Run mortality was further halved when limits on hydro project operation were set using a combination of rainfall events and eel run timing factors. As a strategy for consistently reducing run mortality on an annual basis, suspending generation for a seven day period during the most probable time of peak downstream movement was as unreliable as normal project operation, and was less than half as reliable as limiting hydro project operations on dates encompassing 25–75% of the cumulative eel catch (~ 30 days). Our simulations provide guidance when modification of hydroelectric project operations is being considered as a mitigative tool for downstream passage of eels, and in identifying critical areas for future research.

Introduction

Passage of adult downstream migrant eels *Anguilla* at hydroelectric projects and other barriers has become an issue of concern to management agencies, due to recent population declines of several species (Haro et al. 2000). Mortality of eels due to turbine passage (hereafter turbine mortality) has been estimated in many cases to be greater than 25%, due to the large size of adult eels (EPRI 1999). Further, rates of turbine-induced injuries can be as high as 50% for small eels (22–83 cm TL; Berg 1986) and up to 100% for large eels (> 70 cm TL; Montén 1985). Turbine mortality and injury of eels is also related to design, size, and speed of turbine runners (EPRI 2001).

Reduction of turbine mortality by conventional screening and guidance methods has been problematic. Screens with meshes less than 1.5 cm (Therrien 1999), or trash rack spacings of 2 cm or less coupled with approach velocities less than 0.5 m/s (Adam and Schwevers 1997) are presently the only mechanical barrier types known to be effective for excluding eels from intakes. Behavioral barriers such as light arrays (Hadderingh et al. 1992) have shown some promise in reducing entrainment for eels, but their performance has been variable, depending on environmental conditions and hydraulic characteristics (e.g., turbidity, water velocities) of the site.

One strategy that has been considered for reducing numbers of eels entrained at hydro projects involves altering generation schedules to suspend turbine operation and provide spill flows during times of peak downstream migration (EPRI 1999). Because characteristics of run timing of migratory (silver-phase) eels are rarely well known, an altered generation approach may require generation to be suspended for extended periods during the fall run. Although suspending generation for the entire duration of the migration of silver phase eels (as long as three months) may be an economical and flexible approach for some hydro projects (compared to structural modifications), for larger or peak-demand projects it may not be cost effective or considered a reasonable option.

A clearer understanding of migratory cues for silver phase eels and their run characteristics would help to refine this strategy by targeting periods when downstream migration events are most likely. This could be achieved by identifying environmental conditions, such as rainfall or peaks in flow, when pulses of downstream movement are most likely to occur. Boubée et al. (2001) observed that rainfall triggers were highly correlated with pulses of downstream movements of New Zealand eels, and could be used to alter project operation economically, with significant reduction in overall mortality. Movements of European eels also appear to be related to flow and temperature (Vøllestad et al. 1986), but relationships are less clear for American eels (Euston et al. 1997).

In this paper, we use a six-year dataset of silver-phase eel catches from a small Maine river to analyze the relationship between run timing and environmental variables and assess the effects of three different operational scenarios at a simulated hydroelectric project on eel mortality. Our analysis aims to illustrate the potential of modifying project operation to reduce turbine mortality of eels.

Methods

We obtained daily records of total eel catch (total pounds of eels/d) from an anonymous fisherman who operated an eel weir on a small Maine river during a period from 17 August to 25 October for six consecutive years (1992–1997). We transformed total eel catch in pounds of eels per day to the total number of eels captured per day using length distributions and length-weight relationships for Maine eels given by Oliveira and McCleave (1999). The fisherman also recorded rain events (rain, no rain) on a daily basis. Daily total precipitation data were obtained from the NOAA National Climatic Data Center for a rainfall gauge approximately 15 km west of the weir site. Daily river flow data were obtained from a USGS gauge approximately 6 km downstream of the weir. River temperature data were unavailable for the watershed.

We tested for correlations between mean river flow for the period of weir operation and: timing of run initiation (date of 5% cumulative eel descent, sensu Vøllestad et al. 1986); duration of the initial part of the run (number of days between 5 and 25% cumulative eel descent); and size of the run (total number of eels captured) (partial correlation, controlling for annual effect; Spearman's ρ; Sokal and Rohlf 1995). We also tested for correlations between daily catch and daily river flow and daily gauge precipitation. Number of eels captured on days when rain events (as recorded by the fisherman) occurred was compared to number captured on days when no rain event occurred (Mann–Whitney U-test; Sokal and Rohlf 1995). We generated a cumulative distribution function (CDF) for each of the six years of eel catch data and then randomly sampled the CDF's to produce a combined cumulative distribution function that was representative of the six sampling years (see Whalen et al. 2000). The combined CDF was used to estimate dates of 25, 50, and 75% cumulative passage of the run.

Parameters for the hydroelectric dam mortality simulation were based on typical projects in the northeastern United States built on watersheds of a scale similar to the watershed upstream of the eel weir. Total drainage area was set to 375.5 km^2 (145 mi^2), and maximum project flow was 5.66 m^3/s (200 ft^3/s) approximating the mean annual flow for the river system. Minimum spills were set at 0, 0.25, 0.5, 1.0, 2.0, and

4.0 m³/s (bracketing U. S. Fish and Wildlife Service guidelines of 0.2–0.5 ft³/s for every square mile of drainage; Larsen 1980; Kulik 1990). The simulation emulated a run-of-the-river type project with no storage capacity. Daily inflows to the simulated project were divided into project generation flows that subjected eels to turbine mortality, and into spill flows that subjected eels to mortality caused by passage through sluice gates or spillways. It was assumed that eels exhibited no behavioral preference for turbine or spill routes, and passage route was randomly distributed with respect to flow.

We created six project operation scenarios as follows: *Normal Operation*, simulated mortality occurring under normal project operations having no operational mitigative procedures to reduce eel mortality; *Rain Event Suspension*, simulated mortality incurred by suspending hydro generation and directing all flow to spill on days when rain events occurred (as noted in the weir catch dataset); *25–75% Cumulative Passage Suspension*, simulated mortality incurred by suspending all generation and directing all flow to spill between dates typically encompassing 25–75% of the cumulative eel passage; *Seven Day Suspension*, simulated mortality incurred by suspending generation and directing all flow to spill for a seven day period centered on the expected date of 50% cumulative eel passage; combination of Rain Event Suspension with 25–75% Cumulative Passage Suspension; and combination of Rain Event Suspension with Seven Day Suspension.

Effects of Operation Scenario On Run Mortality Estimations

The total flow for a given day (USGS gauge data) was apportioned to generation flow and spill flow. If total flow was less than minimum spill, spill flow equaled total flow, and generation flow was zero. Total flow in excess of minimum spill was allocated to generation flow, up to the maximum project flow (5.66 m³/s). Total flows in excess of maximum project flow were added to the spill flow.

The daily number of eels lost to turbine mortality was calculated by multiplying the proportion of total flow passing through the turbines by the number of eels passed that day, which was then multiplied by an assumed turbine mortality rate of 25% (EPRI 1999; 2001). Daily number of eels lost to spill mortality was calculated similarly, using proportion of spill flow and an assumed spill mortality rate of 2% (based on spill mortality data for other fish species; Heisey et al. 1995). Daily generation and spill losses were summed to generate the total number of eels lost per day.

Daily values of number of eels lost were summed over the 70 day run period and divided by the total number of eels observed for the run to derive an estimate of the proportion of the entire run lost to turbine and spill mortality (hereafter termed run mortality). The other operation scenarios utilized steps similar to Normal Operation, with modifications as described above.

Large inter-annual variability occurs in eel migration timing. Failure to consider this variability when developing an operational mitigation alternative for eel passage could adversely affect survival. We determined the inter-annual variability of each operational scenario by calculating the standard deviation (SD) of arcsine transformed percent run mortality for each scenario and each spill rate over the six simulation years (Sokal and Rohlf 1995). We considered run mortality incurred by operational scenarios having higher SDs to be less predictable between years because, in a practical sense, they have a greater likelihood of resulting in a run mortality value different from the modeled average. Calculating the SD also enabled us to compare the unpredictability of implementing each operational scenario relative to the absolute reduction in mortality the scenario returned relative to Normal Operation. We also tested for differences among operational scenarios with a two-way Analysis of variance (ANOVA) on the combined effects of scenario and spill rate on arcsine transformed run mortality.

Results

Dates of 5 and 25% cumulative catch for silver phase American eels in the Maine river varied between years (Figure 1). The 5% cumulative catch date ranged between 18 August and 10 September, while the 25% cumulative catch date ranged from 27 August to 26 September. Dates associated with 25 and 75% mean cumulative passage as estimated from the combined CDF were 10 September and 6 October, respectively (Figure 2), and the estimated date of 50% passage was 23 September. Based on this 50% passage midpoint, the start and end dates for the

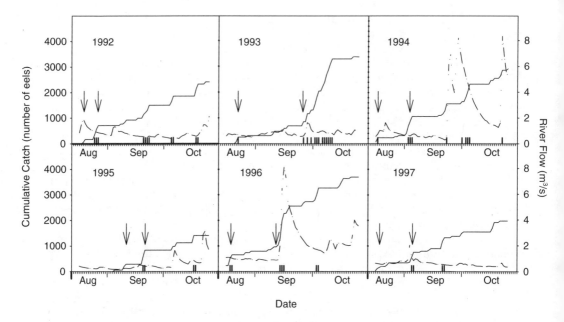

Figure 1. Time series of cumulative catch (estimated number of eels, solid line) and flow (USGS gauge data, dashed line) from the Maine eel weir 1992–1996. Vertical bars (no scale) indicate days when significant rain events were recorded by the weir fisherman. Arrows indicate dates of 5 and 25% cumulative catch.

Seven Day scenario were 20 and 26 September, respectively. Size of runs varied from an estimated total catch of 1409 eels (1995) to 3699 eels (1996).

We found no significant correlations between mean flow during the run and date of 5% cumulative eel capture (Spearman's ρ, $p = 0.15$), number of days between 5% and 25% cumulative eel capture ($p = 0.87$), or size of the run (total number of eels per year; $p = 0.11$). However, statistical power of these tests was low, resulting from low sample size (six year dataset). There was a positive correlation between daily proportion of the annual run and daily flow within years (Spearman Partial Correlation; $p < 0.01$), but no significant correlation between daily proportion of the annual run and daily gauge rainfall ($p = 0.74$). However, more eels were captured on days when rain events were recorded by the fisherman than on days when no rain event was recorded (Mann–Whitney U; $p < 0.001$).

Run mortality was significantly different between operation scenarios (ANOVA; $p < 0.001$) and spill rates ($p < 0.001$). Mortality generally

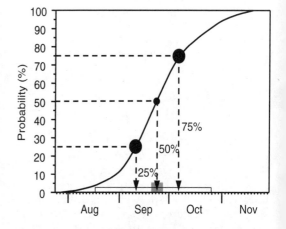

Figure 2. Combined cumulative distribution function generated from random sampling the six years of silver phase eel weir catch data. Estimates for 25 and 75% cumulative catch (dashed lines) correspond to dates of 10 September and 6 October, respectively. Open bar on x-axis spans the period of weir operation; shaded bar denotes a seven day period centered on the date of 50% cumulative catch.

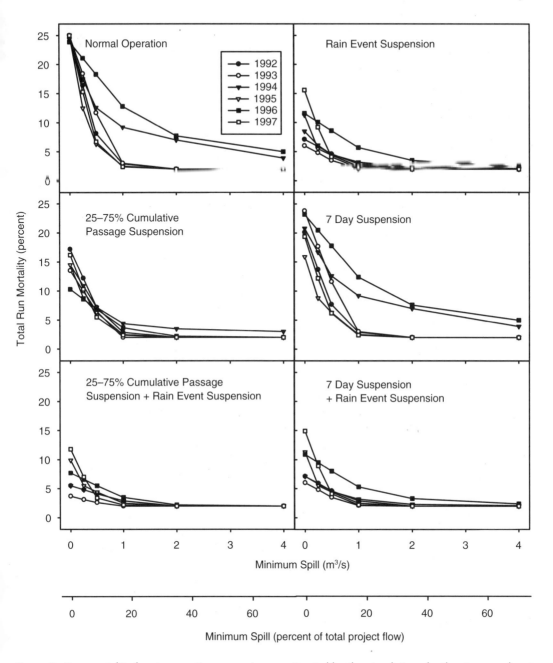

Figure 3. Run mortality for six operation scenarios as estimated by the simulation, for the six annual weir catch datasets. Simulations were run for 5.66 m³/s maximum generation flow and 0.25, 0.5, 1.0, 2.0, and 4.0 m³/s minimum spill conditions. Turbine and spill mortalities were set at 25 and 2 %, respectively.

decreased with increasing spill (Figure 3), but the difference between spill rates was significant only for spill rates ≤ 1 m³/s (Scheffé's Test; $p < 0.05$). Under the Normal Operation scenario, mortality was generally higher in high flow years (1994 and 1996), when most of the flow was directed through the turbines. This relationship was similar for the Seven Day Suspension scenario, but less pronounced in the other operational scenarios. Average run mortality of eels under

25–75% Cumulative Passage Suspension and Rain Event Suspension scenarios was approximately two thirds to one half that of Normal Operation, respectively (Table 1). Combining the Rain Event Suspension with the 25–75% Cumulative Passage Suspension and Seven Day Suspension approximately halved each scenario's average run mortality rate (Table 1).

The simulation results predict that variability of run mortality between years was greatest under the Seven Day Suspension scenario, returning a SD comparable to that for the Normal Operation scenario and more than two-fold greater than that returned for the 25–75% Cumulative Passage Suspension, the least variable scenario (Table 1). Adding the Rain Event Suspension to the Seven Day Suspension nearly halved the SD, while no reduction in annual variability was achieved by combining Rain Event Suspension with the 25–75% Cumulative Passage Suspension.

Discussion

Run timing of silver phase eels in the Maine river was highly variable between years, and daily movements and dates of initiation of the run were not strongly correlated with daily records of rainfall. However, significant eel movements did appear to be triggered by river flow and local rain events (as recorded by the fisherman), with up to an estimated 20% of the run passing in a single day during one rain event (i.e., 100 lbs or 285 fish on 17 December, 1995). Local rain events were not strongly associated with flow data from the USGS gauge 6 km downstream, suggesting that eels may have initiated downstream migration in response to very low amounts of rainfall or environmental triggers other than flow or rainfall (e.g., rainfall-induced olfactory cues). The influence of mean flow during the run on run size and timing was equivocal, but the low number of sample years (six) precludes meaningful assessment of these relationships.

Given the results of our analysis of the effects of river flow and rainfall on silver eel migration timing, we feel these are potential factors for consideration in modifying hydro project operations for protection of downstream migrant eels. Including rainfall as a decisional factor in planning suspension of project operations decreased the run mortality of both the Seven Day Suspension and 25–75% Cumulative Passage Suspension scenarios, and substantially minimized the variability in run mortality for the Seven Day Suspension scenario. Clearly, the limitation of including rainfall as a decisional factor is that silver eel migration, while cued in some respect by rainfall and peaks in flow, is not dependent upon rain occurring (see Vøllestad et al. 1986). Depending on the size of a hydro project relative to river flows and basin area (see below), augmenting predetermined operational modification periods by including rainfall as a decisional criterion may have some merit. Other proximate environmental factors we did not address, such as lunar phase, may also hold promise as a decisional tool for restricting hydro project operations to benefit silver eel

Table 1. Mean annual mortality for each operation scenario and spill condition for the 1992 to 1996 seasons; standard deviations (SD) in parentheses. RES = Rain Event Suspension; 25–27% CPS = 25–75% Cumulative Passage Suspension; 7-DS = 7-Day Suspension. SD is calculated from arcsine square root transforms of mortality estimates for each year. Mean mortality and SD for all spill conditions are shown in the rightmost two columns.

Operation scenario	Minimum Spill (m^3/s)						Mean mortality	Mean SD
	0.00	0.25	0.50	1.00	2.00	4.00		
Normal Operation	24.7 (.01)	16.9 (.08)	10.6 (.14)	5.5 (.18)	3.8 (.14)	2.8 (.07)	10.7	0.10
7-DS	20.5 (.07)	14.9 (.12)	10.4 (.15)	5.4 (.18)	3.8 (.14)	2.8 (.07)	9.6	0.12
25–75% CPS	14.4 (.07)	10.1 (.04)	6.6 (.03)	3.0 (.05)	2.3 (.04)	2.2 (.03)	6.4	0.04
RES	10.1 (.12)	6.9 (.08)	5.0 (.08)	3.1 (.07)	2.3 (.04)	2.1 (.01)	4.9	0.07
7-DS + RES	9.6 (.11)	6.7 (.08)	4.9 (.07)	3.0 (.06)	2.3 (.03)	2.1 (.01)	4.8	0.06
25–75% CPS + RES	7.3 (.12)	5.3 (.06)	4.0 (.05)	2.7 (.03)	2.0 (.01)	2.2 (0)	3.9	0.05

downstream passage (Euston et al. 1997). Without question, further research and analyses are needed to investigate the effect of proximate environmental factors on eel migration and how they may be potentially used as decisional criteria to further refine hydro project operation modifications to benefit silver eel passage.

Based on the results of the simulations, run mortality was dependent on the magnitude and timing of both flow and spill. Under conditions of low river flows and minimal or no spill, fish were directed primarily through the turbines, but few fish were migrating, so the number of fish killed was low and their contribution to run mortality was small. As river flow increased, more fish emigrated, but because all but the highest flows in this simulation were below maximum project flows, these fish were still entrained into the turbines, substantially increasing the contribution to run mortality. When flows exceeded maximum project flows, fish were passed via spill, reducing the proportion of fish killed by turbines on a given day. However, this condition occurred in only one simulation year (1994) for only two days (22 and 23 October); as such, our simulation represents conditions where river flows during the run are usually less than maximum project flows. Increasing the proportion of spill relative to flow through the turbines reduced run mortality in every operation scenario; however, reduction in run mortality became less significant as minimum spill exceeded 1 m^3/s (20%) of the maximum project flow.

The degree to which periods of suspension of operation consistently match peak downstream movements of eels on an annual basis depends on the inter-annual variability of the migration timing and the length of the period of suspension. For silver eels in this small Maine river (that exhibited substantive inter-annual variability in run timing), the Seven Day Suspension scenario failed to provide a consistently protective alternative to the ~ 30 day period of operation suspension incurred by the 25–75% Cumulative Passage Suspension. It is evident from our simulation that scenarios involving short periods of suspension of operations are less likely to consistently bracket times of peak downstream movement of eels from year to year.

Ultimately, however, the specifications of a hydroelectric project operational modification for silver eel mortality mitigation must be balanced with other factors, including energy losses, project economics, and overall goals for protection of eels and other fishes in river systems (EPRI 1999). Our simulation model, being flow-based, could be modified and expanded to include economic costs of various operation suspension scenarios in terms of lost generation. From a predetermined starting point, simulations could be repeated at increasingly narrower operation suspension intervals to establish acceptable breakpoints between run mortality, variability in mortality, and project economics. Such an approach would provide a powerful and useful tool for developing balanced and feasible eel entrainment mitigation options at hydroelectric projects. Highly hydro-regulated rivers also pose the significant challenge of cumulative passage mortality (McCleave 2001). In addition to our previous recommendations, we suggest that operational modifications will be most effective and beneficial to eels if they are coordinated among hydro projects. Without a coordinated approach, gains in eel entrainment mitigation at one project could be easily nullified by a lack of mitigation at subsequent projects downstream.

Because migration timing of eel runs may not be similar between watersheds or geographic locations, knowledge of site-specific run timing for downstream migration is important for targeting periods when suspension of generation will have a consistent and maximal reduction of turbine mortality. Characteristics of run timing may also vary within a watershed. Small or headwater streams may exhibit flow patterns that are very sensitive to rainfall in both magnitude and duration. The results presented here use data from a watershed of this type. In contrast, peaks in eel runs in larger or mainstem rivers may be "damped" or more prolonged, and less sensitive to environmental factors than smaller streams. Rivers with regulated flow, storage capacity, or peak generation schedules will also exhibit complex flow patterns that may not be tightly linked to rainfall events. Timing of eel runs in such large or regulated river systems have not been extensively characterized, and require additional study.

In order to improve the potential for operational mitigative techniques as solutions for downstream eel passage, several areas of future research are identified. These include: a better understanding of migration timing (especially annual variation in timing); identification of the influence of environmental factors on run timing and individual eel behavior; assessment of effects of river/project size on run characteristics; tests of predictive model assumptions (e.g., estimates of turbine/spill mortality, passage route proportional to flow); and refinement of designs for predictive models (e.g., to account for passage of a fixed number of migrants). The simulation described here is provided only as a demonstrative tool for assessment of potential effects of changes in project operation on run mortality of downstream migrant eels. In order for this model to accurately simulate passage at a particular project, it requires refined estimates of the parameters listed above; once achieved, the predictive power of the model could be improved to better define operation schedules that result in significant and consistent reductions in passage mortality.

Acknowledgments

We thank the Maine Department of Marine Resources and the anonymous eel fisherman for providing weir catch data. John Warner of the U. S. Fish and Wildlife Service and Brandon Kulik of Kleinschmidt USA provided data on scaling of parameters for the simulated hydro project. Glenn Çada, Terry Euston, Brandon Kulik, Steve Railsback, and Ned Taft reviewed earlier drafts of the manuscript. The views represented herein are those of the authors and do not represent the official views and policies of their respective agencies.

References

Adam, B., and D. U. Schwevers. 1997. Behavioral surveys of eels (*Anguilla anguilla*) migrating downstream under laboratory conditions. Institute of Applied Ecology, Neustader Weg 25, 36320 Kirtorf-Wahlen, Germany.

Berg, R. 1986. Fish passage through Kaplan turbines at a power plant on the River Neckar and subsequent eel injuries. Vie et Milieu 36:307–310.

Boubée, J., C. P. Mitchell, B. L. Chisnall, D. W. West, E. J. Bowman, and A. Haro 2001. Factors regulating the downstream migration of mature eels (*Anguilla* spp.) at Aniwhenua Dam, Bay of Plenty, New Zealand. New Zealand Journal of Marine, and Freshwater Research 35:121–134.

EPRI (Electric Power Research Institute). 1999. American eel (*Anguilla rostrata*) scoping study: a literature and data review of life history, stock status, population dynamics, and hydroelectric facility impacts. Report Number TR-111873. Palo Alto, California.

EPRI (Electric Power Research Institute). 2001. Review and documentation of research and technologies on passage and protection of downstream migrating catadromous eels at hydroelectric facilities. Report Number 1000730. Palo Alto, California.

Euston, E. T., D. D. Royer, and C. Simons. 1997. Relationship of emigration of silver American eels (*Anguilla rostrata*) to environmental variables at a low head hydro station. Waterpower 97:549–558.

Hadderingh, R. H., J. W. Stoep, J. W. Vander, and J. W. P. W. Habracken. 1992. Deflecting eels from water inlets of power stations with light. Irish Fisheries Investigations Series A (Freshwater) 36:80–89.

Haro, A., W. Richkus, K. Whalen, A. Hoar, W.-D. Busch, S. Lary, T. Brush, and D. Dixon. 2000. Population decline of the American eel: implications for science and management. Fisheries 25(9):7–16.

Heisey, P. G., D. Mathur, and E. T. Euston. 1995. Fish injury and mortality in spillage and turbine passage. Waterpower 95(2):1416–1423.

Kulik, B. H. 1990. A method to refine the New England aquatic base flow policy. Rivers 1(1):8–22.

Larsen, H. N. 1980. New England flow regulation policy. Memorandum to Area Manager, New England Area Office, from Regional Director, Region 5, U.S. Fish and Wildlife Service, Newton Corner, Massachusetts.

McCleave, J. D. 2001. Simulation of impact of dams, and fishing weirs on reproductive potential of silver-phase American eels in the Kennebec River Basin, Maine. North American Journal of Fisheries Management 21:592–605.

Montén, E. 1985. Fish and turbines: fish injuries during passage through power station turbines. Vattenfall, Stockholm, Sweden.

Oliveira, K., and J. McCleave. 1999. American eel biology: population structure and reproductive potential. Pages 24–38 *in* Eel and Elver Progress Report, May 1999. Stock Enhancement Division, Maine Department of Marine Resources, Augusta, Maine.

Sokal, R. R., and F. J. Rohlf. 1995. Biometry: the principles and practices of statistics in biological research. 2nd edition. Freeman, New York.

Therrien, J. 1999. Suivi environemental de la centrale hydroélectrique du barrage La Pulpe sur la rivière Rimouski en 1998. Rapport réalizé par Groupe-conseil Génivar inc. pour Rimouski Hydroélectrique Inc.

Vøllestad, L. A., B. Jonsson, N. A Hvidsten, T. F. Naesje, O. Haraldstad, and J. Ruud-Hansen. 1986. Environmental factors regulating the seaward migration of European silver eels, *Anguilla anguilla*. Canadian Journal of Fisheries and Aquatic Sciences 43:1909–1916.

Whalen, K. G., D. L. Parrish, M. E. Mather, and J. R. McMenemy. 2000. Cross-tributary analysis of parr to smolt recruitment of Atlantic salmon (*Salmo salar*). Canadian Journal of Fisheries, and Aquatic Sciences 57:1607–1616.

Evaluation of Angled Bar Racks and Louvers for Guiding Silver Phase American Eels

STEPHEN V. AMARAL*

Alden Research Laboratory, Inc., 30 Shrewsbury Street, Holden, Massachusetts 01520, USA
(508) 829–6000, ext. 415, Fax (508) 829–5939
*Corresponding author: amaral@aldenlab.com

FREDERICK C. WINCHELL

Alden Research Laboratory, Inc., 30 Shrewsbury Street, Holden, Massachusetts 01520, USA
(508) 829–6000, ext. 473, Fax (508) 829–5939
Email: winchell@aldenlab.com

BRIAN J. MCMAHON

Alden Research Laboratory, Inc., 30 Shrewsbury Street, Holden, Massachusetts 01520, USA
(508) 829–6000, ext. 447, Fax (508) 829–5939

DOUGLAS A. DIXON

Electric Power Research Institute, 3412 Hillview Avenue, Palo Alto, California 94304, USA
(804) 642–1025, Email: ddixon@epri.com

Abstract.—The ability of downstream migrant silver phase American eels *Anguilla rostrata* to guide along various configurations of angled bar racks (25- and 50-mm clear spacing) and louvers (50-mm clear spacing) was evaluated in a laboratory flume. Guidance tests were conducted with bar racks and louvers angled at 45 and 15 degrees to the approach flow at velocities of 0.3 m/s to 0.9 m/s. A full-depth bypass was used for all tests. Guidance efficiency was calculated by dividing the number of fish recovered from the bypass by the total number recovered downstream (bypass and entrainment combined). Mean guidance efficiency (MGE) with the 45-degree, 25-mm bar rack ranged from a low of 56.8% at 0.6 m/s to a high of 65.9% at 0.9 m/s. MGE of the 45-degree, 50-mm bar rack decreased from 72.7% at 0.3 m/s to 54.5% at 0.9 m/s, while MGE of the 45-degree louver ranged from a low of 34.9% at 0.3 m/s to a high of 61.9% at 0.6 m/s. Guidance efficiency was considerably higher for tests with the 15-degree structures, exceeding 88% at all velocities during tests with a solid bottom overlay placed over the lower 30 cm of each structure. During tests without the overlay in place and at a velocity of 0.6 m/s, guidance efficiency of both bar racks and louvers decreased to 83.3 and 60.7%, respectively. The estimated guidance efficiencies indicate that angled bar racks and louvers have potential for diverting American eels away from hydro intakes, particularly if a shallow angle is employed (e.g., 15 degrees to the approach flow). However, we believe our estimates of guidance efficiency are higher than would be experienced at an actual intake due to the full depth bypass, the limited depth of the flume, and the short length of each rack configuration that we evaluated.

Introduction

Recent declines in American and European eel populations have generated a great deal of interest in the impacts of human activities on the various lifestages of both species (EPRI 1999; Haro et al. 2000a). In particular, many resource agencies in the United States have begun to examine the effects of hydroelectric projects on the upstream and downstream movements of American eel. The current technology and knowledge for passing elvers and young yellow phase eels upstream of hydro projects is more advanced than what is available and known for safely passing silver eels downstream. Horizontally angled bar racks and louvers are being considered as primary measures for diverting silver eels away from turbine intakes and towards downstream bypasses at some North American hydro projects. However, the ability of these types of structures to effectively guide silver eels has not been determined.

Angled bar racks and louvers have been prescribed or installed at many hydro sites in the Northeast U.S. (Cada and Sale 1993; EPRI 1998), mostly for the purpose of guiding anadromous fishes (Atlantic salmon smolts, juvenile clupeids) to a downstream bypass. Most installations to date have involved the application of angled bar racks with narrow spacing (25 to 50 mm clear spacing). Although results from field evaluations have been mixed, there have been enough installations to indicate that when such facilities are properly designed they have potential to divert a large portion of the targeted fish. Consequently, angled bar racks have been prescribed or recommended for use as a guidance measure for silver eels despite a lack of evidence that they can effectively guide this species, and despite the fact that American eel are very different with respect to morphology, migration size, swimming capabilities, and behavior compared to other diadromous fish species for which these diversion structures have been designed.

There is very limited information on the ability of the catadromous American eel to guide along angled bar racks and louvers. In the one field study conducted in the U.S., no eels were collected in head race nets downstream of a 45-degree, 25-mm spaced bar rack installed at the Stillwater Hydroelectric Project on the Stillwater River in Maine (Barnes-Williams 1998). Twenty-five American eels were collected in the bypass during this study; these fish accounted for about 15% of the total bypass catch and were the second most abundant species collected from the bypass. However, the collection nets used in this study did not sample the entire flow passing through the powerhouse (or even one unit) and net collection efficiencies were not estimated for outmigrating eels. Laboratory studies investigating European eel guidance with bar rack screens and louvers have been conducted in Germany (Adam et al. 1999) and demonstrated that achieving eel guidance with these types of structures may be difficult. Adam et al. (1999) reported that silver European eels exhibited no searching behavior when encountering the bar rack and louver arrays that were evaluated and that impingement occurred at approach velocities greater than 0.5 m/s.

In response to the lack of information on American eel guidance with angled diversion structures, we conducted a two-year laboratory study designed to answer basic questions related to the potential for these technologies to be successfully applied at hydro projects past which silver eels must migrate. The goal of our study was to generate baseline data on the effectiveness of angled bar racks and louvers in guiding American eels to a bypass. To achieve this goal, the objectives of our study program were to collect quantitative and qualitative data on the guidance and behavior of silver eels as they approach and either guide along or pass through several different bar rack and louver configurations at approach velocities typically found in hydro project forebays.

Methods

Test Fish

Approximately 200 to 300 American eels were obtained from commercial fishermen in Maine for testing during each year of the study. The fishermen collected migrating adult eels with fyke nets located at a lake outlet during fall in 1999 and 2000. We believe greater than 90% of the fish that we received were silver eels based on external coloration and gonad condition of several fish that were internally examined. The average total length of the eels that were evaluated was 558 mm (SD = 46 mm; range = 151–697 mm) in 1999 and 569 mm (SD = 76 mm; range = 410– 781 mm) in 2000.

Test Facility

Fish guidance tests were conducted in an indoor flume constructed for the specific purpose of testing fish under realistic and controlled conditions with various fish passage and protection technologies. The entire test flume was about 24.4 m long and the section in which fish passage tests were conducted was about 1.7 m wide by 2.1 m deep. Major components of the test facility included an upstream fixed isolation screen, the bar rack/louver array, a collection net and pen downstream of the bar rack/

louver, and a bypass with an inclined screen leading to a collection box. A bow thruster powered by a 150-hp electric motor was used to recirculate water at the selected test velocities. The flume was readily adapted to test different rack angles, bar angles, bar spacings, water velocities, and bypass configurations.

We used bar racks with slats that were oriented perpendicular to the axis of the support structure and louver slats that were perpendicular to the approach flow (Figure 1). The test flume was configured with the bar rack and louver arrays angled 45 degrees to the flow in 1999 (Figure 2) and angled 15 degrees to the flow in 2000 (Figure 3). A surface release box located upstream of the guidance structures was used to introduce fish in 1999 (Figure 2), and a pressurized injection system with a submerged exit was used in 2000 (Figure 3). The flow rate in the flume ranged from about 0.8 m³/s at an approach velocity of 0.3 m/s to 2.5 m³/s at an approach velocity of 0.9 m/s. Water depth in the test flume was maintained between 1.65 and 1.72 m depending on the bar rack and louver configuration being evaluated (e.g., because the louvers produced a greater head differential, particularly at the higher approach velocities, lower water depths were required to ensure that the flume flow did not pass over the top of the slats).

The bar rack and louver arrays (i.e., slats and cross members) were supplied by Hydrothane Systems of Pembroke, Massachusetts. Hydrothane bar racks are fabricated with a high-density polyethylene material and are rounded at the ends to improve hydraulic efficiency. The bar slats measured 12.5 mm wide by 100 mm deep and were separated by spacers and held together with four horizontal cross members (50 mm in diameter). The height of the bar slats was 1.8 m. The 45-degree bar rack and louver arrays comprised a single section of slats that had a horizontal length of 2.0 m. The 15-degree structures were constructed in three sections, two were 1.4 m in length and one was 2.0 m long, for a total guidance structure length of 4.8 m.

The downstream end of each array terminated at a bypass entrance that extended the full depth of the water column and was 15.2 cm wide. The bypass received about 10 to 12% of the total flume flow. A wedge-wire screen angled at 16 degrees from the flume bottom passed most of the bypass flow while guiding fish to a bypass collection box. This screen was 5 m in length, extending from 0.6 m downstream of the bypass entrance to the collection box. The collection system for fish entrained through the bar racks and louver array consisted of a net that tapered about 6 m to a funnel that delivered fish into a collection pen.

Approach velocities were measured at three depths along vertical transects at several locations upstream of and along the angled arrays. Water velocities also were measured at three depths along a vertical transect located at the bypass entrance. Approach velocities in the channel upstream of the angled arrays were typically within 10 to 15% of target velocities (i.e., 0.3, 0.6, and 0.9 m/s). Bypass entrance velocities for the 45-degree arrays were slightly less than the velocities in front of the slats at the downstream end. For the 15-degree arrays, bypass entrance velocities generally were about the same or slightly higher than the velocities at the downstream end of each array. We were able to achieve bypass velocities that were about 1.2 to 1.5 times higher than the approach velocity for tests with the 45-degree structures and about 1.6 to 2.0 times higher for tests with the 15-degree arrays. A bypass approach velocity ratio of 1:1.5 has been shown to produce favorable hydraulic conditions for effective guidance with louvers (Ducharme 1972).

Test Parameters

The primary bar rack and louver design parameters that we evaluated included bar slat angle (i.e., bar rack or louver orientation; Figure 1), bar slat spacing (25 and 50 mm for the bar racks

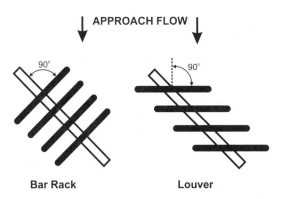

Figure 1. Orientation of bar slats for angled bar racks and louvers. The structures depicted are angled at 45 degrees to the approach flow.

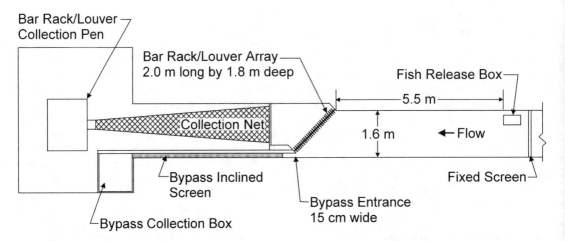

Figure 2. Plan view of fish testing facility configured for tests with the bar rack and louver arrays angled at 45 degrees to the approach flow.

and 50 cm for the louvers), approach velocity (0.3, 0.6, 0.75, and 0.9 m/s), and bar rack and louver angle to the approach flow (45 degrees in 1999 and 15 degrees in 2000; Table 1). In addition to these parameters, a solid bottom overlay that covered the lower 30 cm of each guidance structure was used in most tests conducted in 2000 with the 15-degree structures. Tests at 0.75 m/s were only conducted with the 45-degree louver array. Higher velocity (i.e., 0.9 m/s) resulted in stress on test facility components due to relatively high head losses created by this configuration.

The 45-degree angle was initially selected for evaluation because it has been frequently prescribed or recommended by resource agencies for guiding downstream migrant anadromous fish at hydro projects and is being considered for riverine species as well. After reviewing the results from the 1999 tests with American eel and other fish (Amaral et al. 2001), we made several changes to the design of the testing facility in 2000 in an attempt to increase guidance efficiency. A bottom release system, which was used for all species tested in 2000, was added primarily to accommodate the behavior of benthic fish

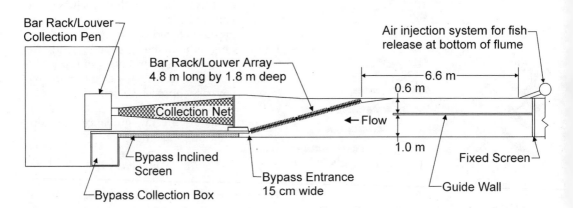

Figure 3. Plan view of fish testing facility configured for tests with the bar rack and louver arrays angled at 15 degrees to the approach flow.

Table 1. Bar rack and louver configurations and approach velocities that were evaluated for guidance efficiency with American eel.

Guidance structure	Angle to flow	Clear spacing (mm)	Bottom overlay	Approach velocity (m/s)
bar rack	45°	25	no	0.3, 0.6, 0.9
bar rack	45°	50	no	0.3, 0.6, 0.9
bar rack	15°	50	yes	0.3, 0.6, 0.9
bar rack	15°	50	no	0.6
louver	45°	50	no	0.3, 0.6, 0.75
louver	15°	50	no	0.3, 0.6, 0.9
louver	15°	50	no	0.6

(two sturgeon species) that were also being evaluated (Amaral et al. 2002). The release of fish near the bottom of the flume was prompted by observations in 1999 that indicated lake sturgeon released at the surface were approaching the bar racks and louvers high in the water column where their swimming ability may have been reduced. This was not the case with American eel, which were observed in close proximity to the flume bottom as they approach the guidance structures, regardless of where they were released. The bottom overlay placed over the lower 30 cm of the 15-degree bar racks and louvers also was designed to take advantage of the location and behavior of bottom-oriented fish, including American eel. Fish moving downstream near the flume bottom were expected to follow the overlay towards the bypass. In addition, a guide wall was also installed for 2000 tests and was designed to deliver fish to the upstream end of the guidance structures (Figure 3). The guide wall was used to minimize the potential for fish to move across the flume and follow the wall opposite the release location downstream to the bypass without interacting with the bar racks or louvers. Finally, although not chosen for American eel specifically, the shallower angle (15 degrees) of the bar rack and louver arrays used in 2000 tests was selected based on past research that had shown effective guidance with other fish species at this angle (EPRI 1986, 1998).

Testing Procedures

We released fish on the side of the flume opposite the bypass to maximize the potential for interaction with the bar racks and louver. For the evaluation of the 45-degree structures in 1999, fish were introduced from a surface release box during tests at approach velocities of 0.3 and 0.6 m/s. Fish were introduced directly into the flume where the release box was located for tests at 0.75 m/s (louvers) and 0.9 m/s (bar racks) because the release box was not stable at approach velocities greater than 0.6 m/s. We used an air-injection system for introducing fish to the flume through the bottom release system at all approach velocities during 2000 tests. This system introduced fish into the flume through a 15-cm diameter pipe with an exit that was about 8 cm off the flume bottom.

The guidance efficiency of American eel was quantitatively evaluated by determining the percentage of fish released upstream of the bar racks and louvers that were recovered from the bypass collection box and the bar rack/louver collection pen over a specified time period. Three trials were conducted per approach velocity for most bar rack and louver configurations that were evaluated. In 1999, we released distinctly marked groups of 15 fish per velocity at 15-minute intervals (i.e., 45 fish were released within a 30 minute period; one velocity tested per night). In 2000, we conducted one trial (15 fish per trial) for each velocity on each night of testing to increase the independence of replicate velocity trials. Because eels are sensitive to light and typically migrate during nighttime hours, we released the first group of fish on each test night between 1700 and 1800 hrs. All tests with the 15 and 45-degree angled structures were conducted with naive fish that were tested only once (i.e., fish were not re-tested). Test velocities were maintained during a trial until all or most fish were recovered from the downstream collection locations. Greater than 90% of released fish were recovered during all tests with the 45-degree structures (Table 2) and during most tests with the 15-degree structures (Table 3). At the end of each test the following sequence of activities was followed: (1) the flow was shut off; (2) screens were dropped behind the guidance structures

Table 2. Summary of results from American eel guidance trials with bar racks and louvers angled 45 degrees to the approach flow.

Approach velocity (m/s)	Number of trials (N)	Number of fish released	Number of fish entrained	Number of fish bypassed	Total recovered	Percent recovery	Mean guidance efficiency (%) (SE)
			45° Bar rack with 25-mm spacing				
0.30	3	45	15	28	43	95.6	64.8 (8.0)
0.60	3	45	19	25	44	97.8	56.5 (7.0)
0.90	3	45	15	29	44	97.8	65.9 (12.0)
			45° Bar rack with 50-mm spacing				
0.30	3	45	12	32	44	97.8	72.5 (5.0)
0.60	3	45	19	26	45	90.0	57.8 (4.0)
0.90	3	45	20	24	44	97.8	53.3 (3.0)
			45° Louver (50-mm spacing)				
0.30	3	45	28	14	42	93.3	33.3 (2.0)
0.60	3	45	16	26	42	93.3	62.1 (4.0)
0.75	3	45	24	20	44	97.8	45.4 (4.0)

and in front of the bypass to prevent fish from moving out of or into these areas; (3) overhead lights were turned on; and (4) all un-recovered fish were removed from upstream areas. For each time interval, recovered fish were measured for length, marks were identified, and recovery location was recorded.

We also released groups of American eel during daytime hours to videotape behavior in the vicinity of the bar racks and louvers. The videotapes were used to qualitatively assess behavior as fish encountered the bar racks and louvers. Guidance efficiencies were not estimated for these releases. We videotaped fish guidance for at least one release group at each velocity. Video of fish interacting with the 45-degree structures was recorded from two underwater black and white Sony cameras (model HUM-351/352/354)

Table 3. Summary of results from American eel guidance trials with bar racks and louvers angled 15 degrees to the approach flow. The clear spacing of all 15 degree bar rack and louver configurations was 50-mm.

Approach velocity (m/s)	Number of trials (N)	Number of fish released	Number of fish entrained	Number of fish bypassed	Total recovered	Percent recovery	Mean guidance efficiency (%) (SE)
			15° Bar rack with bottom overlay				
0.3	3	45	2	39	41	91.1	95.1 (3.0)
0.6	3	45	2	38	40	88.9	95.2 (24.0)
0.9	3	45	5	40	45	100.0	88.9 (4.0)
			15° Bar rack without bottom overlay				
0.6	2	30	4	20	24	80.0	83.3 (0.00)
			15° Louver with bottom overlay				
0.3	3	45	5	39	44	97.8	88.7 (4.0)
0.6	3	45	2	39	41	91.1	95.2 (2.0)
0.9	3	45	4	37	41	91.1	90.3 (2.0)
			15° Louver without bottom overlay				
0.6	3	45	14	21	35	77.8	60.7 (4.0)

that were mounted on bar slats near the water surface. These cameras provided complete coverage for the lower two-thirds of the guidance arrays and the bypass entrance. Two cameras were also mounted on the 15-degree bar rack and louver array. These cameras were positioned in the center of the two upstream rack sections; a third underwater camera was mounted on the wall leading to the bypass and was aimed downstream at the bypass entrance. In addition to the underwater cameras, a Sony hand-held camera recorded fish movements through a full-depth plexiglass window that extended from about 1 m downstream of the bypass entrance to about 2 m upstream of the entrance.

Video images were recorded during nighttime guidance efficiency tests in 2000 to determine the percent of fish that moved downstream along the wall leading to the bypass. A Sony handheld camera positioned across from the downstream end of the guide wall was submerged in a waterproof cylinder and operated in nightshot mode (i.e., a built in infrared light was used to illuminate the area directly below the camera) to capture movements of fish moving upstream and downstream at this location. The counts of downstream moving fish at this location were used to assess the number of guided fish that may have followed the wall downstream to the bypass without having interacted with the guidance structures.

Data Analysis

The guidance efficiency (E) of each release group (i.e., replicate trial) was calculated using the following formula:

$$E = R_b/(R_b + R_e);$$

where R_b is the number of fish bypassed (i.e., recovered from the bypass collection box) and R_e is the number of fish entrained (recovered in the collection net downstream of the guidance arrays). Mean guidance efficiencies were calculated from the replicate trials that were conducted with each species, guidance structure configuration, and approach velocity. This method of estimating guidance efficiency does not account for fish that encounter the bar racks but do not enter the bypass or pass through the guidance structures (i.e., failed attempts at passage), or fish that do not move downstream after they are released. Also, we did not quantify the number of attempts or approaches individual fish made before they either were bypassed or entrained. Although the number of attempts fish make before they pass downstream through an available route is an important measure of guidance efficiency, we could not effectively see eels under the nighttime conditions during which tests were conducted. Similarly, we could not determine if fish that remained upstream during an entire test had made any attempts to pass downstream or simply did not move downstream after release. Therefore, we decided to exclude non-recovered fish (i.e., fish that did not pass downstream through the bypass or guidance structure) from the estimation of guidance efficiency because we could not determine if fish remaining upstream was due to failed passage or lack of movement.

Results

Mean guidance efficiency of American eel varied among the three rack configurations and three approach velocities that were evaluated with the 45-degree angled arrays (Table 2). Mean guidance efficiency was greater than 50% for tests with the two bar rack configurations (i.e., 25- and 50-mm clear spacing) at all three velocities and for tests at 0.60 m/s with the louver array (Table 2). Mean guidance efficiency did not demonstrate any trend related to velocity among the three angled configurations that were evaluated. Eel guidance with the 25-mm spaced bar rack was similar among all three velocities, ranging from a low of 56.8% at 0.6 m/s to a high of 65.9% at 0.9 m/s. Guidance efficiency decreased with increased velocity for tests with the 50-mm spaced bar rack. During louver tests, guidance was greatest at 0.6 m/s and lowest at 0.3 m/s (Table 2).

Recovery rates for the combined releases (i.e., three groups of 15 fish) at each test velocity with the 45-degree arrays ranged from 93.3 to 100.0% (Table 2). The majority of eels recovered from each location typically were collected within the first hour after release. Recovery time decreased with increasing velocity, resulting in tests at 0.3 m/s being run longer than tests at 0.6, 0.75 and 0.9 m/s. Also, the time it took to recover more than 90% of the fish during louver tests at 0.3 and 0.6 m/s was greater than it was during tests with the 25- and 50-mm spaced bar racks.

American eel guidance increased substantially for tests with the 15-degree structures when the bottom overlay was in place, exceeding 90% at velocities of 0.3 and 0.6 m/s with the bar rack and at velocities of 0.6 and 0.9 m/s with the louvers (Table 3). During tests at 0.6 m/s without the bottom overlay, guidance efficiency of American eel was 83.3% for the bar racks (about 12% less than with the overlay) and 60% for the louvers (35% less than with the overlay). Recovery rates of fish released during tests with the 15-degree arrays were generally lower than those observed with the 45-degree structures. However, about 90% or more of released fish were typically recovered during tests with the 15-degree arrays. Also, between 65 and 98% of recovered fish were collected within the first hour of each test.

Nighttime video observations of eel movements along the wall leading to the bypass during tests with the 15-degree arrays demonstrated that few eels were exhibiting this behavior, particularly at approach velocities of 0.6 and 0.9 m/s. The average number of eels observed moving downstream along the bypass wall was 4.6, 0.6, and 0.3 for trials at 0.3, 0.6, and 0.9 m/s, respectively.

Daytime observations of American eel movement in the vicinity of the bar rack and louvers provided some useful insights regarding eel behavior under daylight conditions. Visual observations revealed that American eel approached the various array configurations while in close contact with the flume floor and walls. Eels that encountered the bar racks and louvers usually made contact with the bar slats whether they were facing upstream or downstream. This contact often startled the fish as if they were unaware that the bar rack was present, even when fish contacted the structures with their heads first. After contact, eels either turned and quickly moved back upstream, moved slowly along or in and out of the slats, passed through the slats, or guided to the bypass. Several eels were observed "resting" along the bar rack structure, often moving in and out of slat openings or along the upstream face of the racks. Although some eels entered the bypass after contacting the bar racks or louvers, some fish moved downstream along the wall leading to the bypass. These fish typically would enter the bypass without having encountered the rack structures. Many of these eels also were observed swimming out of the bypass after entering it.

Discussion

The results of our study indicate that angled bar racks and louvers have potential for diverting downstream-migrating silver eels to a bypass, particularly structures angled at 15 degrees to the approach flow. Up to 72% of recaptured eels were diverted to the bypass during tests with the 45-degree bar racks and louvers, and up to 95% were diverted with the 15-degree arrays. With a few exceptions, we observed relatively high guidance efficiencies at all approach velocities (i.e., guidance was not considerably less at the highest velocity tested compared to guidance at the lowest velocity), demonstrating that American eels have potential to be effectively guided at the range of velocities that typically occur at most hydro sites (i.e., velocities between 0.3 and 0.9 m/s).

Our observations of eel guidance with the angled bar racks and louvers was in stark contrast to the results of a similar laboratory study conducted in Germany with European eels. Adam et al. (1999) evaluated eel passage with 90- and 15-degree bar racks (20-mm spacing) and 15-degree louvers (100-mm spacing) in a flume and observed no guidance to a bypass. Adam et al. (1999) also reported that eels were unable to free themselves after impinging on the guidance structure slats when approach velocities were greater than 0.5 m/s. During daytime release trials with the 45-degree arrays, we observed eels laying against the slats, moving in and out of the slats, and leaving the slats to move upstream at all velocities that were evaluated, including 0.9 m/s. The large difference in guidance rates between the two studies may be related to differences in light conditions, bypass configuration, and flume depth. All of our guidance efficiency tests were conducted at night when we could not observe eel behavior, whereas all the tests conducted by Adam et al. (1999) were performed during daytime hours. Silver eels are believed to be very sensitive to light during their outmigration and generally do not move during daylight hours (Haraldstad et al. 1985; Vøllestad et al. 1986; EPRI 2001). It is possible that daytime behaviors observed during both studies were not representative of behaviors that occur under low-light conditions when downstream movements typically occur.

Although guidance rates differed, daytime observations of eel behavior as they encountered the guidance structures during the study conducted by Adam et al. (1999) were similar to our observations. In particular, during both studies, most eels were observed striking the bar rack and louver slats with their heads, after which they quickly turned 180 degrees and moved back upstream. Also, Adam et al. (1999) reported that eels would move in and out of the slats, which was observed during our study as well. Because of the light sensitivity of silver-phase eels, we cannot assume that the daytime behaviors that were observed are representative of what occurred during nighttime guidance tests. All fish released during the daytime hours appeared to have no limitations in swimming ability at any of three velocities evaluated. Eels were able to move quickly upstream, even at 0.9 m/s.

Our analysis of guidance efficiency for the various guidance structure parameters demonstrated that the 15-degree angle with the overlay in place produced the highest guidance efficiencies. Slat orientation (i.e., bar rack vs. louver) did not appear to be a significant factor affecting guidance rates for the 15-degree structures, probably because at this angle the difference in the angle of the slats is small and similar flow conditions are produced. Conversely, at an angle of 45 degrees, guidance was considerably greater for the bar rack slat orientation than it was for the louvers slat orientation, at least at the lowest and highest approach velocities. Trials with the two spacings (25 and 50 mm) that were evaluated with the 45-degree bar racks indicated that the smaller spacing may be preferable at higher velocities, whereas guidance efficiencies were similar at the two lower velocities tested. Previous research with louvers at shallow angles (15 to 30 degrees) has indicated that wider spacings (up to 30 cm) than we evaluated may be acceptable for some species (Ducharme 1972).

Although guidance efficiencies were over 50% for most tests with the bar racks and louvers, there are differences between the design of the structures that we evaluated in the laboratory and the design of structures that would be installed at an actual hydro site. We evaluated guidance with a full-depth bypass, whereas bypasses at many hydro projects are located at the surface. The ability of eels to locate surface bypasses is not known, although there is some evidence that when eels encounter a hydro project they will search for a suitable downstream passage route (Haro et al. 2000b). The length of the bar rack and louver arrays that we evaluated were 2.0 m (45-degree structures) and 4.8 m (15-degree structures) in length. These are shorter than the length of typical field installations. Guidance efficiencies may be lower if fish must negotiate a much longer length of bar racks or louvers to reach a bypass.

The distribution of fish across the width of the flume may have produced overestimates of guidance efficiencies if a large portion of the eels were attracted to the wall leading to the bypass. Daytime observations revealed that eels moving along this wall generally would continue into the bypass without having encountered the bar rack or louver structures. However, the location of eels as they moved downstream during most nighttime guidance tests is not completely known due to our inability to observe fish under unlighted conditions. Our observations upstream of the bypass during nighttime tests with the 15-degree arrays indicated that few fish were following the wall leading to the bypass, especially at approach velocities of 0.3 and 0.6 m/s. Therefore, the potential for eels following the wall leading to the bypass and then entering the bypass without interacting with the bar racks, was probably minimal and did not lead to inflated estimates of guidance efficiency. The considerably higher guidance efficiencies observed with the 15-degree arrays also supports this conclusion because more eels would have been able to enter the bypass in this manner (i.e., no interaction with bar racks or louvers) without the guide wall in place during tests with the 45-degree structures.

The results of our study provide an important foundation for the continued development of structural fish passage technologies that are designed to guide silver American eels to a downstream bypass. Bar racks and louvers angled at 15 degrees to the approach flow demonstrated considerable potential for effectively guiding silver American eels to a bypass, whereas the potential for successful application of 45-degree structures is less favorable. Although 15-degree structures appear to be biologically effective, this angle would result in a very long facility that could be difficult to install and maintain at small hydro sites (both from an engineering and economic standpoint). Future field studies will be important to verify that the results from laboratory tests are reasonably representative of the guidance efficiencies that can be expected at most hydro projects and to determine which design parameters (e.g., slat spacing, approach velocity, slat orientation, and use of a bottom overlay) may be most important for successful application at various types of water intakes.

Acknowledgments

The research was sponsored by EPRI (Electric Power Research Institute) of Palo Alto, California. We thank Hydrothane, Inc. for providing the bar racks and louvers and Alex Haro of the Conte Anadromous Fish Research Center for assisting with the design of the video system and for providing lighting equipment. Dave Michaud, Chris Tomichek, Corrie Rose, Jim Fossum, Tom Thuemler, Kurt Newman, Kevin McGrath, and Ned Taft all provided valuable input to the design and performance of this study. We also thank Rolf Hadderingh, Beate Adam, and Alex Haro for their insightful reviews of the original manuscript.

References

Adam, B., U. Schwevers, and U. Dumont. 1999. Beitrage zum Schutz abwandernder Fische, Verhaltensbeobachungen in einem Modellgerinne. Bibliothek Natur & Wissenschaft Band 16, p. 1– 63. Verlag Natur & Wissenschaft, Solingen, Germany.

. Amaral, S. V., J. L. Black, B. J. McMahon, and D. A. Dixon. 2002. Evaluation of angled bar racks and louvers for guiding lake and shortnose sturgeon. Pages 197– 210 in Van Winkle, W., P. Anders, D. H. Secor, and D. A. Dixon, editors. Biology, Management, and Protection of North American Sturgeon. American Fisheries Society, Symposium 28, Bethesda, Maryland.

Amaral, S. V., B. J. McMahon, J. L. Black, F. C. Winchell, and D. A. Dixon. 2001. Fish guidance efficiency of angled bar racks and lovers. In Waterpower XII, Advancing Technology for Sustainable Energy. HCI Publications, Inc., Kansas City, Missouri.

Barnes-Williams Environmental Consultants, Inc. 1998. Downstream Fish Passage Monitoring Study, Final Report, Stillwater Hydroelectric Project (FERC # 4684) Stillwater, New York. Prepared for Newric Hydro, Inc.

Cada, G. F., and M. J. Sale. 1993. Status of fish passage facilities at nonfederal hydropower projects. Fisheries 18(7):4–12.

Ducharme, L. J. A. 1972. An application of louver deflectors for guiding Atlantic salmon (*Salmo salar*) Smolts from Power Turbines. Journal of the Fisheries Research Board of Canada 29(10):1397– 1404.

EPRI (Electric Power Research Institute). 1986. Assessment of downstream migrant fish protection technologies for hydroelectric application. Prepared by Stone & Webster Engineering Corporation, EPRI Report 2694-1, Palo Alto, California.

EPRI (Electric Power Research Institute). 1998. Review of downstream fish passage and protection technology evaluations and effectiveness. Prepared by Alden Research Laboratory, Inc., EPRI Report TR-111517, Palo Alto, California.

EPRI (Electric Power Research Institute). 1999. American Eel (*Anguilla rostrata*) scoping study: A literature and data review of life history, stock status, population dynamics, and hydroelectric impacts. Prepared by Versar Inc., EPRI Report No. TR-111873, Palo Alto, California.

EPRI (Electric Power Research Institute). 2001. Review and Documentation of Research and Technologies on Passage and Protection of Downstream Migrating Catadromous Eels at Hydroelectric Facilities. Prepared by Versar, Inc., EPRI Report TR-1000730, Palo Alto, California.

Haraldstad, O., L. A. Vøllestad, and B. Jonsson. 1985. Descent of European silver eels, *Anguilla anguilla* L., in a Norwegian watercourse. Journal of Fish Biology 26:37–41

Haro, A., W. Richkus, K. Whalen, A. Hoar, W. D. Busch, S. Lary, T. Brush, and D. Dixon. 2000a. Population Decline of the American Eel. Fisheries 25(9):7–16

Haro, A., T. Castro-Santos, and J. Boubée. 2000b. Behavior and passage of silver-phase eels, *Anguilla rostrata* (leSueur), at a small hydroelectric facility. Dana 12:33–42.

Vøllestad, L. A., B. Jonsson, N. A. Hvidsten, T. F. Naesje, O. Haraldstad, and J. Ruud-Hansen. 1986. Environmental factors regulating the seaward migration of European silver eels, *Anguilla anguilla*. Canadian Journal of Fisheries and Aquatic Sciences 43:1909–1916.

Review of Research and Technologies on Passage and Protection of Downstream Migrating Catadromous Eels at Hydroelectric Facilities

WILLIAM A. RICHKUS*

Versar, Inc. 9200 Rumsey Road, Columbia, Maryland 21045, USA
*Corresponding author: richkuswil@versar.com

DOUGLAS A. DIXON

Electric Power Research Institute, 3412 Hillview Avenue, Palo Alto, California 94304, USA

Abstract.—Mortality of migrating catadromous eels at hydroelectric facilities is an impact that may be contributing to stock declines of several *Anguilla* species. We present a review of existing information on downstream migratory behavior and existing fish passage technologies related to minimizing turbine entrainment and mortality of downstream migrating eels. Studies suggest that approaching or reaching sexual maturity is a necessary, but not a sufficient condition for migration to ensue, with water temperature, precipitation, flow and moon phase triggering migration in most watersheds. Once migration is initiated, eels appear to move downstream at a rate consistent with flow velocity. Movement patterns are often significantly altered when obstacles, such as dams and hydroelectric facilities, are encountered. Downstream migrating silver eels appear not to use visual cues, but physically "bump into" barriers; eels encountering physical obstacles tend to show a "startle" response and move quickly back upstream, as opposed to initiating search behaviors to find a way around the obstacle. Of the various behavioral technologies examined (e.g., light, sound, water jets, bubble curtains, electric fields) for their effectiveness on influencing the behavior of downstream migrating eels, only light and infrasound (< 100 Hz) have demonstrated some level of effectiveness. Physical barriers have potential for use, but most likely only in smaller river systems and at smaller projects. Attraction of migrating eels to non-turbine bypass routes requires that a substantial portion of river flow be diverted through a bypass. Complete project shutdown during eel migration periods provides 100% protection of migrants, but can be very costly because of lost power generation. The accuracy of predictions of when pulses of migrants may pass any individual hydroelectric facility, based on statistical correlations, is generally low. Trap and transport ensures that all eels captured will avoid turbine passage and associated mortality. The applicability of many of the findings from previous and ongoing research to resolving eel passage issues at projects on large rivers is unknown.

Introduction

Stocks of catadromous *Anguillid* eels are in apparent decline in many areas of the world (Haro et al. 2000a; Richkus and Whalen 2000; EIFAC/ICES 2001). The reasons for stock declines are generally unknown, but may include oceanic influences, pollution, over-fishing, natural predation and disease, and direct (turbine mortality) and indirect (changes in habitat availability) impacts caused by hydroelectric and other water resource development projects (Haro et al. 2000a; Richkus and Whalen 2000; Tatsukawa and Matsumiya 1999; Moriarty and Dekker 1997; Castonguay et al. 1994a, b). Declines in eel abundance have triggered management concerns on the part of domestic (state and federal) and international fisheries managers. These concerns have led to the development of an interstate American eel *Anguilla rostrata* management plan in the U.S. (ASMFC 2000) and development of management recommendations by the European Inland Fisheries Advisory Commission (EIFAC) and International Council for Exploration of the Sea (ICES) Working Group on Eels (EIFAC/ICES 2001). Concerns about the decline in eel abundance and the development of management initiatives intended to reverse the decline have elevated interest in potential sources of anthropogenic impacts on eel populations (in particular, the effects of hydroelectric facilities) and methods to mitigate the effects. The EIFAC/ICES Working Group recommendations (EIFAC/ICES 2001) include a call for resolutions to fish passage problems at obstructions that contribute to mortality of migrating silver eels. The Working Group identified a need to document and describe cost-effective technologies to allow safe and effective downstream passage.

This paper is abstracted from a report prepared for EPRI that summarizes the most current research and existing technologies related to minimizing turbine entrainment of downstream migrating eels, directing them to downstream passage facilities, and/or successfully passing eels around hydroelectric projects (EPRI 2001). Such information is considered to be of value to owners and operators of hydroelectric facilities located in watersheds that support eel populations, as well as to the agencies responsible for regulating those facilities.

Methods

Potential sources of relevant literature and data were identified by contacting individuals involved in eel research and management, conducting bibliographic searches of existing and new publications, searching through "gray" literature (i.e., unpublished studies and reports), accessing bibliographic search services, and submitting inquiries via an internet eel work group (EIFAC/ICES Working Group on Eel; anguilla@rivo.wag-ur.nl; http://www.rivo.wag-ur.nl/mailman/listinfo/anguilla). Researchers participating in this Internet discussion group, which most likely includes all major eel researchers around the world, were solicited via e-mail for information on any current or ongoing projects relevant to downstream migration and hydroelectric impacts and mitigation measures.

Documents that were identified as potentially having useful information were acquired and reviewed for relevance to the objectives of the EPRI report. Categories of information sought during the literature search and acquisition phase of the project included: downstream migratory behavior; turbine mortality and factors influencing its magnitude; and potential mitigation measures and methodologies.

The information presented herein was drawn from studies of sexually mature adult (i.e., "silver") American eel, European eel *A. anguilla*, and freshwater eels *A. australis* and *A. dieffenbachii* of New Zealand, Australia and other countries in that geographical region. Most researchers believe that these species of the genus *Anguilla* all exhibit similar migratory behavior, and that findings related to one species may be generally applicable to other species.

Downstream Migratory Behavior

A comprehensive and detailed understanding of the behavior of downstream migrating eels is critical to the identification, design and implementation of measures intended to reduce their passage through hydroelectric turbines and the mortality that results from such passage. Knowledge of the factors that cause eels to initiate their downstream migration and that affect their patterns of movement (temporally as well as spatially within the water body) is needed to develop effective downstream passage strategies that could reduce mortality rates at hydroelectric projects. Knowledge of how eels move downstream and how they react to various stimuli during such movement can contribute to the design of devices and structures that might divert eels from turbine entrances and/or attract them to alternative passage routes.

Seasonal Movement Patterns

The downstream migration of partially or fully mature eels is commonly viewed as a seasonal phenomenon, with peak migration occurring in the fall (Haro et al., this volume; Winn et al. 1975; Haraldstad et al. 1984; Vøllestad et al. 1986; Mitchell and Chisnall, 1992; Aoyama et al. 2000). An exception to this general seasonal pattern of fall migration occurs in the St. Lawrence River. In this large river, migration from Lake Ontario and Lake Champlain may begin as early as late spring (e.g., May) and continue through late summer, peaking in mid-summer (McGrath et al., this volume[1]). Despite this summer movement pattern that occurs upstream, migration to the ocean from the St. Lawrence River estuary occurs over a brief period of only about one

---

[1] McGrath, K. J., J. Bernier, S. Ault, J.D. Dutil, and K. Reid.

month in the fall (Caron et al., this volume), a seasonal pattern more similar to that seen in smaller rivers and streams. Facey and Van Den Avyle (1987) suggest that emigration from well-inland lakes may begin earlier than from waters nearer the ocean.

Although the majority of eel outmigrations in smaller rivers and streams may occur within a three-month fall period, the pattern of runs within this period can vary substantially from year to year (Vøllestad et al. 1986; Haro et al., this volume). The findings of many studies suggest that approaching or reaching sexual maturity is a necessary, but not sufficient condition for migration to ensue, with environmental factors constituting actual migration triggers, resulting in migration occurring in pulses. An exception to this pattern may exist in the upper reaches of large river systems, such as the St. Lawrence, where downstream migration appears to occur on a relatively continuous basis, independent of expected environmental triggers (McGrath et al., this volume*). While a fall migration pattern is typical, silver eels have been found migrating well into winter and even spring (Facey and Helfman 1985; Euston et al. 1997, 1998; Stevens et al. 1996).

Migration Triggers and Short-term Movement Patterns

The variability of silver eel movement within seasonal migratory periods is reflected in highly punctuated patterns of downstream movement. Numbers of silver eel moving downstream may differ markedly from day-to-day, with long daily stretches of no movement interspersed with short peaks of extensive movement (Lowe 1952; Winn et al. 1975; Wippelhauser et al. 1998). The three primary categories of environmental variables identified in the literature as influencing short-term patterns of silver eel downstream migration, individually or in concert with each other, include: (1) water temperature; (2) river or stream discharge, water level and/or precipitation events; and (3) light intensity, including moon phase (Lowe 1952; Haraldstad et al. 1984).

Water temperature has been identified in many studies as having a dominant influence on when migratory activity occurs. Vøllstad et al. (1986) reported that over a nine-year period, the majority of silver European eels in the River Imsa migrated at temperatures between 9 and 12°C. However, the migration patterns shown in the upper, non-tidal St. Lawrence River and the Richlieu River are contrary to the patterns shown elsewhere, since peak migration occurs in summer months when water temperatures are highest and often above 20°C (Kleinschmidt Associates 2001; Richard Verdon, personal communication).

River discharge, water level, and precipitation are variables that are often inextricably related and all have potential to act as triggers for migratory pulses of eels. Silver eel migrations in Europe, North America, and New Zealand have often been shown to coincide with periods of increased discharge (Lowe 1952; Mitchell 1995; Winn et al. 1975; Vøllestad et al. 1986; Euston et al. 1997, 1998; Haro et al., this volume).

Increases in discharge in unregulated waters during the fall season are always associated with precipitation events. However, precipitation events cause changes in a number of environmental variables in addition to water level and flow rates (i.e., temperature and chemistry), and changes in these variables could also affect eel behavior. Tesch (1977) notes that eel movement in large rivers sometimes occurs at the time of a storm but before the precipitation from that storm causes changes in river height or flow. Charles Mitchell and Associates (Mitchell 1995) established that the migration of silver eels in a New Zealand watershed tends to be in advance of the flood wave associated with precipitation events, although very closely tied to the precipitation events themselves. Haro et al. (this volume) report that for eel runs in a small river in Maine over a six year period, significantly more eels were captured on days when rain events occurred than on days when no rain occurred. Tesch (1977) suggests that a change in water level, in and of itself, is unlikely to be a migratory cue or trigger for eels, but that river flow rate (i.e., velocity) has a major influence on migration. However, Vøllestad et al. (1986) and others have documented that while peaks in flow in the fall may cause significant migratory release, they may also be accompanied by little movement. Also, silver eels have been found to move downstream during periods of low and decreasing discharge (Haraldstad et al. 1984; Euston et al. 1997, 1998). Most migration in the upper St. Lawrence River occurs during summer low flow periods, where flow shows limited variability and there are no flow pulses (Kleinschmidt Associates 2001). The location at which eel migration is being monitored and the size of the water body where sampling occurs can greatly influence the migration patterns observed, with the episodic migratory pulses being most evident in smaller rivers and streams. The lack of consistency of findings suggests that more information is needed to resolve differences between studies and to describe the discharge/migration relationship.

Light intensity and lunar phase play a major role in establishing short-term and diurnal patterns of silver eel migration. Winn et al. (1975) describe a migration pattern in a small North American stream that demonstrates the relationships among precipitation and moon phase and moon rise in stimulating migration pulses. Tesch (1977) notes that silver eels are caught in European commercial fisheries mostly from an hour after sunset until one and a half-hours before sunrise. Eighty-five percent of downstream migrating eels tracked using sonic tags in the St. Lawrence River approached the Beauharnois hydroelectric project at night (Desrochers 2001). Most existing literature support the view that silver eel migration is almost entirely nocturnal.

Over a large spatial scale, Lowe (1952) found that peaks in movement of silver eel corresponded to periods of the new moon. Several accounts reviewed by Tesch (1977) indicate the strong relationship between silver eel migration and moon phase. Winn et al. (1975) reported primary downstream migration occurs after rains, but also observed downstream migration during the third and fourth lunar quarters when rain was not a factor. The lunar influence on eel migratory activity may be exerted in two ways: endogenous lunar cycles; or, inhibition of movement by moonlight. A number of studies have suggested that eels exhibit an endogenous lunar cycle of activity (Jens 1952; Boëtius 1967; Hain 1975; Edel 1976). Numerous additional studies have also demonstrated the very significant inhibitory impact that light has on eel activity, even when intensity levels are as low as moon light. Haraldstad et al. (1984) suggest that migration activity is probably adapted to occur at periods with low light intensity, and not to lunar phases per se. Vøllestad et al. (1986) reported that silver eels usually started to migrate shortly after sunset and stopped when the moon appeared above the horizon and illuminated the river, a pattern consistent with that observed by Winn et al. (1975).

Hvidsten (1985) presented the results of an extensive multiple correlation analysis of factors that were believed to have potential to influence silver eel migration from the River Imsa in Norway. Ten variables were included in this analysis: moon phase; average daily water flow; change in water flow; water temperature; positive and negative change in water temperature; rainfall between 7 a.m. and 7 p.m.; cloud cover; barometric air pressure; positive and negative change in air pressure; and, wind. Parameters explaining a statistically significant portion of total variation in eel movement, in order of importance, were rainfall, changed water flow, moon phase, and air pressure.

Post-trigger Migration Patterns

Limited literature is available from which to characterize the way in which eels continue their downstream migration once it is initiated and the rate at which they travel. Tesch (1977) suggests that silver eels drift downstream in the middle depths of rivers, based on the fact that catches of nets set on the river bottom are lower than those set in mid-water. He also indicates that they have frequently been seen to drift in groups. Tesch (1997, cited in Behrmann-Godel 2000) found that acoustic-tagged silver eels in the Elbe and Weser Rivers moved steadily downstream, after daylight, in the middle of the stream (i.e., always in the main current of the river), and that their downstream velocity was equal to or less than the river water velocity (i.e., suggesting passive drift). Preliminary analyses of 62 acoustically tagged silver eels on the St. Lawrence River suggests that most movement occurs at night, that most of the eels are moving at a rate slightly faster than prevailing currents, and that movement through the hydroelectric dam occurs relatively quickly with little perceived delay (Kevin McGrath, personal communication). In the Elbe, tagged silver eels covered at least 30 km a day (Lühmann and Mann 1958). Behrmann-Godel (2000) stated that the rate of downstream passage of their acoustic-tagged eels varied considerably, with individual eels requiring from 1 to 18 days to move from point of release downstream several kilometers to a hydroelectric project where their monitors were located. In the Shenandoah River, Virginia, acoustic-tagged silver eels required an average of 21 days to move from a release point to a hydroelectric dam 4.8 km downstream (RMC 1995). Desrochers (2001) reported that eels migrating in the St. Lawrence River exhibited an average migration rate of 10 km/day, and passed the Beauharnois Dam within an average of 31 minutes of reaching the project, thus showing no delays.

Several radio tracking studies that have been conducted with outmigrating silver eels provide detailed information on the complex movement patterns that they exhibited after their release and when in proximity to a dam. Haro and Castros-Santos (1997) and Haro et al. (2000b) reported that eels exhibited variable behavior upon release. Some showed considerable activity and entered the forebay area, whereas others moved between near-field zones but then became sedentary. Thirteen of 25 eels tagged in 1996 and 1997 entered the forebay zone at least once, with 10 of the 13 passing the project via the turbines. Eels that were active in the vicinity of the station spent from 1 to 41 minutes in the forebay and ventured into the forebay up to nine times.

In a study in which eels were tracked as they approached a dam and after they passed downstream, Durif et al. (this volume) found that 15 eels stopped less than 24 hours after their downstream movement was initiated and that their downstream movement did not begin until 1 to 26 days later, when environmental triggers occurred. Watene et al. (this volume) reported that one New Zealand silver eel tracked using radio-telemetry continued a migration initiated at night into daylight hours, but upon encountering a dam after daylight, the eel moved back upstream and remained sedentary until night time. Lowe (1952) suggests that the migratory activity of eels triggered by an initial stimulus, such as a flood event, does not last throughout the whole journey to the sea, but that the majority of the eels probably need reactivating several times. She also suggests that the cumulative effect of light as an inhibiting factor may be responsible for such a phenomenon, such that the migratory urge is reduced after several days of exposure to regular daylight during downstream migratory movement.

Behavior During Downstream Migration

The manner in which silver eels behave during their downstream migration (e.g., spatial movement patterns, both horizontal and vertical; swimming behavior; body orientation) affects how they may respond to obstacles encountered during their migration and how and where potential diversion devices should be placed and oriented to be most effective. No detailed in situ observations of downstream swimming behavior of American eels have been reported in the literature. Observations of silver-phase European eels in an experimental flume indicate eels exhibit three forms of downstream behavior: passive drift, controlled drift, and active downstream movement (Adam and Schwevers 1997; Adam et al. 1997). Passive drift was characterized by the absence of rheotropic behavior and almost no swimming activity, which resulted in net downstream displacement of the eel. When colliding with an obstacle, the drifting eels exhibited a "startle" response, and "sprinted" upstream against the flow. At medium velocities (0.2–0.3 m/s), the eels let themselves drift to the bed of the flume, which was smooth. It appears unlikely that such behavior would occur in natural river systems with uneven bottoms. At higher velocities, the eels ceased the passive drift, converting to controlled drift. All of these observations were made under lighted conditions, to allow visual observation.

Adam et al. (1997) noted that silver eels moving downstream in the experimental channel always collided with objects (smaller yellow eels appeared to be able to see and avoid objects). After an intense collision, eels showed the "startle" response, sprinting upstream. After a moderate collision, eels generally settled in front of the threshold. In flume studies using silver American eels, Amaral et al. (this volume) made similar observations of eel responses to structures, in this case bar racks and louvers. Eels that encountered the objects usually made contact with them, regardless of whether they were moving downstream headfirst or tail-first. Upon contacting the structure, they moved back upstream, rapidly if hitting with their head but more slowly if making first contact with their caudal fin.

Protection of Downstream Migrating Eels

Study findings summarized in EPRI (2001) show that turbine entrainment injury and mortality rates for eels may be as low as 9%, but more often appear to be higher. Mortality estimates for some European hydro projects have been on the order of 50% and in New Zealand, for the short fin and long fin eels which are much larger than the American and European eel species, they have approached 100% at some facilities. Mortality estimates such as these suggest that successfully deterring eels from entering turbines and bypassing them downstream could substantially enhance the number of eels that eventually reach the ocean from watersheds in which there is substantial hydroelectric development.

The key components of a successful operational scenario for downstream passage include both a suitable deterrent from turbine entry as well as an available bypass to attract and/or facilitate the passage of eels downstream (Hadderingh et al. 1992). Deterrents to turbine entry can be categorized as either behavioral (e.g., light, electricity, sound, bubble or water jet screens) or mechanical (e.g., louvers, angled bars, and screens). Also, means of attracting eels to bypass structures may, in some instances, be useful in the absence of deterrents to turbine entry. Other means of reducing eel turbine entrainment include modification of plant operations and the capture and downstream transport of migrating eels.

Behavioral Barriers

Light. Controlled and in-situ experimentation indicate that both American and European eels exhibit an avoidance reaction to light, although the extent of avoidance varies among the studies. Lowe (1952) describes the results of extensive studies of light as a barrier to migrating silver eels in laboratory flumes and at six different field sites in England during the 1940s. While the types of light arrangements investigated varied substantially, diversion percentages in many individual studies were high (e.g., 72% and 92% of captured eels being taken in the dark, non-illuminated side at two test sites). Diversion effectiveness was low during flood events, when water was turbid, and low diversion was observed under some specific light arrangements. This work was performed in relatively small rivers, in which the entire width of the river could be trapped.

More recent studies, both in the field and the laboratory, have produced inconsistent results. Hadderingh (1982) reported that the effectiveness of illumination in reducing eel entrainment at a power plant was relatively low for juvenile yellow eels (21% average), but much higher for silver eels (54% average). Hadderingh et al. (1992) reported on subsequent studies conducted under both controlled and field conditions. In the laboratory, light intensities of from 1 to 10 Lux induced a significant avoidance response (64% to 90% avoidance). During studies conducted at the Bergum power station, no deflection of yellow eels was found and only 6% deflection of silver eels occurred. Additional studies with lamps above water, as well as below, resulted in deflection of 51% of yellow eels and 25% of silver eels (note the difference from the 1982 work). Hadderingh and Potter (1995) and Hadderingh and Smythe (1997) showed that light was effective in deflecting eels in the experimental flume, even at illumination as low as 0.007 Lux and regardless of the type of lamp used. They also noted that the effectiveness of light for diversion was related to water velocity, with the diversion effectiveness decreasing with higher water velocities. Hadderingh and Smythe (1997) recommend that light barriers be placed at an obtuse angle to a flow, so that when the eels react and deflect from the angled line of lights, they are directed toward a bypass entrance. They recommend that the angle between the light barrier and flow direction be small to enhance diversion effectiveness. Hadderingh et al. (1999) reported additional positive light diversion results from studies with silver eels. In the experimental flume, 65% of silver eels could be deflected with a fluorescent lamp at an illumination level of about 0.003-0.005 Lux.

In contrast with these promising results, Adam and Schwevers (1997) found silver eels exhibited no avoidance reaction to strobe lights when tested in an experimental channel. They attributed the failure of light to induce an effect to changes in the spectral sensitivity of the silver eel eye; that is, the shift in spectral sensitivity toward rod vision of blue wavelengths that occurs during metamorphosis (see Pankhurst 1982). Halsband (1989) attempted to guide silver eels away from hydroelectric facility turbines using a light barrier of 20 mercury vapor lamps, but was apparently unsuccessful. Therrien and Verreault (1998) reported low percentage effectiveness of a light diversion array at a small hydroelectric facility on the Rimouski River, a tributary of the St. Lawrence River.

Sound: In a review of the use of sound to control fish behavior, Popper and Carlson (1998) concluded that most experiments testing the usefulness of sound signals have produced ambiguous results, except when ultrasound has been used to control some clupeid species. They concluded that too little is known about the suitability of various signals for controlling fish behavior, and they presented no findings related to eels. Sand et al. (2000) reported results of studies using infrasound (specifically, 11.8 Hz) to evoke avoidance responses in migrating European silver eels. A highly significant reduction (43%) in the number of eels entering river sections ensonified with infrasound occurred, while numbers in the section along the opposite river bank, which was not exposed to infrasound, increased by 44 to 52%.

Water Jets and Air Bubbles: Adam and Schwevers (1997) tested the response of eels to water jet and air-bubble curtains. Eels showed only an initial avoidance response to these types of barriers, but then swam through them once they habituated to the test conditions. Schultze (1989) found that eels in an experimental flume responded to a bubble curtain/strobe light combination at first with an avoidance reaction, but quickly acclimated and subsequently showed no response. Based on these studies, water jet and air bubble barriers appear to be ineffective in altering silver eel migration patterns.

Electrical Fields: Eels are known to be very sensitive to electric fields, as evidenced, for example, by their susceptibility to electro-shocking sampling procedures. Despite this sensitivity, a difficulty encountered with the use of electricity for guiding eels along spillways or deterring them from turbine entry is charging an electrical field with sufficient voltage to dissuade eels from moving through an area without charging the field to a degree as to stun eels when they approach the field. Gleeson (1997) described the use of an electrical field that was developed for diverting eels from passage over a dam and noted that eels that approached this device too rapidly were often stunned and would sink and be moved downstream by current, resulting in a lack of diversion in the direction of their downstream movement. Hadderingh and Jansen (1990) reported results for experiments that evaluated the effectiveness of electric screens for diverting fish under laboratory and field conditions. Effectiveness was as high as 75% in some tests, but as low as -38% in others (i.e., 38% more eels were taken in the portion of the river with the electric field). They concluded from the results of the field studies that the reliability of the electric screen was too uncertain for practical application in algae-rich waters. Also, they indicated that installation of a screen of this type at a large power plant or hydroelectric intake would be very expensive, making it impractical.

Physical Barriers (Angled Bar Racks, Louvers, and Screens)

Difficulties in developing effective structural barriers that are designed to eliminate or reduce eel entrainment result from the fact that eels may try to pass through such guidance structures after making contact (EPRI 2001). Adam and Schwevers (1997) discuss detailed observations of European silver eel behavior in response to angled vertical bar racks and flow velocity in an experimental channel. Eels moving downstream often collided with the racks and exhibited no lateral searching behavior in front of angled bar screens before physically touching the structures. In tests conducted at higher water velocities, eels were observed to physically force their way through the bars after being impinged initially. Based on positive guidance, Adams and Schwevers (1997) suggested that a 25-degree angled screen may be used in combination with a surface collection device to bypass eels downstream. Observations of impingement at various flow velocities during this study clearly indicated that maintaining approach velocities less than 50 cm/s for angled screens may be necessary to minimize impingement of European eels.

Amaral et al. (this volume) presented results of two years of study of the effectiveness of angled bar racks and louvers at diverting silver American eels. Guidance efficiency with a 50-mm bar rack declined from a high of 72.7% at 0.3 m/s to a low of 54.5% at 0.9 m/s. Efficiency also increased markedly with the structures at a 15-degree angle to flow, exceeding 90% at velocities of 0.3 and 0.6 m/s with bar racks. However, for the 15-degree angle structures, a solid bottom overlay was attached to the lower 30 cm of the bar racks and the louvers to improve the guidance of bottom-oriented species such as eels. The authors cautioned that their experimental facility employed a full-depth bypass and relatively short lengths of bar racks and louvers, conditions unlike what might be installed at actual hydroelectric facilities and conclude that field tests are required to get a more accurate assessment of diversion potential of these devices for eels.

Adam and Schwevers (1997) also investigated the use of wedge-wire screens for eel diversion. They reported that the response of the eels was related to velocity, with a "startle" response occurring when flow velocity was less than 0.7 m/s. Also, eels did not impinge on the screens even at velocities greater than 1.0 m/s. Schultze (1989) concluded that some variation of wedge wire screen barriers was the type of diversion structure most likely to be effective for eels. He found that while eels demonstrated a "startle" response to the screen when first encountering it, they were not impinged on the smooth surface at velocities that exceeded their swimming ability (> 0.7 m/s), but were instead pushed up the angled screen into a bypass structure perpendicular to the flow, and did not suffer any injury. Based on these observations, Schultze (1989) proposed that wedge-wire screens should be installed at an angle of 40 degrees for effective eel guidance.

By-pass Facilities and Induced Flows

While a number of studies conducted in the field have shown that eels will use bypass facilities (Haro et al. 2000b; Schultze 1989; Durif et al., this volume; Rathcke 1993), controlled studies have not been conducted in which location, dimension, and flow rates of bypass facilities were modified and eel passage documented in a rigorous statistical fashion. Berg (1995) suggested that an efficient eel bypass should have entrance velocities that are similar to those at a project's intake trash racks. Additionally, Berg (1995) concluded that bypass flows should be at least 5% of total river discharge (some studies have suggested that bypass flows should be 50% of river discharge for effective passage).

Altered Generation Schedule

Reducing or ceasing generation may be a viable mitigation measure for safe downstream passage of eels in some circumstances. The costs of ceasing generation may be less than the cost of installing physical structures, such as bar racks or screens, particularly if nighttime migratory behavior of eels coincides with off-peak rate schedules (Gilbert and Wenger 1996). Ceasing generation during the hours of highest downstream eel passage has been the preferred and least-cost alternative method for avoiding impacts at four dams on the South Fork of the Shenandoah River, Virginia (Gilbert and Wenger 1996). However, knowing when peak passage periods may occur has proven to be problematic and a considerable challenge.

Euston et al. (1997) carried out regression modeling in an attempt to develop a migration prediction model based on the influence of a range of environmental variables (dissolved, oxygen, water temperature, plant load, river flow, river flow squared, water temperature-river flow interaction term, and number of days from full moon) on the migratory pattern of eels at a hydroelectric project in Virginia. The highest percentage of variation explainable by a three variable model was 19.8%. Based on this finding, they concluded that a meaningful predictive model could not be developed to establish when plant shutdowns should occur. Multivariate correlation analyses conducted by Hvidsten (1985) resulted in regressions explaining from 9% to 68% of variance of silver eel migration from the River Imsa over seven years for which data were analyzed. In five of the seven years, the regression model R^2 exceeded 46%. These results were significantly better than those of Euston et al. (1997), possibly because 10 variables were incorporated into the correlation analyses and the number of years for which data were available was much higher, or because of differences in environmental conditions.

Haro et al. (this volume) also developed a statistical model to predict eel migration in a small river in Maine. They found no significant correlation between daily flow and daily catch, mean annual flow and 5% cumulative eel descent, or mean annual flow and number of days between 5% and 25% cumulative eel descent. However, they did establish probable dates of 5% and 75% of run passage and a strong correlation between outmigrations and precipitation events. After incorporating these findings into a simulation model, they concluded that a reduction in total passage mortality at a hydroelectric project on the small river in which this work was conducted was possible. Simulated reductions under different scenarios were as great as 50% (e.g., from estimated mean mortality of 10.7% under all operating scenarios to 4.8%). However, they point out that peaks in eel runs in larger or mainstream rivers and in systems where flow is regulated may be less predictable. In these instances, benefits of plant operation changes would be much more difficult to predict.

Because numerous factors may contribute to triggering and sustaining silver eel migration runs, it is likely that the estimates of run timing derived from modeling efforts may prove to have a high degree of uncertainty. Altering the operations of a hydroelectric facility on the basis of uncertain predictions could result in either insufficient protection provided to migrating eels, or excessive cost due to plant shutdown in excess of the period required to provide the degree of protection sought.

Trap and Transport

Because of the potential for high cumulative mortality rates experienced by migrating eels that must pass several hydroelectric facilities, collection of silver eels at upriver sites and their transport to downstream release sites may be an effective mitigation alternative. Mitchell and Chisnall (1992) suggested a trap and transport program as the only practical means of ensuring survival of downstream migrant eels in the Upper Rangitaiki River system in New Zealand. Sistenich (1998) also describe a trap-and-transport program for European eels in a German river. This program employs an intense fishery for migrating silver eels above four hydroelectric projects and their release at the mouth of the Moselle River where it enters the Rhine. In 1997, 1,500 kg of silver eels, over 90% female, were transported in this program, with 2,000 kg being transported in 1998, 3,418 kg in 1999, and 4,600 kg in 2000.

Conclusions

The literature that exists on the migratory behavior of the various *Anguilla* species documents some general behavioral patterns that appear to be common to all (e.g., triggering of migration by rain and runoff events, aversion to light). Unfortunately, the same body of literature documents the high degree of variability shown by eels in expressing these general patterns and the fact that much about their behavior is still unknown. As a result, the identification of measures that could be employed at any given hydroelectric facility to ensure effective eel protection remains extremely difficult. The most comprehensive data and information on eel behavior, as well as on their response to various types of mitigation measures, are those that have been acquired in relatively small river systems or streams and in experimental chambers. This is because accurate sampling and documentation of behavior is difficult or impossible in large waters. As a consequence, the applicability of many of the findings summarized here to resolving eel passage issues at projects on large rivers is unknown. However, some general conclusions can be drawn from the available literature.

Nighttime plant shutdown over the entire period of potential silver eel migration (as defined by observations on the seasonal timing of out-migration or by single or multiple environmental triggers of migration) may be the most protective mitigation alternative, but also the potentially most expensive due to lost generation. While the predictions of pulses of eel migration derived from most statistical models may have a high degree of uncertainty, such models may have sufficient reliability to be useful for timing hydroelectric project shutdowns in small river systems in order to protect downstream migrants. The use of such models in these circumstances could reduce the required period of plant shutdown, and thus provide a cost-effective means of protecting migrants, if sufficient data are available from which to construct a relatively reliable model.

Four of the potential mitigation measures reviewed showed some degree of promise, these include angled light arrays, mechanical barriers, infrasound, and bypass flows (alone or in combination with diversion devices). In all instances, the measures were shown to be effective only under laboratory conditions or at relatively small projects. Therefore, their efficacy in natural conditions at projects varying widely in size cannot be predicted. Because each technology has specific advantages and disadvantages, site-specific factors are likely to play a key role in establishing which of the measures would be most appropriate for any individual project. Selection of an appropriate mitigation measure for eel passage must also account for potential consequences to species other than eel. For example, light barriers, while appearing to be a potentially effective diversion technology for eels under many circumstances, might adversely affect the migration of other species, such as anadromous clupeids.

Many of the findings we have discussed are from on-going or preliminary studies and/or have not been peer reviewed. In many instances, follow-up studies to further explore promising approaches or to more accurately characterize eel behavior of particular relevance to successful mitigation would be of great value for enhancing the protection of migrating eels.

Acknowledgments

The primary funding for this project was provided by EPRI (Electric Power Research Institute) of Palo Alto, California. Additional funding support was provided by Allegheny Energy, Dominion Generation, Duke Energy, Exelon Generation Company, Hydro-Quebec, New York Power Authority, Ontario Power Generation and the U.S. Department of Energy—Hydropower Program. Valuable information and review comments on the initial EPRI report were provided by Glenn Cada, Charles Simons, Terry Euston, Greg Pope, Ron Sheehan, Bob Graham, Kevin McGrath, Scott Ault, Richard Verdon, Alex Haro, and Jean-Denis Dutil. We also express our appreciation for the helpful comments on the original draft of this manuscript provided by Steve Amaral and Kevin McGrath.

References

Adam, B., and D. U. Schwevers. 1997. Behavioral Surveys of Eels (*Anguilla anguilla*) Migrating Downstream Under Laboratory Conditions. 36320 Kirtorf-Wahlen, Germany: Institute of Applied Ecology.

Adam, B., U. Schwevers, and U. Dumont. 1997. Beiträge zum Schutz abwandernder Fische—Verhaltens-beobachtungen in einem Modellgerinne. Bibliothek Natur & Wissenschaft Band 16. Verlag Natur & Wissenschaft, Solingen, Germany.

Aoyama, J., S. Watanabe, T. Miyai, S. Sasai, M. Nishida, and K. Tsukamoto. 2000. The European Eel, *Anguilla anguilla* L., in Japanese Waters. Dana 12:1–5

Atlantic States Marine Fisheries Commission (ASMFC). 2000. Interstate Fishery Management Plan for American Eel. Fishery Management Report No. 36. Washington, D.C.

Behrmann-Godel, J. 2000. Telemetrische Untersuchung der Herbstlichen Wanderung der Blankaale (*Anguilla anguilla* L.) in der Staustufe Trier [Telemetric Investigation of the Fall Travel of the Silver Eel in the 'City of Trier' Compartment of the Moselle River]. Final report for the Eel Conservation Initiative, Rheinland-Pfalz/ RWE Energie AG.

Berg, R. 1995. Gedanken zur Problematik "Turbinenschaden an Aalen" Arb. Dtsch. Fischerei-Verband 64:21–39

Boëtius, J. 1967. Experimental Indication of Lunar Activity in European Silver Eels, *Anguilla anguilla* (L.). Meddr. Danm, Fisk. -og Havunders 6:1–6

Castonguay, M., P. V. Hodson, C. M. Couillard, M. J. Eckersley, J. –D. Dutil, and G. Verreault. 1994a. Why is recruitment of the American eel, *Anguilla rostrata*, declining in the St. Lawrence River and Gulf? Canadian Journal of Fisheries and Aquatic Science 51:479–488.

Castonguay, M., P. V. Hodson, C. Moriarty, K. F. Drinkwater, and B. M. Jessop. 1994b. Is there a role in the ocean environment in American and European eel decline? Fisheries Oceanography 3:197–203.

Desrochers, D. 2001. Étude de la Migration des Anguilles (Anguilla rostrata) du Saint-Laurent et Passe Migratoire à Anguille au Barrage de Chambly. Report prepared by Milieu inc. for Hydro-Québec, Montreal, Québec, Canada

Edel, R. K. 1976. Activity Rhythms of Maturing American Eels, *Anguilla rostrata*. Marine Biology 36:283–289

Electric Power Research Institute(EPRI). 2001. Review and Documentation of Research and Technologies on Passage and Protection of Downstream Migrating Catadromous Eels at Hydroelectric Facilities. EPRI Report 1000730, Palo Alto, California.

European Inland Fisheries Advisory Commission and International Council for the Exploration of the Sea (EIFAC/ICES). 2001. Report of the Eleventh Session of the Joint EIFAC/ICES Working Group on Eels, Silkeborg, Denmark (20–24 September, 1999). EIFAC Occasional Paper No. 34. Report prepared for the Food and Agriculture Organization of the United Nations, Rome, Italy.

Euston, E. T., D. D. Royer, and C. L. Simmons. 1997. Relationship of Emigration of Silver American Eels (*Anguilla rostrata*) to Environmental Variables at a Low Head Hydro Station. Proceedings of the International Conference on Hydropower, 5–8 August 1997, Atlanta, Georgia.

Euston, E. T., D. D. Royer, and C. L. Simmons. 1998. American Eels and Hydro Plants: Clues to Eel Passage. Hydro Review August: 94–103.

Facey, D. E., and G. S. Helfman. 1985. Reproductive Migration of American Eels in Georgia. Proceedings of the Annual Conference of Southeast Association of Fish and Wildlife Agencies 39:132–138

Facey, D. E., and M. J. Van Den Avyle. 1987. Species Profiles: Life Histories and Environmental Requirements of Coastal Fishes and Invertebrates (North Atlantic) – American Eel. U.S. Fish and Wildlife Service Biological Report 82 (11.74) and U.S. Army Corps of Engineers, TR EL-82-4.

Gilbert, B., and B. Wenger. 1996. Entrainment Mortality of American Eels at Two Hydroelectric Plants in Virginia. Problems and Solutions. U.S. Fish and Wildlife Service, American Eel Passage Workshop Transcripts, 31 July 1996, Hadley, Massachusetts.

Gleeson, L. 1997. Use of Electric Eel Passage. Presented at the Fish Passage Workshop (sponsored by Alden Research Laboratory, Inc., Conte Anadromous Fish Research Center, Electric Power Research Institute, and Wisconsin Electric), 6–8 May 1997, Milwaukee, Wisconsin.

Hadderingh, R. H. 1982. Experimental Reduction of Fish Impingement by Artificial Illumination at Bergum Power Station. Int. Revue Ges. Hydrobiol. 67(6):887–900

Hadderingh, R. H., and M. R. De Potter. 1995. Reduction of Fish Mortality at Power Stations. Proceedings POWER-GEN '95 Europe, 16–18 May 1995, Amsterdam, Netherlands.

Hadderingh, R. H., and H. Jansen. 1990. Electric Fish Screen Experiments Ender Laboratory and Field Conditions. Pages 266–280 in I.G. Cowx, editor. Developments in Electric Fishing. Blackwell Science Publications, Fishing News Books, Oxford, England.

Hadderingh, R. H., and A. G. Smythe. 1997. Deflecting Eels from Power Stations with Light. KEMA Environmental Services, Presented at the Fish Passage Workshop, 6–8 May 1997, Milwaukee, Wisconsin.

Hadderingh, R. H., J. W. Van Der Stoep, and J. M. P. M. Habraken. 1992. Deflecting Eels from Water Inlets of Power Stations with Lights. Irish Fisheries Investigations Series A (Freshwater) 36:78–87

Hadderingh, R. H., G. H. F. M. Van Aerssen, R. F. L. J. De Beijer, and G. Van Der Velde. 1999. Reaction of Silver Eels to Artificial Light Sources and Water Currents: An Experimental Deflection Study. Regulated Rivers: Research & Management 15(4):365–371

Hain, J. H. W. 1975. The Behaviour of Migratory Eels, *Anguilla rostrata*, in Response to Current, Salinity, and Lunar Period. Helgolander wiss. Meeresunters 27:211–233

Halsband, E. 1989. Nur durch elektrischen Impulsstrom ist das Scheuchen und Leiten von Fischen wirksam möglich. Der Fischwirt 39(10):76–79

Haraldstad, Ø., L.A. Vøllestad, and B. Jonsson. 1984. Descent of European Silver Eels, *Anguilla anguilla* L., in a Norwegian Watercourse. Journal of Fish Biology 26:37–41

Haro, A., and T. Castro-Santos. 1997. Downstream Migrant Eel Telemetry Studies, Cabot Station, Connecticut River, 1996. Conte Anadromous Fish Research Center, Turner's Falls, Massachusetts.

Haro, A., W. Richkus, K. Whalen, A. Hoar, W. Dieter-Busch, S. Lary, T. Brush, and D. Dixon. 2000a. Population Decline of the American Eel: Implications for Research and Management. Fisheries 25(9):7–16

Haro, A., T. Castro-Santos, and J. Boubée. 2000b. Behavior and Passage of Silver-phase Eels, *Anguilla rostrata* (LeSueur), at a Small Hydroelectric Facility. Dana 12:33–42

Hvidsten, N.A. 1985. Yield of Silver Eels and Factors Affecting Downstream Migration in the Stream Imsa, Norway. Reprinted from Institute of Freshwater Research, Drottningholm. National Swedish Board of Fisheries, Report No. 62, pp. 75–85.

Jens, G. 1952. Über Den Lunaren Rhythmus Der Blankaalwanderung. Archiv für Fischereiwissenschaft 4:94–110.

Kleinschmidt Associates. 2001. New York Power Authority, St. Lawrence-FDR Power Project (FERC NO. 2000): 2000 Survey of American Eel in the Tailwater of the International St. Lawrence Power Project. Draft Final Report, June 2001. White Plains, New York.

Lowe, R. H. 1952. The Influence of Light and Other Factors on the Seaward Migration of the Silver Eel, *Anguilla anguilla* L. Journal of Animal Ecology 21:275–309.

Lühmann, M., and H. Mann. 1958. Wiederfänge Markierter Elbaale vor der Küste Dänemarks. Archiv für Fischereiwissenschaft 9:200–202.

Mitchell, C. P. 1995. Trapping the Adult Eel Migration at Aniwhenua Power Station—Investigation No. 1940. Charles Mitchell & Associates, Rotorua, New Zealand.

Mitchell, C. P., and B. L. Chisnall. 1992. Problems Facing Migratory Native Fish Populations in the Upper Rangitaiki River System. New Zealand Freshwater Fisheries, Rotorua. Miscellaneous

Report No.119. Document prepared for Bay of Plenty Electric Power Board.

Moriarty, C., and W. Dekker, editors. 1997. Management of the European eel. Marine Institute of Fisheries, Bulletin No. 15, Dublin, Ireland.

Pankhurst, N. W. 1982. Relation of Visual Changes to the Onset of Sexual Maturation in the European Eel, *Anguilla anguilla* (L.). Journal of Fish Biology 21:127–140

Popper, A. N., and T. J. Carlson. 1998. Application of Sound and Other Stimuli to Control Fish Behavior. Transactions of the American Fisheries Society 127:673–707

Rathcke, P. C. 1993. Untersuchung uber die Schadigungen von Fischen durch Turbine und Rechen im Wasserkraftwerk Dringenauer Muhle (Bad Pyrmont). Arb. Dtsch. Fischerei-Verband 59:37–74

Richkus, W., and K. Whalen. 2000. Evidence for a Decline in the Abundance of the American Eel, *Anguilla rostrata* (LeSueur), in North America Since the Early 1980s. Dana 12:83–97

RMC Environmental Services. 1995. Report on Studies to Evaluate American Eel Passage. Prepared by RMC Environmental Services for Allegheny Power Service Corporation. RMC Environmental Services: Drumore, Pennsylvania.

Sand, O., P. S. Enger, H. E. Karlsen, F. Knudsen, and T. Kvernstuen. 2000. Avoidance Responses to Infrasound in Downstream Migrating European Silver Eels, *Anguilla anguilla*. Environmental Biology of Fishes 57:327–336

Schultze, D. 1989. Versuche zur Ermittlung von Turbinenschaden an Aalen am Kraftwerk Wahnhausen. Arb. Dtsch. Fischerei-Verband 47:13–24

Sistenich, H. P. 1998. Wasserkraftnutzung im Spannungsfeld okonomischer und okologischer Anforderungen. Pp. 18–27, Fische und Turbinen: Schutz der Fische vor Verletzungen in den Kraftwerksanlagen staugeregelter Flusse. Symposium in Veitschochheim, 16–17 November 1998. Bezirk Unterfranken, Fachberatung fur Fischerei & Aalschutzinitiative des Landes Rheinland-Pfalz und der RWE Energie AG, Essen, Germany.

Stevens, G., K. Webb, and G. C. Crommentt. 1996. Entrainment Mortality and Use of Bottom Openings During Downstream Passage of American Eels at Two Hydroelectric Plants in Maine: A Cooperative Solution. US Fish and Wildlife Service, American Eel Passage Workshop Transcripts, 31 July 1996, Hadley, Massachusetts.

Tatsukawa, K., and Y. Matsumiya. 1999. Conservation and management of eel stocks in Japan. Gekkan Kaiyo, Gougai 18:1–8.

Tesch, F. 1977. The Eel [English translation of German language original, Der Aal, published in 1973]. London: Chapman and Hall.

Therrien, J., and G. Verreault. 1998. Évaluation d'un Dispositif de Dèvalaison et des Populations d'Anguilles en Migration dans la Rivière Rimouski. Ministére de l'Énvironnement et de la Faune, Québec, Québec, Canada.

Vøllestad, L. A., B. Jonsson, N. A. Hvidsten, T. F. Naesje, Ø. Haraldstad, and J. Ruud-Hansen. 1986. Environmental Factors Regulating the Seaward Migration of European Silver Eels, *Anguilla anguilla*. Canadian Journal Fisheries Aquatic Science 43:1909–1916

Winn, H. E., W. A. Richkus, and L. K. Winn. 1975. Sexual Dimorphism and Natural Movements of the American Eel, *Anguilla rostrata*, in Rhode Island Streams and Estuaries. Helgoländer wiss. Meeresunters. 27:156–166

Wippelhauser, G. S., L. Flagg, J. McCleave, J. Moring, K. Oliveira, J. Brockway, M. Cieri, and L. Daniels. 1998. Maine Department of Marine Resources, Stock Enhancement Division, Eel and Elver Progress Report, February 1998, Augusta, Maine.

DATE DUE

SCI QL 637.9 .A5 I67 2003

International Symposium
Biology, Management, and

Biology, management, and
protection of catadromous